10th Asia Pacific Structural Engineering and Construction Conference (APSEC 2018)

IOP Conference Series: Materials Science and Engineering
Volume 513

Langkawi, Malaysia
13 – 15 November 2018

ISBN: 978-1-5108-8635-3
ISSN: 1757-8981

Printed from e-media with permission by:

Curran Associates, Inc.
57 Morehouse Lane
Red Hook, NY 12571

Some format issues inherent in the e-media version may also appear in this print version.

Printed by Curran Associates, Inc. (2019)

For permission requests, please contact the Institute of Physics at the address below.

Institute of Physics
Dirac House, Temple Back
Bristol BS1 6BE UK

Phone: 44 1 17 929 7481
Fax: 44 1 17 920 0979

techtracking@iop.org

Additional copies of this publication are available from:

Curran Associates, Inc.
57 Morehouse Lane
Red Hook, NY 12571 USA
Phone: 845-758-0400
Fax: 845-758-2633
Email: curran@proceedings.com
Web: www.proceedings.com

Table of contents

Volume 513

10th Asia Pacific Structural Engineering and Construction Conference 2018

13–15 November 2018, Langkawi, Malaysia

Accepted papers received: 8 March 2019
Published online: 25 April 2019

Preface

Papers

IOP Conf. Series: Materials Science and Engineering **513** (2019) 011001 doi:10.1088/1757-899X/513/1/011001

Preface

The construction industry has accepted innovation and safeguarded great extent of sustainability by creative applications in design and construction practices. A sustainable structure or construction practice implicates that it is judiciously deployed to challenges faced by modern construction industry. As a matter of fact, the innovation in this field has kept developing or evolving to meet the expectation of ever more demanding design requirements in addition to persistent needs for change and improvement in performance. The problems faced contemporarily by the construction industry may not necessarily identical as in the future. Nevertheless, innovation and developments should not be ceased in order to offer cost effective sustainable solution that can change these challenges into opportunities.

The School of Civil Engineering, Faculty of Engineering, Universiti Teknologi Malaysia (UTM), organised this International Conference to discuss the challenges as well as innovations in regards to sustainable construction industry, which including new materials, technologies, environmental concerns and economic factors to enhance its excellence. This is done by gathering Engineers, designers, researchers, scientists and scholars from different countries all around the world to celebrate the excellence in sustainable structure and construction practices. This conference dealt with key technical under the theme of "Sustaining the world with better structures and construction practice". Over 65 papers were presented by authors during this Conference and these are compiled in a single volume conference proceeding.

The event was co-organized with Construction Research Institute of Malaysia (CREAM). The organisation is gratefully acknowledged for their indispensable support. Organising a Conference is immense and substantial. Assistances and helps from all those involved are gratefully acknowledged, particularly the members of Technical and Supporting Committee for their supports from start to finish; the Authors and the Session Chairs for their invaluable contributions.

This Proceeding has been completed via camera-ready copy printed from the digital manuscripts submitted by the authors and editing has been limited to the least, where it was considered unavoidable.

Tan Sri. Datuk. Prof. Dr. Ir. Ts. Tr. Ar. Noor Nabilah Sarbini

IOP Conf. Series: Materials Science and Engineering **513** (2019) 011002 doi:10.1088/1757-899X/513/1/011002

Peer review statement

All papers published in this volume of *IOP Conference Series: Materials Science and Engineering* have been peer reviewed through processes administered by the proceedings Editors. Reviews were conducted by expert referees to the professional and scientific standards expected of a proceedings journal published by IOP Publishing.

IOP Conf. Series: Materials Science and Engineering **513** (2019) 012001 doi:10.1088/1757-899X/513/1/012001

Review on the method for carbon footprint calculation of highway development

Che Muhammad Fatihi Hafifi Che Wahid[1], Muhammad Rozaid Ramli[1], Eeydzah Aminudin[1*], Zainura Zainon Noor[2,3], Monh Neardey[1], Rozana Zakaria[1], Mohd Rosli Hainin[1], Rosli Mohamad Zin[1], Mohd Khairul Idham Mohd Satar[1], Muhammad Naqiuddin Mohd Warid[1], Muhd Zaimi Abd. Majid[1] and Chai Chang Sa'ar[4]

[1]School of Civil Engineering, Faculty of Engineering, Universiti Teknologi Malaysia, 81310 Johor Bahru, Johor, Malaysia
[2]Center for Environmental Sustainability and Water Security (IPASA), Universiti Teknologi Malaysia, 81310 Johor Bahru, Johor, Malaysia
[3]School of Chemical Engineering, Faculty of Engineering, Universiti Teknologi Malaysia, 81310 Johor Bahru, Johor, Malaysia
[4]University of Reading Malaysia, Educity, 79200 Iskandar Puteri, Johor

[*]Corresponding author: eeydzah@utm.my

Abstract. Highway infrastructure emits large quantities of carbon dioxide over their entire life cycle including emission from production of raw materials and also emission from construction, operation, maintenance, and rehabilitation of the roads. The past emission trends of carbon dioxide from these sectors are investigated through their sources while mitigation and abatement strategies are suggested. This paper brings together, for the first time, a systematic review of the carbon footprint calculator of 21 case study highway from 8 different countries were investigated. This review focuses on method for calculate carbon footprint of highway development, through the synthesis of the overall outcomes of these studies, to identify whether using excel tool or Life Cycle Assessment (LCA) software achieve better performance and easier to calculate the emission from highway development. It is observed that using excel tool is better to calculate and analysis the carbon footprint from many sources and stage compare than calculate using LCA software. Therefore, understanding the relationships between methods for calculate carbon footprint life cycle energy consumption and GHG emissions are critically important for highway agencies to evaluate their annual greenhouse gas emission reports.

1. Introduction

As Malaysia transforms into a high-income nation, the national development strategy must be in line with the megatrends of the world, especially climate change. Therefore, Malaysia has already signatory and pledged to cut national carbon emission intensity by 45 % by 2030 based on 2005 emission levels. Align with The National Green Technology Policy greenhouse gases emission was introduced in order to ensure sustainable development approach can significantly progress besides providing the major improvement, especially in construction sector.

Carbon footprint technology is still evolving, especially with regard to the accuracy of estimations and transparency of calculations methods, as a means of benchmarking and comparison. Hence, there is a need to explore, in more detail, the consequences of personal choices before setting a normative concept of responsible behaviour toward emissions reductions [1].

Therefore, this study focuses on the comparison of carbon emission calculations by either using excel or software. An overview of the case studies from previous research is presented in Table 1. Some attributes such as country, date, type of LCA approach used and stage duration of scale analysis of previous research. The most common scale of analysis is production and construction process. The most common LCA method is studies focusing on create excel calculator tool are also frequent.

Table 1. Carbon Footprint quantification studies on road project.

Authors	Country	LCA Method		Scale of analysis				
		Crete Excel	Use LCA software	*P	*D	*C	*O	*M
Venmathy et al., 2015 [2]	Malaysia	/			/	/	/	/
White et al., 2010 [3]	U.S	/		/		/		
Santero et al., 2011 [4]	U.S	-	-	/		/	/	/
Kucukvar & Tatari, 2012 [5]	U.S		/*P-LCA	/		/		
Loijos et al., 2013 [6]	U.S		/*Gabi	/		/	/	/
Barandica et al., 2013 [7]	Spain	/				/	/	/
Wang, 2016 [8]	China	/		/		/		
Sreedhar et al., 2016 [9]	India	/		/	/	/	/	
Kamyab et al., 2015 [10]	Malaysia	/				/		
Thives & Ghisi, 2017 [11]	Brazil	/		/		/	/	/
Ghazy, et al., 2016 [12]	Egypt.		/*LCA-	/		/	/	/
Rahman, et al., 2017 [13]	Saudi A.	/	based			/	/	

*P-Production, *D-Design, *C-Construction, *O-Operation, *M-Maintenance

2. Carbon footprint calculation

Before undertaking the study in Malaysia, understanding of the definition of carbon footprint and the way calculations are made is to be understood first. According to Ramachandra, carbon footprint is the total amount of GHG emission caused direct and indirect by an individual, event, organization, and product, expressed in equivalent tons of carbon dioxide (t CO_2e) [14].

Embodied Emissions is GHG emissions which the sum of all the energy required to produce any goods or services, considered as if that energy was incorporated or 'embodied' in the product itself. Carbon footprint consists of its initial embodied energy, its recurrent embodied energy and its operational energy over its lifetime. So, in order estimate the carbon footprints of a product, it is crucial to take into consideration the embodied energy of the product and also the transportation factor. With this, the total amount of carbon footprints caused by the product over the life stage can be estimated.

2.1. Scope

First step is essential to understand all possible sources of direct and indirect CO_2 emissions during the full life cycle period. There are many researchers quantifying the sources of carbon emission by divided into three scopes according to the GHG Protocol models [15]. Direct emissions Scope 1 are released from sources within the organizational boundaries of the entity being inventoried. Besides that, indirect emissions are consider in Scope 2 are released from sources outside of the organizational boundaries of the entity being inventoried but are a consequence of the energy purchases of the entity. Lastly, optional indirect emissions Scope 3 are a broad category that covers all other releases that are an indirect consequence of the entity's operations.

However, the ability of highway utilities to effect significant emission reductions or sequestration of GHGs may require the inclusion of projects impacting Scope 3 and Embodied Energy Emissions or other projects outside of their own boundaries. Scope 3 is direct emission, GHG emissions which the sum of all the energy required to produce any goods or services, considered as if that energy was incorporated or 'embodied' in the product itself. For example, possible sources GHG emitters on highway development are discussed in table 2 - 4 below.

Table 2. Sources in scope 1 during full life cycle period of highway.

Sources	Design Stage	Construction Stage	Operation & Maintenance Stage
Fleet Vehicle	/	/	/
On-site construction by machine used		/	/

Table 3. Sources in scope 2 during full life cycle period of highway.

Sources	Design Stage	Construction Stage	Operation & Maintenance Stage
Electricity (power plant that generates the electricity consumed by the entity)	/	/	/

Table 4. Sources in scope 3 during full life cycle period of highway.

Sources	Design Stage	Construction Stage	Operation & Maintenance Stage
Material	/	/	/
Staff Commuting	/	/	/
Water	/	/	/
Waste	/	/	/
Transportation	/	/	/
Supplier		/	/

2.2. Emission factors of emission sources

Carbon Footprint can be measured by estimating the amount emission emitted by multiplying activity data (such as the amount of fuel used) with relevant emissions conversion factors. Emission factors will be identified according to the data availability and cited based on the reference as show in Table 5 - 9 below.

Table 5. Emission factor for utilities.

Activity	Emission Factor	Unit	Sources
Electricity (Peninsular)	0.000694	t CO_2/kWh	SEDA, [16]
Electricity (Sabah & Wilayah Persekutuan Labuan)	0.000536	t CO_2/kWh	SEDA, [16]
Water	0.000376	t CO_2/m^3	Shimizu et al., [17]
Diesel	0.002672	t CO_2/liter	Defra, [18]
	0.003176	t CO_2/liter	Zainab et al., [19]
	0.002564	t CO_2/liter	Venmathy, [2]
	0.000185	t CO_2/km	Venmathy, [2]
Petrol	0.002322	t CO_2/liter	Defra, [18]
	0.002233	t CO_2/liter	Venmathy, [2]
	0.000201	t CO_2/km	Venmathy, [2]

Table 6. Emission factor for vehicle.

Activity	Emission Factor	Unit	Sources
Motorcycle	0.000107	t CO_2/km	Defra, [18]
Car	0.000216	t CO_2/km	Defra, [18]
Hybrid Car	0.000120	t CO_2/km	Defra, [18]
4WD	0.000183	t CO_2/km	Defra, [18]

Table 7. Emission factor for transportation of material.

Activity	Emission Factor	Unit	Sources
Dump Truck	0.001100	t CO_2/km	Sreedhar et al., [9]
Truck (14 ton)	0.001100	t CO_2/km	Sreedhar et al., [9]
Truck standard 16-24 tons	0.000282	t CO_2/ton-km	White et al., [3]
Lorry	0.000846	t CO_2/km	White et al., [3]
LMV (Goods)	0.000914	t CO_2/km	Sreedhar et al., [9]
LMV (Passenger)	0.000460	t CO_2/km	Sreedhar et al., [9]

Table 8. Emission factor for material.

Activity	Emission Factor	Unit	Sources
Sand	0.0000025	t CO_2/kg	White et. al., [3]
Concrete	0.1055000	t CO_2/t	White et al., [3]
	0.0001340	t CO_2/kg	Chen et al., [20]
Crusher run	0.0000026	t CO_2/kg	White et. al., [3]
Wet-mix Roadbase	1.1400000	kg CO_2/ t	Sreedhar et al., [9]
Prime coat	0.0001600	t CO_2/kg	Wang et al., [21]
	0.0205000	kg CO_2/m^2	Sreedhar et al., [9]
Dense Bitumen Macadam (DBM)	0.0786000	t CO_2/t	Sreedhar et al., [9]
	0.0238000	t CO_2/t	White et. al., [3]
	0.0000630	t CO_2/kg	Chen et al., [20]
Binder course	0.0786000	t CO_2/t	Sreedhar et al., [9]
	0.0238000	t CO_2/t	White et. al., [3]
	0.0000630	t CO_2/kg	Chen et al., [20]
Tack coat	0.0001600	t CO_2/kg	Wang et al., [21]
	0.0205000	kg CO_2/m^2	Sreedhar et al., [9]
Wearing course	0.0786000	t CO_2/t	Sreedhar et al., [9]
	0.0238000	t CO_2/t	White et. al., [3]
	0.0000630	t CO_2/kg	Chen et al., [20]
Steel	0.0046700	t CO_2/kg	Sreedhar et al., [9]

Table 9. Result of carbon emission per unit activity from previous study.

Activity	Emission Factor	Unit	Sources
Clearance for road construction	0.006560	t CO_2/m^3	Sreedhar et al., [9]
Excavation with excavator	0.000539	t CO_2/m^3	Sreedhar et al., [9]
Construction of road sub-base	0.001140	t CO_2/ t sub base	Wang et al., [21]
Rolling of layers, for one layer	0.000102	t CO_2/m^2 compacted surface	Sreedhar et al., [9]
Prime coat	0.0000205	t CO_2/m^2 applied surface	Sreedhar et al., [9]
Construction of road base course(unbound)	0.0011400	t CO_2/t roadbase	Sreedhar et al., [9]
Tack coat	0.0000205	t CO_2/m^2 applied surface	Sreedhar et al., [9]
Paving of asphalt Layers, for one layer	0.0000965	t CO_2/m^2 applied surface	Sreedhar et al., [9]
Rolling of asphalt layers, for one layer	0.0001300	t CO_2/m^2 applied surface	Sreedhar et al., [9]

2.3. Calculation approach

According to report by National Corporate GHG Reporting Programmed for Malaysia, there are several approaches which could be taken in order to calculate carbon footprint [22]. For example calculations approach as show in table 10. Based on GHG Protocol, an organization should use the most accurate calculation approach available which is appropriate for their reporting. If it is not possible to calculate emissions from known activity data, the organization needs to estimate its emissions and extrapolate on the basis of known activity data.

Table 10. The calculation approach for calculating GHG Emissions [23].

No.	Calculation Approach	Description
1	Direct Measurement	- Monitor GHG concentration and flow rate - May be expensive and difficult to be implemented.
2	Stoichiometric Calculation	- Measure which elements enter and leave the system
3	Estimate emissions	- Most common approach for calculating GHG emissions - Apply documented emissions factors to known activity data from the organizations.

3. Comparison on the method whether using excel tool or Life Cycle Assessment (LCA) software

There are various tools for conducting LCA or for supporting the different phases and applications of LCA. Table 11 includes a list of existing LCA-related tools. Some main characteristics of the tools are included assessment on carbon footprint related to highway development, following the Carbon Footprint approach and IPCC guidelines for GHG emission assessment, consider all phase of development and activity, Carbon Offset, and result divided into scope. The summary of result from 21 case study as show in figure 1 below.

As also discussed in this report, there are many LCA methods and tools but not many is tailored for the highway development sector. Some difficulties in applying LCA in to highway development activity include the complexity of the methodology and lack of understanding.

Table 11. Comparison on the method.

Tool	Name	A	B	C	D	E	Reference
Excel Spreadsheet / Web Base	Highway England Carbon Tool	*	*	*			https://www.gov.uk/government/publications/carbon-tool
	Highways Agency Carbon Calculation For MAJOR PROJECTS	*	*				https://www.gov.uk/.../Major_Projects_HA_Carbon_Calculation_Spreadsheet.xls
	Highways Agency Carbon Calculation For DBFOs	*	*				https://www.gov.uk/government/.../DBFO_HA_Carbon_Calculation_Spreadsheet.xls
	Carbon Footprint Calculator (develop by UTP)	*	*				Rozana Kasbon, et al., [24]
	Carbon Footprint Calculator Highway India	*	*				Sreedhar S. et al., [9]
	Defra / DECC's GHG Conversion Factors, UK	*	*			*	http://carboncalculator.direct.gov.uk/index.html
	Carbon Trust Standard Carbon footprint spreadsheet	*	*		*		https://www.carbontrust.com/.../carbon-trust-standard-carbon-footprint-spreadsheet.xls
	Direct Emissions	*	*				www.wri.org/sites/default/files/direct_emissions.xls
	Carbon Footprint	*	*		*		https://www.carbonfootprint.com/

							calculator.aspx
	IESR Kalkulator Jejak Karbon, Indonesia	*	*				http://www.iesr.or.id/kkv3/
	Resurgence Calculator	*	*				https://www.resurgence.org/resources/carbon-calculator.html
	Climatecare Carbon Calculator	*	*		*		https://climatecare.org/calculator/
Software	GaBi	*	*	*			http://www.gabi-software.com
	GEMIS	*	*	*			http://www.iinas.org/news de.html, accessed: 21.02.2014
	openLCA	*	*	*			http://www.openlca.org
	SimaPro	*	*	*			http://www.pre-sustainability.com/software, accessed: 21.02.2014
	Umberto	*	*	*			http://www.umberto.de/en, accessed: 21.02.2014
	Spectrum IEEE	*					https://spectrum.ieee.org/static/carbon-calculator
	bp Carbon Calculator	*					http://www.bp.com/iframe.do?categoryId=9027929&contentId=7050956
	Safeclimate Calculator	*					http://www.safeclimate.net/calculator/
	MyFootprint Calculator	*					http://www.myfootprint.org/
Number of tool		21	17	6	3	1	

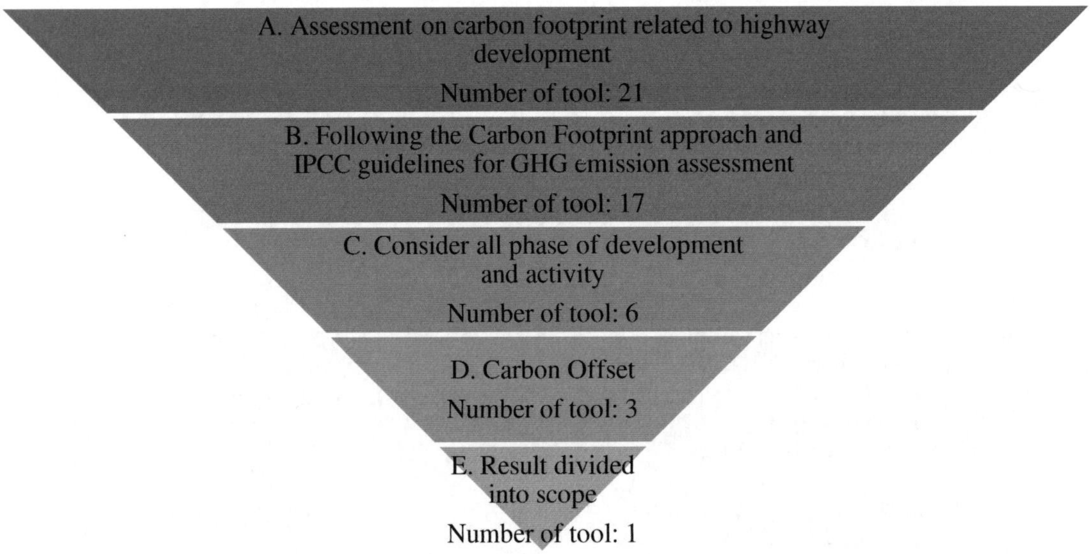

Figure 1. Application of carbon footprint calculation on highway development.

It is observed that using excel tool are better performance and easier to calculate and analysis the carbon footprint from many sources and stage compare than calculate using LCA software. Therefore, understanding the relationships between methods for calculate carbon footprint life cycle energy consumption and GHG emissions are critically important for highway agencies to evaluate their annual greenhouse gas emission reports.

4. Conclusion and recommendation

The Carbon Footprint Calculator can be established and implemented in Malaysia which can be the reference and guideline in the sustainable development of Green Highway. Besides, it also can be used as a highway development decision support tool in Malaysia.

In addition, the Carbon Footprint Calculator is developed specifically for the Malaysian tropical weather, environment, cultural and social needs. Thus, this research will be well contributed to the benefits of the nation especially towards the nation carbon reduction which believed to be achieve 45% reduction by the year 2030.

References

[1] Whitmarsh L and Neill S O 2010 Green identity , green living ? The role of pro-environmental self-identity in determining consistency across diverse pro-environmental behaviours *J. Environ. Psychol.* **30** 305–314
[2] Venmathy S, Zainura Z N, Mohd Fadhil M D, Che Hafizan C H 2015 Assessment of SAJ's Carbon Footprint SAJ doi:10 2172/875800
[3] White P, Golden J S, Biligiri K P and Kaloush K 2010 Modeling climate change impacts of pavement production and construction *Resour Conserv. Recycl.* **54** 776–782
[4] Santero N J, Masanet E and Horvath A 2011 Life-cycle assessment of pavements Part I : Critical review *Resour. Conserv. Recycl.* **55** 801–809
[5] Kucukvar M and Tatari O 2013 Ecologically based hybrid life cycle analysis of continuously reinforced concrete and hot-mix asphalt pavements *Transp. Res. Part D* **17** 86–90 (2012)
[6] Loijos A, Santero N and Ochsendorf J 2013 Life cycle climate impacts of the US concrete pavement network *Resour. Conserv. Recycl.* **72** 76–83
[7] Barandica J M, Fernández-Sánchez G, Berzosa Á ,Delgado J A and Acosta F J 2013 Applying life cycle thinking to reduce greenhouse gas emissions from road projects *J. Clean. Prod.* **57** 79–91
[8] Wang X, Duan Z, Wu L and Yang D 2015 Estimation of carbon dioxide emission in highway construction: A case study in southwest region of China *J. Clean. Prod* **103** 705–714
[9] Sreedhar S, Jichkar P and Biligiri K P, 2016 Investigation of Carbon Footprints of Highway Construction Materials in India *Transp. Res. Procedia.* **17** 291–300
[10] Kamyaba H, Jeng S L, Tayebeh K, Wai S H, Rahmalan A, Haslenda H, Ho C W, Ali K, Chew T L 2015 Greenhouse gas emission of organic waste composting: A case study of universiti teknologi malaysia green campus flagship project *J. Teknol.* **4** 113–117
[11] Thives L P and Ghisi E 2017 Asphalt mixtures emission and energy consumption: A review *Renew. Sustain. Energy Rev.* **72** 473–484
[12] Ghazy M, Abdallah A, Basiouny M and Saad M 2016 Life cycle assessment of flexible pavement construction *Br. J. Appl. Sci. Technol.* **12** 1–17
[13] Rahman S M, Khondaker A N, Hasan M A and Reza 2017 I Greenhouse gas emissions from road transportation in saudi arabia - a challenging frontier *Renew Sustain. Energy. Rev.* **69** 812–821
[14] Ramachandra T V, Aithal B H and Sreejith K 2015 GHG footprint of major cities in india *Renew. Sustain. Energy Rev.* **44** 473–495
[15] Harangozo G and Szigeti C 2017 Corporate carbon footprint analysis in practice – with a special focus on validity and reliability issues *J. Clean Prod.* **167** 1177–1183
[16] SEDA 2018 CO_2 avoidance as 2018 Available at: http://www seda gov my
[17] Shimizu Y, Dejima S and Toyosada K 2012 The CO_2 emission factor of water in japan *Water (Switzerland)* **4** 759–769
[18] Defra 2012 Methodology paper for emission factors *Methodol. Pap. Emiss. Factors* 1–85 doi:v 1 2 1 final
[19] Zainab A H and Afizah A 2016 Carbon footprint of road pavement rehabilitation : case study in sungai petani , kedah *Int. J. Appl. Environ. Sci.* **11** 1285–1302

IOP Conf. Series: Materials Science and Engineering **513** (2019) 012001 doi:10.1088/1757-899X/513/1/012001

[20] Chen J, Zhao F, Liu Z, Ou X and Hao H 2017 Greenhouse gas emissions from road construction in china: A province-level analysis *J. Clean. Prod.* **168** 1039–1047
[21] Wang X, Duan Z, Wu L and Yang D 2015 Estimation of carbon dioxide emission in highway construction: A case study in southwest region of china *J. Clean. Prod.* **103** 705–714
[22] NRE M 2017 *National Corporate GHG Reporting Programme for Malaysia Ministry of Natural Resources and Environment (NRE) Malaysia* **36**
[23] Natalia H 2009 A study of carbon footprinting techniques, best practices review and the application of carbon footprinting analysis for kruger products limited (Royal Roads University)
[24] Kasbon R, Jalil N A, Mazlan E M and Mahamad S 2011 Carbon footprints calculator for construction industry with case- based reasoning *2nd Int. Conf. Netw. Inf. Technol.* **17** 286–293

IOP Conf. Series: Materials Science and Engineering **513** (2019) 012002 doi:10.1088/1757-899X/513/1/012002

Effects of craftsmen supervisors project management skills on construction project success in Nigeria

Ibrahim Ibrahim Inuwa[1*], Silas Joseph[2] and Mohammed Mukhtar Musa[1]

[1]Department of Quantity Surveying, Abubakar Tafawa Balewa University, Bauchi-Nigeria
[2]Silas Joe & Partners Consulting Firm, Abuja-Nigeria

[*]Corresponding author: iiinuwa@atbu.edu.ng

Abstract. In Nigeria there is shortage of competent craftsmen due to project management skills (PMS) deficiency in their quality of training. This study investigates the effect of craftsmen supervisors PMS on construction project success in Nigeria, with a view to improving on craftsmen supervisors' productivity. Disproportionate stratified random sampling technique was used to administer 132 questionnaires to project managers (PM). Data were analysed using descriptive and inferential statistics. Result shows that craftsmen supervisors PMS requirements are relevant and positively relates to construction project success. This implied that PMS can improve craftsmen supervisors productivity, as well as bridge the gap between project management approach (PMA) and construction operations execution. The skills will as well enhanced construction supervision and operations productivity to attain project success. Hence the study recommends PMS to be integrated in craftsmen supervisors training in Nigeria.

1. Introduction

Monuments such as Angkor temples of Cambodia, the giant course way of Ireland, the Great Wall of China, and the pyramids of Giza show results of PMA in the technical competence of craftsmen supervisors in managing the construction operations [1] [2] [3] [4]. Craftsmen are contractors operatives who contribute skilfully with their hands in the practical realization of a construction project, while craftsmen supervisors are experience construction craftsmen that have been upgraded to line managers and authorized to directly manage other craftsmen operations while reporting to a higher ranking manager on the activities on construction site [5] [6]. Ness [7] draws the attention of the construction industry on neglect of the huge reservoir of practical, experiential and tacit knowledge for the advancement of its craftsmen to managerial position. Chindo *et al.* [8] observed that the absence of a framework for skills training in construction is one of the major challenges in the training of craftsmen in the Nigerian construction industry (NCI). Usman *et al.* [6] asserted that periodic training of construction craftsmen is a pre-requisite for improving on their productivity.

According to Kerzner [5], construction craftsmen supervisors (line managers) must understand the principle of project management since they are responsible for recruiting operatives, dispatching tasks to operatives, as well as responsible for overseeing operatives activities in a project. Thus, Kerzner [5] stressed that understanding the principles of project management is a necessity for craftsmen supervisors to provide visible support and commitment for the construction process. However, in

Nigeria, craftsmen training has been inadequate and declined to 42% [9]. Moreover, the few craftsmen trained lacked quality craftsmanship that match the demand of the construction industry [10]. This trend resulted in the inability of the craftsmen to keep abreast with the technological development in the NCI [11] [12]. A review of studies conducted on craftsmen training in the NCI fall short of identifying PMS requirements for the training of craftsmen, and the effect of craftsmen PMS on construction project success (see: [13] [14] [8] [11] [6] [12] [8] [10]). Likewise, a review of the curricula use by craftsmen training institutions in Nigeria are deficient of PMS integration in their training modules [15] [16] [17]. Hence this study investigates the effect of craftsmen supervisors PMS on construction project success in Abuja, Nigeria, with a view to improving on craftsmen supervisors productivity. To achieve its aim the following objectives were outlined: to identify and assess craftsmen supervisors PMS requirements, and to determine the relationship of craftsmen supervisors PMS and construction project success

2. Literature review

2.1. Theoretical construct

This study adopted project management theory (PMT) propounded by Koskela and Howell [18]. Project management is the art of directing and coordinating human and material resources through the life of a project by using modern management techniques to achieve pre-determined goals of scope, cost, time, quality and participant satisfaction [1]. Koskela and Howell [18] sub divided PMT into: theory of project and theory of management (see Table 1). The theory of project have three concepts: transformation, flow, and value (TFV). While the theory of management comprised of three models: planning, execution, and control [19].

Table 1. Project management theoretical constructs [19].

Subject of Theory		Concepts in Theory
1. Project		Transformation
		Flow
		Value generation
2. Management		Management-as-planning
	Planning	Management-as-organizing
		Classical communication theory
	Execution	Language/action perspective
		Thermostat model
	Control	Scientific experimentation model

2.1.1. Theory of project

i. Transformation

Transformation is concern with how inputs are converted into output [20]. This enables a project to be realize in an optimal manner and tasks executed in an optimal sequence through identification of the various tasks involve and their sequential relationship, curtailing uncertainty on requirements and tasks, and efficient project resource scheduling [19]. Transformation process portrays an overview of the operational structure of the project, thus provides management with the technical basis for answers to economic, time-scheduling, quality, quantity and organization questions [21].

ii. Flow

The flow concept is about eliminating non-value adding activities and integrating value adding activities throughout the transformation process [22]. Flow is achieved through the improvement of project lead time, variability reduction, flexibility, and transparency [22]. Thus, the flow concept in construction is contextualized into: adding and non-value adding activities, even flow, variability, and preconditions to work tasks in construction (see [22]).

iii. Value

The value concept is concern with how to integrate customer/end user requirements in the design and construction process of a construction project [22]. Thus, the concept of value in project management requires a manager to integrate customer/end-user optimum requirements in their construction design and execution; it takes into cognizance the project inputs, tasks, and sequence of operations throughout the transformation path [20].

2.1.2. Theory of management

i. Theory of planning

The theory of planning is subdivided into: management-as-planning and management-as-organizing. Management-as-planning theorized that planning a project has a managerial part and an effector part [18]; managerial part focus on creation and revision of plan (planning), while the effector part is the implementation of the resultant plan. Management-as-organizing optimally aimed at assembling the necessary resources into a cohesive structure in accordance with the project plan requirements [23].

ii. Theory of execution

The theory of execution is modelled into: dispatching model, and the language/action perspective model. The dispatching model conceptualized that, managerially, execution is about assigning (dispatching) tasks to work stations and, this is regarded as the classical communication theory [19]. Classical communication model focus on passing instructions to operatives without considering their effective understanding of the instructions passed. The short comings of the classical communication theory is complemented with the language/action perspective model, which emphasize a two-way communication and commitment, instead of the mere one-way communication [19]. Thus, communication of tasks dispatched to work stations must be comprehensive to the operatives; they should be feedback mechanisms that will convey the operatives understanding of the instruction passed.

iii. Theory of control

The theory of control consist of: thermostat model and the scientific experimental model (Koskela & Howell 2002a). Thermostat model conceptualized that in the project production process, there is a process to control, a unit for performance measurement, a standard of performance, and a controlling unit [19]. The model enables a production process to be monitored and controlled (Inuwa 2014). While the scientific experimental model focus on studying the causes of deviation in project production performance in order to proffer solution [19].

2.2. Craftsmen supervisors PMS requirements

A skill is the ability to do something well; expertise [24]. PMS for craftsmen supervisors are skills that give craftsmen supervisors the optimal ability to carry out their supervisory task in line with construction project management processes [10]. PMS requirements for craftsmen supervisors can be acquired through the application of project management's knowledge, skills, and techniques [1]. Training craftsmen supervisors on the application of project management entails giving them teaching and practices on the application of project management in order to bring them to the desired standard of behaviour, efficiency or physical condition for optimal realization of their supervisory role in the execution of construction project [25]. This study explore PMT to identify craftsmen supervisors PMS requirements. The project transformation process informs project manager of the overview of the operational structure of the project, thus provides management with the technical basis for answers to economic, time-scheduling, quality, quantity and organization questions. These subsequently guide the acquisition of skills on: design comprehension, resource scheduling, time scheduling, tasks execution effectiveness, and tasks execution effectiveness enhancing [21] [19]. The project flow concept is achieved through the display of skills that enable adequate resource organization, timely project completion, and adequate pre-activity identification. Thus, informs the development of construction craftsmen supervisors skills on: optimal workmanship attainment, task execution effectiveness, and task preconditions identification [22]. The value concept entails managers to optimize workmanship that ensures meeting the value requirement of the project stakeholders [1]. Hence craftsmen supervisors need skills on optimal workmanship attainment. Understanding the concept of planning in the theory of management enables craftsmen supervisors acquire skills on: task execution planning, task execution implementation, and task execution management [23] [18] [22]. The theory of execution enhances the craftsmen supervisors communication skill in dispatching tasks to work stations [20]. While the theory of control enhances craftsmen supervisors skill on how to monitor and control the performance of operatives in the construction project production process [20].

2.3. Relationship of craftsmen supervisors PMS and construction project success

Kerzner [5] acknowledged that PMA allows the arrangement of work flow and project coordination horizontally and vertically, thus resulting in extensive planning and coordination. Moreover the PMA enables work to be organized across the various functional groups that work with each other. This results in improved coordination and communication among employees and managers, as well as in project success through high level productivity, efficiency, and effectiveness [5]. Though several studies have given different approach to what constitute success of a project [5] [26]. Kerzner [5] reported that the success of a project entails meeting budget, timeliness, ensuring quality performance, and meeting customers' satisfaction. Besides [5], other factors that constitute project success are environmental and safety regulations, and delivering the productivity for the manufacturing operations. Khosravi and Afshari [26], identified the following as project success criteria: time performance, cost performance, quality performance, health, safety and environment of the project, and client satisfaction. Thus, this study adopted the project success criteria reported by Kerzner [5], and Khosravi and Afshari [26] to determine the effects of craftsmen supervisors PMS on construction project success. The success criteria are: project cost effectiveness (PCE), attainment of project timeliness (APT), attainment of quality workmanship (AQW), achieving customers/end users requirements (ACEUR), adherence to project specifications (APS), curtailing construction materials waste (CCMW), curtailing construction site accidents (CCSA), ensuring environmental friendliness (EEF), project quality attainment (PQA), less/absence of construction project rework (LACPR), and smooth construction activities execution and integration (SCAEI).

3. Methodology

3.1. Questionnaire design and administration

This study used the three higher levels of cognitive domain of Bloom's revised taxonomy of educational objectives (analysing, evaluating & creating) (see [27] to review PMT propounded by Koskela and Howell [18] in identifying craftsmen supervisors PMS requirements, and other relevant literature to identify construction project success factors in relation to craftsmen supervisors PMS. Subsequently, designed a questionnaire to collect data on: respondents' demographic profile, craftsmen supervisors PMS requirements, and project success in relation to craftsmen supervisors PMS. PM and seasoned academicians in construction project management were used to pre-test the questionnaire. Internal consistency testing of the questionnaire using Cronbach alpha record 0.84, 0.83, and 0.83 reliability values for respondents' profiles, relevance of craftsmen supervisors PMS, and effects of craftsmen supervisors PMS on construction project success respectively. These portray the questionnaire to be reliable and error free [28]. Likert scales of 1-5 were used to assess the two study construct on the: relevance of craftsmen supervisors PMS requirements, and the effects of craftsmen supervisors PMS on construction project success. The scale interpretation for relevance of PMS are: 1-irrelevant, 2-least relevant, 3-fairly relevant, 4-relevant, and 5-very relevant. While the scale for construction project success are: 1-ineffective, 2-least effective, 3-moderately effective, 4-effective, and 5-highly effective. The study population is heterogeneous comprising of 200 accessible construction professionals that are statutorily certified by the Nigerian government to practice project management: architects, builders, engineers, and quantity surveyors [29]. These set of professionals conform to Chitkara's [3] recommendation that construction PM should be persons with atleast a degree or its equivalent in engineering, architecture, and quantity surveying. A sample size of 132 PM was established using Krejcie and Morgan [30] table. Hence 132 questionnaires were administered to the respondents using disproportionate stratified random sampling technique. The heterogeneous nature and the uneven distribution of the study population informs the adoption of disproportionate stratified random sampling technique. The survey was conducted in Abuja the administrative capital of Nigeria because it has the highest concentration of experience construction PM [31] [32]. The survey records 44% (58) valid response rate: architects (arc) (14), builders (bldr) (9), engineers (engr) (25), and quantity surveyors (qs) (10). Respondents are from the public (83.8%) and private (16.2%) sectors of the Nigerian economy respectively. This reflect the divides in the Nigerian economy which is a mix economy (Inuwa, 2014).

3.2. Method of data analysis
Data from the survey were analysed using statistical package for social sciences (SPSS). Percentages and frequencies were used to analyse respondents' profile, while mean scores were used to rank craftsmen supervisors PMS requirements identified. ANOVA with post hoc and Spearman's rank order correlation coefficient were used to test hypotheses mirroring the study objectives, and to determine the relationship of craftsmen supervisors PMS requirement and project success respectively.

4. Results and discussions

4.1. Respondents profiles
Respondents' educational qualifications are: certificate (1.7%), HND (15.5%), PGD (19%), BSc/B.Tech (25.9%), MSc (36.2%), and PhD (1.7%). Despite being certified as PM, their core construction disciplines are rooted either in architecture (24.1%), building (15%), engineering (43.1%), or quantity surveying (17.2%). About 98% of the respondents hold managerial positions in their respective organisations: top management (47%), middle management (28%), lower management (22.4%), and supervisory staff (line managers) (2%). Respondents have an average of 15 years' experience in the construction industry. These results revealed that most of the respondents' are core construction PM that are qualified and experience to respond to an enquiry of this nature and as such, assist the study in ensuring that all the variables that have an influence on the correctness of the data have been analysed. By implication, this ensures that the respondents used for the study are the most appropriate sample [28] [3].

4.2. Craftsmen PMS requirements
Table 2 depicts responses on craftsmen supervisors PMS requirements and their level of significance using ANOVA with post hoc at 95% confidence level. Design comprehension skill (DCS) was ranked very relevant (4.33-4.50) by all the respondents. Total mean ($T\overline{X}$) of the respondents ranked DCS relevant (4.47). Optimal workmanship attaining skill (OWAS) was ranked by all respondents but 'qs' as relevant (4.21-4.40). Quantity surveyors ranked OWAS very relevant (4.80). However OWAS was averaged by all the respondents as relevant (4.36). Task dispatching communication skill (TDCS) was ranked relevant (4.20-4.40), with the respondents $T\overline{X}$ of 4.22 showing it to be relevant. Task execution effectiveness enhancing skill (TEEES) ranked relevant (3.89-4.40) by all the respondent which also corresponds with their $T\overline{X}$ (4.26) as relevant. Task execution implementation skill (TEIS) ranked relevant (3.92-4.07) by all respondents, as well as their total mean ($T\overline{X}$=3.98). The individual groups ranking on task execution monitoring skill, task execution planning skill, task precondition identification skill, task resource organizing skill and task resource scheduling skills were relevant PMS requirement for craftsmen supervisors training respectively (see table 1). The ANOVA in table 2 shows that there is no statistical significant difference in the mean ranking of respondents on the PMS

requirement. Their *p* value exceed 0.05 ($p>0.05$), thus the null hypothesis that there is no statistically significant difference in the respondents mean ranking on PMS requirements for craftsmen supervisors training was accepted. The PMS identified assists craftsmen supervisors in the planning, organising, monitoring and controlling of all aspects of their supervisory role in a construction project as well as, sharpened their motivational skills on their subjects to improve productivity that results in achieving construction project success. Skills identified are in conformity with the factors identified by Alinaitwe *et al.* [33], John *et al.* [10], and Tunji-Olayeni *et al.* [34] as necessary for enhancing construction craftsmen productivity. However their studies were not focused on identifying craftsmen supervisors PMS requirements.

Table 2. Responses on PMS requirements for craftsmen supervisors training.

S/N	PMS	Respondents mean scores								
		arc	bldr	engr	qs	$T\bar{X}$	SD	F	Sig.	LS
1	Design comprehension skill	4.50	4.33	4.48	4.50	4.47	0.60	0.17	0.92	NS
2	Optimal workmanship attaining skill	4.21	4.00	4.40	4.80	4.36	0.72	0.39	0.08	NS
3	Task dispatching communication skill	4.29	4.00	4.20	4.40	4.22	0.84	0.38	0.77	NS
4	Task execution effectiveness enhancing skill	4.21	3.89	4.36	4.40	4.26	0.81	0.87	0.46	NS
5	Task execution implementation skill	4.07	4.00	3.92	4.00	3.98	0.93	0.08	0.97	NS
6	Task execution monitoring skill	4.29	3.89	3.72	4.50	4.02	0.96	2.17	0.10	NS
7	Task execution management skill	4.00	3.89	4.00	4.40	4.05	0.91	0.62	0.61	NS
8	Task execution planning skill	3.86	3.67	3.84	4.00	3.84	0.91	0.20	0.89	NS
9	Task precondition identification skill	3.79	3.56	4.08	4.10	3.93	0.92	0.96	0.42	NS
10	Task resource organizing skill	3.79	4.00	3.88	4.50	3.98	1.02	1.13	0.35	NS
11	Task resource scheduling skill	4.21	4.20	3.88	4.50	4.07	0.88	1.48	0.23	NS

Note: $T\bar{X}$ total mean, SD- standard deviation, F-F ratio, LS- Level of significance, NS-Not significant, S-significant

4.3. Effects of craftsmen supervisors PMS on construction project success

Table 3 shows the result of Spearman's rank order correlation coefficient to determine whether there is a relationship between craftsmen supervisors PMS requirement and construction project success. The result shows the Spearman's rank order coefficient to be 0.731 ($r= 0.731$); thus indicating that there is a positive relationship between PMS and construction project success. When the coefficient was squared and multiplied by 100 ($r^2 \times 100$), the product is 53.44%; this figure is termed the coefficient of determination [28]. The coefficient of determination is a statistic that measures how much an independent variable explains the variance in the dependent variable [35]. This infers that craftsmen supervisors PMS explained 53.44% variances that determine construction project success. Moreover the null hypothesis tested at 99% confidence level stating that there is no positive relationship between craftsmen supervisors PMS requirement and construction project success was rejected, because the computed *p* value is less than 1 ($p<0.01$) [35]. Thus craftsmen supervisors PMS is shown to have a positive influence on construction project success. The construction project success criteria used were adopted from Kerzner [5], and Khosravi and Afshari [26]. However this finding contrast with their study by relating craftsmen supervisors PMS with project success criteria. Moreover it implies that the more craftsmen supervisors acquire PMS, the better it is for construction project success.

IOP Conf. Series: Materials Science and Engineering **513** (2019) 012002 doi:10.1088/1757-899X/513/1/012002

Table 3. Effects of craftsmen supervisors PMS on project success.

Variables			Inferences			
S/N	X (PSF)	Y (PMSRF)	Correlation (ρ)	Coefficient of determination	LS (p)**	Decision on H_0
1	ACEUR	PMSRF	0.598	35.76	0.00	rejected
2	APS	PMSRF	0.748	55.96	0.00	rejected
3	APT	PMSRF	0.684	46.79	0.00	rejected
4	AQW	PMSRF	0.826	68.23	0.00	rejected
5	CCMW	PMSRF	0.655	42.90	0.00	rejected
6	CCSA	PMSRF	0.614	37.67	0.00	rejected
7	EEF	PMSRF	0.547	29.92	0.00	rejected
8	LACPR	PMSRF	0.573	32.83	0.00	rejected
9	PCE	PMSRF	0.608	36.97	0.00	rejected
10	PQA	PMSRF	0.522	27.25	0.00	rejected
11	SCAEI	PMSRF	0.559	31.25	0.00	rejected
12	PSF	PMSRF	0.731	53.44	0.00	rejected

Key: PSF-project success factors, PMSRF-project management skills requirement factors, LS-level of significance, H_0 – null hypothesis, **- correlation is significant at the 0.01 level (1-tailed)

5. Conclusions and recommendations

Studies have shown that the quantity of craftsmen training in Nigeria has declined and their quality is poor compare to other developing and developed countries. This resulted in the NCI performance to be dismal due to poor productivity. Literature reviewed show that PMT is an area that needs to be explores to bridge the gap between craftsmen supervisors (line managers) and the operatives executing construction operations on sites to attain construction project success. Thus this study identified craftsmen PMS requirements and their effects on construction project success. Subsequently, a structured questionnaire was designed and administered to certify construction PM in Abuja using stratified disproportionate random sampling technique. Data were analysed using descriptive and inferential statistics. All the craftsmen PMS requirements identified were ranked relevant, and their PMS are found to be positively related to construction project success. This implied that craftsmen supervisors PMS could bridge the gap between PMA and construction operations execution. Moreover craftsmen supervisors PMS can enhance construction supervision and operations productivity to attain construction project success. Hence the study recommends PMS to be integrated in craftsmen supervisors training in Nigeria.

References

[1] N J Smith 2002 *Engineering Project Management* 2 edn. ed. Oxford: Blackwell Science Ltd.

[2] L Ireland 2006 *Project Management: Past, Present, Future* Available: www.osun.org.

[3] K K Chitkara 2012 *Construction Project Management: Planning, Scheduling & Controlling* New Delhi: Tata McGraw-Hill

[4] E M Awe , P Stephenson and A Griffith 2013 *Impact of Vocational Ttraining on Skilled Labour Shortage within the Nigerian Construction Industry* vol. 2 no. 3 pp. 156-162

[5] H Kerzner 2000 A*pplied Project Management: Best Practices on Implementation* New York: John Wiley & Sons, Inc.

[6] N D Usman, I I Inuwa and A I Iro 2012b The Influence of Unethical Professional Practices on the Management of Construction Projects in North Eastern Sates of Nigeria *International Journal of Economics Development Research and Investment,* pp. 3(2), 124-129

[7] K Ness 2009 Not Just About Bricks: the invisible building worker in *In: Dainty, ARJ (Ed.) Proceedings of the 25th Annual ARCOM Conference. 7-9 September, 2009*, Nottingham, UK 1: 635-644, 2009

[8] P G Chindo, A D Ibrahim, Y M Ibrahim and Y G Musa-Haddary 2015 Assessing The Performance of Nigerian Trained Construction Craft Skills in *Proceedings of the 2nd Research Conference of the Nigerian Institute of Quantity Surveyors*, Akure-Nigeria

[9] Africa Economic Brief (AEB) 2015 Closing South Africas skilled Worker Gap: higher

[10] I B John, O J Faremi and O S Lawal 2016 Construction Craftsmen Skill Needs and Training in the Nigerian Construction Industry *Lagos Journal of Environmental Studies* **8** pp. 72-84

[11] A O Onuka and K O Ajayi 2012 Effects of Manpower Development on Workers Job Performance *European Journal of Educational Studies* **4** pp. 423-433

[12] P O Adewale, A B Siyanbola and S O Siyanbola 2014 Building Craftsmen Skill Development and Nigeria's Vision 20: 20-20: Imperative & Daunting Challenges *International Journal of Vocational and Technical Education* **6** pp. 36-42

[13] J S Dantong 2007 Training of Construction Craftsmen in the Nigerian Construction Industry MSc Thesis, University of Jos, Nigeria

[14] N Aniekwu and C A Ozochi 2010 Restructuring Education, Training & Human Resouce Development in the Nigerian Counstruction Industry *Journal of Sciences & Technology Education Research* pp. 1(5), 92-98

[15] NBTE 2001 National Diploma & Higher National Diploma Curriculum, Kaduna-Nigeria: NBTE

[16] Federal Ministry of Labour-FML, 2006, Labour Trade Test Curriculum, Abuja-Nigeria: FML

[17] Universal Basic Education Board (UBEB) 2008 Curriculum of Technical College, Abuja: UBEB

[18] L Koskela and G Howell 2002a The Theory of Project Management: Explanation to Novel Methods in *Proceedings IGLC-10*, Gramado-Brazil

[19] L Koskela and G Howell 2002bThe Theory of Project Management-Problem and Opportunity in *Working Paper. VTT Technical Research Centre of Finland & Lean Construction Institute*

[20] I IInuwa 2014 Project Planning In Construction Procurement: The Case of Nigerian Indigenous Contractors PhD Thesis: Jomo Kenyatta University of Agriculture and Technology, Nairobi-Kenya

[21] H Knoepfel 1992 Theory and Practice of Project Management in Construction *International Journal of Project Management,* pp. 10 (4), 243-251

[22] K Kraemer, G Henrich, L Koskela and M Kagioglou 2014 How Construction Flows Have Been Understood in Lean Construction [Online]. Available: http://www.laurikoskela.com.

[23] J W Weiss and R K Wysocki 1992 5-Phase of Project Management: A Practical Planning & Implementation Guide, Cambridge, Massachusetts: Perseus Books

[24] Oxford Dictionary of English 2010, Oxford Dictionary of English, Oxford: Oxford University Press.

[25] · I I Inuwa and S U Kunya 2015 Project Management Theory Exploration for Indigenous Contractors' Project Planning in Nigeria *International Journal of Applied Science and Technology* **5** pp. 60-72

[26] S Khosravi and H Afshari 2011 A Success Measurement Model for Construction Projects Singapore

[27] G Guthrie 2010 *Basic Research Methods: An Entry to Social Science Research* New Delhi: SAGE Publications India Pvt Ltd

[28] J Pallant 2011 S*PSS Survival Manual: A Step by Step to Data Analysis Using SPSS* New South Wales:Allen & Unwin

[29] D Ekundayo, C Jewell and O A Awodele 2013 Executive Project Management Structure and the Challenges Facing its Adoption in the Nigerian Construction *International Journal of Architecture, Engineering and Construction,* pp. 2(3),158-169

[30] R Krejcie and D W Morgan 1970 Determining of Sample Size for Research Activities *Educational & Psychological Measurement,* pp. 30, 607-610

[31] J O Ameh and K T Odusami 2010 Nigerian Building Professionals' Ethical Ideology and Perceived Ethical Judgement *Australian Journal of Construction Economics and Building,* pp. 10(3),1-13

[32] A Oladapo 2006 The Impact of ICT on Professional Practice in Nigerian Construction Industry *The Electronic Journal on Information System in Developing Countries,* pp. 1-19

[33] H M Alinaitwe, J A Makwali and B Hansson 2007 Factors Affecting the Productivity of Building Craftsmen-Studies of Uganda *Journal of Civil Engineering and Management* **13** pp. 169-176

[34] P F Tunji-Olayeni, T O Mosaku, O I Fagbenle and L M Amusan 2016 Project Management Competencies of Indigenous Contractors in Nigeria *African Journal of Built Environment Research* pp. 1 (1) 49-58

[35] J Pallant 2011 *SPSS Survival Manual: A Step by Step to Data Analysis Using SPSS* New South Wales: Allen & Unwin

[36] J P Lewis 2007 F*undamentals of Project Management 3rd edn.* New York: AMACOM

[37] A Roberts and W Wallace 2004 P*roject Management, Edinburgh*: Pearson Education

[38] Y Wang and E D Gibson 2008 A Study of Preproject Planning and Project Success The 25th International Symposium on Automation & Robotics in Construction, Vilnius

[39] S K Bokinni 2005 Skill Acquisition and Development for Craftsmen and Artisans *The Professional Builder, Journal of the Nigerian Institute of Building,* pp. 100-111

[40] O F Femi 2014 Building Construction Technician Training: It's Relevance to Modern Construction Industry in Nigeria *International Journal of Technology Enhancement and Emerging Engineering Research* **2** pp. 58-68

[41] C C Nwachukwu and F I Emoh 2011 Building construction project management success as a critical issue in Real Estate Development and Investment *American Journal of Social & Management Sciences* **2** pp. 56-75

[42] "The Balance," 11 February 2017. [Online]. Available: http://www.thebalance.com.

[43] W Y Oso and D Onen 2011 Research proposal and report writing: A handbook for beginning researchers, Nairobi: the jomo kenyatta foundation

[44] A I Iro, I I Inuwa and J S Dantong 2013 Investigation Into Contractors Craftsmen Training in the Nigerian Construction Industry *International Journal of Engineering Research and Technology* pp. 2(1),1-6

IOP Conf. Series: Materials Science and Engineering **513** (2019) 012003 doi:10.1088/1757-899X/513/1/012003

Building projects delivery challenges on remote sites in northern Nigeria

Auwalu Sani Shehu[1], Yakubu Ibrahim[1] and Ibrahim Ibrahim Inuwa[1*]

[1]Department of Quantity Surveying, Faculty of Environmental Technology, Abubakar Tafawa Balewa University (ATBU), Bauchi-Nigeria

*Corresponding author: iiinuwa@atbu.edu.ng

Abstract. Most building projects sited on remote sites in Nigeria fail due to less attention given to logistics. Hence, this paper identifies and assesses challenges faced by such projects, as well as strategies for curbing them. Literature review, unstructured interview, and personal observation were employed to design a structured questionnire for the study. A total of 117 questionnaires were administered to building procurement professionals through disproportionate stratified random sampling technique in northern Nigeria. Subsequently, descriptive and inferential statistics were used to analyze data obtained. Six challenges and ten strategies were identified. The challenges were assessed to be severe; with resource scarcity, predetermined designs, and poor accessibility found to be extremely severe. While maintaining retainer-ship with suppliers, outsourcing work force, integrating local community in works, stacking materials before starting works, and subletting labour were ranked highly effective strategies for curbing the challenges. The study recommends: appropriate attention on project site logistics; government provision of socio-economic infrastructures in rural areas; and the adoption of effective curbing strategies where challenges seem inevitable.

1. Introduction

The concern of key construction project stakeholders globally (clients, consultants and contractors) is to deliver a project that meets time, cost, quality, and other requirements that define project success [1,2,3,4]. The quest for attaining construction project success in a developing country, in contrast with a developed country, can be made more difficult if the project is located on a remote site, due to inadequate socio-economic infrastructures that ensure quality, timeliness, and budget effectiveness [1] [2]. This agrees with Ofori [1] and Callahan *et al.* [3] assertion that attainment of construction project success in developing countries is increasingly difficult, due to the absence of the necessary external factors that are attributable to site location. Thus, resulting in projects time and cost overruns, and quality being compromised. In Nigeria, logistic problems associated with project location on remote sites according to Shehu [4], attracts little attention. The objectives of this paper therefore, is to identify and assess challenges faced by counstruction projects on remote sites in northern Nigeria, as well as strategies for curbing them. In other to achive its objectives, this study sought to answer the following questions: what are the challenges faced by building projects on remote sites in northern Nigeria; and what are the strategies for curbing them?

2. Review of literature

2.1. Nigeria as a developing country

Going by the UNDP [5] report, there are over 100 countries globally that are referred to as developing countries; Nigeria inclusive. A developing country is a country with an economy characterised by poor level of industrialisation, less income per capita, low life expectancy, low rate of literacy, and poor standard of living; such a country is also referred to as a developing economy or an underdeveloped country. The Human Development Index (HDI) determine a country's level of development. The HDI is a statistic that aggregates a country's level of economy/development in a composite index; it consider level of industrialisation, income per capita, life expectancy (health measure), literacy level, and standard of living [5]. Thus by implication, Nigeria as a developing country lacks the adequate infrastructures that support industrialisation, capital formation and investment potentials, health, education, and agriculture. Hence, their economic deficiency constitutes logistical problem that adversely affects the attainment of project success especially on a remote site.

2.2. Challenges faced by building projects located on remote sites

Remote construction sites are sites that are situated far from the main centres of population; they are in most cases difficult to access [4] [6]. By their characteristics, most remote sites are found in developing countries [7]. This is virtually because of the level of under development and the concentration of development in urban centers [8]. Consequently, leads to paucity of construction resources in the rural communities. It follows therefore that the remoter the location the higher the scarcity of construction resources and the costlier the construction development. The features of a remote site have adverse disadvantages on the realization of a project. These disadvantages pose severe challenges to project success, and they include: poor communication; poor accessibility; distance; expensive site situations; pre-determined designs; and resource scarcity [13,14,15,16].

2.2..1 Poor communication

According to Sidawi [9], remote construction project site exists in many regions throughout the world. One of the problem faced on such site, is in managing the communication chains in implementing such projects. Poor communication on construction site will obviously affect project planning, as well as project process integration [9]. This will leads to misinterpretations and miscommunications of project results and needs issues. Szava *et al.* [10] revealed that communication problem makes it difficult to remobilize workers back to remote sites to effect remedial works after the expiration of defects liability period. This poses a problem that usually results in arbitration and in some cases to ligation as the owner perceives the project as being uncompleted.

2.2.2. Poor accessibility

Remote sites are mostly characterized by poor accessibility, hence making the contractor incurs more cost that is transferred to client as site overheads. In most of the remote sites in northern Nigeria, access roads are either absent or in poor conditions. The situation in most cases calls for the construction of access roads; in some cases leading to delay in project execution and cost overrun. Thomas [11] confirmed that poor accessibility leads to high cost of labour, money, the technology and methods to be used in construction. This is because in most remote sites adhering to a strict construction method may mean adopting the use of certain equipment and personnel which in turn means increasing the scope of logistic supports (preliminaries) required to achieve that [12]. Generally, accessibility of any site commands the cost of achieving project goals set at the brief and design stages [9]. It is therefore pertinent to know the level of accessibility of a site to the human and material resources needed to be mobilized for the actualization of the project at that stage.

2.2.3. Distance

Szava *et al.* [10] reported that the overall increase in the cost of construction in remote locations is compounded by the rising cost of transportation which increases progressively. The impact of higher transport costs is felt where human and material resources need to be transported from central regions to remote site where important construction projects are sited [13]. Moreover, fuel price which is another component of transport cost increases accordingly. Szava *et al.* [10] further reported that labour cost shot up to 60% of the total construction cost in remote areas which may result in pay hesitant and consequent resentment of people to work in remote sites. Similarly, Sidawi [9] pointed out distance-specific problems occasioned by remote sited projects to:

i. Delay in achieving project goals due to mistakes in construction, delay in project time-table, personnel safety issues, poor quality in construction, low productivity of workers
ii. Using unskilled workers due to unavailability of skilled workers ;
iii. Low productivity due to inaccessibility to specified materials on time and shortage of construction equipment; and
iv. Unavailability of supervisors when needed.

In some situations contractors have had to contend with carrying out works using the obtainable skills with very little input from appointed supervisors due to distance and the fact that there may be other projects in near places to supervise at the same time [14]. In Nigeria, many projects were completed in remote locations without the predetermined supervision due to distance of the project location.

2.2.4. Expensive site situations

Most construction sites in remote areas are small and medium sized, in cases where such projects are large and gigantic, the logistic supports allowed in the project plan takes care of the infrastructure challenges, for instance; access roads, water supply, electricity, communication and resident supervision and accommodation [15]. However, where projects are small and medium size, such provisions to take care of the logistic support cannot be incorporated in the project plans as in many situations such logistic requirements are costlier than the actual project itself. On remote sites where operatives cannot be shuttled every day due to distance, the need to provide accommodation becomes inevitable. Moreover, welfare facilities such as mess room, convenience, and other social amenities will be needed to provide a suitable working condition. Contractors handling construction projects located on remote sites in Nigeria find it difficult to address issues emerging from poor logistics consideration due to inappropriate attention given to logistics. Thus, putting them in a dilemma as to whether they are doing business or running a charity. Usually, the preliminaries allowed in such project cannot take care of the logistic support requirement of the projects, hence a contractor has to make do with substandard materials and poor workmanship, which eventually result in project failure. In addition [10], the isolation and long transport distances have a compounding effect on all other logistics issues, thereby increasing normal risk. This means that any unforeseen events, conditions, or even small mistakes like miscalculating the amount of materials needed, or forgetting a tool, can have substantial consequences.

2.2.5. Predetermined designs

Most projects (e.g. dispensaries, schools, etc.) sited on remote sites in Nigeria, which are sponsored by public and development partners, adopt prototype designs. Such projects faced serious challenges because the design adopted failed to consider the dynamics (logistics) of each site [4]. Consequently, neglecting the uniqueness of each project [21,22]. Accordingly, when the uniqueness of a project manifest itself, it becomes a problem that was not foreseen and planned for, thus resulting in excessive variations, litigation, or eventually in abandonment.

2.2.6. Resource scarcity
Construction resource encompasses both human, material, and equipment and plants. It is very evident that construction projects sited on remote sites experience resource scarcity especially in a developing country [5,15,14,17]. In most remote sites in Nigeria, lack of skilled operatives compelled contractors to engage substandard operatives, thus contributing to resource management difficulty. The transporting and maintenance of plants needed on site is certainly a challenge too, and frustrates projects on remote sites in the attainment of success [12,13,23].

3. Methodology

3.1. Study area
This study was conducted in the north-eastern and north-western geopolitical zones of Nigeria. The two zones constitute slightly more than one-third of Nigeria's 36 States (13 states). They occupy a land size of 512, 572.06 sq. km representing almost 55% of Nigeria's total land size of 937,052.16 sq. km and accommodate a population size of over 55 million people, representing 39% of Nigeria's total population of over 140 million people [16]. In the north-eastern zone, Bauchi and Gombe States were selected as a sub-study area, while Kano, Katsina, and Jigawa States were selected in the north-western zone to complement the research study area. These States were selected because of their large population and concentration of construction activities in their respective geo-political zones [24,25,26].

3.2. Research design, population, sample size and sampling technique
This study's objectives inform the adoption of an exploratory (interviews & literature review) and descriptive (observation & questionnaire survey) research design [24,27]. Findings from literature review, interviews and physical observations on remote sites were used to design the study questionnaire. The population for this study comprised of a set of core building procurement professionals: architects; engineers; and quantity surveyors. They were drawn from 75 identified remote sites in the north-eastern and north-western geo-political zones of Nigeria. Hence, given a population size of 225 professionals from 75 remote sites. Subsequently Krejcie and Morgan [17] table was used to arrive at a sample size of 144. Disproportionate stratified random sampling technique was employed to administer questionnaire; this technique was used to capture all the sites, as well as to proportionately address the dispoportionate numbers of respondents on the various sites identified [18].

3.3. Method of data analysis
This study used Microsoft excel to conduct descriptive and inferential statistics analyses on the data obtained from questionnaire responses. A 5-point Likert scale of severity (scale: 5-extremely severe; 4-severe; 3-moderately severe; 2-least severe; & 1-not severe) and relative effectiveness (scale: 5-highly effective; 4-effective; 3-moderately effective; 2-least effective; & 1-ineffective) index (see: eqn. 1 & 2) were used to rank the severity of challenges faced by projects on remote sites and the relative effectiveness of the strategies for curbing the challenges respectively. The formula used for the computation of the ranked severity index and relative effectiveness index are given as [4]:

$$Severity\ Index(SI) = \sum {}^{IF}/_{N} \qquad (1)$$

- *Range for SI:* 0-1.44 not severe; 1.45-2.49 least severe; 2.45-3.49 moderately severe; 3.45-4.49 severe; 4.50-5.00 extremely severe.

$$Relative\ Effective\ index\ (REI) = \sum_{i=1}^{n} {}^{aixi}/_{3} \qquad (2)$$

- *Range for EI*: 0-1.49 ineffective; 1.45-2.49 least effective; 2.45-3.49 moderately effective; 3.45-4.49 effective; 4.50-5.00 highly effective.

4. Results and discussions

Table 1 depicts that a total of 117 valid questionnaires were returned out of a total of 144 distributed; representing 81.25% response rate. This response rate is higher than other studies carried out in the construction industry. For instance: Usman *et al.* [19] recorded a response rate of 55.25%; Iro *et al.* [20] received a 55.25% response rate; Ibrahim [21] recorded a 47% response rate; [22] recorded a 25.4% response rate; Olatunji [23] obtained a 33.5% response rate, Crafford [24] received a response rate of 22.8%. Hence, making it more appropriate for generalization of the results on the study population [35,36]. Table 1 also shows that all the respondents' used for the study are specialized in core building procurement professions [25] and about 89% of the respondents' have more than six years working experience in the construction industry. This indicates that all the respondents' are knowledgeable and aware of the problem being investigated.

Table 1. Respondents' profile & questionnaire administration.

Questionnaire response rates				Respondents' organizations' profile					
Distribution				Profile			Profile		
respondent	out (no.)	back & valid (no.)	% response rates	educational specialization	Freq	%	working experience (years)	Freq	%
Architects	53	40	75.47	Architecture	40	46.43	≤ 5	13	11.43
Engineers	48	35	72.92	Engineering	35	33.57	6-10	60	51.42
Q/Surveyors	43	42	97.67	Q/Surveying	42	30.00	>10	34	37.15
Total	**144**	**117**	**81.25**	**Total**	**117**	**100**	**Total**	**117**	**100**

Source: Field survey (2015)

Table 2 portrays the respondents' ranking of the relative severity and relative effectiveness index of the severity of challenges building projects faced on remote sites in northern Nigeria, and the effectiveness of the strategies for curbing them respectively. All the identified challenges were ranked within the range of 3.9714-4.7140, revealing that all the challenges are experienced at a severe level; with resource scarcity, predetermined designs, and poor accessibility ranked as extremely severe challenges.

Table 2. Ranking of relative Severity and relative effectiveness index.

Ranking of relative severity index (RSI) of project challenges on remote sites				
S/N	Challenges	RSI	Severity	Rank
1	Resource Scarcity	4.7140	Extremely severe	1st
2	Predetermined designs	4.6857	Extremely severe	2nd
3	Poor accessibility	4.5444	Extremely severe	3rd
4	Distance	4.0286	Severe	4th
5	Poor communication	4.0286	Severe	4th
6	Expensive site situation	3.9714	Severe	5th
Ranking of relative effectiveness index (REI) of strategies for curbing challenges				
S/N	Strategies	REI	effectiveness	Rank
1	Maintaining retainer-ship with suppliers	4.6123	highly effective	1st
2	Outsourcing work force	4.6142	highly effective	2nd
3	Integrating local community in works	4.6211	highly effective	2nd
4	Stacking Materials before starting works	4.5122	highly effective	3rd
5	Subletting labour	4.5001	highly effective	4th
6	Partnering with material suppliers	4.4911	effective	5th
7	Pay increase to worker	4.1190	effective	6th
8	Subletting part of the works	3.9523	effective	7th
9	Sacrificing Profit to achieve delivery	2.2315	Least effective	8th
10	Working slowly as materials arrived	2.2215	Least effective	9th

Source: Field survey (2015)

Table 2 shows that maintaining retainer-ship with suppliers, outsourcing work force, integrating local community in works, stacking materials before starting works, and subletting labour strategies are highly effective in curbing challenges faced by building projects located on remote sites in northern Nigeria. Partnering with material suppliers, pay increase to workers, and subletting part of the works are ranked to be effective strategies in curbing challenges encountered by building projects sited on remote sites in northern Nigeria, while sacrificing profit to achieve delivery and working slowly as materials arrived are least effective strategies. The findings of this study are in accord with previous studies on the identified projects challenges on remote sites and strategies for curbing them [13,14,15,17,18,19,37,38]. However, this study is in contrast with previous studies in view of its assessment of the severity of challeneges faced by building projects sited on remote sites in northern Nigeria, as well as on the assessment of the strategies for curbing them.

5. Conclusion and recommendation

This paper identified and assesses challenges faced by building projects on remote sites in northern Nigeria, as well as strategies for curbing them. The study employed exploratory and descriptive survey designs. Six challenges on remote sites were identified to be severe; with resource scarcity, predetermined designs, and poor accessibility assessed to be extremely severe. Ten strategies were identified as strategies for curbing the projects challenges on remote sites, with the following strategies found to be highly effective; maintaining retainer-ship with suppliers, outsourcing work force, integrating local community in works, stacking materials before starting works, and subletting labour. This paper recommended that: projects on remote sites should be considered on their unique site logistics; government should provides basic socio-economic infrastructures that support economic development in rural areas; and where challenges seem inevitable, effective strategies for curbing such challenges should be adopted.

References

[1] O. O. Aina and A. B. Wahab 2011 An Assessment of Buildability Problems in the Nigerian Construction Industry, *Global Journal of Researches in Engineering* 11 (2), version 1.0, 43-52.

[2] A. Ogwueleka 2011 "The Critical Success Fcators Influencing Project Performance in Nigeria," *International Journal of Management Science and Engineering Management* 6(5) 343-349.

[3] R. Takim and A. Akintoye 2002 Performance Indicators for Successful Construction Project Performance *in Association of Researchers in Construction Management Conference*, University of Northumbria.

[4] P. A. Bowen, K. A. Hall, P. J. Edwards and R. G. Pearl 2000 Perceptions of Time, Cost & Quality Management on Building Project *Australian Journal of Construction Economics & Building* **2**(2), 48-56.

[5] G. Ofori 2014 Nature of Construction Industry, its Needs & its Development: A Review of Four Decades of Research *in CIB W107 Conference*, Lagos, 2014.

[6] RICE 2006 Construction Economy Report No. 47. Research Institute of Construction and Economy, Tokyo.

[7] A. Callahan 2012 Achieving Superior Delivery of Capital Projects. Denver: Accenture. 2012.

[8] A. S. Shehu 2015 Strategies for Successful Building Projects Delivery on Remote Sites in Nigeria MSc Thesis: Abubakar Tafawa Balewa University, Bauchi-Nigeria, 2015.

[9] UNDP, "UNDP Human Development Reports," 4 October 2015. [Online]. Available: hdr.undp.org.

[10] Oxford Dictionary of English, Oxford Dictionary of English, Oxford: Oxford University Press, 2010.

[11] C. Plessis 2007 "A Strategic Framework for Sustainable Construction in Developing Countries.," Construction Management and Economics **25**(1) 67 – 76.
[12] Accenture 2012, Developing Strategies for effective Delivery of Capital Projects., Texas: Accenture, 2012.
[13] B. Sidawi 2012 Management Problems of Remote Construction Project and Potential IT Solutions: A case for Saudi Arabia *Journal of Information Technology in Construction*, ISSN 1874 – 4753, 103 – 120.
[14] J. McIllwraith 2005 Dig Deep to fill skills shortages – Resources Boom: Special Report.," The Australian, Australia.
[15] A. M. M. W. B. and. W. G. Szava 2007 The Cost of Housing In remote Indigenous communities from Northern Territory Construction Industry.," Alice Springs, Centre for Appropriate Technology.
[16] S. Olander and A. Landin 2005 Evaluation of Stakeholder Influence in Implementation of Construction projects. *International Journal of Project Management* **23**(4) 321 – 328.
[17] H. Thomas 2002 "Construction Practices in Developing Countries *Journal of Construction Engineering and Management* **128**(1) 1 – 7.
[18] A. O. Ralph and R. Iyagba 2012 Factors Affecting Contractor Performance: A Comparative Study of Non Value-Adding Activities in Nigeria and Indonesia *Journal of Emerging Trends in Economics and Management Sciences (JETEMS)* **3**(5) 2141-7024.
[19] S. Scally 2011 "Remote Area Building and Sustainable Development," Online journal Environmental Design Guide. ISSN: 1442 – 5017, retrieved via www.environmentdesignguide.com.au retrieved on 27/07/2012, 2011.
[20] N. Gil, Tommelein, I.D., and Ballard, G. 2012 Theoretical Comparison of Alternative Delivery Systems for Projects in Unpredictable Environments Journal of Construction Management and Economics, p. 495 – 508.
[21] K. K. Chitkara, Construction Project Management: Planning, Scheduling & Controlling, New Delhi: Tata McGraw-Hill, 2012.
[22] H. Kerzner, Applied Project Management: Best Practices on Implementation, New York: John Wiley & Sons, Inc., 2000.
[23] G. E. Gibson, G. O. Migliaccio and J. T. O'Connor 2007 Changing Project Delivery Strategy: An Implementation Framework *Journal of Civil and Environmental Engineering* p. 205 – 215.
[24] I. I. Inuwa, *Project Planning In Construction Procurement: The Case of Nigerian Indigenous Contractors* PhD Thesis: Jomo Kenyatta University of Agriculture and Technology, Nairobi-Kenya, 2014.
[25] N. D. Usman, I. I. Inuwa and A. I. Iro 2012 "The Influence of Unethical Professional Practices on the Management of Construction Projects in North Eastern Sates of Nigeria," International Journal of Economics Development Research and Investment **3**(2) 124-129.
[26] J. O. Ameh and K. T. Odusami 2010 Nigerian Building Professionals' Ethical Ideology and Perceived Ethical Judgement *Australian Journal of Construction Economics and Building*, **10**(3) 1-13.
[27] W. Y. Oso and D. Onen, Writing Research Proposal and Report: A Handbook for Beginning Researchers, Rsvd edn., Nairobi-Kenya: The Jomo Kenyatta Foundation, 2011.
[28] R. Krejcie and D. W. Morgan 1970, Determinig of Sample Size for Research Activities *Educational & Psychological Measurement* **30** 607-610.
[29] C. Dawson, Practical Research Methods: A user-friendly guide to mastering Research Techiques & Projects, Oxford: howtobooks, 2002.

[30] A. I. Iro, I. I. Inuwa and J. S. Dantong 2013 Investigation Into Contractors Craftsmen Training in the Nigerian Construction Industry *Internation Journal of Engineering Research and Technology* **2**(1) 1-6.
[31] A. D. Ibrahim 2011 *Investigating Into the Knowledge Requirements of Nigerian Quantity Surveyors to Meet Future Challenges in 25th Biennial Conference of The NIQS* (pp. 7-18), Abuja, Nigeria, 2011.
[32] F. A. Emuze 2011 "Performance Improvement in South African Construction," PhD Thesis NMMU, South Africa, 2011.
[33] A. A. Olatunji 2010 "Influences on Construction Project Delivery Time," PhD Thesis NMMU, South Africa, 2010.
[34] G. J. Crafford, "Clients' Views on Construction and Design Team Competencies," PhD. Thesis: NMMU, South Africa, 2007.
[35] A. A. Sambo, Statistical Principles for Research in Education and Social Sciences, Zaria-Nigeria: S. Asekome & co, 2008.
[36] G. Guthrie, Basic Research Methods: An Entry to Social Science Research, New Delhi: SAGE Publications India Pvt Ltd, 2010.
[37] S. and B. B. McAnulty, "Construction Challenges in Remote Australian Locations. UK." in 26th Annual Association of Researchers in Construction Management Conference Leeds, 2010.
[38] M. Datta 2009, "Challenges Facing the Construction Industry in Developing Countries," *Journal of Science and Environmental Research* 78 – 89.
[39] F. Bennett 2003 The Management of Construction Projects: A Life Cycle Approach, Oxford: Butterworth-Heinnemann.
[40] R. L. D. N. R. and. U. S. Fellows 1990 Construction Management in Practice. London.: Longman.

IOP Conf. Series: Materials Science and Engineering **513** (2019) 012004 doi:10.1088/1757-899X/513/1/012004

Strength and water absorption properties of lightweight concrete brick containing expanded polystyrene and palm oil fuel ash

A Suraya Hani[1*], M Y Nurain Izzati[2], S Shahiron[2], A Sallehuddin Shah[2], O Mohamad Hairi[1], A Mohd Sufyan[1], A J Mohamad Luthfi[1], J Zalipah[2], W A Mohamad Nor Akasyah[3] and K Nurul Amirah[1]

[1]Department of Civil Engineering, Faculty of Engineering Technology, Universiti Tun Hussein Onn Malaysia, Pagoh, Johor, Malaysia
[2]Faculty of Civil and Environmental Engineering, Universiti Tun Hussein Onn Malaysia, Batu Pahat, Johor, Malaysia
[3]Melayu Jati Enterprise, Batu Pahat, Johor, Malaysia

[*]Corresponding author: suraya.uthm@gmail.com

Abstract. This paper is focusing on strength and water absorption properties of lightweight concrete brick containing Expanded Polystyrene (EPS) and Palm Oil Fuel Ash (POFA) as partial replacement materials of fine aggregates and cement respectively. EPS has been chosen as lightweight aggregates material due to its characteristic which is extremely light. Meanwhile, POFA has been chosen as cement replacement due to the cementitious characteristic which could act as the binder in the concrete mixture. The replacement percentage of EPS is 0%, 20%, 30%, 40% and 50% whilst the replacement percentage of POFA is 0%, 5%, 10%, 15%, 20% and 25%. The brick properties that have been investigated in this study are density, water absorption and compressive strength. Based on the experimental results, the density and compressive strength of the brick is decreased as the percentage of the replacement increased. Meanwhile, for water absorption, it was found that the percentage of water absorption of brick was increased as the percentage of POFA increased, and it was decreased as the percentage of EPS increased. Based on the findings, the properties obtained has satisfied the requirement where the brick density for lightweight should be less than 1680 kg/m^3 and the strength for load bearing and non-load bearing brick is 11.7 MPa and 3.45 MPa for each individual unit. Meanwhile, for water absorption, the percentage of water absorption of brick should be less than 12% [1,2]. From this study, it was found that, the replacement of sand and cement by EPS and POFA give significant impact towards density, strength and water absorption performance of concrete brick.

1. Introduction

Due to rapid growth of population in Malaysia, Malaysian government is focusing on low and medium cost housing project as stated in the Eleventh Malaysia Plan (2016-2020) in the third strategic thrusts where government is uplifting B40 households towards middle-class society by increasing the provision of affordable housing. However, in the effort of increasing the number of housing, government strictly aware on the sustainable issues especially in the construction industries. Therefore, due to high demand in housing and sustainability awareness, construction industries and Malaysia researchers has collaborates in the production of sustainable building materials such as lightweight concrete brick.

Abundance of research on the production of lightweight concrete brick has been conducted [3-5]. One of the most preferable lightweight aggregates materials that have been widely used is EPS. This is due to its characteristic which is extremely light. EPS has significantly contribute to the reduction of brick density [3-13]. However, based on findings by previous researches, EPS concrete or EPS brick has low in strength [3-13]. Due to this, researchers have extend their research in improvising the EPS concrete brick strength with the addition of pozzolanic materials such as fly ash, rice husk ask and silica fume [14-20]. The ability of these pozzolanic materials in enhancing the strength of brick or concrete has been proven. For instance, in a study conducted by Ling & Teo [3] has shown that, replacement of cement by 10% of Rice Husk Ash (RHA) has increased the brick strength from 13 MPa (normal brick) up to 17 MPa. In other study, by Sadrmomtazi et al [14], 10% of silica fume has increased the strength of concrete from 43 MPa to 47.6 MPa. It has proven that, optimum replacement percentage by pozzolanic materials could enhance the strength performance of brick or concrete.

Therefore, for this study another pozzolanic material which is POFA has been used as cement replacement in the production of lightweight concrete brick. POFA is an agricultural waste produced by the palm oil industries. Malaysia has known as one of the biggest palm oil producer in the world. The vast production of POFA by these industries has become one of the environmental issues in Malaysia. This is because this waste could not be used as fertilizer. Therefore, from time to time, the generation of this waste has increase the landfill area. In order to overcome this problems, researcher has extended their research on POFA as cement replacement [20-24]. Hence, this study has focus on production of concrete brick containing EPS and POFA with different percentage of replacement. Among the brick properties that has been tested are density, compressive strength and rate of water absorption.

2. Materials and brick sample preparation

2.1. Materials

In this study, raw materials that have been used for the production of brick are Ordinary Portland Cement (OPC), river sand, Palm Oil Fuel Ash (POFA), Expanded Polystyrene (EPS), tap water and super plasticizer.

2.1.1. Ordinary Portland cement (OPC)

The classification of OPC used in this study is type 1 based on the ASTM C150 [25]. The supplier for this cement is Tasek Corporation Berhad which certified by SIRIM. Cement need to be stored in an air tight container to prevent any moisture contact. The chemical composition of OPC has shown in Table 1.

Table 1. Chemical Composition of OPC and POFA.

Chemical Composition	OPC	POFA
Silicon Dioxide (SiO_2)	14.6	55.20
Aluminium Oxide (Al_2O_3)	3.95	4.48
Ferric Oxide (Fe_2O_3)	3.46	5.44
Calcium Oxide (CaO)	57.1	4.12
Potassium Oxide (K_2O)	0.51	2.28
Magnesium Oxide (MgO)	1.62	2.25
Sodium Oxide (Na_2O)	-	0.1
Sulfur Trioxide (SO_3)	3.43	2.25

2.1.2. Palm Oil Fuel Ash (POFA)

POFA was obtained from one of Palm Oil mill in Batu Pahat, Johor. The ash was initially dried in oven at 105°C for 24 hours in order to remove the moisture content. After dried, POFA was sieved by 300 μm sieve passing. Table 1 shows the chemical composition of POFA obtained from the XRF analysis.

2.1.3. Fine Aggregates (sand)

Fine aggregates that has been used is river sand. The sieve analysis of sand was conducted in accordance to ASTM C136 [26]. The sand was dried in the oven at 105°C±5 for 24 hours to ensure the moisture content is constant. Then, the sand was sieved through 2.36 mm sieve opening.

2.1.4. Expanded Polystyrene (EPS)

The size of EPS used in this study was between 1.18 mm to 2.36 mm. EPS used was collected from ST Polyfoam Industries Sdn. Bhd which located in Batu Pahat, Johor.

2.2. Brick sample preparation

For this study, 30 types of concrete brick samples were prepared. The percentage of EPS is 0%. 20%, 30%, 40% and 50% meanwhile the percentage of POFA is 0%, 5%, 10%, 15%, 20% and 25%. Water cement ratio for this study is 0.5 and super plasticizer is 8ml for every 1 kg binder. The mix proportion of bricks as shown in Table 2.

Table 2. Mix proportions (kg/m^3).

Samples	Cement	POFA	Sand	EPS
E0P0 (C)	495.34	-	1484.72	-
E20P0	495.34	-	1186.81	3.98
E30P0	495.34	-	1038.19	5.98
E40P0	495.34	-	888.88	7.97
E50P0	495.34	-	744.44	9.96
E0P5	470.13	25	1484.72	-
E20P5	470.13	25	1186.81	3.98
E30P5	470.13	25	1038.19	5.98
E40P5	470.13	25	888.88	7.97
E50P5	470.13	25	744.44	9.96
E0P10	445.14	50	1484.72	-
E20P10	445.14	50	1186.81	3.98
E30P10	445.14	50	1038.19	5.98
E40P10	445.14	50	888.88	7.97
E50P10	445.14	50	744.44	9.96
E0P15	420.83	75	1484.72	-
E20P15	420.83	75	1186.81	3.98
E30P15	420.83	75	1038.19	5.98
E40P15	420.83	75	888.88	7.97
E50P15	420.83	75	744.44	9.96
E0P20	395.83	100	1484.72	-
E20P20	395.83	100	1186.81	3.98
E30P20	395.83	100	1038.19	5.98
E40P20	395.83	100	888.88	7.97
E50P20	395.83	100	744.44	9.96
E0P25	370.53	125	1484.72	-
E20P25	370.53	125	1186.81	3.98
E30P25	370.53	125	1038.19	5.98
E40P25	370.53	125	888.88	7.97
E50P25	370.53	125	744.44	9.96

*E = EPS, P = POFA, C = control brick

IOP Conf. Series: Materials Science and Engineering **513** (2019) 012004 doi:10.1088/1757-899X/513/1/012004

3. Test methods

3.1. Hardened brick density
In determining the density of the brick samples, the test was conducted accordance to BS EN 12390-7 [27] in which determination of the brick mass will be as-received condition. The measurement of density was conducted at the age of 7 and 28 days. Three samples have been prepared for every mix proportion.

3.2. Compressive strength test
Compressive strength test for this study was conducted according to the ASTM C140-11a [28]. For this test, three brick samples have been prepared for every mix proportion and all the samples tested are full-sized. Any excess moisture or any particles must be removed from the brick surface and the loading plate of the machine before conducting the test. All the brick samples was cured by air curing process in the laboratory until the day of test. For this test, the samples were also tested at the age of 7 and 28 days.

3.3. Water Absorption Test
Water absorption test was conducted in order to determine the percentage of the water absorption by the brick. The test was accordance to the BS 1881: 122 [29]. Initially, the brick samples were dried in the oven at $100^{o}C \pm 5^{o}C$ for 24 hours. The samples were then left to cool down before been immersed in the water tank for another 24 hours. The weight of samples were taken before and after the samples were immersed in the tank in order to determine the percentage of the water absorption. This test was conducted at the age of 28 days.

4. Results and discussion
As already mentioned earlier for density and compressive strength test the tests were conducted at the age of 7 and 28 days whilst for water absorption test, the test was conducted at the age of 28 days. Every types of brick use constant water cement ratio which is 0.5. Following sub sections discuss the outcomes from this study.

4.1. Hardened brick density
Density of 30 samples were determined as shown in Figure 1. Based on the figure, generally the density of the brick samples decreased as the percentages of the replacement materials increased. At 7 days, density of control brick (E0P0) is 2150 kg/m^3. As the percentage of POFA increased, the brick samples shows slight reduction in density where for each increment of 5% POFA about 2 to 4% of brick density has reduced. For E0P25, where the replacement of POFA is 25% (maximum replacement), the brick density is 1864.29 kg/m^3 which is 13.3% lesser than normal brick. This has shown the contribution of POFA in the reduction of brick density.

Besides, EPS as the lightweight aggregates materials also contribute in the reduction of the brick samples. The contribution of EPS in the reduction of the brick is more significant comparing to POFA. For instance at 7 days, density of E20P0 is 1864 kg/m^3. This shows that, 20% of EPS has reduced the brick density up to 13%. For E30P0, E40P0, and E50P0 where the percentage of EPS is 30%, 40% and 50% respectively, it was found that the brick density was decreased from 17.3% up to 32.2% from normal brick. This has proven that, the characteristic of EPS which is extremely light give a vast impact towards the reduction of brick density.

The reduction of the brick density continuous to decrease as the percentage of replacement materials increased. Based on the figure, density of brick with maximum replacement of materials which is E50P25 is 1385.71 kg/m^3 which is 36% lesser than normal brick. From this it can conclude that, the density of the brick has achieve lightweight brick density as stated in ASTM C90 [1] where for lightweight brick, the density should less that 1680 kg/m^3.

Alongside materials replacement, brick age also contribute in the reduction of density. As can be seen, density of brick at 28 days is slightly lower that brick density at 7 days. For example, for brick sample E0P25 the density at 7 days is 1864.29 kg/m^3 whilst the density at 28 days is 1850 kg/m^3. From this it can be seen that, the density was reduced about 0.8%. Another example, for brick E50P0 the density at 7 days is 1457.14 kg/m^3 and the density at 28 days is 1435.71 kg/m^3. The reduction of density is about 1.5%. Perhaps, the reduction of brick density was due to loss of water content throughout the air curing process.

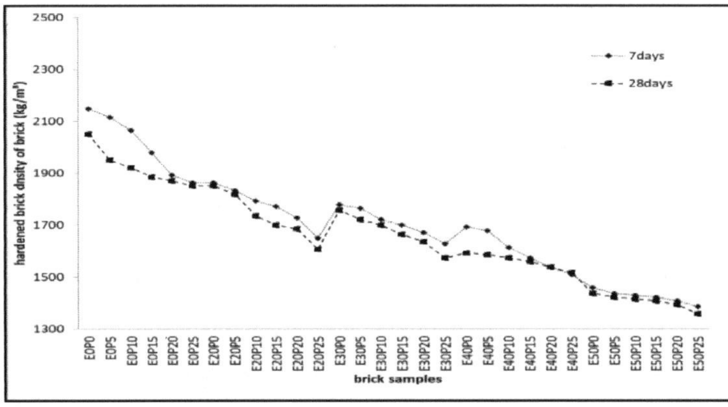

Figure 1. Hardened brick density of bricks.

4.2. Compressive strength

Figure 2 shows brick strength of 30 different types of brick. Based on the figure shown, it can be seen that, overall strength of the brick decrease as the percentage of replacement materials increase. However, it also can be observed that, the brick strengths are fluctuate. The irregular rise and fall of strength is affected by the percentage of replacement materials.

From figure 2, the brick strength of control brick (E0P0) at 7 days is 25.7 MPa. It was observed that brick strength was continuously decreased as the percentage of POFA increase. For example, the strength of E0P25 is 7.4 MPa. The reduction of the strength is about 71% from the normal brick. However, it was found that, brick with 10% of POFA (E0P10) has slightly increase the brick strength comparing to 5% POFA brick (E0P5). The same finding was found at 28 days, where brick with 10% of POFA (E0P10) has increased the brick strength. In addition, at 28 days, brick of E0P10 has higher strength than normal brick where the strength E0P10 is 31 MPa meanwhile, E0P0 is 27.4 MPa. This has shown that, 10% of POFA replacement could enhance the brick strength.

Similarly with density, the existing of EPS has contribute a great impact towards the brick strength. Compressive strength of all brick specimens containing EPS was lower as compared to the control brick. For instance at 28 days, the strength of control brick was 27.4 MPa. The strength was decreased as the percentage of EPS increase. For E20P0 the strength was 23.3MPa, for E30P0 the strength was 21.6 MPa, for E40P0 the strength was 12.6 MPa and lastly for E50P0 the strength was 11.7Mpa. 50% replacement of EPS has reduced the brick strength up to 57%. The great reduction in strength was due to the characteristic of EPS with is extremely low in strength which is almost zero in strength Miled et al [10].

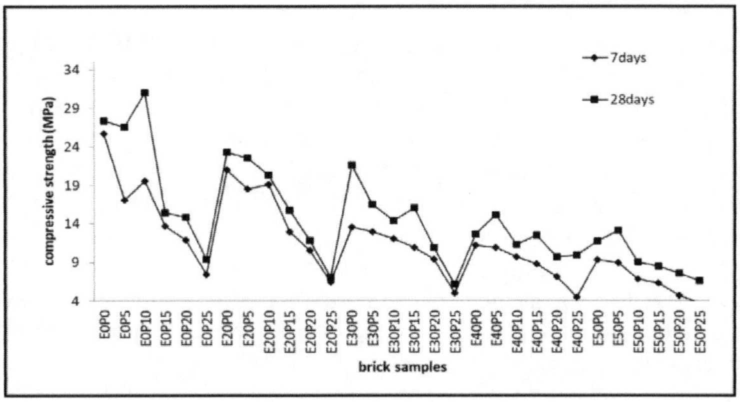

Figure 2. Compressive strength of bricks.

4.3. Water Absorption

Figure 3 shows the percentage of water absorption of 30 different types of brick. As can be seen, water absorption of E0P0 (control brick) was 11.23%. The percentage of water absorption was increased as the percentage of POFA increased. For specimen E0P5 which is 5% replacement of POFA, the water absorption of brick has increased about 1.6% from the control brick. The following specimens which are E0P10, E0P15, E0P20 and E0P25 show that the percentage of water absorption was increased about 0.5 to 0.7% for every 5% increment of POFA.

However, opposite findings was observed where the percentage of water absorption of brick decreased as the percentage of EPS increased. Brick samples with EPS had lower percentage of water absorption as compared to control brick. For instance, the percentage of water absorption for E20P0 was 6.95% which was 4.28% less than E0P0. Percentage of water absorption continuous to decrease where percentage of water absorption of brick samples E30P0, E40P0 and E50P0 were 6.1%, 5.82% and 4.37% respectively.

This can be conclude that, the increment for percentage of water absorption as the percentage of POFA increased was due to the characteristic of POFA which was coarser in structure and highly absorbent material. Meanwhile, the reduction of water absorption with the increment of EPS was due to non-absorbent characteristic of EPS.

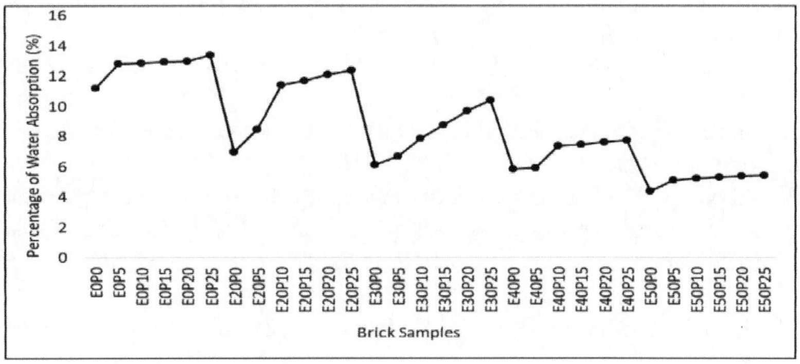

Figure 3. Water absorption of bricks.

5. Conclusion

- The density of the brick was affected by the replacement of POFA and EPS in the brick. As the percentage of replacement materials increase, the density of the brick decrease. In this study, maximum replacement of materials is 50% of EPS and 25% of POFA has reduced the density of the brick up to 34% from normal brick. This indicates that, EPS and POFA are highly potential materials for the production of lightweight.

- The mechanical properties of brick which is compressive strength is also highly affected by the existing of EPS and POFA in the brick. Based on the findings, generally the replacement of POFA and EPS has reduced the brick strength. However, it was found that, replacement of 10% of POFA showed an increment in brick strength. This has proven the potential of POFA as pozzolanic material to enhance the strength of brick.
- For water absorption properties, the findings shows that, as the percentage of POFA increase, the percentage of water absorption increase. This is due to the coarser structure of POFA which contributed to high formation of porosity in the brick. Meanwhile, opposite findings was observed when the replacement of EPS increase, the percentage of water absorption decrease. This is believed due to characteristic of EPS which is non-absorbent that helps in reduction of water absorption.

Acknowledgement

The authors gratefully acknowledge the Universiti Tun Hussein Onn Malaysia and Ministry of Higher Education for providing the financial support through the Fundamental Research Grant Scheme (FRGS), Vot 1618 and Postgraduate Research Grant (GPPS), Vot U77.

References

[1] ASTM C90 -11a. *Standard Specification for Loadbearing Concrete Masonry Units*. ASTM Masonry Standard for the Building Industry 7th Edition 2012
[2] ASTM C129 - 11. *Standard Specification for Nonloadbearing Concrete Masonry Units*. ASTM Masonry Standard for the Building Industry 7th Edition 2012
[3] Ling I H and Teo D C L 2011 Properties of EPS RHA lightweight concrete bricks under different
curing conditions. *Construction and Building Materials 25*(8) p 3648
[4] Hernández-Zaragoza J B, López-Lara T, Horta-Rangel J, López-Cajún C, Rojas-González E, García-Rodríguez F J, and Adue J 2013 Cellular concrete bricks with recycled expanded polystyrene aggregate. *Advances in Materials Science and Engineering*
[5] Veiseh S and Yousefi A A 2003 The use of polystyrene in lightweight brick production. *Iranian Polymer Journal 12 p 323*
[6] Bouvard D, Chaix J M, Dendievel R, Fazekas A, Létang J M, Peix G and Quenard D 2007 Characterization and simulation of microstructure and properties of EPS lightweight concrete. *Cement and Concrete Research* **37**(12) p 1666
[7] Sayadi A A, Tapia J V, Neitzert T R and Clifton G C 2016, Effects of expanded polystyrene (EPS)
particles on fire resistance, thermal conductivity and compressive strength of foamed concrete. *Construction and Building Materials* **112** p 716
[8] Tamut T , Prabhu R, Venkataramana K and Yaragal S C 2014 Partial replacement of coarse aggregates by expanded polystyrene beads in concrete. *Int. J. Res. Eng. Techno* **3**(02) p 238
[9] Cook D J 1973 Expanded polystyrene beads as lightweight aggregate for concrete [J]. *Precast Concrete* **4** p 691
[10] Miled K, K Sab and R Le Roy 2007 Particle size effect on EPS lightweight concrete compressive
strength: experimental investigation and modelling *Mechanics of Materials* **39**(3) p 222
[11] Xu Y, Jiang L, Xu J. and Li Y 2012 Mechanical properties of expanded polystyrene lightweight aggregate concrete and brick. *Construction and Building Materials* **27**(1) p 32-38
[12] Tang W C, Lo Y and Nadeem A B I D 2008 Mechanical and drying shrinkage properties of structural-graded polystyrene aggregate concrete. *Cement and Concrete Composites* **30**(5) p 3-409.

[13] Laukaitis A, Žurauskas R and Kerien J 2005 The effect of foam polystyrene granules on cement composite properties. *Cement and Concrete Composites* **27**(1) p 41

[14] Sadrmomtazi A, Langeroudi M M, Fasihi A and Haghi A K 2009. *An investigation on effect of using PP fibers and different cementitious materials on mechanical properties of EPS concrete*
Proceeding of 3rd International Conference on Concrete & Development p 1035

[15] Herki A B, Khatib J M and Negim E M 2013 Lightweight concrete made from waste polystyrene
and fly as *World Applied Sciences Journal* **21**(9) p 1356

[16] Babu K G and Babu D S 2003 Behaviour of lightweight expanded polystyrene concrete containing silica fume *Cement and Concrete Research* **33**(5) p 755

[17] Babu K G and Babu D S 2004 Performance of fly ash concretes containing lightweight EPS aggregates. *Cement and concrete composites* **26**(6) p 605

[18] Babu D S, Babu K G and Wee T H 2005 Properties of lightweight expanded polystyrene aggregate concretes containing fly ash. *Cement and Concrete Research* **35**(6) p 1218

[19] Katkhuda H, Hanayneh B and Shatarat N 2009 Influence of silica fume on high strength lightweight concrete. *World Academy of Science, Engineering and Technology* **58** p 781

[20] Adnan S H, Abadalla M A S, Jamellodin Z 2017 The mechanical and physical properties of concrete containing polystyrene beads as aggregate and palm oil fuel ash as cement replacement material *Construction Research Journal* p 114

[21] Munir A 2015 Utilization of palm oil fuel ash (POFA) in producing lightweight foamed concrete
for non-structural building material. *Procedia Engineering* **125** p 739

[22] Ahmad M H, Omar R C, Malek M A, Noor N M and Thiruselvam S 2008 *Compressive strength of palm oil fuel ash concrete* Proceedings of the international conference on construction and building technology, Kuala Lumpur, Malaysia p 297

[23] Awal A A and Hussin M W 2011 Effect of palm oil fuel ash in controlling heat of hydration of concrete. *Procedia Engineering* **14** p 2650

[24] Awal A S M A and Nguong S K 2010. *A short-term investigation on high volume palm oil fuel ash (POFA) concrete.* Proceedings of the 35th Conference on our World in Concrete and Structure p 185

[25] ASTM C150/C150M-11. *Standard Specification for Portland Cement.* ASTM Masonry Standard
for the Building Industry 7th Edition 2012

[26] ASTM C136-06. *Standard Test Method for Sieve Analysis of Fine and Coarse Aggregates.* ASTM
Masonry Standard for the Building Industry 7th Edition 2012

[27] BS EN 12390-7. *Testing Hardened Concrete. Density of Hardened Concrete.* British Standard 2009

[28] ASTM C140- 11a. *Standard Test Method for Sampling and Testing Concrete Masonry Units and*
Related Units. ASTM Masonry Standard for the Building Industry 7th Edition 2012

[29] BS 1881: part 122. *Method for Determination of Water Absorption.* British Standard 1983

IOP Conf. Series: Materials Science and Engineering **513** (2019) 012005 doi:10.1088/1757-899X/513/1/012005

Strength and water absorption properties of lightweight concrete brick

M Y Nurain Izzati[1], A Suraya Hani[2*], S Shahiron[1], A Sallehuddin Shah[1], O Mohamad Hairi[2], J Zalipah[1], A H Noor Azlina[1], W A Mohamad Nor Akasyah[3] and K Nurul Amirah[2]

[1]Faculty of Civil and Environmental Engineering, Universiti Tun Hussein Onn Malaysia, Batu Pahat, Johor, Malaysia
[2]Department of Civil Engineering, Faculty of Engineering Technology, Universiti Tun Hussein Onn Malaysia, Pagoh, Johor, Malaysia
[3]Melayu Jati Enterprise, Batu Pahat, Johor, Malaysia

[*]Corresponding author: suraya.uthm@gmail.com

Abstract. This study presented the strength and water absorption properties of lightweight concrete brick containing Expanded Polystyrene (EPS) as the material replacement for sand. EPS has been widely used by the construction industries in the production of lightweight building materials such as wall panel and lightweight concrete blocks. Hence, in this study, EPS has been used as the replacement material in the production of lightweight concrete brick. Replacement of sand by EPS was based on volume. The replacement percentage of EPS was 0%, 20%, 30%, 40% and 50%. The brick properties that have been investigated in this study were density, water absorption and compressive strength. Based on the experimental results, the density and compressive strength of the brick was decreased as the percentage of the replacement increased. Meanwhile, for water absorption properties, it was found that water absorption of the brick was decreased as the percentage of EPS increased. However, the properties obtained has satisfied the requirement where the brick density for lightweight should be less than 1680 kg/m^3 and the strength for load bearing and non-load bearing brick is 11.7 MPa and 3.45 MPa for each individual unit. Meanwhile, for water absorption, the percentage of water absorption of brick should be less than 12% [1,2]. From this study, it was found that, the replacement of sand by EPS give significant impact towards strength and water absorption performance of concrete brick.

1. Introduction

Concrete brick or concrete masonry is well known building materials used in many countries all over the world. In Malaysia, the demand of the concrete brick for housing construction has increase day by day especially in urban area which the population rate is rapidly increased. Due to high demand, the rate of housing prices also increased. The high prices of housing are due to the increasing in the total cost of construction and limitation of raw materials. Hence, the lightweight building materials such as lightweight concrete brick could benefits the economy and society because it will minimise the construction cost in terms of manufacturing, transportation and handling cost. With that it could reduce the housing prices.

IOP Conf. Series: Materials Science and Engineering **513** (2019) 012005 doi:10.1088/1757-899X/513/1/012005

One of the most preferable lightweight material is Expanded Polystyrene (EPS). EPS consist of 98% of air and 2% of polystyrene. Due to this, EPS has a very low density which could contribute in the reduction of building materials mass. Abundance of studies has shown that EPS has significantly contribute to the reduction of building materials density [3-10, 12-15, 17-22].

Other than density, rate of water absorption also among significant parameter that should take into consideration in the production of brick. As compared to conventional brick such as clay brick, the rate of water absorption of EPS brick was found to be lower than any other types of conventional bricks. This is due to the non-absorbent characteristic of EPS. High rate of water absorption could affects the performance of concrete in term of strength. For that reason, EPS is preferred as replacement material for improving the water absorption performance of brick. On the other hand, replacement materials by EPS has reduced the strength performance of the building materials [3-5]. However according to the findings in this study, the strength of concrete bricks has achieved the requirement strength for load and non-loading brick where the minimum strength requirement for load and non-load bearing brick are 11.7MPa and 3.45MPa respectively [1,2].

Therefore, in this study, the brick performance was investigated to identify the effects of different percentage of EPS in the concrete brick and also make a comparison with the normal concrete brick.

2. Materials and brick sample preparation

2.1. Materials

In this study, raw materials that have been used for the production of brick are Ordinary Portland Cement (OPC), river sand, Expanded Polystyrene (EPS), tap water and super plasticizer.

2.1.1. Ordinary Portland cement (OPC)

The classification of OPC used in this study is type 1 based on the ASTM C150 [23]. The supplier for this cement is Tasek Corporation Berhad which certified by SIRIM. Cement need to be stored in an air tight container to prevent any moisture contact. The chemical composition of OPC has shown in Table 1.

Table 1. Chemical Composition of OPC.

Chemical Composition	OPC
Silicon Dioxide (SiO_2)	14.6
Aluminium Oxide (Al_2O_3)	3.95
Ferric Oxide (Fe_2O_3)	3.46
Calcium Oxide (CaO)	57.1
Potassium Oxide (K_2O)	0.51
Magnesium Oxide (MgO)	1.62
Sodium Oxide (Na_2O)	-
Sulfur Trioxide (SO_3)	3.43

2.1.2. Fine Aggregates (sand)

Fine aggregates that has been used is river sand. The sieve analysis of sand was conducted in accordance to ASTM C 136 [24]. The sand was dried in the oven at 105°C±5 for 24 hours to ensure the moisture content is constant. Then, the sand was sieved through 2.36 mm sieve opening.

2.1.3. Expanded Polystyrene (EPS)

The size of EPS used in this study was between 1.18 mm to 2.36 mm. EPS used was collected from ST Polyfoam Industries Sdn. Bhd which located in Batu Pahat, Johor.

2.2. Brick sample preparation

For this study, five (5) types of concrete brick samples were prepared which are C, E20, E30, E40 and E50. Each types have different percentage of EPS which are 0%, 20%, 30%, 40% and 50% respectively. Water cement ratio for this study is 0.5 and super plasticizer is 8ml for every 1 kg binder. The curing age for the specimens were at 7 and 28 days. The mix proportion of bricks as shown in Table 2.

Table 2. Mix proportions (kg/m^3).

samples	cement	sand	EPS
C	495.34	1484.72	-
E20	495.34	1186.81	3.98
E30	495.34	1038.19	5.98
E40	495.34	888.88	7.97
E50	495.34	744.44	9.96

*E = EPS, C = control brick

3. Test methods

3.1. Hardened brick density

In determining the density of the brick samples, the test was conducted accordance to BS EN 12390-7 [25] in which determination of the brick mass will be as-received condition. Three samples have been prepared for every mix proportion.

3.2. Compressive strength test

Compressive strength test for this study was conducted according to the ASTM C140-11a [26]. For this test, three brick samples have been prepared for every mix proportion and all the samples tested were full-sized. Any excess moisture or any particles must be removed from the brick surface and the loading plate of the machine before conducting the test. All the brick samples were cured by air curing process in the laboratory until the day of test.

3.3. Water Absorption Test

Water absorption test was conducted in order to determine the percentage of the water absorption by the brick. The test was accordance to the BS 1881: 122 [27]. Initially, the brick samples were dried in the oven at 100°C ± 5°C for 24 hours. The samples were then left to cool down before been immersed in the water tank for another 24 hours. The weight of samples were taken before and after the samples were immersed in the tank in order to determine the percentage of the water absorption. This test was conducted at the curing age of 28 days only.

4. Results and discussion

As mentioned earlier for density and compressive strength, the tests were conducted at the age of 7 and 28 days whilst for water absorption test, the test was conducted at the age of 28 days only. Every types of brick use constant water cement ratio which is 0.5. Following sub sections discuss the outcomes from this study.

4.1. Hardened brick density

Figure 1 shows the density of five (5) types of concrete brick. Based on the figure it can be seen that, density of the brick samples decreased as the percentages of the EPS increased. At 7 days, density of control brick (C) is 2150 kg/m^3 meanwhile the density of E20, E30, E40 and E50 are 1864 kg/m^3, 1778.57 kg/m^3, 1692.86 kg/m^3, and 1457.14 kg/m^3 respectively. This shows that, 20% of EPS has reduced the brick density up to 13%. For E30, E40, and E50 it was found that the brick density was decreased from 17.3% up to 32.2% from normal brick. This has verified that, EPS has significantly reduce the brick density due to its characteristic which is low in density. Similar finding was found by Mulla & Shelake [8] where replacement of EPS in the concrete mix has reduced the brick density about 29% to 40% from the conventional brick. Meanwhile, according to latest research by Tayal et al [21], where 30% of EPS has reduced the concrete density up to 37% from normal concrete.

The reduction of the brick density continuous to decrease as the percentage of replacement materials increased. According to ASTM C90 [1], density for lightweight masonry brick should be less that 1680 kg/m^3. Therefore, from Figure 1, it can be observed that density of brick E40 and E50 has achieved the lightweight density as stated in the ASTM C90 [1].

Besides percentage of materials replacement, brick age similarly contribute in the reduction of density. As can be seen, density of brick at 28 days is slightly lower than brick density at 7 days. For example, for brick sample E20 the density at 7 days is 1864 kg/m^3 whilst the density at 28 days is 1850 kg/m^3. From this it can be seen that, the brick density was reduced about 0.8% from day 7 to day 28. Another example, for brick E50 the density at 7 days is 1457.14 kg/m^3 and the density at 28 days is 1435.71 kg/m^3. The reduction of brick density was about 1.5%. Possibly, the reduction of brick density was due to loss of water content throughout the air curing process.

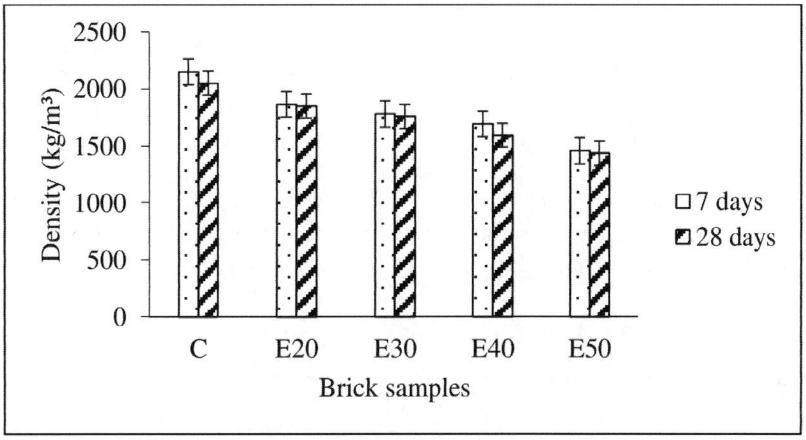

Figure 1. Hardened brick density of bricks.

4.2. Compressive strength

Figure 2 shows compressive strength of each types of brick. Based on figure shown, it can be seen that, overall strength of the brick decrease as the percentage of EPS increase. For instance, the brick strength of control brick (C) at 7 days is 25. 7 MPa whilst for E20, E30, E40 and E50 the brick strength were 21 MPa, 13.5 MPa, 11.2 MPa and 9.3 MPa respectively. It can be seen that, 50% of EPS has reduced the brick strength up to 64% from the normal brick. The same observation was found by Tamut et al [6] where in their studies the replacement of aggregates by EPS was 5%, 10%, 15%, 20%, 25% and 30%. According to their findings, the compressive strength of the concrete was reduced up to 55% as compared to the normal concrete. This has shown that, the replacement of sand by EPS also significantly reduced the brick strength. The great reduction in strength is due to the characteristic of EPS with is extremely low in strength which is almost zero in strength.

Meanwhile, for brick samples at 28 days, it can be observed that, all types of brick have higher strength as compared to the earlier strength. This was due to hydration process by the cement paste of brick at the later age.

Although EPS has greatly reduced the brick strength, yet it was found that the brick strength of 50% EPS is 13.5 MPa which is above the minimum requirement for load and non-load bearing brick as stated in ASTM C90 and ASTM C129 [1,2].

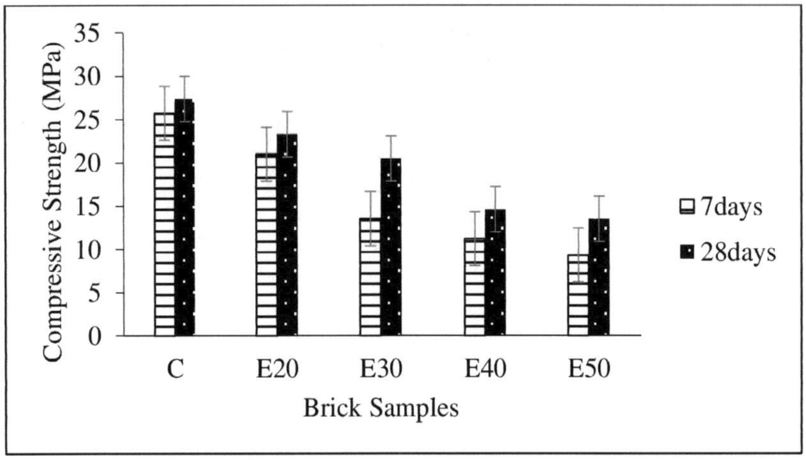

Figure 2. Compressive strength of bricks.

4.3. Water Absorption

Figure 3 shows the percentage of water absorption for each different types of brick. As can be seen, water absorption of control brick (C) is 11.23%. The percentage of water absorption of brick decrease as the percentage of EPS increase. Brick samples with EPS had lower percentage of water absorption comparing to control brick (C). For instance, E20 has water absorption of 6.95% which is 4.28% less than C. Percentage of water absorption continuous to decrease where water absorption of brick samples E30, E40 and E50 are 6.1%, 5.82% and 4.37% respectively. The same observation was found by Babu et al [19] where in their study on the effect of moisture migration of concrete containing EPS. They have found that, concrete containing EPS show lower water absorption comparing to normal concrete. This can be conclude that, the reduction of percentage of water absorption with the increment of EPS was due to non-absorbent characteristic of EPS.

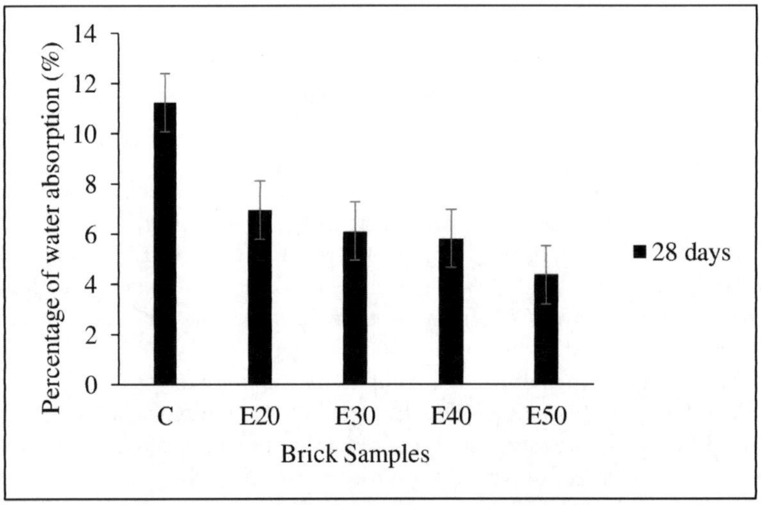

Figure 3. Percentage of water absorption of bricks.

5. Conclusion

- For density, it can be seen that EPS has significantly reduced the brick density where, as the percentage of replacement materials increase, the density of the brick decrease. In this study, maximum replacement of materials is 50% of EPS has reduced the density of the brick up to 30% from normal brick. This indicates that, EPS is a highly potential materials for the production of lightweight concrete brick.
- Strength properties of brick also highly affected by the existing of EPS in the brick. Based on the findings, generally the EPS has reduced the brick strength. However, it was found that, maximum replacement of EPS which is 50% has strength of 13.5 MPa which is above the minimum requirement for load and non-load bearing brick as stated in ASTM C90 [1] and ASTM C129 [2]. This has proven that, the replacement of EPS did not compromise the brick strength of brick.
- For water absorption properties, the findings shows that, as the replacement of EPS increase, the percentage of water absorption decrease. This is believed due to characteristic of EPS which is non-absorbent that helps in reduction of water absorption.

Acknowledgement
The authors gratefully acknowledge the Universiti Tun Hussein Onn Malaysia and Ministry of Higher Education for providing the financial support through the Fundamental Research Grant Scheme (FRGS), Vot 1618 and Postgraduate Research Grant (GPPS), Vot U77.

References

[1] ASTM C90 -11a. *Standard Specification for Loadbearing Concrete Masonry Units* ASTM Masonry Standard for the Building Industry 7th Edition 2012

[2] ASTM C129- 11. *Standard Specification for Nonloadbearing Concrete Masonry Units* ASTM Masonry Standard for the Building Industry 7th Edition 2012

[3] Ling I H and Teo D C L 2011 Properties of EPS RHA lightweight concrete bricks under different curing conditions *Construction and Building Materials* **25**(8) 3648-3655

[4] Hernández-Zaragoza J B, López-Lara T, Horta-Rangel J, López-Cajún C, Rojas-González E, García-Rodríguez F J, and Adue J 2013 Cellular concrete bricks with recycled expanded polystyrene aggregate *Advances in Materials Science and Engineering*

[5] Veiseh S and Yousefi A A 2003 The use of polystyrene in lightweight brick production *Iranian Polymer Journal* **12** 323-330

[6] Bouvard D, Chaix J M, Dendievel R, Fazekas A, Létang J M, Peix G and Quenard D 2007 Characterization and simulation of microstructure and properties of EPS lightweight Concrete *Cement and Concrete Research* **37**(12) 1666-1673

[7] Sayadi A A, Tapia J V, Neitzert T R and Clifton G C 2016 Effects of expanded polystyrene (EPS) particles on fire resistance, thermal conductivity and compressive strength of foamed Concrete *Construction and Building Materials* **112** 716-724

[8] Tamut T, Prabhu R, Venkataramana K and Yaragal S C 2014 Partial replacement of coarse aggregates by expanded polystyrene beads in concrete *Int. J. Res. Eng. Technol.* **3**(02) 238-241

[9] Cook D J 1973 Expanded polystyrene beads as lightweight aggregate for concrete *J. Precast Concrete* **4** 691-693

[10] Mulla A and Shelake A 2016 Lightweight Expanded Polystyrene Beads Concrete *Int J Res Advent Technol.* 17

[11] Sadrmomtazi A, Langeroudi M M, Fasihi A and Haghi A K 2009 *An investigation on effect of using PP fibers and different cementitious materials on mechanical properties of EPS concrete* In Proceeding of 3rd International Conference on Concrete & Development, 1035-1044

[12] Tang W C, Lo Y and Nadeem A B I D 2008 Mechanical and drying shrinkage properties of structural-graded polystyrene aggregate concrete *Cement and Concrete Composites* **30**(5) 403-409
[13] Ranjbar M M and Mousavi S Y 2015 Strength and durability assessment of self-compacted lightweight concrete containing expanded polystyrene *Materials and Structures* **48**(4) 1001-1011
[14] Xu Y, Jiang L, Xu J and Li Y 2012 Mechanical properties of expanded polystyrene lightweight aggregate concrete and brick *Construction and Building Materials* **27**(1) 32-38
[15] Herki A B, Khatib J M and Negim E M 2013 Lightweight concrete made from waste polystyrene and fly ash *World Applied Sciences Journal* **21**(9) 1356-1360
[16] Nekooie M A, Goodall H and Mohamad M I March 2015 *Selection of sustainable materials for lightweight concrete* In Proceedings of the Institution of Civil Engineers-Engineering Sustainability **168** 159-172
[17] Babu K G and Babu D S 2003 Behaviour of lightweight expanded polystyrene concrete containing silica fume *Cement and Concrete Research* **33**(5) 755-762
[18] Babu K G and Babu D S 2004 Performance of fly ash concretes containing lightweight EPS aggregates *Cement and concrete composites* **26**(6) 605-611
[19] Babu D S, Babu K G and Wee T H 2005 Properties of lightweight expanded polystyrene aggregate concretes containing fly ash *Cement and Concrete Research* **35**(6) 1218-1223
[20] Babu D S, Babu K G and Tiong-Huan W 2006 Effect of polystyrene aggregate size on strength and moisture migration characteristics of lightweight concrete. *Cement and Concrete Composites* **28**(6) 520-527
[21] Adnan S H, Abadalla M A S, Jamellodin Z 2017 The mechanical and physical properties of concrete containing polystyrene beads as aggregate and palm oil fuel ash as cement replacement material *Malaysian Construction Research Journal* 114
[22] Tayal A, Gupta G, Choudhary P, Tomar T, Kumar N and Mittal S 2018 Lightweight Concrete using Recycled Expanded Polystyrene Beads
[23] ASTM C150/C150M-11. *Standard Specification for Portland Cement.* ASTM Masonry Standard for the Building Industry 7th Edition 2012
[24] ASTM C136-06. *Standard Test Method for Sieve Analysis of Fine and Coarse Aggregates.* ASTM Masonry Standard for the Building Industry 7th Edition 2012
[25] BS EN 12390-7. *Testing Hardened Concrete. Density of Hardened Concrete.* British Standard 2009
[26] ASTM C140- 11a. *Standard Test Method for Sampling and Testing Concrete Masonry Units and Related Units.* ASTM Masonry Standard for the Building Industry 7th Edition 2012
[27] BS 1881: part 122. *Method for Determination of Water Absorption.* British Standard 1983

IOP Conf. Series: Materials Science and Engineering **513** (2019) 012006 doi:10.1088/1757-899X/513/1/012006

Behaviour of kenaf fibre reinforced composite

Mst. Sadia Mahzabin[1*], Lim Jee Hock[1], Lim Siong Kang[1] and Ehsan Nikbakht Jarghouyeh[2]

[1]LKC Faculty of Engineering and Science, Department of Civil Engineering, Universiti Tunku Abdul Rahman (UTAR), Jalan Sungai Long, Bandar Sungai Long, Cheras, 43000 kajang, Selangor, Malaysia.
[2]Civil Engineering, Engineering Campus, Universiti Teknologi Petronas (UTP)

[*]Corresponding author: sadia@utar.edu.my

Abstract. This study presents the comparison on behavior of kenaf fibre composite and normal concrete. Use of natural fibre in construction industry is of tremendously interest in the past few decades. This study focus on investigate the material characteristic and the structural behavior of kenaf fibre composite compare with normal concrete. The percentages of kenaf fibre composite were calculated depend on the weight of concrete. Experimental investigation and comparison on behaviour such as compression test, splitting tensile test, slump test, water absorption test and 4 point loading beam test were discussed in this study. These studies found that kenaf fibre composite have better behaviour than normal concrete. Kenaf fibre composite has higher load capacity as compare with normal concrete. Moreover, these show that kenaf fibre composite are capable to behave better than normal concrete in all aspect when act as structural member.

1. Introduction

In the recent time, natural fibres have become increasingly popular in construction industry. This is because of its benefits such as renewability, low density, and high specific strength [1]. Natural fibres such as coconut fibres (coir), sugarcane bagasse fibres, bamboo fibres, and kenaf fibres, are introduced to the construction industry to use as the replacement of the traditional reinforce fibre for concrete. Traditional reinforce fibres such as glass fibres, steel fibres, and carbon fibres have add in great mechanical properties to the concrete, but these fibres will also causing some environmental issue due to their non-degradability. Besides that, these fibres are relatively expensive when compare with the natural fibres.

Among the natural fibres mention above, kenaf fibre has been give an account of possess great tensile and flexural strength [2]. Tensile and flexural strength of the fibre are key criteria to take into consideration if the fibre is planning to be used as reinforced fibre in concrete. The excellent properties of kenaf fibre result in a new concrete technology “kenaf fibre reinforced concrete”. As literally, kenaf fibre reinforced concrete is the concrete that reinforce by kenaf fibre. Natural fibre composites generally have high specific strength, biodegradability, and low density [3]. According to Nishino [4], the mechanical strength and thermal properties of kenaf composites are higher than other natural fibre composites. Hence, kenaf fibre is more suitable to be used as reinforce fibre to the concrete as compare with other natural fibres. Moreover, kenaf is relatively commercially available and economically efficient among other natural fibre reinforcing material [5].

Natural fibre as reinforcement in concrete to replace the traditional reinforces fibre is of tremendous interest as natural fibre offer a more economical and environmental friendly in enhancing "green" building. "Green" building has received huge attention as the environmental issues are become more seriously in recent decade. Construction industry is playing an important role in this issue as the construction industry is one of the majors' contributors to the environmental issues. Construction industry has been using steel bars and synthetic fibres to reinforce the structure concrete. Huge amount of energy is required to produce steel bar and lots of pollutants are release during the process of producing energy and steel bars. The synthetic fibres will also induce some environmental issues due to their non-degradability.

Due to the disadvantages of the steel bars and synthetic fibres, natural fibres have become the alternative material for the construction industry. The use of natural fibres will reduce the application of steel bars and synthetic fibres. Hence, it is reducing the impact of construction industry toward environment.

From the past research, kenaf fibre is seen to be better than other natural fibres. The tensile strength of kenaf fibre has revealed to be 930MPa, which is higher than other natural fibres [6]. Kenaf fibre is capable to improve the structural properties. However, study of kenaf fibre combined with concrete has yet been published. Hence, it is necessary to investigate the hidden benefits of kenaf fibre composite so as to improve the properties of kenaf fibre composite. As we all know, the concrete is generally good in compression, but weak in tension. Therefore, the present research intended to study the shear behaviour of kenaf fibre composite and investigate the differences between normal concrete and kenaf fibre reinforced concrete.

The application of kenaf fibre composite can bring many benefits to the society. First and foremost, kenaf fibre composite can reduce the huge cost of using conventional building material. Kenaf fibre is relatively cheaper as compare to the conventional building material such as steel bars and synthetic fibres. The use of kenaf fibre can enhance the tensile strength of the concrete which will reduce the amount of steel bars require to reinforce the structure. Besides that, kenaf fibre will also replace the synthetic fibres as reinforce fibre in concrete. Hence, this will result in a lower cost of construction in term of materials. Lower cost of construction means the property can be retail at lower price.

The application of kenaf fibre (natural fibre) will replace the synthetic fibres. Synthetic fibres will cause some environmental issues due to their non-degradability. Moreover, some synthetic fibres are produced from non-renewable resource such as iron. In the other hand, kenaf fibre is renewable and biodegradable. Moreover, the planting of the kenaf fibre will also provide jobs for people. Overall, the result is a cleaner, healthier and better living environment for mankind.

In this research, the testing of the kenaf fibre composite will be carried out to investigate the kenaf fibre composite by adding 0.40% and 0.45% of kenaf fibre to act as structure member. The testing of the normal concrete will also carry out concurrently. This is to compare the behaviour of the kenaf fibre composite with normal concrete. Kenaf fibre composite and normal concrete will be cast into cube size and cylinder size. Compression test and tensile splitting test will be carried out to test the concrete cube and cylinder. An initial study on mix proportion of kenaf fibre composite will be conduct to gain the desirable characteristic strength. At the same time, the normal concrete with target strength of 30 MPa is design. All concrete specimens will go through the water curing session to optimise the properties of the specimens. Meanwhile, few cube specimens will be cast to perform the cube test to determine the compressive strength achieved. The concrete compression machine will be used to carry out this testing.

2. Methodology

This chapter was focused on the method and procedure to carry out this research. The materials used in this research also expounded. The trial mix proportion to obtain desire concrete strength was to be investigated. The dimension of the concrete cubes and cylinders were verified as well. Slump test, compression test, tensile splitting test, water absorption tests and beam test were carried out to fulfil the research objectives.

2.1. Materials

The basic materials used to produce the concrete include Ordinary Portland Cement (OPC), fine aggregate, coarse aggregate, Kenaf fibre and water.

2.2. Kenaf Fibre

Raw kenaf fibre were cut into 5cm length and treat chemically. The purpose of making kenaf fibre into 5 cm length was for ease of mixing and dispersion. Kenaf fibre were treated with 0.1 mol of aqueous sodium hydroxide (NaOH) which was produced by dissolve of 40 gm of NaOH in 1 L of distilled water for 2 hours. The purpose of this treatment is to remove any impurities on the kenaf fibre and chemically modify the surface of kenaf fibre. The volume of fibre added in the composite are 0.40% and 0.45% by weight of concrete. The sample pictures of kenaf fibre in 5cm and kenaf fibre under treatment of 0.1mol of NaOH solution are show in Figure 1 and Figure 2.

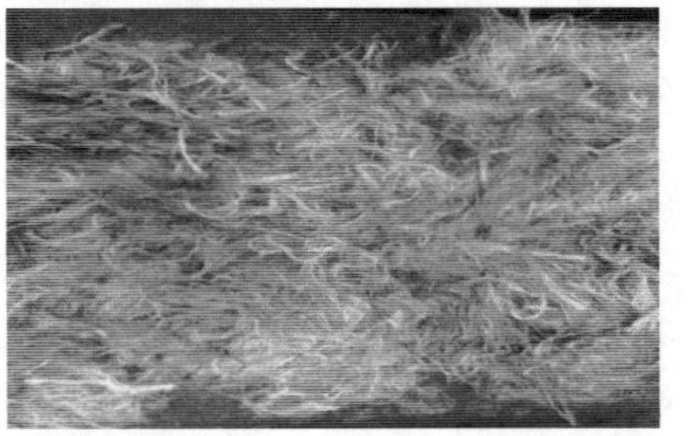

Figure 1. Kenaf fibre in 5cm length.

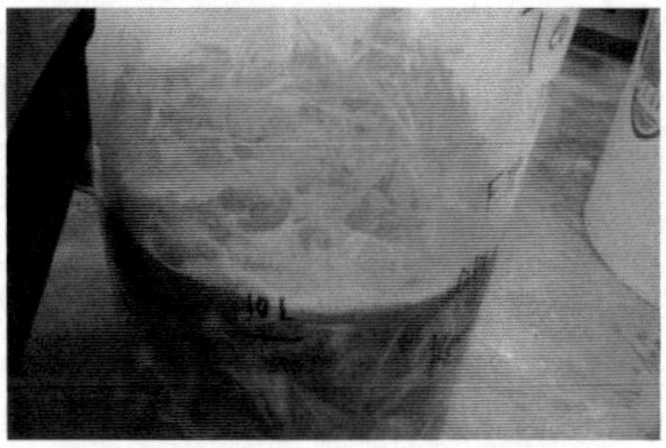

Figure 2. Kenaf fibre under treatment of 0.1mol of NaOH solution.

2.3. Initial mixture proportion

In order to achieve the optimum strength of the concrete, trials are conducted on the mixing proportion by varying the amount of the materials such as water to cement ratio (w/c), cement, water, sand and aggregate. Tables 1 show the mixing proportion used in the beginning to obtain the optimum concrete strength for normal concrete (NC) and kenaf fibre Composite (KFC).

Table 1. Initial mixture proportion.

Name	w/c	Cement (kg/m^3)	Water (kg/m^3)	Fine Aggregate <2.36mm	Fine Aggregate <4.5mm	Coarse Aggregate <10mm	Coarse Aggregate <20mm
NC	0.47	340	160	258	258	460	925
0.40% KFC	0.47	340	160	258	258	460	925
0.45% KFC	047	340	160	258	258	460	925

2.4. Mixing procedure

The procedure of making and curing of concrete done in this research is fully complies with ASTM C 192 [7]. Mixing procedure is a critical criterion to produce an effective mixing and obtain a trustworthy result. Procedure and duration of mixing is essential to develop a homogenous mix. Different mixing proportion has different duration of mixing. It is very important to take into account of the usage of chemical admixture, water-to-cement ratio and sequence.

First of all, measure the weight of each material to ensure the mix proportion is follow. After that, all materials will put into concrete mixer for mixing. The coarsest material will pour into the concrete mixer first and the finest material will pour into concrete mixer lastly. The sequence will be coarse

IOP Conf. Series: Materials Science and Engineering **513** (2019) 012006 doi:10.1088/1757-899X/513/1/012006

aggregate, fine aggregate, OPC, and kenaf fibre. After these materials are mixes together in concrete mixer, weighted water is pour into concrete mixer to mix together with the materials mention above.

2.5. Casting and curing

Casting procedure will be following the ASTM C 31[8] standard. Free water will be maintained on the surface of specimen during curing. Concrete cubes, cylinders and beams will be cure in water tank for 7 days and 28 days.

2.6. Compression test

Concrete compression test is performed in fully complies with ASTM C 109 [9]. The purpose of this test is to study the effect of kenaf fibre on the compressive strength of concrete. Concrete compressive strength at 7 days and 28 days will be recorded for both normal reinforced concrete and kenaf fibre reinforced concrete. The cube mould with 100mm x 100mm x 100mm dimension is used.
Concrete cube will be placed in the compression machine and apply increasing load on it until failure. The maximum load applied before concrete cube reach failure is recorded for both normal reinforced concrete and kenaf fibre reinforced concrete.

2.7. Tensile splitting test

Concrete cylinder specimens will be used to perform tensile splitting test as complies with ASTM C 496 [10]. The purpose of this test is to study the effect of kenaf fibre on the tensile strength of concrete. Concrete tensile strength at 7 days and 28 days will be recorded for both normal reinforced concrete and kenaf fibre reinforced concrete.
Concrete cylinder specimens will be placed in equipment before place into compression machine and apply increasing load on it until failure. The maximum load applied before concrete cylinder reach failure is recorded for both normal reinforced concrete and kenaf fibre reinforced concrete.

2.8. Water absorption test

Water absorption test of concrete is performed as complies with ASTM C 1585[11]. The purpose to carry out water absorption test is to investigate the rate of absorption or sorptivity of the water by hydraulic cement concrete.

2.9. The 4-point loading beam test

For the concrete beam samples, two beams of kenaf fibre reinforced concrete with different kenaf fibre contents and one control sample with normal weight concrete (NWC) would be required. In order to obtain a more accurate result of the kenaf fibre reinforced lightweight concrete, 3 similar samples of the same kenaf fibre content would be needed. Thus, a total of 9 samples would be required. All the beams were subjected to a modify 4-point loading test. The beam's reinforcements and beams are shown in the figures 3 and 4 below.

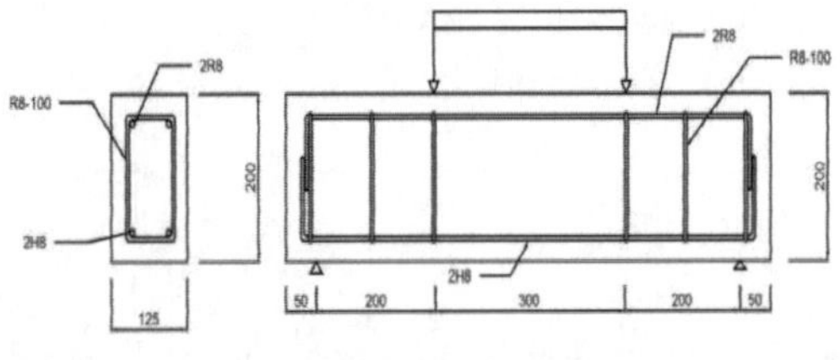

Figure 3. Reinforcement layout of beam sample.

Figure 4. Concrete beam produced.

3. Results and discussion

This chapter present the data obtained from series of testing and discussion on the data and other observation from testing.

3.1. Specimens properties and observation

Table 2 shows the 7 days and 28 days compressive strength of normal concrete (NC) and kenaf fibre composite (KFC).

Table 2. 7 and 28 days compressive strength.

Reference Name	7-day Strength (N/mm^2)	Average 7-day Strength (N/mm^2)	28-day Strength (N/mm^2)	Average 28-day (N/mm^2)
0.4% KFC	25.23	24.49	32.64	32.68
	23.56		31.98	
	24.68		33.41	
0.45% KFC	23.12	22.18	33.68	32.28
	20.97		31.46	
	22.45		31.70	
NC	22.85	23.69	32.56	33.84
	24.32		34.85	
	23.90		34.11	

Table 2 shows that kenaf fibre composites have gained higher 7 days strength than normal concrete but the 28 days strength is similar. The normal concrete has followed the typical strength-gain curve which shows that the 7 days strength is around 65% of 28 days strength. However, kenaf fibres composite does not exatly follow the typical strength-gain curve. The 7 days strength of kenaf fibre composite is varies from 70% to 75% of 28 days strength. Besides that, crack pattern of the concrete cube of normal concrete and kenaf fibre composite is also different. Concrete cube of normal concrete has many large cracks develop on it and also more fragment. On the other hand, concrete cube of kenaf fibre composite behave oppositely with concrete cube of normal concrete. Figure 5 and Figure 6 show the concrete cube of normal concrete and kenaf fibre composite after compression test. Table 3 shows the 7 days and 28 days tensile splitting strength of normal concrete and kenaf fibre composite.

Figure 5. Concrete cube of normal concrete after compression test.

Figure 6. Concrete cube of kenaf fibre composite after compression test.

Table 3 shows that the 7 days and 28 days tensile splitting strength of normal concrete is lower than kenaf fibre composite. Besides that, splitting crack of normal concrete also appears to be larger than kenaf fibre composite. This is because of the kenaf fibre help the concrete to withstand the tensile splitting force by holding the concrete together. Kenaf fibre will fail to hold the concrete together after the tensile splitting force exceeds the kenaf fibre tensile strength. Figure 7 and Figure 8 show the kenaf fibre inside concrete cylinder specimen and the splitting crack of the kenaf fibre composite cylinder specimen. Table 4 shows the 7 days and 28 days density of normal concrete and kenaf fibre composite.

Table 4 shows that both kenaf fibre composites density are lower than the normal concrete density. This is because of the kenaf fibre occupy a space in the concrete. Density of the concrete can be greatly affected by the total quantity of cement and aggregate. When kenaf fibre is added, it will cause

IOP Conf. Series: Materials Science and Engineering **513** (2019) 012006 doi:10.1088/1757-899X/513/1/012006

less aggregate contain in the concrete. Moreover, kenaf fibre has a lower density than aggregate. Hence, the density of the kenaf fibre composite will be lower than normal concrete. Table 5 show the water absorption of normal concrete and kenaf fibre composite.

Table 3. 7 days and 28 days tensile splitting strength.

Reference Name	7-days Strength (N/mm^2)	Average 7-days Strength (N/mm^2)	28-days Strength (N/mm^2)	Average 28-days Strength (N/mm^2)
	1.59		1.98	
NC	1.55	1.55	2.12	2.07
	1.52		2.10	
	2.23		3.56	
0.4% KFC	2.50	2.37	3.49	3.58
	2.38		3.69	
	2.14		3.16	
0.45% KFC	2.10	2.13	3.21	3.28
	2.15		3.47	

Figure 7. Kenaf fibre inside concrete cylinder specimen.

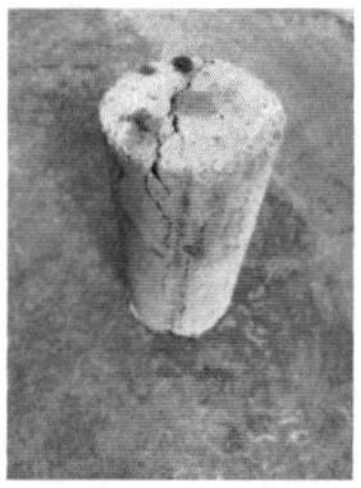

Figure 8. Kenaf fibre composite cylinder specimen.

Table 4. 7 days and 28 days density.

Reference Name	7-day Density (kg/m^3)	Average 7-day Density (kg/m^3)	28-day Density (kg/m^3)	Average 28-day Density (kg/m^3)
	2500		2530	
NC	2380	2427	2390	2447
	2400		2420	
	2360		2350	
0.4% KFC	2320	2340	2336	2335
	2340		2320	
	2260		2340	
0.45% KFC	2300	2260	2360	2320
	2220		2260	

Table 5 shows that both kenaf fibre composites have absorb more water than normal concrete. Moreover, water absorption increase as the kenaf fibre contains increase. According to Tope Moses [2], the increased water absorption may be attributed to the increase in void and pore with the inclusion of kenaf fiber in the concrete. Hence, as the kenaf fibre increase, more void and pore are present in the concrete. Besides that, kenaf fibre also has higher water absorbability. Table 6 shows the slump value for normal concrete and kenaf fibre composite.

Table 5. Water absorption test.

Reference Name	Water Absorption (g)	Average Water Absorption (g)	Water Absorption (%)	Average Water Absorption (%)
	60		2.61	
NC	80	73	3.42	3.18
	80		3.51	
	100		4.42	
0.4% KFC	90	90	3.60	4.01
	80		4.01	
	120		5.31	
0.45% KFC	120	130	5.31	5.82
	150		6.85	

Table 6. Slump value.

Reference Name	Slump Value (cm)
NC	15.5
0.40% KFC	5.3
0.45% KFC	5

Table 6 shows that the slump value of kenaf fibre composite is much smaller than normal concrete. Slump value is one of the indicators for the workability of concrete mixture. The higher the slump value, the better the workability of the concrete mixture. The reduction of slump value in kenaf fibre composite may cause by the high water absorbability of kenaf fibre. Because of most water was absorbed by the kenaf fibre, less water are remain for the concrete mixture. Hence, the slump value of kenaf fibre composite will be greatly reduced.

3.2. Structure test

Structure test will be carried out once the beam specimens reach the target compressive strength which is 30MPa, not matter whether the beam specimens has reach 28 days. The compressive strength will be determined by compression test of concrete cube cast by the same batch of concrete mixture. Theoretical design moment and experimental ultimate moment of the beam specimens is shows in Table 7.

Table 7. Moment comparison.

Reference Name	Theoretical Design Moment (kNm)	Ultimate Aplied Load (kN)	Experimental Ultimate Moment (kNm)	Capacity Ratio
		130	13	1.91
NC	6.8	130	13	1.91
		137	13.7	2.01
		137	13.7	2.01
0.40% KFC	6.8	138	13.8	2.03
		139	13.9	2.04
		140	14	2.06
0.45% KFC	6.8	142	14.2	2.09
		144	14.4	2.12

Table 7 shows that the experimental ultimate moment of normal concrete is 91%, 91% and 101% (average 94.33%) higher than theoretical design moment. For 0.45% KFC, the experimental ultimate moment is 106%, 109% and 112% (average 109%) higher than theoretical design moment. For 0.40% KFC, the experimental ultimate moment is 101%, 103% and 104% (average 102.67%) higher than theoretical design moment. Besides that, 0.45% KFC has taken the highest load among these proportions.

4. Conclusion

The required results are finally obtained after a series of laboratory works. The testing to investigate the physical properties of kenaf fibre is water absorption test. It was found out that the kenaf fibre composite will absorb more water than normal concrete. Besides that, as the kenaf fibre increase, the water absorb by the composite also increase. This is because of the increase in void and pore with the inclusion of kenaf fibre in concrete. Moreover, kenaf fibre also has higher water absorbability. Because of the reasons mention above,

The compressive strength of the kenaf fibre composites were determined to be 32.68 MPa and 33.84 MPa for 0.40% and 0.45% kenaf fibre composite respectively. The targeted compressive strength before the laboratory works were carried out is 30 MPa. The results obtained have fulfilled the targeted compressive strength. The result shows that all concretes are capable to take more load after exceed theoretical design moment. 0.45% KFC averagely takes 9.67kN more load than normal concrete before yielding.

Acknowledgments

The authors acknowledge the financial support provided by the Universiti Tunku Abdul Rahman (UTAR) through the funding (IPSR/RMC/UTARRF/2016-C1/M2) for carrying out this study.

References

[1] Ochi S, 2007 Mechanical properties of kenaf fibers and kenaf/PLA composites. *Mechanics of Materials*. **40** (4-5) 446–452.

[2] Moses O, Samson D and Waila O 2015 Compressive strength characteristics of kenaf fibre reinforced cement mortar. *Advances in Materials*. 6-10.

[3] Raman Bharath V R, Ramnath B V and Manoharan N, 2015 Kenaf Fibre Reinforced Composites: A Review. *Kenaf Fibre Reinforced Composites: A Review*. 10.

[4] Nishino T, Hirao K, Kotera M, Nakamae K and Inagaki H, 2003 Kenaf reinforced biodegradable composite. *Composites Science and Technology* **63** 1281-1286.

[5] Saba N, Paridah M and Jawaid M, 2015 Mechanical properties of kenaf fibre reinforced polymer composite: A review. *Construction and Building Materials* **76** 87–96.

[6] Akil H, 2011 Kenaf fiber reinforced composites: A review. *Materials & Design* **32**(8-9) 4107–4121.

[7] ASTM C 192 - C192M - 16. Making and Curing Concrete Test Specimens in the Laboratory.

[8] *ASTM C31 / C31M - 18a*. Standard Practice for Making and Curing Concrete Test Specimens in the Field.

[9] ASTM C109 / C109M - 16a. Standard Test Method for Compressive Strength of Hydraulic Cement Mortars (Using 2-in. or [50-mm] Cube Specimen.

[10] ASTM International 2017 *ASTM C496 / C496M-17, Standard Test Method for Splitting Tensile Strength of Cylindrical Concrete Specimens.* West Conshohocken, PA: ASTM International.

[11] ASTM C1585 – 13. Standard Test Method for Measurement of Rate of Absorption of Water by Hydraulic-Cement Concretes

IOP Conf. Series: Materials Science and Engineering **513** (2019) 012007 doi:10.1088/1757-899X/513/1/012007

Effect of steel fibers on self-compacting concrete slump flow and compressive strength

N Majain[1*], A B A Rahman[1], R N Mohamed[1] and A Adnan[1]

[1]Department of Structures and Materials, School of Civil Engineering, Faculty of Engineering, Universiti Teknologi Malaysia, 81310, Johor Bahru, Malaysia

[*]Corresponding author: nelly.majain1979@graduate.utm.my

Abstract. Ever since Self-Compacting Concrete (SCC) was introduced, various attempts have been made to further enhance its quality and robustness. The addition of steel fibers into the SCC mix is found to have increased the hardened properties of concrete. However, it is also acknowledged that the addition of steel fibers into the fresh SCC mix poses a negative effect on the workability which may cause segregation and bleeding. Thus, some modifications are required on the mix proportions to obtain a good flowability without bleeding mixes. In this study, four types of mixes were prepared which comprising of normal concrete (NC), SCC and Self-Compacting Concrete with Steel Fibers (SCCSF) with two volume fractions of still fibers 0.5% and 1.0%. The results revealed that an increased amount of steel fibers in the SCC mix improve the compressive strength of concrete but reduced the slump flow of the fresh SCC.

1. Introduction

The application of Self-Compacting Concrete (SCC) in the construction industry has gained wider attention ever since it was introduced in 1988 by Okamura [1]. Since then, it has been used for various types of structural elements such as bridges, slabs, precast beams, and tunnel segment. Undisputedly, SCC has many advantages over conventional concrete because it can flow and compact in a mould or formwork under its own weight without the need for vibration. Because of this superior quality, adequate compaction can be achieved even in a congested reinforcement and restricted areas. Basically, as stated in "The European Guidelines for Self-Compacting Concrete"[2], there are several testing methods that need to be performed in order to assess the SCC mixes. The result of the testing provides information on the flowability, viscosity, passing ability and segregation resistance of SCC at fresh state. Consequently, some desired mechanical properties are expected on the hardened state of SCC.

Various attempts have been made to further enhance the quality of SCC because just like normal concrete, it is also susceptible to tensile crack and flexural failure. The study of Self-Compacting Concrete with Steel Fibers (SCCSF) for reinforced concrete structure has gained wider attention due to the benefits of having more ductile and higher residual tensile strength as compared to conventional concrete [3]. Besides, depending on the structures, the use of steel fibers can reduce the required amount of conventional steel reinforcement while maintaining the satisfactory performance of the structure [4].

However, the addition of steel fibers into SCC may affect the fluidity of the materials. For instance, the increased in fibers volume fraction and aspect ratio will decrease the slump flow and passing ability, while increasing the flow time of SCCSF [5]. Often, a high content of steel fibers is difficult to be evenly distributed in a mix. Nevertheless, a good distribution is necessary in order to achieve optimum benefits of the fibers. Meanwhile, the effect of aspect ratio is different whether it is small or large values. Prior research has proven that the use of fibers with small aspect ratio can increase the workability of SCCSF [5]. In relation to strength, when the larger aspect ratio is being utilized, the strength performance will be increased. Hence, caution must be exercised when selecting the aspect ratio of steel fibers because this parameter has a major effect on both workability and strength properties of concrete. Accordingly, the development of optimum mix proportions for SCCSF is imperative since consideration needs to be given for both fresh and hardened properties. For that reason, this study attempts to investigate the effect of selected steel fibers on the workability of SCC and how it can enhance the hardened properties of concrete especially in mitigating crack propagation.

2. Experimental program

The experimental program was organized into two main parts. The first part is the mixing of NC, plain SCC, and SCC using a different volume fractions of steel fibers; SCCSF-0.5% and SCCSF-1.0%. The second part is the testing of fresh and hardened states for all concrete mixes.

2.1. Materials

2.1.1. Normal concrete and self-compacting concrete

In these mixtures, the cement used was the Ordinary Portland Cement (OPC) CEM I 42.5N according to European standard EN 197-1 [6]. The type of aggregates used were natural aggregates and consist of sand (0.15 – 4.75 mm) as fine aggregates and gravel (5 – 10 mm) as coarse aggregate, with relative densities of 2700 kg/m^3 for each type. For SCC mixes, in order to have an increased paste volume and a good cohesion, fly ash class F produced by coal-burning electric utilities obtained from Tanjung Bin Power Plant, Johor was added. The density of the fly ash is 2100 kg/m^3 and it replaces 30% of the total cement content. To achieve high flowability in SCC, superplasticizers or high range water reducing admixture is needed and vital. In this study, Sika Viscocrete-2044 was used and it is the third generation of polycarboxylate ether-based superplasticizer.

2.1.2. Steel fibers

In this research, the STAHLCON Steel Fibers HE 0.55/35 (hooked end steel fibers) as shown in figure 1 were used with two different percentages of fibers volume fractions, V_f = 0.5% and 1.0%. The properties and dimensions of the hooked end steel fibers are as given in table 1.

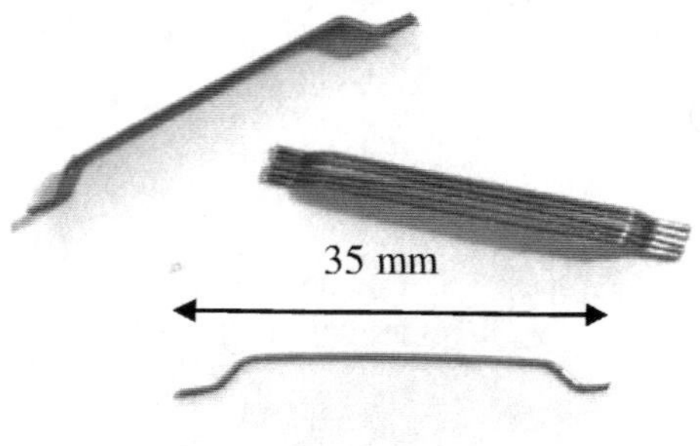

Figure 1. STAHLCON steel fibers HE 0.55/35.

Table 1. Properties of hooked end steel fibers.

Component	Value
Diameter (mm)	0.55
Length (mm)	35
Aspect ratio (L/D)	63.6
Density (kg/m^3)	7850
Tensile Strength (MPa)	1200
Elastic Modulus (GPa)	205

2.2. Mix design

2.2.1. Mix design proportions
The mix design proportions for both NC and SCC were adopted from a previous researcher with some modification made [7]. The NC design was based on the Department of Environment (DOE) method while the SCC mix design was based on the recommendation by "The European Guidelines for Self-Compacting Concrete" issued by the European Federation. The mixes were designed to develop a 28-days characteristics strength of 40 MPa. The materials mix proportions are summarized in table 2. As for the superplasticizer, the supplier's recommended dosage for SCC is within 1-2 % by weight of powder. The w/c ratio is kept as 34% in order to have the good flowability without bleeding.

Table 2. NC and SCC materials compositions (1 m^3 Volume).

Component	Content (kg/m^3)	
	NC	SCC
Cement	444.44	416.82
Fly Ash	-	178.64
Fine Aggregate	812.93	783.14
Coarse Aggregate	954.30	729.00
Water	213.33	200.00

2.2.2. Mixing procedure
The mixing procedure for SCC and SCCSF follows the method suggested by previous researchers [8] and [9]. For SCC, the first step is to mix the coarse aggregate and fine aggregate together for 30 seconds. In the second step, half of the total volume of water was added and the mixing continues for a further 1 minute. Next, the powder (cement and fly ash) and half of the remaining water were added into the mixture for another 2 minutes. After that, the remaining water and superplasticizer were added progressively. Finally, the mixing process continued for another 2 minutes until the mixture became homogeneous. As for the SCCSF, the same steps were applied with the addition of steel fibers during the first step of the mixing.

2.3. Specimen details
Four mixes were prepared in this study consisting of one mixture for normal concrete, one mixture for plain SCC and two mixtures for SCCSF with fibers volume fractions of V_f = 0.5% and 1.0%. For each concrete mixes, six numbers of concrete cube specimens with dimensions of 100 x 100 x 100 mm were prepared for compressive strength tests.

2.4. *Testing*
In this study, the slump flow tests were conducted which is a primary testing used in the field because of the convenience and the ability to show the initial condition of the concrete mixture. This test gives some information on the flowability and segregation of fresh SCC and SCCSF through measurement and visual assessments. As for the hardened state, the cubic specimens were demoulded 24 hours after casting and then cured for 7 days and 28 days in water. Subsequently, the compressive tests were conducted to evaluate the strength of the concrete.

2.4.1. *Slump flow test*
The slump flow test for fresh SCC and SCCSF (V_f = 0.5% and 1.0%) were conducted in accordance with BS EN 12350-8-2010 [10]. Basically, the procedure begins with filling the slump mould in a single layer without compaction and then the mould is raised from the concrete immediately and slowly in a vertical direction. Often, this test is performed with the slump cone upside down to make it easier to fill. The value of the slump flow is the average of the largest slump flow diameter, d_1 and the second diameter d_2 that is perpendicular to the first measured diameter, d_1. Figure 2 presents the slump flow diameter measurement. The acceptance criteria as stated in "The European Guidelines for Self-Compacting Concrete" is between 550 mm and 850 mm.

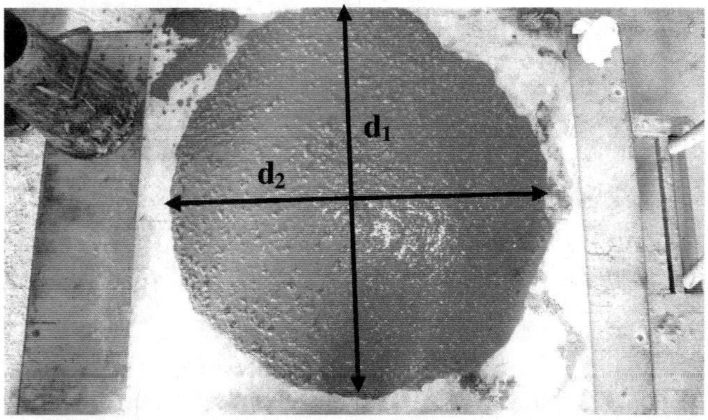

Figure 2. Slump flow measurement.

2.4.2. *Compressive strength test*
The compressive tests were conducted using the NL Compression Machine with a capacity of 3000 kN as shown in figure 3. The test was performed based on the requirements by BS EN 12390-3:2009 [11]. In this test, the strength properties of all cubes specimens at hardened state for all concrete mixes, NC, SCC and SCCSF (V_f = 0.5% and 1.0%) were determined at the age of 7 and 28 days. The test begins with applying load on the cube not exceeding 30% of the design failure load and then followed by a constant and continuous loading until failure at a rate of 0.6 ± 0.2 $N/mm^2.s$. The compressive strength was derived by dividing the maximum compression force to the specimen cross-sectional area.

IOP Conf. Series: Materials Science and Engineering **513** (2019) 012007 doi:10.1088/1757-899X/513/1/012007

Figure 3. Compressive strength test machine.

3. Results and discussion

3.1. Slump flow

The composition of materials are similar for all SCC mixes and the difference between all mixes lies on the fibers volume fractions. Table 3 presents the results of the slump flow for plain SCC and SCCSF with a fibers volume fraction of 0.5% and 1.0%. As shown, the average slump flow for plain SCC is 775 mm diameter which is higher than SCCSF-0.5% and SCCSF-1.0% with average slump flow of 675 mm and 640 mm, respectively. The reason is that concrete without steel fibers can flow easily without any obstruction from the fibers. Figures 4(a) and (b) present the slump flows for both SCC and SCCSF-0.5% which show good flowability without bleeding. However, the slump flow for SCCSF-1.0% shows some segregation and bleeding in the mix as depicted in figure 5(a), though in the previous study conducted [7], with the same mix composition there was no segregation and bleeding for this mix. This could be due to the different type of superplasticizer used in this study. As observed, the slump flow diameter for SCCSF-1.0% is only slightly smaller than the SCCSF-0.5% although it was expected that the diameter is significantly reduced due to the double amount of steel fibers used. Therefore, for SCCSF-1.0%, casting work was repeated with some slight adjustment on the amount of fine aggregate. The volume of fine aggregate was increased by 2% to allow for a higher volume of mortar and hence the dosage of the superplasticizer was also increased from 1.7% to 2%. Figure 5(b) shows the slump flow of SCCSF-1.0% after the adjustment was made. The concrete slump flow is much better compared to the previous mix design and there is no segregation observed. This proves that the content and particle size of aggregates play a vital role and can significantly affect the concrete flow.

Table 3. Slump flow test results.

Mixture	d_1 (mm)	d_2 (mm)	Slump flow average (mm)	Superplasticizer dosage (%)	Remark
SCC	800	750	775	1.5	Satisfactory
SCCSF-0.5%	700	650	675	1.6	Satisfactory
SCCSF-1.0%	650	630	640	1.7	Bleeding & Segregation

* d_1 = Largest slump flow diameter * d_2 = diameter perpendicular to d_1

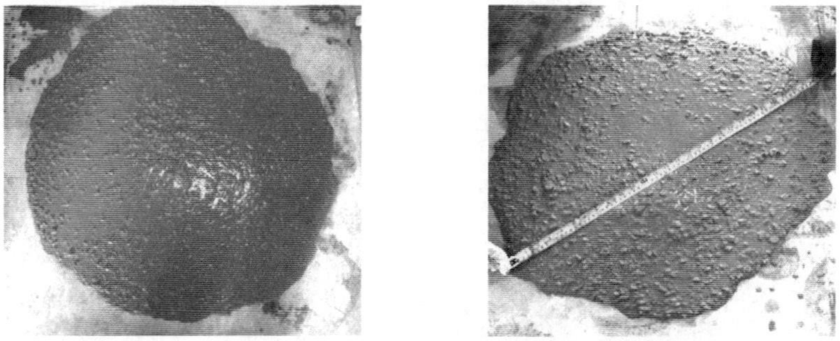

(a) (b)

Figure 4. Slump flow for SCC & SCCSF Mixes (a) SCC. (b) SCCSF-0.5%.

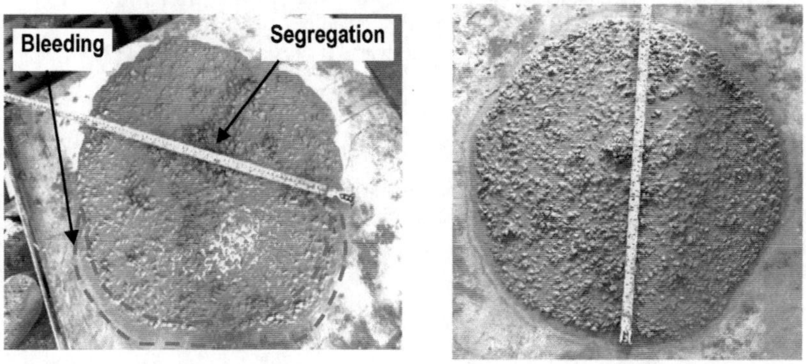

(a) (b)

Figure 5. Slump flow (a) SCCSF-1.0%. (b) SCCSF-1.0% after adjustment of mix.

3.2. Compressive strength

Table 4 presents the results of the compressive strength test for NC, plain SCC, SCCSF-0.5% and SCCSF-1.0% at the age of 7 and 28 days. Generally, the characteristics strength of the concrete achieved the intended strength of 40 MPa at 28 days. In fact, the strength has already been achieved even on the seventh day and evidently, the strength of SCC is much higher than NC by 39% at 28 days. This is due to the low water/cement ratio in SCC which contributes to a higher strength of concrete. The outcome of the test also shows that the strength of SCCSF-1.0% was the highest although at fresh state there were some bleeding and segregation issues. This could be because the presence of steel fibers helped to improve the compressive strength in concrete as mentioned by other previous researchers [12]. However, the influence of steel fibers on the compressive strength is not very substantial as the increment of strength at 28 days between SCC and SCCSF-0.5% is only around 6.6% while between SCCSF-0.5% and SCCSF-1.0% is only 1.4%. The reason is that steel fibers primary contribution are not really in compression but mainly as a crack arrester which restricts the development of cracks. Consequently, the assessment was often made based on the failure mode of the specimens. Figure 6 shows the comparison of the failure mode between SCC, SCCSF-0.5% and SCCSF-1.0%. As can be seen, the plain SCC cube crushed along the perimeters of the cube whereas the SCCSF-0.5% cube showed some crack opening on the surface while for the SCCSF-1.0%, just a hairline crack was observed. This is owing to the fact that steel fibers were able to control the propagation of micro-cracks [13] and bridge the cracks formed during the compressive loading and thus prevent the cubes from crushing.

Table 4. Compressive strength test results.

Mixture	Average Compressive Strength (MPa)	
	7 Days	28 Days
NC	43.71	50.30
SCC	52.94	69.99
SCCSF- 0.5%	53.98	74.61
SCCSF- 1.0%	54.08	75.67

Figure 6. Comparison of Cube Test Failure Mode for SCC, SCCSF-0.5% and SCCSF-1.0%.

4. Conclusion

In conclusion, the addition of steel fibers into the SCC mix may affect the workability and fluidity of the fresh SCC and causes segregation and bleeding in the mixtures. Thus, some modifications on the mix proportion of SCC may be required. Additionally, the inclusion of steel fibers in SCC have slightly increased the compressive strength of concrete and changes the failure modes.

Acknowledgments

The authors would like to express their deepest gratitude to the Ministry of Higher Education Malaysia (MOHE), Universiti Teknologi Malaysia (UTM), and Research Grant Votes No. 14H69 for making this study possible.

References

[1] Okamura H and Ouchi M 2003 Self-compacting concrete *J Adv Concr Technol.* **1**(1), 5–15

[2] European Project G 2005 *The European guidelines for self-compacting concrete* (United Kingdom)

[3] Ding X, Li C, Han B, Lu Y and Zhao S 2018 Effects of different deformed steel-fibers on preparation and fundamental properties of self-compacting SFRC *Constr Build Mater.* **168** 471–81

[4] Aslani F and Nejadi S 2012 Bond characteristics of steel fiber and deformed reinforcing steel bar embedded in steel fiber reinforced self-compacting concrete (SFRSCC) *Cent Eur J Eng.* **2**(3) 445–70

[5] Grunewald S and Walraven J C 2001 Parameter-study on the influence of steel fibers and coarse aggregate content on the fresh properties of self-compacting concrete *Cem Concr Res.* **31**(12) 1793–8

IOP Conf. Series: Materials Science and Engineering **513** (2019) 012007 doi:10.1088/1757-899X/513/1/012007

[6] British Standard Institute 2011 *BS EN 197-1 Cement composition, specifications and conformity criteria for common cements* (London)
[7] Lamide J A, Mohamed R N and Abd Rahman A B 2016 Experimental results on the shear behaviour of steel fibre self-compacting concrete (SFSCC) beams *J Teknol.* **78**(11) 103–11
[8] Paine K A 1998 *Steel fibre reinforced concrete for prestressed hollow core slabs* (Ph.D. Thesis: University of Nottingham)
[9] Mohamed R N 2009 *Shear strength of precast beam-half joints using steel fibre self-compacting concrete* (Ph.D Thesis: The University of Nottingham)
[10] British Standard Institute 2010 *BS EN 12350-8 Testing fresh concrete-Part 8: Self-Compacting Concrete -Slump Flow Test* (London)
[11] British Standard Institute 2009 *BS EN 12390-3 Testing Hardened Concrete-Part 3: Compressive Strength of Test Specimens* (London)
[12] Song P S and Hwang S 2004 Mechanical properties of high-strength steel fiber-reinforced concrete *Constr Build Mater.* **18**(9) 669–73
[13] Rambo D A S, Silva F D A and Toledo Filho R D 2014 Effect of steel fiber hybridization on the fracture behavior of self-consolidating concretes *Cem Concr Compos.* **54** 100–9

Shear strength behaviour of different energy in compaction using consolidated drained triaxial test

A S A Rahman[1*], M J M Noor[1], J Ahmad[1], N H M Zain[1] and M I F Rosli[1]

[1] Faculty of Civil Engineering, Universiti Teknologi MARA, Shah Alam, Selangor, Malaysia

[*] Corresponding author: abdulsamad@salam.uitm.edu.my

Abstract. The concept of compaction has been widely used in the construction activities such as highways embankments, earth dams and others engineering structures where the loose soils must be compacted to increase the unit weights and thus increase the bearing capacity of the subjected soils. However, little intention has been directed towards understanding the influence of compaction energy on the behaviour of shear strength using consolidated drained triaxial test. This research was conducted to study the effect of different energy in compaction for 25 number of blows compare with 40 number of blows using Standard Proctor Test. Results shows that the effect of different energy in compaction is not too significant where the changes is merely about 1° only in effective friction angles. It is because during consolidation stage in triaxial, the effect of different energy has no influence on the specimens. During consolidation stages in triaxial, the soil particles rearrange to each other with different applied pressure thus eliminates the effect of different energy on the specimens due to compaction effort.

1. Introduction

Ground improvement works is necessary before starting construction works at project site especially with weak soil condition. Prior to construction, soil upgrading will be performed to improve its mechanical behaviours in line with the requirements for construction [1]. Soil upgrading can be either by modification, stabilization of soil or both. Rahman et al. [2] stated that soil stabilization is carried out in order to improve the bearing capacity of the soil to accommodate construction needs while soil modification involves the addition of any modifier into the soil to modify the properties of soil such as its strength. The methods available to improve ground soils includes compaction process, pilling work, soil stabilization, vertical drain and load reduction method.

Compaction can be defined as the densification process of unsaturated soil due to decrease in air volume without any changes in volume of water or liquid, by mechanical forces [3]. Compacted residual soil has been widely accepted in tropical or semi-tropical areas as a filling materials where the soil is used as a civil engineering structures such as road based, embankments, slope stability, lateral earth pressure and landfill [4].

Soil compaction was necessary in many engineering works in order to ensure that the soil can sustain loadings from the infrastructure. However, compaction will also decrease the permeability of soil, causing erosion and run off. According to Akshaya [5], when compaction energy is increased the value of the optimum moisture content (OMC) will decrease, while the value of maximum dry density (MDD) will increase due to the increase in soil compaction energy.

There are various geotechnical properties of granitic residual soil that need to be considered in the design of structures to ensure that it is safe and robust [6]. Townsend [7] stated that residual soils are formed from chemical weathering therefore the engineering properties of residual soil depend on climate, parent rock, topography and the time it is formed. Based on Brand [8] and Fookes [9], residual soil is found partially saturated consisting of partly entrapped air and water due to the elevation of the ground from sea level causing the shear strength to vary.

Compaction of granitic residual soil are most important as they are vital in geotechnical design of shallow foundations. The shear resistance is not the main issue but the expansion and reduction of volume due to drying and wetting will affect the compactness of the granitic residual soil [10]. Compactness of the soil will affect the design of shallow foundations, embankments, dams and many other structures while consolidation is the change in volume due to increase of overburden pressure at a prolonged time causing reduction of soil volume and relates to the clay content that exist within the soil.

Residual soils are weathering products of rock that usually formed under unsaturated condition. It is formed through the chemical and mechanical weathering of parent rocks in situ. More than 80% of land in Malaysia is covered by the residual granite soil. The low level of ground water table causes the soil to be unsaturated and when infiltrated by rainwater the moisture content increases a soil because wetter. These soils generally belong to the residual category that may exhibit collapse settlement upon wetting [11]. When the soil in wetted condition i.e. inundation, there are possibilities that the soil is expose to large and sudden reduction in volume and known as collapsible soil. The are two main characteristic of the collapsible soil which is consists of loose cemented deposits and the soil is normally in dry condition. Apart of the two main characteristic of the collapsible soil, it also can sustain with larger vertical stresses and with a little effect in compression. However, when the collapsible soil is in wetted condition, then it will results in larger settlement without increase in vertical stress [12].

As reported by Zhao [13], the dominants minerals that exhibit in the granite are quartz and feldspar where the content of the minerals were at 30% and 65% respectively. Almost every part of the Peninsular Malaysia was covered by residual soil where the colour of the soil is in red due to reaction of natural lateritic and laterite form because of the disintegration of fossils under climatic conditions [9].

Chiu and Ng [14] found that when the process of weathering take place for the granite residual soil, the soil will be in sandy state as sand-sized particles of quartz and partially weathered feldspar. Eventually, the weathered feldspar grain will gradually be transformed into fined-grained clay minerals for sometimes. However, the quartz minerals remain in contact due to its resistance to the weathering and results in having the soil in both sand-sized quartz and clay.

In this research, a compaction process at different energy level was performed to improve the shear strength of granitic residual soil. This was carried out on soil specimens using the consolidated drained triaxial test. Hypothetically, the shear strength should increase proportionally with the higher energy in compaction. Results should show that the process of compaction using 40 number of blows should produce higher effective friction angles in the soil specimen compared to 25 number of blows.

2. Research methodology

A laboratory study was done to investigate the effect of shear strength of granite residual soil when tested with consolidated drained triaxial test with compacted of different specimens with different energy of 25 number of blows and 40 number of blows. There are two types of data collected from the laboratory procedure which is index properties and engineering properties. For index properties test, the test that will be conducted were moisture content, wet and dry sieving, sedimentation and the Atterberg limit test. On the other hand, the consolidated drained triaxial test (CD) was conducted to determine the effective minimum internal friction angle at failure, $\phi'_{\min_f}$, transition shear strength, τ_t

and transition effective stress, $(\sigma\text{-}u_w)_t$ based to the curved surface shear strength envelope framework model. About four (4) different effective stresses were applied at 50, 100, 200 and 300kPa to obtain a series of Mohr circles.

The soil samples were taken at Hutan Lipur Jeram Toi, Kuala Klawang, Negeri Sembilan at coordinates of 2°52'09.1"N; 102°00'50.6"E. Samples were collected at the site location and put in polyethylene bags and bring the soil to the laboratory in UiTM as soon as possible for testing.

About eight (8) cylindrical remoulded specimens with size of 50 mm diameter with 100 mm in height were used for the test. In order to make the remoulded specimens in identical conditions, the same optimum moisture content with maximum dry density were used for 25 number of blows and 40 number of blows consecutively. The aim is to evaluate the effect of different compaction energy in number of blows to the shear strength of the granitic residual soil.

3. Analysis of results

Granitic residual soil was taken from Hutan Lipur Jeram Toi, Kuala Klawang, Negeri Sembilan and classified as silty SAND with 11% of gravel, 60% of sand, 18% of silt and 11% of clay. Moreover, during compaction test, results indicate that the optimum water content for 25 number of blows was at 15% with maximum dry density at 1790 kg/m^3 while for 40 number of blows, the optimum water content was 12.5% with maximum dry density at 1880 kg/m^3. Result reveals that there is a significantly reduction in water content of 17% with an increasing in dry density of 5% when the energy in compaction is increase from 25 number of blows to 40 number of blows. Figure 1 illustrate the optimum moisture content with respect to maximum dry density for the 25 and 40 number of blows.

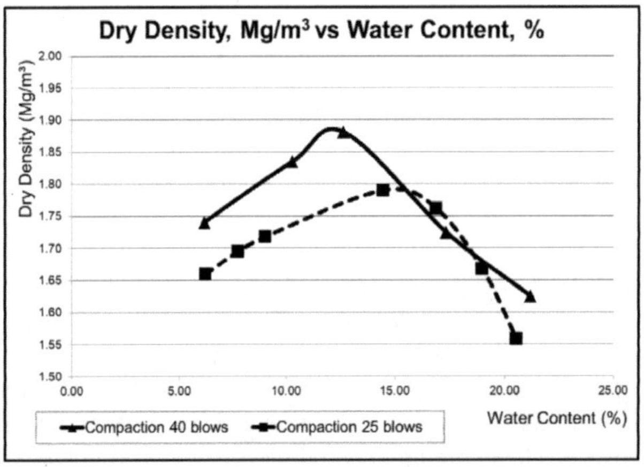

Figure 1. Optimum moisture content against maximum dry density for standard Proctor compaction at different number of blows.

Basically, when the soil is compacted with higher number of blows, it will re-orientate to denser state of packing and thus will reduce the void ratio. This will reduce the ability of water to seep through the soil and results in reduction of water content in the soil. This explain why the water content for 40 number of blows is less compared to 25 number of blows.

Effective internal friction angle of the saturated specimens can be plotted using the stress-strain graph tabulated from consolidated drained triaxial test. Table 1 illustrate the tabulated data for shear strength parameters of 25 number of blows with respect to deviator stress, pore water pressure and cell pressure. Results reveal that the effective friction angle at failure is 37° with transition shear strength, τ_t is 66kPa and transition effective stress, $(\sigma\text{-}u_w)_t$ is 62kPa. In order to determine the transition shear strength, τ_t and transition effective stress, $(\sigma\text{-}u_w)_t$, the non-linear shear strength envelope need to be plotted as shown in Figure 2(b). The values of the transition shear strength, τ_t and transition effective stress, $(\sigma\text{-}u_w)_t$ can be calculated using the plotted Mohr envelope especially at the lower shear-stress value i.e. at 50kPa effective stress Mohr envelope.

Table 1. Effective shear stress parameters at failure for 25 number of blows.

Effective Stress, kPa	Condition of Failure			Shear Strength Parameters		
	Deviator Stress (kPa)	Pore Water Pressure (kPa)	Cell Pressure (kPa)	ϕ'_f	τ_t kPa	$(\sigma\text{-}U_w)_t$ kPa
50	190	400	450	37°	66	62
100	350	400	500			
200	648	400	600			
300	956	400	700			

Figure 2(a) illustrate the deviator stress for each stress-strain curve at maximum for the saturated specimens at 25 number of blows. Nevertheless, the recorded data at peak for the deviator stress are at 190, 350, 648 and 956kPa for an at effective stresses of 50, 100, 200 and 300kPa respectively. Finally, about four (4) failure Mohr envelopes can be drawn using non-linear failure envelope as shown in Figure 2(b).

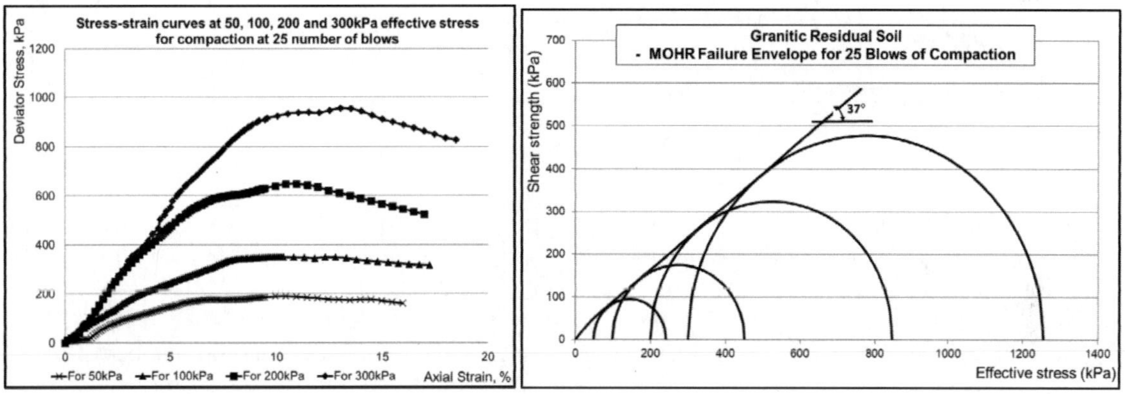

(a) (b)

Figure 2. (a) Illustrate the stress-strain curves for 25 number of blows at 50, 100, 200 and 300kPa effective stress while in Figure 2(b) illustrate the non-linear shear strength envelope at 50, 100, 200 and 300kPa effective stress.

The results of shear strength parameters at 40 number of blows in compaction was recorded and tabulated in Table 2. From the tabulated table, that the effective friction angle at failure **was at** 38° with the transition shear strength, τ_t is 68kPa and transition effective stress, $(\sigma\text{-}u_w)_t$ is 61kPa. On the other hand, Figure 3(a) illustrate the deviator stress at failure for each of the stress-strain curve for the saturated specimens at 40 number of blows. The deviator stress at maximum were recorded as 195, 355, 660 and 995kPa for the stress-strain curves at effective stresses of 50, 100, 200 and 300kPa respectively. Four failure Mohr envelopes were drawn using non-linear failure envelope as shown in Figure 3(b).

Table 2. Effective shear stress parameters at failure for 40 number of blows in compaction.

Effective Stress, kPa	Condition of Failure			Shear Strength Parameters		
	Deviator Stress (kPa)	Pore Water Pressure (kPa)	Cell Pressure (kPa)	ϕ'_f	τ_t kPa	$(\sigma-U_w)_t$ kPa
50	195	400	450	38°	68	61
100	355	400	500			
200	660	400	600			
300	995	400	700			

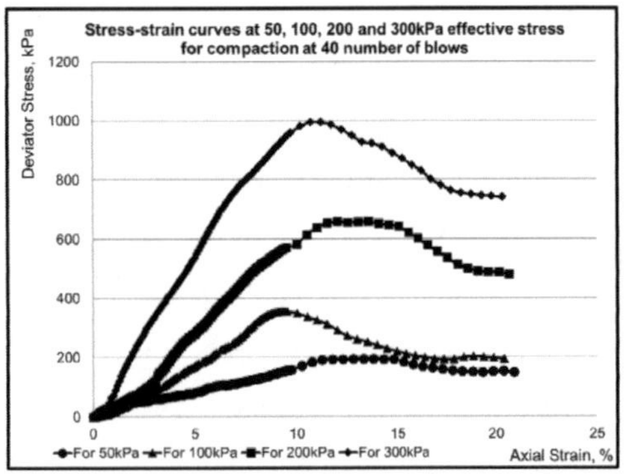

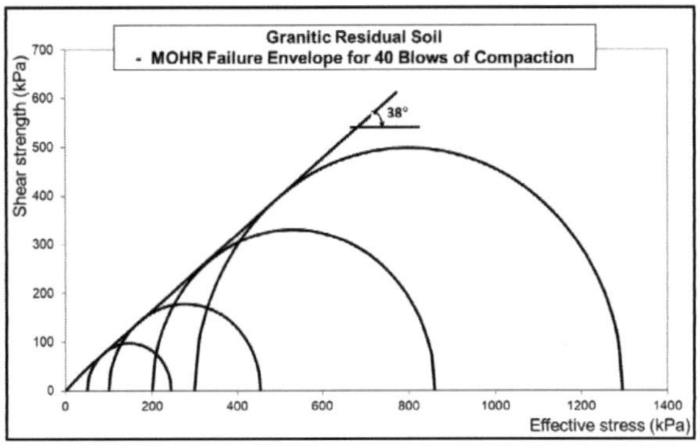

(a) (b)

Figure 3. (a) Illustrate the stress-strain curves for 40 number of blows at 50, 100, 200 and 300kPa effective stress while in Figure 3(b) illustrate the non-linear shear strength envelope at 50, 100, 200 and 300kPa effective stress.

The difference in effective friction angles between 25 number of blows and 40 number of blows was only about 1°, that is between 37° and 38°. This small change in effective friction angle have insufficient influence on the shear strength of the soil. Even though the increase in compaction from 25 to 40 number of blows significantly increase the water content and dry density, it did not give significant contribution to the shear strength. This means that an increasing in number of blows for compaction did not increase the shear strength of the soil at all.

Figure 4(a) shows the combination of recorded maximum deviator stress for each stress-strain curve of the saturated specimens at 25 and 40 number of blows. Figure 4(b) on the other hand shows the combination of failure envelope of 25 and 40 number of blows.

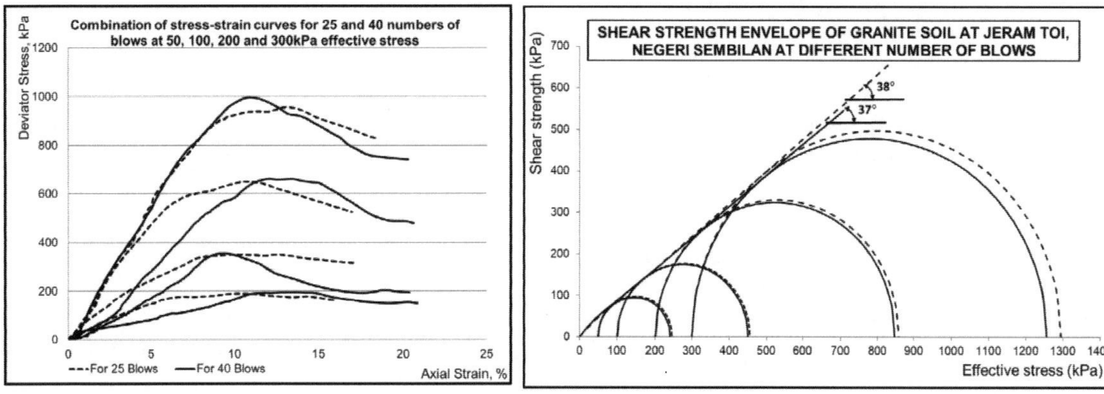

(a) (b)

Figure 4. (a) Shows the combination of stress-strain curves for 25 and 40 number of blows at 50, 100, 200 and 300kPa effective stress while in Figure 4(b) shows the non-linear shear strength envelope at 50, 100, 200 and 300kPa effective stress.

4. Conclusion and recommendation

Based on the results presented, the compaction for 40 number of blows gave a higher value of dry density but lower value water content compared with 20 number of blows. It can be said that the higher energy applied in the compaction would result in better dry density but lower permeability. Different energy in compaction can influence the shear strength of the soil. However, in consolidated drained triaxial test the results of shear strength between 25 and 40 number of blows shows that the values of the effective friction angle is almost the same which is 37° and 38° respectively. The difference in 1° of the shear strength will not give any significant impact on the soil. Even though there is different energy in compaction between 25 and 40 number of blows, the difference in energy did not influence the shear strength of the soil when tested in the consolidated drained triaxial test. This is could be due to the specimens being saturated in consolidated stages, resulting in the soil skeleton re-arranging and repacking itself to a denser state according to the desired effective stress. Therefore, even if the specimens were compacted using different energy i.e. different in number of blows, the shear strength will not increase since the specimens will have to consolidated with the desired effective stress in the consolidation stages. As for the recommendations for further research, studies can be carried out on the impact of different energy in the form of number of blows on the shear strength using direct shear strength method since this this method will not involved any consolidation stages. The results could then be compared with the outcome from the triaxial test.

References

[1] Rahman Z A, Lee J Y Y, Rahim S A , Lihan T and Idris W M R 2015 Application of gypsum and fly ash as additives in stabilization of tropical peat soil *Journal of Applied Science* **15** (7) 1007-1012

[2] Rahman Z A, Ashari H H, Sahibin A R, Tukimat L and Razi I W M 2014 Effect of rice husk ash addition on geotechnical characteristics of treated residual soil *Journal of Agriculture & Environment Sciences* **14** (12) 1368 - 1377

[3] Ghosh R 2013 Effect of soil moisture in the analysis of undrained shear strength of compacted clayey soil *Journal of Civil Engineering and Construction Technology* 4(1) 23-31

[4] Taha M R, Mofiz S A and Hossain M K Mohamad A 2000 Stress-strain behaviour of compacted residual soil in direct shear test P*roceedings of ISRM International Symposium, Melbourne, Australia*

[5] Akshaya K S and Ranjan K M 2015 Effect of compaction energy on engineering properties of fly ash-granite dust stabilized expansive soil *International Journal of Engineering and Technology* **7**1617-1624

[6] Marto K and Kassim F 2003 Characterisation of Malaysian residual soil for geotechnical and construction engineering *Project Report Vote NO:72256 UTM, Malaysia*

[7] Townsend F C 1985 Geotechnical characteristics of residual soils *Journal of Geotechnical Engineering* **111** 77-92

[8] Brand E W 1982 Analysis and design in residual soils *Proceedings of the Conference on Engineering and Construction in Tropical and Residual Soils ASCE, Honolulu, Hawaii* 89-129

[9] Fookes P G 1997 *Tropical residual soils*, 1st. ed. London: The Geological Society London.

[10] Brand E W 1982 Analysis and design in residual soils *Proceedings of the Conference on Engineering and Construction in Tropical and Residual Soils. ASCE, Honolulu, Hawaii* 89-129

[11] Ayadat T and Hanna A M 2011 Assesment of soil collapse prediction methods *International Journal of Engineering* **25** 19-26

[12] Ahmed F A, Yahaya A S and Farooq M A 2006 Characterization and geotechnical properties of Penang residual soils with emphasis on landslides *American Journal of Environmental Sciences* **2** 121-128

[13] Zhao J 1994 *E*ngineering properties of the weathered Bukit Timah granite and residual soils *Regional Conference in Geotechnical Engineering* **94** Melaka, Malaysia

[14] Chiu C F and Ng C W 2014 Relationships between chemical weathering indices and physical and mechanical properties of decomposed granite *Engineering Geology* **179** 76-89

[15] Little A L 1969 The engineering clasification of residual tropical soils *Proceedings 7th International Conference Soil Mechanics and Foundation Engineering, Mexico* **1** 1-10

IOP Conf. Series: Materials Science and Engineering **513** (2019) 012009 doi:10.1088/1757-899X/513/1/012009

A review on mechanism of rainwater in triggering landslide

Shamsan Alsubal[1*], Nasiman bin Sapari[1], Indra S.H. Harahap[1] and Mohammed Ali Mohammed Al-Bared[1]

[1]Department of Civil and Environmental Engineering, Universiti Teknologi PETRONAS (UTP), 32610 Perak, Malaysia

[*]Corresponding author: shamsan2040@gmail.com

Abstract. Landslide could be evaluated as a very serious issue in tropical countries due to its hazardous impact on lives and properties. One of the factors influencing to the development of landslides is the rainfall infiltration which is considered as a major triggering factor. It decreases the matric suction, increases the pore-water pressure and causes the groundwater table to rise. This paper is reviewing the mechanism of rainfall induced slope failure, the effect of rainfall infiltration, soil suction and the rise of groundwater table on slope stability. The results show high rainfall intensity infiltrates initially in a higher rate than soil permeability and develops a perched water table which induces shallow landslide. While, low rainfall intensity for a longer period is more dangerous than heavy rain for a short period. It increases the saturation of the soil and causes the matric suction to dissipate and this triggers landslide. Moreover, the review reveals that the factor of safety increases when the initial soil suction at the ground surface is high. The rise of groundwater increases the pore-water pressure and soften the slope-forming material and leads to landslide.

1. Introduction

Landslide is defined by Gruden [1] as "a mass of rock, debris or earth moving or sliding down on a slope". Werner [2] also defined landside as a geological phenomena that includes a wide range of movement such as slope failure, rock falls, or debris flow, which occurs onshore or offshore. Meanwhile, unsaturated soil is described as the soil that is formed by solid particles, porewater, and air. The degree of saturation (S) that is defined as the volume of air to that of water is less than one; $S<1$ [3, 4]. However, it is important to recognize the fourth phase namely, air-water interface or contractile skin [4]. Unsaturated soils have been of particular concern due to their complex behaviour resulted from their negative pore water pressures [4]. The soil can be described as saturated or unsaturated soil dpending on the climate conditions that plays an important role. The evaporation processes is responsible for the removal of water from the soil or through evapotranspiration by the vegetation cover. The soil is dried by grass, plants and trees by enforcing a tension to the pore-water, most of the trees are able to apply about 1 to 2 Mpa of tension and the tension will be acting in all direction [5, 6]. The areas with annual evaporation from the ground suface more than annual preciptation are known as arid zones [4]. The areas are classified as extremely arid, arid and semi arid zones and they usually have high water table. The climatic changes highly affect the water content of the soil. During precipitation, the porewater pressure increases toward postive values. Therefore, changes occur in the volume and shear strength of the soil [4]. The soil zone above the groundwater table is called Vadose zone, while the zone that is located immediately after the groundwater table and ranged from less than 1 m to approximately 10 m thickness is called capillary fringe and it is almost saturated [5].

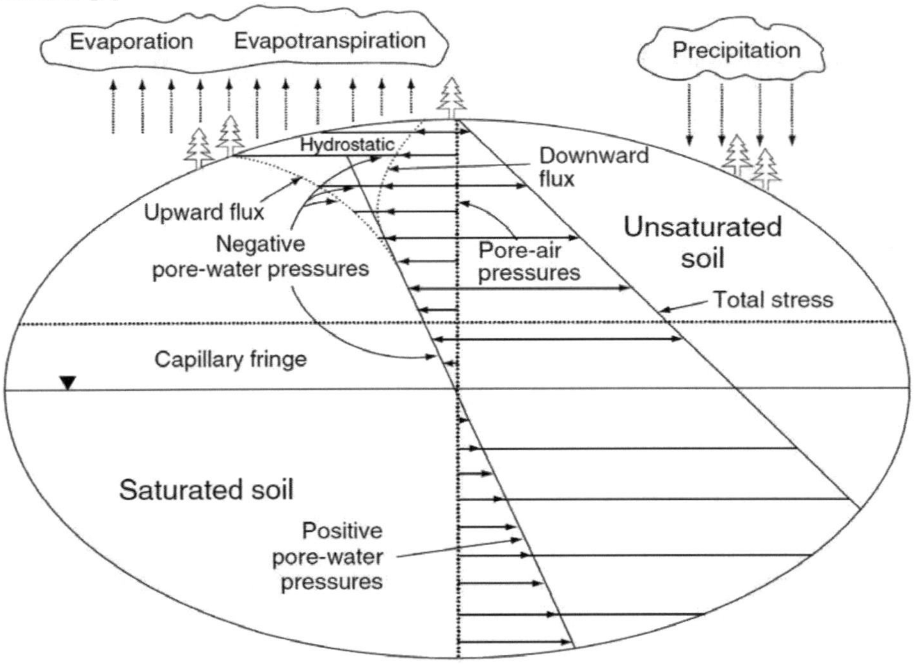

Figure 1. A visualization of saturated/unsaturated soil mechanics based on the nature of the fluid phases [5].

There are many influencing factors prompting the stability of slopes such as the slope forming materials, location of water table, and soil suction. The slope fails due to weakness of the slope forming material that can initially be triggered by the changes in the soil properties [7]. Rainfall is one of the main triggers which makes several variations in the soil properties. It increases the porewater pressure and weaken the strength of the soil. Besides, the saturation of the soil at the first 0.3 to 0.4 m results in the dissipation of soil suction and induces shallow landslide [8]. Rainfall with low intensity for longer period increases the groundwater table and results in a deep-seated landslide, usually after several rainfall events [9].

To the authors best of knowledge, there is no detailed review on the landslide triggering factors in relation to rainfall infiltration, soil suction, and rise of ground water table. Hence, this paper is reviewing those factors and providing a detailed mechanism that explains the landslide failure.

2. Rainfall infiltration

Rock and soil slopes are considered as natural hazardous phenomenon happening frequently all over the globe and the event of rainfall may considered as one of the generating factors [10]. Rainfall infiltration within slope formed by unsaturated soil slope is affected significantly by the permeability of soil [11, 12]. More amounts of rainfall water is infiltrating into the soils forming the slope that are characterized with high permeability causes a reduction in the matric suction of the soil due to saturation [13].

The infiltration of the rainwater into the soil with high permeability is considered higher than the soil permeability at the beginning by as much as 3.5 times and then start to degrade toward the steady state conditions [14]. On the other hand, when the intensity of the rainfall is low, the infiltration is low at the crest of the slope and then rise gradually towards the steady state conditions [14]. For a long period of time, low intensity rain is perhaps more unsafe than the heavy rain happening shortly in causing slope failure due to the reduction of soil suction in the first 20 to 50 cm of the soil [15]. Heavy rainfall for a long period reduces the seepage rate within the soil slope [16]. According to a study performed to show the extend of heavy rain in triggering landslide in Shenzhen, China. The landslide occurred after 13h of the rainfall event because of prolonged rainfall. The high permeability outcrops allowed more rainwater to infiltrate to the granite stone. The rainfall water continued seeping through the fill layer even after the rainfall had stopped, and the ground water table rose eventually inducing the landslide [17].

Furthermore, an experimental study was conducted on the rain infiltration in silty sand soil compacted with different water content and connected to a constant water head of 20 mm. The results showed that the soil compacted with lower water content (8 %) has the highest infiltration rate of 3.25 mm/min while it was 2.15 mm/min and 1.21 mm/min with water content of 15.5% and 24.3%, respectively as shown in Fig. 2 [18]. The infiltration decreased with time towards the steady state conditions (k_s value).

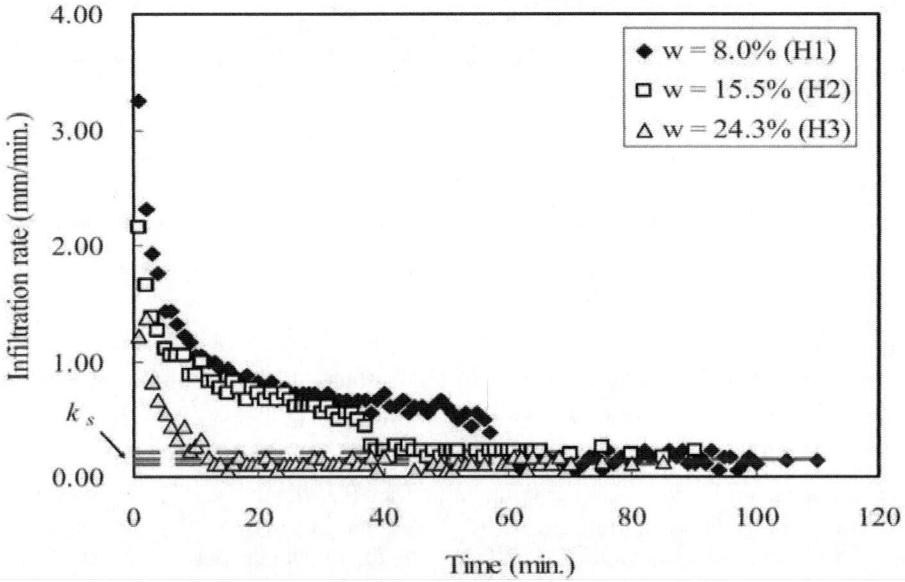

Figure 2. Curves of infiltration for samples exposed to a constant water pressure and compacted at different moisture [18].

The experiment was conducted with soil compacted at similar water content (w = 11.7 %) and exposed to different rain intensity ranging between maximum infiltration (I_{max}) and soil permeability (k_s). The result indicated that before bonding, the infiltration was as high as the rain intensity and it declined towards the steady state value (k_s value) but the time taken to ponding was faster with high rain intensity than it was with low rain intensity due to the fast saturation of the soil specimen [18]. Figure 3 shows the infiltration rate with time.

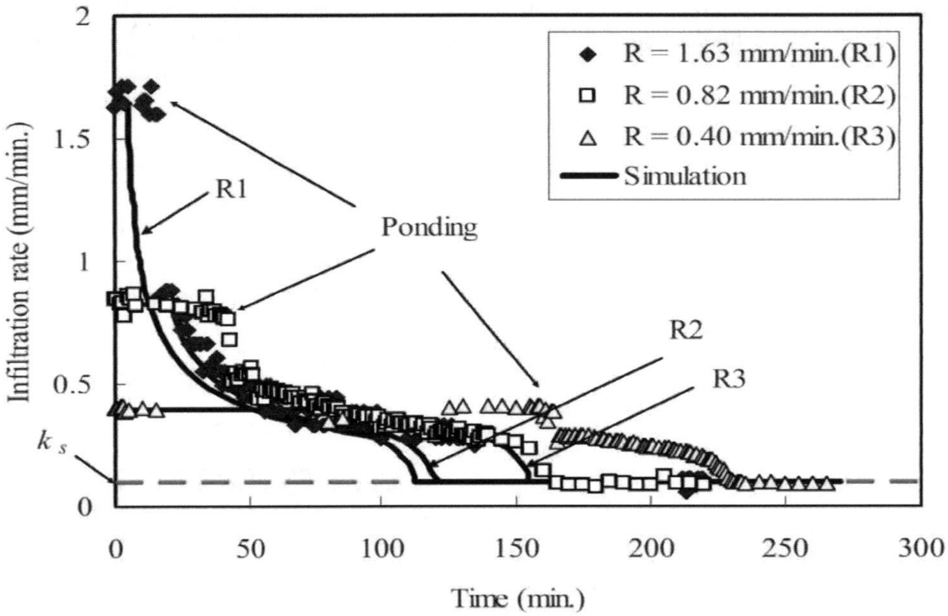

Figure 3. Curves represents infiltration for samples with similar compaction moisture content (w=11.7%) [18].

3. Soil suction

Suction can be described as the state of free energy of soil-water which can usually be measured using its partial vapor pressure. Kelvin's equation described the total suction as follow [5]:

$$\psi = -\frac{RT}{u_{w0}\omega_v}\ln\left(\frac{u_v}{u_{v0}}\right) \tag{1}$$

Where ψ is the total suction in kPa; T is the temperature (k); R is the universal gas constant (J/mol.K); u_{w0} is the specific volume of water (m^3/kg); u_{v0} is the saturation pressure of water vapor over a flat surface of pure water at the same temperature (kPa); ω_v is the molecular mass of water vapor (g/mol); u_v is the partial pressure of pore water vapor (kPa) [5]. In order to give a fixed relationship between relative vapor pressure and total suction, Equation 2 can also be writen at 20° C to as follows:

$$\psi = -135{,}022\, ln\left(\frac{U_v}{U_{v0}}\right) \tag{2}$$

The total suction can evaluated by the following equation:

$$\psi = (u_a - u_w) + \pi \tag{3}$$

Where $(u_a - u_w)$ is the matric suction while (π) is the osmotic suction [5].

The field's matric suction can be measured, the typical air entry value of sandy soils is between 1 and 10 kPa [19]. The residual suction can be estimated to be between 10 and 100 kPa for sand soil [19]. Van Genuchten method can be used to compute the soil suction at field as shown in Equation 4, if the fitting parameters *a, m,* and *n* for the Van Genuchten equation of SWCC are know [20].

$$\psi = \frac{1}{a}\left(\left(\frac{w_S}{w}\right)^{1/m} - 1\right)^{1/n} \tag{4}$$

This Equation is only applicable to evaluate the suction at a range between the residual suction and the air-entry value [19].

The safety condition of the slope can always be measured using the degree of saturation and matric suction that are considered as a significant indicator for safety. High stability can be seen for slopes with high matric suction, while low stability is found in slopes with low matric suction. A single rainfall event at the top part of the soil cover can extremely affect the matric suction and the volumetric water content. However, the deeper parts of the soil will be affected by seasonal trend and unaffected by a single rainfall event [21]. An increase in water content may result in rainwater infiltration that reduces the matric suction in the soil [22-25].The soil suction of soil dissipates due to its saturation during the rainfall events and the failure may result because of the initiation of the positive porewater pressure [8]. At the ground surface, the matric suction is first affected by the direct rainfall [26, 27]. A soil that has a cohesion of 2 kPa may equalize a suction of 7 kPa. Therefore, the factor of safety is 30% higher when the initial soil suction at the ground surface is 50 kPa [28]. Hossain [2010] performed an in-situ soil suction monitoring on a residual granite soil slope. It was found that the average soil suction at the ground surface was 30 kPa. The suction of 30 kPa was equivalent to a cohesion of 9.6 kPa. This is responsible of the increment in the factor of safety from 2.14 to 2.48 [27]. The factor of safety usually increases with the increase of matric suction [29, 30]. Figure 4 shows the increase of factor of safety by increasing the matric suction of the soil slope.

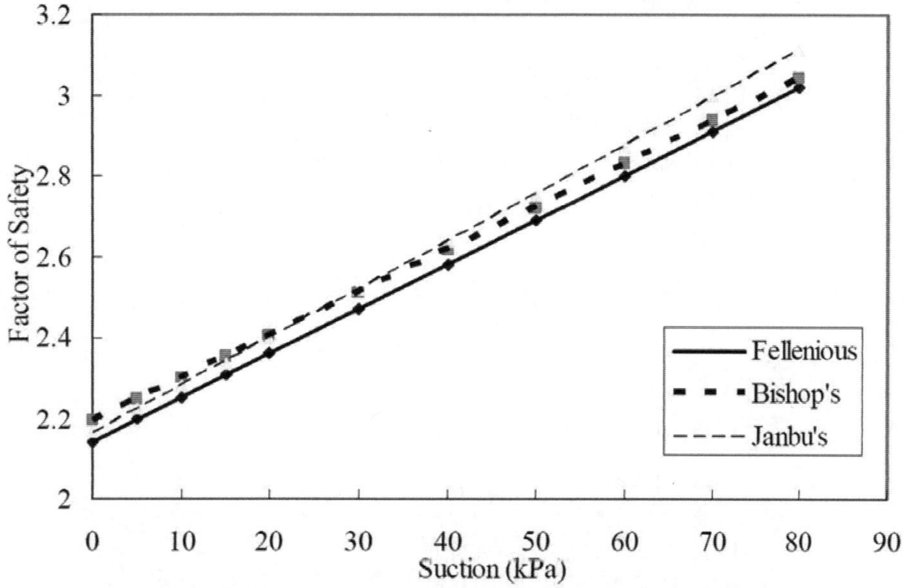

Figure 4. The FOS of a granite soil slope affected by matric suction [27].

Lim [26] conducted an experimental study on an in-situ suction measurement under three different surface conditions; canvas covered surface, grass surface and bare surface. It was found that there was some similarities in the characteristics of the in-situ matric suction variations; the maximum changes of the matric suction occurs near to the ground surface. The variation of the matric suction under the canvas covered section was less significant than that under the grass-covered and the bare sections. The matric suction in the bare slope decreased significantly immediately after the rain storm, while that for grass covered slope decreased immediately after the rainstorms as shown in Figure 5 [26].

IOP Conf. Series: Materials Science and Engineering **513** (2019) 012009 doi:10.1088/1757-899X/513/1/012009

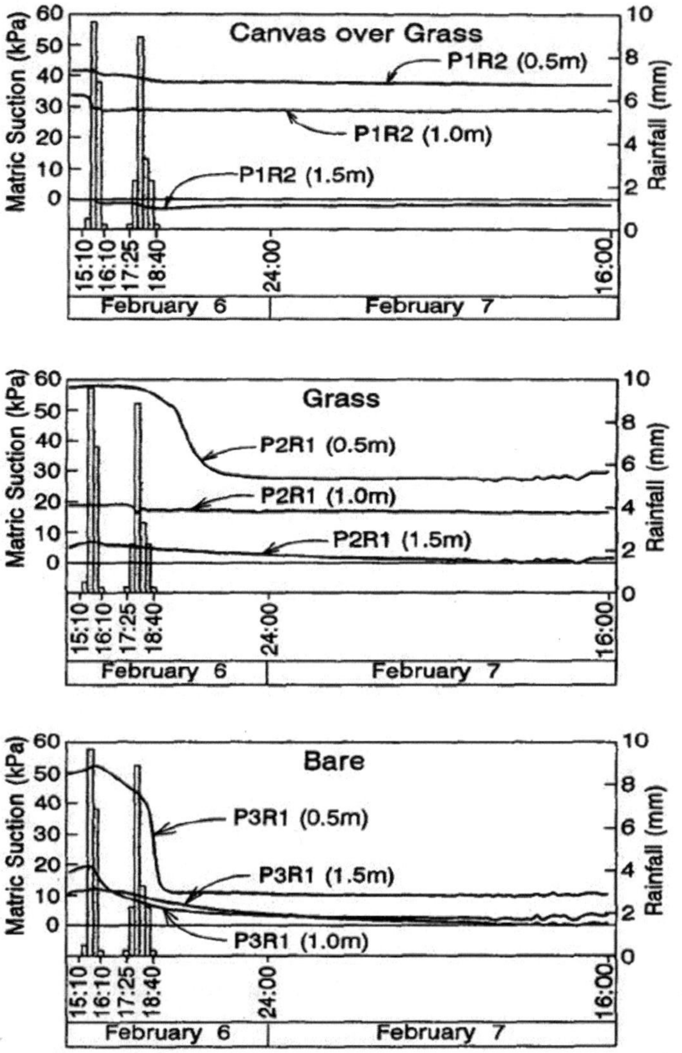

Figure 5. The reduction of matric suction during rainfall at different depth of the soil [26].

4. The rise of groundwater table

The sudden increases in the pore-water pressure of soils due to rainfall softens the slope's forming material and failure of slopes usually occurs [7] . The ground water table rises and results in shallow landslide due to perched water table [11, 31]. Perched water table and sudden increases in the ground water table are a result of the rainstorm [23, 32, 33]. Due to rainfall infiltration, landslides are commonly associated with the advancement of wetting front in the unsaturated soils [24], and the critical factor of safety seemed to be at the same depth as the wetting front [22]. Failure of slopes that have soil with a friction angle less than the slope angel are induced by wetting front [11].

Seepage forces in soils with a low compaction density have a significant effect on seepage velocity [34]. The seepage forces that are proportional to the rainfall intensity and void ratio may causes the formation of landslide. For instance, a reservoir landslide located in Sichuan provinces, China is associated with the reservoir water level drop. The landslide occurred due to the seepage forces caused by the differences between the groundwater level in the slope and the water level of the reservoir [9]. Due to the limitation of soil slope permeability, the drop of the reservoir water level is faster than the groundwater level at the slope. Therefore, the seepage pressure in the slope mass increases and the deformation starts. It was stated that the rise of ground water table within the slope may lead to instability event. Hence, the degradation of the slope may be in a severe mode of failure [35, 36].

4.1. Vibo Valentia Southern Italy landslide

A large landslide occurred at Maierato that caused around 2300 people to be evacuated in addition to high economical losses. The landslide area was 0.3 km^2 with 1.2 km runout and about 10 Mm^3 volume of soil as shown in Fig. 6 [37]. The most possible triggering factor was the increase of pore water pressure due to the rise of groundwater table from the heavy rainfall and the prolonged rainfall one month prior to the landslide occurrence [37, 38]. The last two winters before the landslide were much rainy than the usual within the period from 1919 to 2010 [39]. Figure 7 shows the daily rainfall for the three months prior to the occurrence of the landslide. The cumulative rainfall of the twenty days preceded the landslide was more than 250 mm and around 35 mm the day before the landslide [40].

Figure 6. Post failure photograph of Maierato landslide [39].

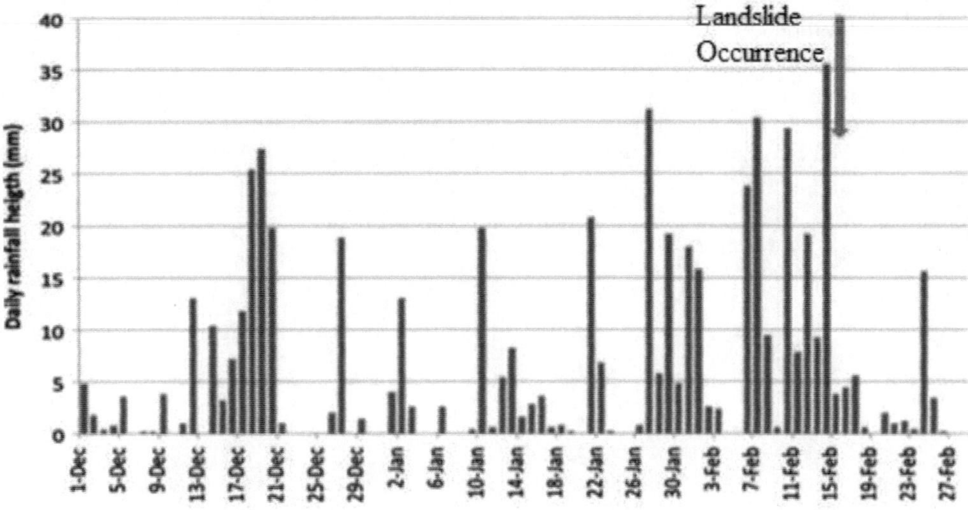

Figure 7. The daily rainfall heights three months prior to the landslide. The red arrow points to the day of landslide [39].

The sliding mass was initiated within Miocene sandstone at a depth of 60 -70 m. The landslide was preceding by a sequence of failures at the toe [39]. The presence of weak and permeable rocks allowed the rain to feed the deep groundwater tables [37, 38]. The rise of groundwater created a high groundwater pressure head, and decreased the effective stresses at the toe, thus inducing failure [40].

A slope stability analysis was performed using finite element method for the calculation of FOS along cross-section A-A (Figure 8). The modeling was carried out using different groundwater levels to recognize the conditions that induced landslide. The result indicated that the factor of safety during the dry conditions was more than three. By using the observed groundwater, the factor of safety decreased by 20%. A further decline of around 20 % is approached considering a rise in groundwater levels of 10 - 15 m. In these ground conditions, the limit equilibrium of the slope stability (Fs = 1) can be achieved when the friction angle of the sandstone is decreased by 50 to 60 % [38].

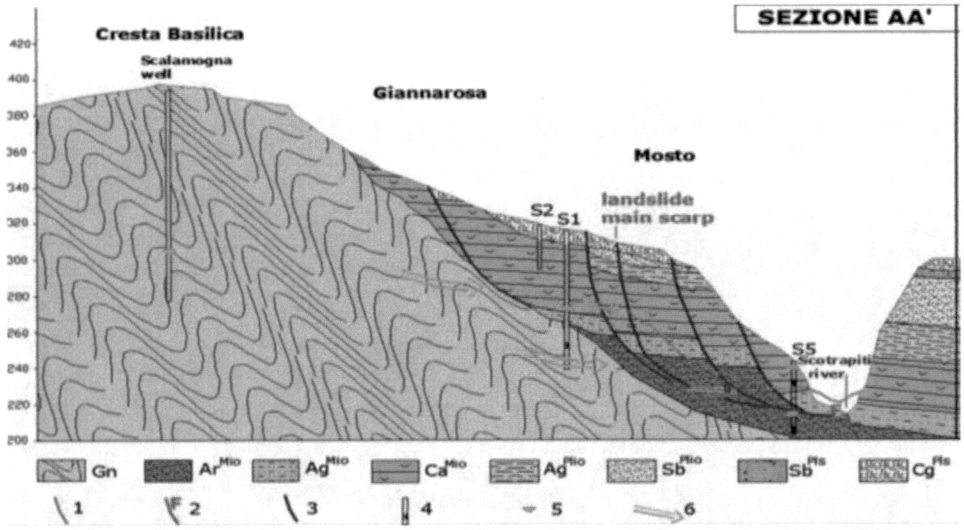

Figure 8. Geological cross-section of the study area before the landslide [38].

4.2. Putrajaya Landslide

A landslide occurred in Precinct 9, Putrajaya, Malaysia caused an evacuation of about 1200 people and buried 27 cars [41]. The failure happened on the western side of a fifty meter hill with a built water tank located at the top. The distance of the tank to the edge of the slope where it failed is 25 m. The area is underlain by graphitic quartz mica schist from Kajang Formation [42]. Figure 9 shows the location of the slope failure.

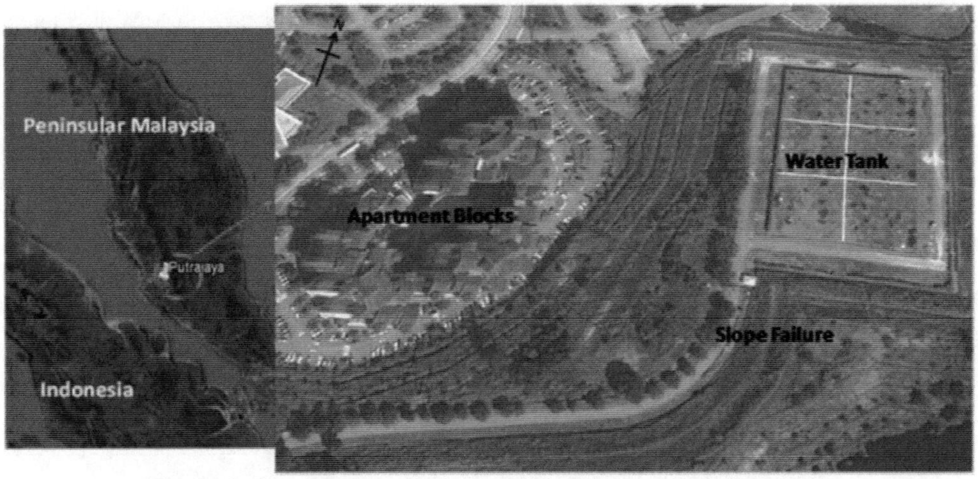

Figure 9. The location of the landslide failure at Putrajaya [42].

A heavy rainfall was recorded in Putrajaya since for two days before the landslide which occurred after the one day of the rainfall [43] and this agrees with the study reported by [44, 45]. Rain gauge located at Precinct 2 recorded a heavy rainfall of 140 mm two days before the landslide as shown in figure 10. The rainfall intensity on the day of the slide was 10 mm. Soil investigation was conducted just after the failure of the slope with a total number of seventeen boreholes. Ismail [2016] also conducted back analysis based on the information obtained from site investigation to identify the causes of landslide. The result indicated that fluctuations in the groundwater due to rainwater infiltration was the main triggering factor of the landslide. The rise of groundwater increased the pore-water pressure at the toe of the slope and reduced the shear strength of the soil. As a result, failure was initiated. It was found that existing groundwater table was high. Based on the measurements of groundwater from boreholes, the groundwater level was at the ground surface at the toe of the slope and 1 m deep at the mid-slope [42].

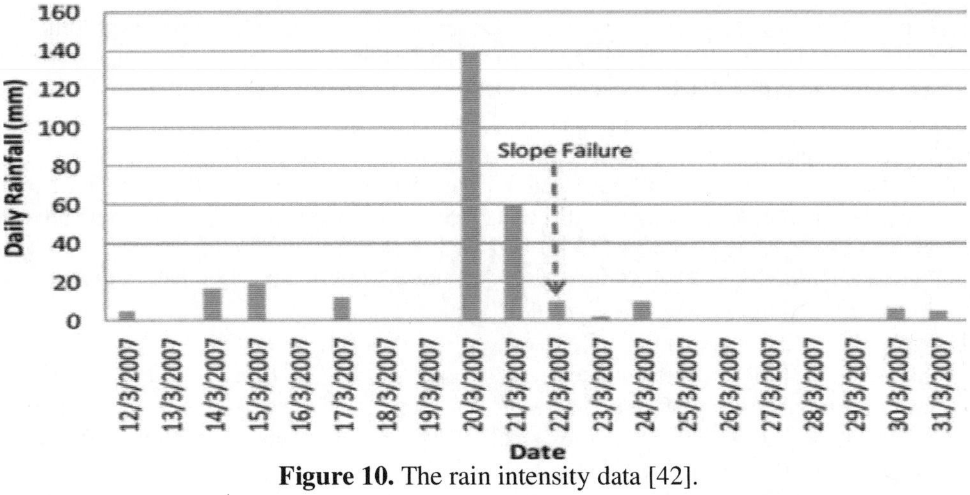

Figure 10. The rain intensity data [42].

5. Conclusion

It can be summarized that the rainfall intensity is one of the factors contributing in triggering hazardous landslides. It seep through the soil resulting the dissipation of the matric suction , increasing the pore-water pressure and reducing the overall strength of the soil. Low rainfall intensity for long time are more hazardous than intensive rainfall for short time. Rainfall infiltrates in a higher rate than soil permeability at the beginning of the rainfall event and drops over time toward the steady state conditions. Heavy rainfall develops a perched water table and the groundwater table rises extensively. The rise of groundwater table is critical once it touches the slip surface of the slope, it causes a deep-seated landslide even though the other part of the slope still dry. Slope failure due to the rise of groundwater table takes one day or more to trigger landslide which is the time taken for the water to infiltrate from the ground surface and the developed perched water tables.

Acknowledgments

The authors are thankful to the Department of Civil and Environmental Engineering, Universiti Teknologi PETRONAS and would like to extend their gratitude to UTP for the generous supports and the provided facilities.

References

[1] D. M. Cruden 1991"A simple definition of a landslide " *Bulletin of Engineering Geology and the Environment.* **43**, pp. 27-29,.
[2] E. D. Werner and H. P. Friedman, 2010 *Landslides: Causes, Types and Effects*: Nova Science Pub Incorporated,.
[3] G. E. Blight 2013*Unsaturated soil mechanics in geotechnical practice*: CRC Press Leiden, the Netherlands.
[4] D. Fredlund and H. Rahardjo 1993 "Introduction to Unsaturated Soil Mechanics," *Soil Mechanics for Unsaturated Soils,* pp. 1-19.
[5] C. W. Ng and B. Menzies 2007. *Advanced unsaturated soil mechanics and engineering*: CRC Press,.
[6] S. A. Taylor and G. L. Ashcroft 1972. *Physical edaphology. The physics of irrigated and nonirrigated soils.*
[7] I. M. Patuti, A. Rifa'i, and K. B. Suryolelono 2017 "Mechanism And Characteristics Of The Landslides In Bone Bolango Regency, Gorontalo Province, Indonesia," *International Journal,* **12** 1-8.
[8] R. P. Orense 2004 "Slope failures triggered by heavy rainfall," *Philippine Engineering Journal,* **25**.
[9] B. Han, S. S. Hou, B. Zhu, L. C. Wang, A. Li, and H. J. Ye 2014 "Deformation Monitoring and Prediction of a Reservoir Landslide in Sichuan Province, China," in *Applied Mechanics and Materials* pp. 2694-2701.
[10] P. De Vita, P. Reichenbach, J. Bathurst, M. Borga, G. Crosta, M. Crozier, *et al.* 1998 "Rainfall-triggered landslides: a reference list," *Environmental Geology* **35**, pp. 219-233.
[11] W. C. Li, L. M. Lee, H. Cai, H. J. Li, F. Dai, and M. L. Wang 2013 "Combined roles of saturated permeability and rainfall characteristics on surficial failure of homogeneous soil slope," *Engineering Geology,* **153** pp. 105-113.
[12] I. Tsaparas, H. Rahardjo, D. G. Toll, and E. C. Leong 2002 "Controlling parameters for rainfall-induced landslides," *Computers and Geotechnics* **29** 1-27.
[13] H. Rahardjo, V. A. Santoso, E. C. Leong, Y. Ng, and C. Tam 2009 "Pore-water pressure characteristics of two instrumented residual soil slopes," in *Proc. of 4th Asia-Pacific Conf. on Unsaturated Soils*, 2009, pp. 333-339.
[14] J. Gasmo, H. Rahardjo, and E. C. Leong 2000 "Infiltration effects on stability of a residual soil slope," *Computers and Geotechnics* **26** 145-165.
[15] A. Galeandro, A. Doglioni, V. Simeone, and J. Šimůnek 2014 "Analysis of infiltration processes into fractured and swelling soils as triggering factors of landslides," *Environmental earth sciences* **71** 2911-2923.
[16] B. Tiwari and S. Caballero 2015 "Experimental Modeling of Rainfall Induced Slope Failures in Compacted Clays," in *IFCEE 2015*, Ed., 2015, pp. 1217-1226.

[17] W. C. Li, F. Dai, Y. Q. Wei, M. L. Wang, H. Min, and L. M. Lee 2015 "Implication of subsurface flow on rainfall-induced landslide: a case study," *Landslides,* pp. 1-15.
[18] A. C. Chiu, W. Zhu, and X. Chen 2009 Rainfall infiltration pattern in an unsaturated silty sand *Journal of Hydrologic Engineering* **14** 882-886.
[19] D. G. Fredlund, H. Rahardjo, and M. D. Fredlund 2012, *Unsaturated soil mechanics in engineering practice*: John Wiley & Sons, 2012.
[20] M. T. Van Genuchten 1980 "A closed-form equation for predicting the hydraulic conductivity of unsaturated soils," *Soil science society of America journal* **44** 892-898.
[21] M. Pirone, R. Papa, M. V. Nicotera, and G. Urciuoli 2015 "In situ monitoring of the groundwater field in an unsaturated pyroclastic slope for slope stability evaluation," *Landslides,* **12** 259-276.
[22] S. Qi and S. K. Vanapalli 2015 "Hydro-mechanical coupling effect on surficial layer stability of unsaturated expansive soil slopes," *Computers and Geotechnics* **70** 68-82.
[23] C. Ng and Q. Shi 1998 "A numerical investigation of the stability of unsaturated soil slopes subjected to transient seepage," *Computers and geotechnics* **22** 1-28.
[24] B. Chae, J. Lee, H. Park, and J. Choi 2015 "A method for predicting the factor of safety of an infinite slope based on the depth ratio of the wetting front induced by rainfall infiltration," *Natural Hazards and Earth System Sciences* **15** 1835-1849.
[25] A. Ali, J. Huang, A. Lyamin, S. Sloan, and M. Cassidy 2014 "Boundary effects of rainfall-induced landslides," *Computers and Geotechnics* **61** 341-354.
[26] T. Lim, H. Rahardjo, M. Chang, and D. G. Fredlund 1996 "Effect of rainfall on matric suctions in a residual soil slope," *Canadian Geotechnical Journal* **33** 618-628.
[27] M. K. Hossain 2010 "Effect of rainfall on matric suction and stability of a residual granite soil slope," *Journal of Dhaka University of Engineering and Technology* **1**.
[28] J. Krahn, D. Fredlund, and M. Klassen 1989 "Effect of soil suction on slope stability at Notch Hill," *Canadian Geotechnical Journal* **26** 269-278.
[29] A. Uchaipichat 2013 "Variation of safety factor with suctions of infinite clay slope under partially saturated condition," *ARPN Journal of Engineering and Applied Sciences,* **8** 166-168.
[30] M. F. Ishak, N. Ali, and A. Kassim 2016 "Tree induced suction on slope stabilization analysis," *ARPN Journal of Engineering and Applied Sciences* **11** 7204-7208.
[31] T. Tsuchida, A. Athapaththu, S. Kawabata, S. Kano, T. Hanaoka, and A. Yuri 2014 "Individual landslide hazard assessment of natural valleys and slopes based on geotechnical investigation and analysis," *Soils and Foundations* **54** 806-819.
[32] A. Leung, H. Sun, S. Millis, J. Pappin, C. Ng, and H. Wong 2011 "Field monitoring of an unsaturated saprolitic hillslope," *Canadian Geotechnical Journal* **48** 339-353.
[33] M. Bordoni, C. Meisina, R. Valentino, N. Lu, M. Bittelli, and S. Chersich 2015 "Hydrological factors affecting rainfall-induced shallow landslides: From the field monitoring to a simplified slope stability analysis," *Engineering Geology* **193** 19-37.
[34] B. Tiwari, D. Tran, B. Ajmera, Y. Carrilo, J. Stapleton, M. Khan, *et al.* 2016 "Effect of Slope Steepness, Void Ratio, and Intensity of Rainfall on Seepage Velocity and the Stability of Slopes," in *Geotechnical and Structural Engineering Congress 2016*, pp. 584-590.
[35] R. A. Abdullah, Q. A. Rosle, M. Al-Bared, N. H. Haron, M. Kamal, and M. Ghazali 2015 "Stability assessment of rock slope at pangsapuri intan, cheras," in *International Conference on Slopes Malaysia. 14th–16th Sept*, 2015.
[36] M. A. M. Al-Bared, R. A. Abdullah, N. Z. M. Yunus, M. Amin, and H. Awang 2015 "Rock slope assessment using kinematic and numerical analyses," *Jurnal Teknologi* **77** 59-66.
[37] P. Gattinoni, L. Scesi, L. Arieni, and M. Canavesi 2012 "The February 2010 large landslide at Maierato, Vibo Valentia, Southern Italy," *Landslides* **9** 255-261.
[38] P. Gattinoni and L. Scesi 2013 "Mechanism and Triggering Factors of the Maierato (VV) Landslide," in *Landslide Science and Practice*, ed: Springer, 2013, pp. 169-176.

[39] A. Doglioni, A. Galeandro, A. Guerricchio, G. Fortunato, E. Guglielmo, M. Ponte, *et al.*, 2010 "Analysis of the rainfall preceding the activation of the large Maierato landslide in 2010," in *Landslide Science and Practice*, ed: Springer, 2013, pp. 107-114.

[40] A. Guerricchio, A. Doglioni, G. Fortunato, A. Galeandro, E. Guglielmo, P. Versace, *et al.*, 2012 "Landslide hazard connected to deep seated gravitational slope deformations and prolonged rainfall: Maierato landslide case history," *Società Geologica Italiana* **21** 574-576.

[41] J. Ahmed, M. Ghazali, M. Mukhlisin, M. Alias, and M. R. Taha 2012 "Effectiveness of horizontal drains in improving slope stability," in *5th Asia-Pacific Conference on Unsaturated Soils 2012*.

[42] M. A. M. Ismail, S. M. Ng, and I. Abustan, "Parametric study of horizontal drains for slope stability measure: A case study in Putrajaya, Malaysia," *KSCE Journal of Civil Engineering,* pp. 1-6.

[43] S. M. Ng, M. A. M. Ismail, C. G. Tan, and R. Yusoh 1999 "Subsurface Characterization Using Seismic Refraction and Borehole Drilling for Slope Stability Study," 1999.

[44] S. Alsubal, N. Bin Sapari, and I. S. Harahap 2018 "Numerical simulation of groundwater rising due to rainfall at far field in triggering landslide," *International Journal Of Advanced And Applied Sciences* **5** 76-86.

[45] S. Alsubal, Sapari, N., and S.H. Harahap 2018, "The Rise of Groundwater Due to Rainfall and the Control of Landslide by Zero-Energy Groundwater Withdrawal System,". *International Journal of Engineering & Technology* **7** p. 6.

IOP Conf. Series: Materials Science and Engineering **513** (2019) 012010 doi:10.1088/1757-899X/513/1/012010

Behaviour of residential structures on problematic ground

T L Ang[1*] and K J Masri[1]

[1]Universiti Teknologi Brunei, Brunei Darussalam

[*]Corresponding author: tingleong.ang@gmail.com

Abstract. Different types of structures behave differently on the effect of differential settlement. In this study, 40 buildings were selected to represent the vast development of national houses. The buildings were selected based on visual observation on cracks and tilted walls. The buildings comprise mostly of piled and unpiled raft foundation supporting framed structures and shear wall structures, except one framed structure with isolated piled foundation. Performance of the structures was compared by using angular distortion. It has been proven that angular distortion can give an indication of the order of magnitude of stresses induced by deformation of the ground. Shear wall structures were found to be more sensitive to vertical ground movement compared to framed structures.

1. Introduction

The response of a building to settlement is dependent on the bearing capacity of the ground and the type and loading condition of the structure. Differential settlement occurred when a structure moves unevenly due to uneven movement of ground. By definition, differential settlement is the differences in vertical displacement of foundations. Uneven settlement at support may result in significant loss of load carrying capacity and thus resulting in damages to various parts of the structure. The ground movement profile is dependent on the soil parameters supporting the foundation and the load applied from the building. The said profile could be concave, convex or the building tilted on one side.

Differential settlement is a common cause of cracks in buildings and it may result in abandonment of the building, depending on the severity of the damages. The remedial actions are usually difficult and require careful analysis of the ground movement. Correct diagnosis of such a problem will avoid unnecessary spending on inappropriate remedial measures.

Differential settlement can cause a structure to rotate or distort causing large strains within structural element and thus a high potential of damages depending on the difference in magnitude of settlement between two points under a structure and the various aspects of the foundation and structural design. The degree of damages can be range from architectural damages such as sticking doors and windows, cracks in wall panels, tiles and plaster, uneven floors to severe cracks in structural elements which could affect serviceability of the structure.

2. Diagnosing differential settlement cracks

Any damages on architectural elements, particularly on infill masonry panels, can usually give a good indication of initiation of differential settlement, then damages to structural elements follows. However, it is always necessary to first verify the causes of crack whether it is due to foundation

IOP Conf. Series: Materials Science and Engineering **513** (2019) 012010 doi:10.1088/1757-899X/513/1/012010

settlement, shrinkage and incompatibility of materials, chemical reactions, thermal movement or the design. It is essential to investigate all possibilities of the causes of the cracks, and finally consider if the crack is a cause of ground movement after having ruled out the others.

In principle, buildings constructed of concrete and bricks are weak in tension. The tension within the material is caused by elongation. When one side of brick panel settles, the distortional effect causes the diagonal of the panel at the settled support to be lengthened and stretched. The resultant tension force pulls the panel apart causing the material to crack. Thus, the crack would be at right angle to the direction of tension. This is the first step to be done to identify damages due to differential settlement. Generally, diagonal cracks extending from openings edges within brick panel can be commonly observed due to presence of structural weak links at the edges of the openings.

The cracks due to differential settlement are likely to initiate at corner of door and window openings or weak structural connections, and it may initiate at higher level of the structure because tendency of horizontal displacement is higher at top level. Cracks in wall panel are usually diagonal, with the horizontal crack measurement normally wider than vertical [1]. This is because the associated rotational movement of structure from differential settlement makes the cracks wider horizontally and the gravity tends to close up the vertical displacement.

3. Angular distortion in structures

The allowable settlement is defined as the acceptable magnitude of settlement that will not affect serviceability of a structure. The determination of tolerance to ground movement by using angular distortion has been proven to be effective by other researchers.

Differential settlement between supports causes distortion and distress to that particular section of the structure. A differential settlement of 50 mm over a span of 4 m may result in greater damage compared to longer span experiencing the similar differential settlement magnitude. This is because the resulted distress is greater in the shorter span.

Angular distortion is the ratio of differential settlement between two points and the horizontal distance apart, less the tilt. Ground settlement under a building is not necessarily a straight line. Hogging or sagging curvature generally exists in related to settlement. Boone used the concept of defining vertical curve in surveying to determine the deflection angle of ground profile [2]. A load bearing wall with symmetrical hogging ground movement profile is illustrated in Figure 1.

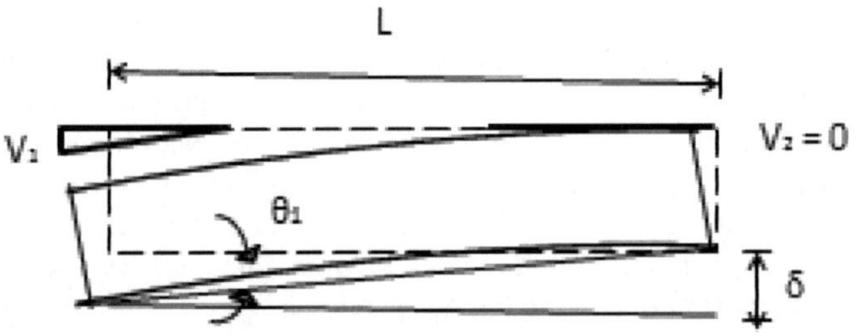

Figure 1. Geometry and notation for deformation of load bearing wall due to settlement (after Boone).

The angular distortion between two points:

$$\beta = V_1 = \theta_1 + \delta/L \tag{1}$$

The radius of curvature is large enough that the angle θ1 between the tangent and the chord is significantly small, therefore

$$\beta = \delta/L \tag{2}$$

The relationship is applied to the series of level readings taken around compound of the representative structures to provide an indication of pattern of vertical movement at the part of structure support.

Skempton and MacDonald [3] concluded in their study on response of conventional frame structures and masonry load bearing walls founded on sand or clay that an angular distortion limit of 1/150 for structural damages, 1/300 for cracking in wall panels and with a factor of 1.5, a limit of about 1/500 was suggested as the allowable value to avoid differential settlement damage.

Polsin and Tokar [4] adopted quite similar approach using 'slope' and 'relative deflection' between two adjacent supports. Their findings agree with Skepmton and MacDonald's [3] on framed structures, taking additional accounts on structural stiffness on Length to Height ratio and the type of building. The proposed limits were 1/200 for structural damage and 1/500 for wall panel. Day [5] studied on behavior of timber framed structure and the proposed limits are 1/100 and 1/300 for structural and wall panel damage limits. March and Thoeny [6] did a study specifically on damages of slab-on-grade foundation of timber framed structures and concluded a proposed limit of 1/230 before the structures experiencing low levels of damage due to settlement.

Table 1 summarized the proposed limits from literature review above. The damaging limits for wall panels were found quite consistent across reinforced concrete, steel and timber framed structures. However, damaging limits to structural elements within timber structures was found comparatively higher as compared to reinforced concrete and steel frame structures due to flexibility of its structural connections.

Table 1. Proposed limits summary by other researchers.

Type of Damage	Reinforced Concrete & Steel Frame		Timber Frame	
	Skempton & MacDonald	Polsin & Tokar	Day	Marsh & Thoeny
Structural	1/150	1/200	1/100	1/230
Wall panel	1/300	1/500	1/300	-

Different type of structures can behave differently on the effect of differential settlement based on the type of design, construction method, type of foundation and capacity of soil underneath it. This paper is intended to explore the tolerable limits of shear wall structures built on problematic soil such as peat, using angular distortion as the key parameter and compare response of such structure to ground movement against framed structures.

4. The scope of study

The scope of this research is to study the difference between the behaviors of framed structure and shear wall dominant structure under the effect of differential settlement. The research is aimed to explore the sensitivity of framed and shear wall structures to differential settlement through field data collection work from the existing low-cost national development houses.

Based on the preliminary survey conducted, 5 sites from the low cost national housing development in Brunei were selected for the study to investigate the performance of shear wall structures and framed structures [7]. A total of 40 buildings were selected as representative houses to produce qualitative data of the structures in response to differential settlement. The 40 representative buildings selected from the 5 sites as shown in Table 3.2, based on observed cracks and visually tilted walls.

4.1. Site 1

The framed structure houses were built on sand replaced peat ground and founded on 200 mm thick waffle raft with primary beams of 750 mm depth and secondary beams of 600 mm depth, with end bearing piles. Identical damages were observed at front and rear brick panel walls with some of the cracks penetrated through the walls (see Figure 2).

4.2. Site 2

Similar to Site 1, this site is situated on sand replaced peat ground. The peat up to depth ranging from 4 m to 6 m was removed and backfilled with sand by hydraulic method to its designed platform level which is 1 m to 1.5 m higher than the original ground level. The self-compacted sand thickness ranged from 4.7 m to 5.3 m and designed with minimum SPT of 15 to provide sufficient bearing capacity for the thick raft foundation.

Table 2. List of selected structures.

Site	Type of house	Type of structure	No of house
1	Terrace 8 units	Framed structure / piled raft	3
	Terrace 6 units	Framed structure/ piled raft	4
2	Semi detached	Shear wall structure / unpiled raft	20
3	Terrace 6 units	Shear wall structure / unpiled raft	9
	Semi detached	Shear wall structure / unpiled raft	2
4	Terrace 6 units	Shear wall structure / piled raft	1
5	Detached	Framed structure / piled foundation	1
	Total		40

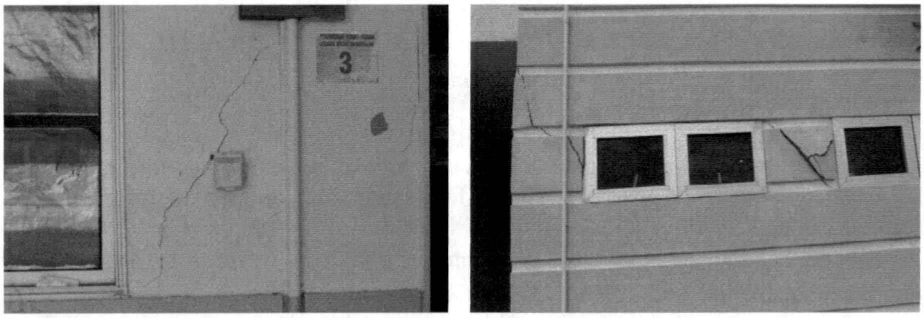

(a) (b)

Figure 2. Cracks observed from site 1 structure (a) Front brick panel, and (b) Rear brick panel.

The semi-detached structures are designed using the concept of tunnel formwork with 100 mm thick shear walls as the dominant structural elements, with transverse beams on first floor and roof level as a tie to the walls.

The buildings are founded on raft of thickness ranging from 150 mm to 300 mm. The 150 mm thickness raft slab is for exterior apron and car porch, 250 mm generally on the main structure and 300mm thick to incorporate punching shear exerted from concrete shear wall. The raft slab is reinforced with two layers of BRC wire mesh. Similarly, the shear walls are reinforced with one layer of BRC wire mesh placed in the middle of the walls. A T10 bar is added at wall edge to act as anchorage reinforcement. The shear walls were casted in situ monolithically with first floor slabs.

The structures do not demonstrate much architectural damages. The brick panels were well confined between shear walls and lateral beams and no evidence of cracks or distortions around the openings. Identical concrete wall cracks initiated around the roof beam connections at concrete gable walls were commonly observed, with some of it in severe condition (see Figure 3). In addition to that, some of the buildings can be seen tilted.

(a) (b)

Figure 3. Cracks observed from site 2 structure (a) Front edge of concrete gable wall, and (b) Rear edge of concrete gable wall.

4.3. Site 3

The site is situated on a cut-and-fill ground. The structure is dominated by precast shear walls with 150 mm thick main exterior walls and 135 mm thick for internal walls. The 150mm thick walls are reinforced with two layers of BRC and two T10 bars added to the wall edge whereas the 135mm thick walls with one layer of BRC placed in the middle of the wall and a T10 bar at the edge. All walls are spliced with equally spaced T15 bars from in-situ structure. The structures are founded generally on 250mm thick raft with 300mm thick at the wall locations. The terrace houses were built using semi-modular structures on cut and fill ground. Massive cracks were observed on concrete walls, slabs and brick panels (see Figure 4).

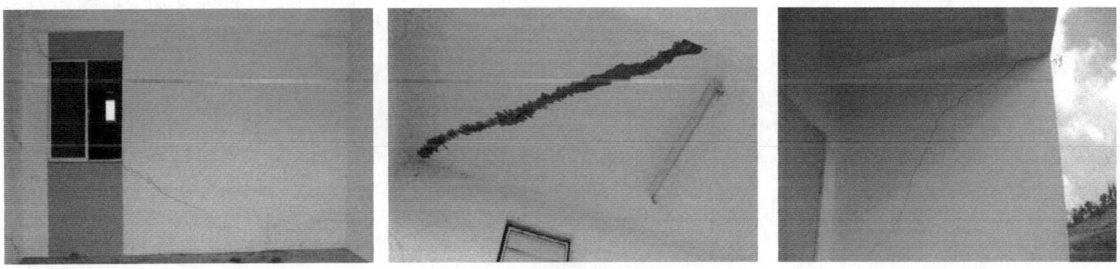

(a) (b) (c)

Figure 4. Cracks observed from site 3 structure (a) Brick panel, (b) Slab, and (c) Concrete wall.

4.4 Site 4

The site is located on a peaty ground treated with prefabricated vertical drains. The selected structure is situated next to approximately 4 m high slope. The foundation design and general layout of the structural elements are generally similar to the structure described in site 1 but adopted shear walls as the dominant structural element. Thickness of the walls is 150 mm. Similar damages as site 1 were observed at rear brick panels of the building (see Figure 5).

4.5 Site 5

This conventional framed structure is a detached residential building located at untreated peaty ground and founded on isolated piles. Severe damage of 15mm crack width and 130mm depth were observed extended from column at the main entrance to ground floor slab (see Figure 6).

IOP Conf. Series: Materials Science and Engineering **513** (2019) 012010 doi:10.1088/1757-899X/513/1/012010

Figure 5. Cracks observed from site 4 structure.

Figure 6. Wall panel crack from site 5 structure.

5. Data Collection Methodology

The research commenced with on-site field data collection by taking reduced levels around perimeter of selected buildings with assistance of professional surveyor. The observed damages were documented in terms of building, location, nature of the damages and photographed. A number of houses at Site 1, 2 and 3 were observed with enormous defects. Readings of reduced levels of all around perimeter of the selected houses and measurement of tilt of walls were collected. Terzaghi [8] suggested that readings of at least 15 points of a defected building shall be referred to describe amount of ground movement. In this research, it was decided to take reduced levels from at least 16 points. The points were selected consistently at support points like columns and walls around exterior perimeter of similar buildings. Additional readings were taken as and where required depending on location of defects and obstructing features. The designed structural layout of the houses at Site 1, 2, 3 and 4 were quite similar (Figure 7). Each unit of houses consist 2 bays at the front, 3 bays at the rear and 3 bays on the side. The bay spans ranging from 3.4 to 5 m.

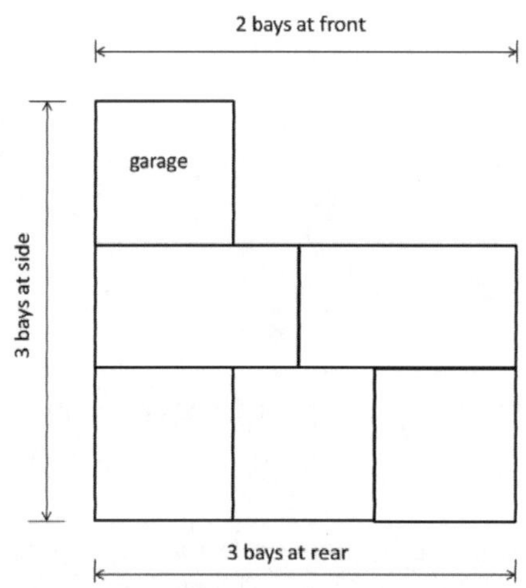

Figure 7. Typical house layout of site 1, 2, 3 and 4.

6. Data Collection Results

Table 3 represents the summarized data obtained from the representative buildings.

6.1. Relationship between maximum angular distortion, δ/L and maximum differential settlement, Δ

The maximum angular distortion δ/L and maximum differential settlement Δ shown in Table 3 was plotted to investigate the relationship following the method of Skempton and MacDonald, and Day. The graph in Figure 8 shows the resulted relationship for framed and shear wall structures on raft foundation.

Table 3. Summary of data.

Site	Block	Building Type[a]	Type of Structure[b]	Max differential settlement Δ(mm)	Max angular distortion δ/L (arch)	Max angular distortion δ/L (struc)	Pattern of settlement profile
1	C10	T8	FRP	29	1/364		sagging
1	C11	T6	FRP	23	1/494		sagging
1	C12	T8	FRP	55	1/312		hogging
1	C23	T8	FRP	22	1/405		sagging
1	C26	T6	FRP	40	1/364		hogging
1	C25	T6	FRP	42	1/564		sagging
1	C24	T6	FRP	36	1/425		hogging
5	EH1	D	FP	36	1/158		hogging
2	B1	SD	SR	30		1/142	hogging
2	B2	SD	SR	48		1/89	hogging
2	B3	SD	SR	13	1/531	1/200	hogging
2	B4	SD	SR	37		1/84	hogging
2	B5	SD	SR	11	1/643	1/325	hogging
2	B6	SD	SR	11	1/650	1/409	hogging
2	B7	SD	SR	24		1/188	hogging
2	B8	SD	SR	16	1/531	1/289	sagging
2	B9	SD	SR	15	1/450	1/173	sagging
2	B10	SD	SR	18	1/544	1/250	hogging
3	A1	T6	SR	27		1/152	hogging
3	A2	T6	SR	43	1/256	1/79	hogging
3	A3	T6	SR	73	1/235	1/45	sagging
3	A4	T6	SR	130		1/32	sagging
3	A5	T6	SR	95		1/147	sagging
3	A6	T6	SR	45	1/333	1/100	sagging
3	A7	T6	SR	60	1/200	1/68	hogging
2	BH3	SD	SR	13	1/531	1/346	sagging
2	BH2	SD	SR	11	1/636		sagging
2	BH4	SD	SR	26		1/163	sagging
2	BH1	SD	SR	10	1/700		sagging
2	BH5	SD	SR	11	1/409		hogging
2	BH6	SD	SR	14	1/778	1/321	sagging
2	BH7	SD	SR	34		1/144	sagging
2	BH8	SD	SR	43		1/163	sagging
2	BH9	SD	SR	27	1/425	1/339	sagging
2	BH10	SD	SR	13		1/377	sagging
3	AHS1	SD	SR	29	1/513	1/117	hogging
3	AHS2	SD	SR	27	1/567	1/89	hogging
3	AHT2	T6	SR	37	1/567	1/92	hogging
3	AHT1	T6	SR	40	1/520	1/65	hogging
4	DH1	T6	SRP	37	1/450	1/94	sagging

[a] Type of building as follows: T8 - 8 unit terrace, T6 - 6 unit terrace, SD - semi-detached, and D - detached house.

[b] Type of structure and foundation as follows: FP - framed structure with isolated piled foundation, FRP - framed structure with piled raft, SR - shear wall structure with raft, and SRP - shear wall with piled raft.

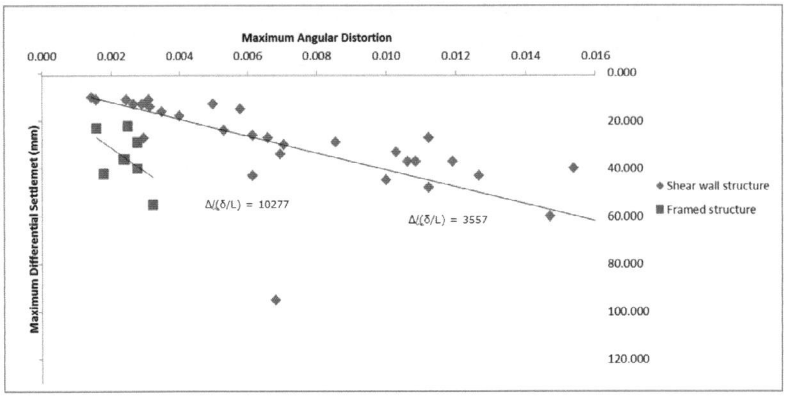

Figure 8. Maximum differential settlement vs maximum angular distortion for framed and shear wall structures on raft foundation.

The line of best fit for framed structures on raft foundation comprises of 7 blocks, shows relation of $\Delta/(\delta/L) = 10277$ in millimeters which is equivalent to 405 in inches. Comparing to the relationship established by Skempton and MacDonald (1956) of $\Delta/(\delta/L) = 350$ in inches (see Figure 9) which had been referred by Day (1990) and Marsh and Thoeny (1999), the result is comparatively lower. Skempton and MacDonald (1956) proposed $\delta/L = 1/300$ as the limit for cracking in brick panel. By substituting this value into the relationship $\Delta = 350(\delta/L)$, the limiting differential settlement for crack in brick wall likely to occur is 30mm. This limit corresponded to the data presented for site 5 where damage on brick wall was detected at differential settlement of 36mm and angular distortion of 1/158.

However, the relationship established by Skempton is not applicable in the case of framed structure on raft foundation. It is probably due to the effect of long raft. Using the relationship determined from Figure 8, and applying the δ/L value in the relationship with reference to the lowest value from block C5 of 1/564, the differential settlement limit Δ=10277/564 = 18 mm for crack in brick wall to likely occur.

In the case of shear wall structures comprising of 32 blocks, the relationship is $\Delta/(\delta/L) = 3557$ in millimeters which is equivalent to $\Delta/(\delta/L) = 140$ in inches. It can be seen that shear wall structures is likely more sensitive to differential settlement compared to framed structures. From Table 3, taking the lowest value of $\delta/L = 1/778$ of Block BH6 from site 2 for wall panel crack to appear, the limiting value of $\Delta = 3557/778 = 5$ mm. Similarly, for structural crack limits, using the lowest value 1/409, the limiting value of $\Delta = 3557/409 = 9$ mm. These limiting values agree with the data.

The reason why the shear wall structures are highly sensitive to differential settlement is likely due to the design of flexible raft slab, walls and first floor slab where brc wire mesh were used as the primary reinforcement.

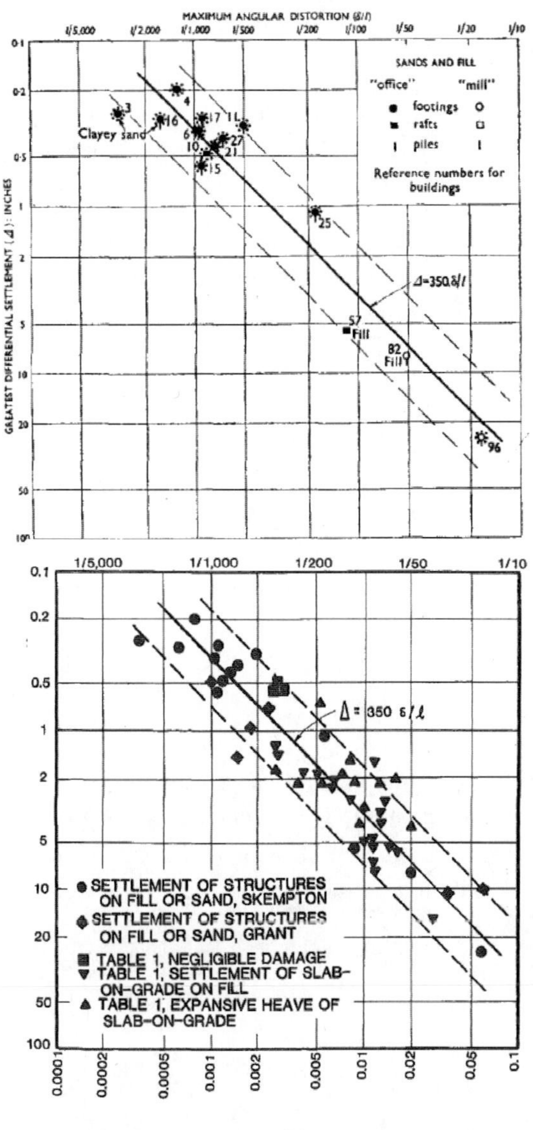

(b)

Figure 9. Maximum differential settlement versus maximum angular distortion (a) Extracted from Skempton and MacDonald, 1956 and (b) Extracted from Day, 1990.

6.2. Comparison between cast-in-suit and precast shear wall structures

A comparison is made on effects of method of construction of shear wall using relationship of δ/L and Δ as shown in Figure10. The gradient of line of best fit for both cases are quite close which is $\Delta = 3404(\delta/L)$ for cast-in-situ structures and $\Delta = 3385(\delta/L)$ for precast. It shows that the method of construction contributes less effect in the difference of severity of damages related to differential settlement.

6.3. Comparison *on effects of length of building of shear wall structures*

All the shear walls are two stories low rise structures and designed and constructed in the almost similar manner of structural details. The E/G of the buildings selected in this study is assumed to be consistent for all buildings since similar concrete grades are specified for all buildings. Considering the difference in length between semi-detached and 6 units terrace buildings, the sensitivity of the buildings to differential settlement is investigated as shown in Figures 11 and 12, using the relationships of max δ/L against max Δ and δ/L against crack width, respectively.

The gradient of line of best fit Δ/(δ/L) in Figure 11 are 3404 and 3233 for semi-detached houses and terrace houses respectively. The 6 units terrace is having 3 times the length of semi-detached buildings. The result reveals that terrace buildings are only 5% more sensitive to differential settlement compared to semi-detached buildings. Both semi-detached (SD) and terrace structures (TS) had shown the same curve of best fit in Figure 12. It shows that the longer buildings do not show much difference in sensitivity to differential settlement compared to shorter building with the same height of shear wall structures.

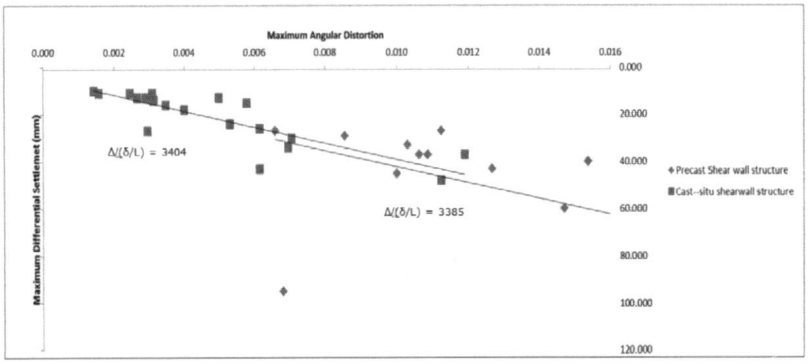

Figure 10. Comparison on effects from method of construction of shear wall structures using relationship between maximum differential settlement and max angular distortion.

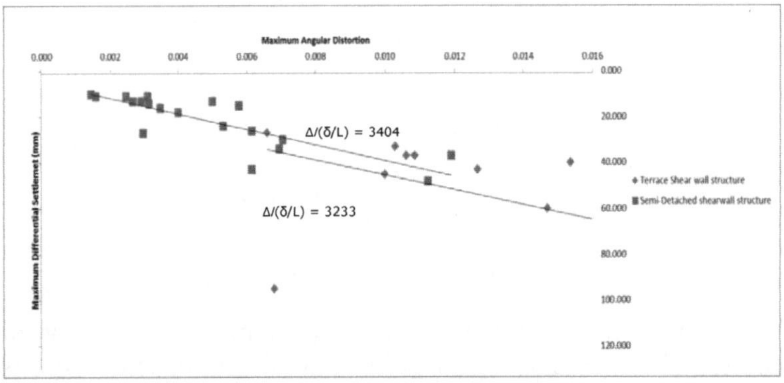

Figure 11. Comparison on effects of length of building of shear wall structures using relationship between maximum differential settlement and maximum angular distortion.

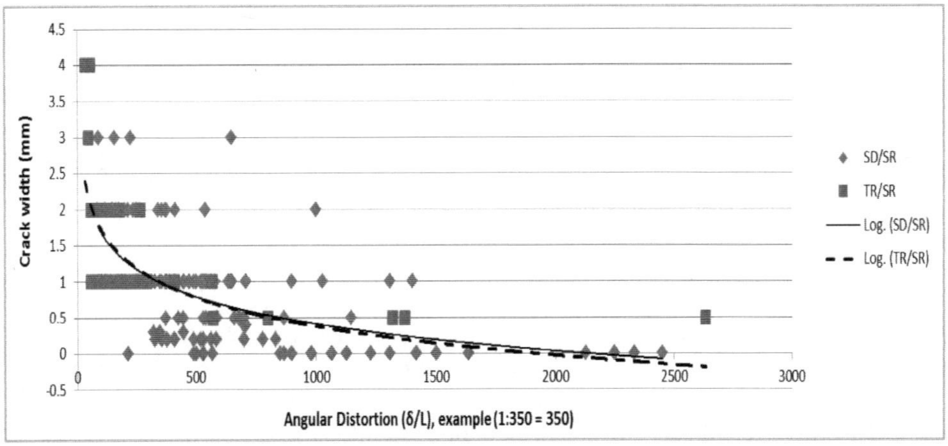

Figure 12. Comparison on effects of length of building of shear wall structures using relationship between angular distortion vs crack width.

7. Conclusion

The performance of framed structures and shear wall dominant structures on raft foundations under influence of differential settlement was examined under this study. Most of the structures are founded on raft, except one framed structure on isolated piles. The studies were made based on evidence of damages during field surveying works.

Proposed tolerable limits by other researchers cannot be simply referred to the structure with similar design concept. The ground condition and type of foundation are key factors to consider. The angular distortion limit for framed structure on peat is found far below the limits proposed by other researchers shown in Table 1.

The framed structures were found to be more sensitive to hogging ground deflection than sagging. The piled raft shear wall structures have shown almost the same response for both sagging and hogging deflection. A relation of $\Delta/(\delta/L) = 10277$ in millimeters (equivalent 405 in inches) was obtained for framed structures. Comparing to $\Delta = 350(\delta/L)$ in inches by Skempton's and MacDonald (1956) on framed structure, the structures undertaken for this study appear to be more sensitive to the ground movement. This is probably due to the effect of complex displacement of long raft.

The damages of unpiled raft structures appear to be inconsistent with respect to both hogging and sagging deflection. The data suggested a more sensitive relationship of $\Delta = 3557(\delta/L)$ in millimeters (equivalent to $\Delta = 140(\delta/L)$ in inches). This is probably because of the design concept that the raft slab and walls are using brc mesh wire as the primary reinforcement. The proposed tolerable angular distortion limits and the maximum differential settlement before cracking occurred are shown in Table 4.

The result of relationship for shear wall structure under cast-in-site and precast construction method was compared and found to be quite close which is $\Delta = 3404(\delta/L)$ for cast-in-situ structures and $\Delta = 3385(\delta/L)$ for precast. This reflects that methods of construction do not contribute much difference in effects of differential settlement.

The effects on L/H ratio on shear wall structures were also investigated and found that buildings which are 3 times longer with the same height only 5% more sensitive to the shorter buildings.

Table 4. Proposed tolerable angular distortion limit and maximum differential settlement.

Type of structure and foundation	Architectural element		Structural element	
	Tolerable angular distortion limit	Maximum differential settlement (mm)	Tolerable angular distortion	Maximum differential settlement (mm)
Framed structure with piled raft	1/560	18	-	
Shear wall structure with unpiled raft	1/780	5	1/140	9

Acceptability of settlement of foundations should be determined uniquely for stability and integrity of a structure based on tolerable movement of their foundation and soil underneath. The relationship $\Delta = 3557(\delta/L)$ is therefore proposed to set as a reference solely for shear wall structures with raft foundation with the concept of structural design for the buildings undertaken for this research.

References

[1] Holland M 2012 *Practical Guide to Diagnosing Structural Movement in Buildings* (Wiley-Blackwell)
[2] Boone S J 1996 Ground movement related building damage *J. Geotech. Engrg.* **122** p 886-896
[3] Skempton A W and MacDonald D H 1956 The allowable settlements of building *Structural and Building Division Meeting 10 May 1956 (Structural Paper* No.50) p 727-768
[4] Polshin D E and Tokar R A 1957 Maximum allowable non-uniform settlement of structures *4th International Conference of soil mechanics and foundation engineering* **1** p 402-405
[5] Day R W 1990 Differential movement of slab-on-grade structures *ASCE Journal of Performance of Constructed Facilities* **4** p 236-241
[6] Marsh E T and Thoeny S A 1999, Damage and distortion criteria for residential slab-on-grade structures, *ASCE Journal of Performance of Constructed Facilities* **13** p 121-127
[7] Ang T L and Masri K J 2018 *Preliminary studies on damages of different type of structures under effect of differential settlement and the proposed research* 6th Annual International Conference on Architecture and Civil Engineering (ACE 2018) p 141-151
[8] Terzaghi K 1935, The actual factor of safety in foundations *Struct. Eng.* **13** p 126-161

IOP Conf. Series: Materials Science and Engineering **513** (2019) 012011 doi:10.1088/1757-899X/513/1/012011

Calculation of vertical bearing capacity factor N_γ of strip footing by FEM

Phuor Ty[1*], Indra S H Harahap[1] and Ng Cheng Yee[1]

[1]Civil and Environmental Engineering Department, UniversitiTeknologi PETRONAS,32610 Bandar Seri Iskandar, Malaysia.

[*]Corresponding author:ty168.utp@gmail.com

Abstract. This paper presents the calculation of vertical bearing capacity factor N_γ of strip footings, which is generally representative of the footing of the buildings, by using the finite element algorithm based elasto-viscoplastic solution. The soil behaviour is idealized as elastic perfectly plastic satisfying the Mohr Coulomb yield criterion with only non-associated flow rules because it tends to treat the actual behaviour of soil in which value of its frictional angle is always different from that of dilative angle. The total bearing resistance is computed using the superposition assumption in Terzaghi classical equation of the three basic components. The comparison of the present solutions with the previous published works including experimental and numerical studies have been presented and discussed. The results reveal in close agreement with the existing published data. In the nutshell, by applying the non-associated soil flow rule representing the real behaviour of soil, the conservation of the superposition assumption is confirmed. Therefore, it could conclude that the results obtaining from these techniques may be confidently applied to a wide range of soil bearing resistance problems.

1. Introduction

One amongst the most highly interesting areas in geotechnical engineering for researchers and practicing engineers is the computation of bearing capacity of various footings such as strip, circular, ring and conical footings. Generally, the calculation of bearing capacity is based on the superposition assumption proposed by Terzaghi, in which the contribution of different loading and soil strength parameter such as cohesion, friction angle, surface surcharge and self-weight, expressed in the form of non-dimensional bearing capacity factors;N_c, N_q and N_γ are added [1]. Based on the used solution techniques, both analytical and numerical solutions to bearing capacity problem can be classified into the following categories: the limit equilibrium method, the upper bound plastic limit analysis method, method of characteristics, finite difference method and finite element method. In recent year, finite element method has been widely used to compute the bearing capacity of strip, circular, ring and conical footings and its pre- failure behaviour because it is capable of combining all the parameters into a single problem. Even though analytical expressions derived from the theory of plasticity provide the bearing capacity factors;N_c and N_q that remain unchanged from the variation in the roughness of the footing-soil interface [2,3,4,5], there is a considerable difference in the magnitudes of N_γ between the smooth and rough footing [2,3,4,5,6]. Consequently, there are several methods have been used to calculate N_γ, including the stress field method [7], the partial plastic stress field method [8], the limit equilibrium method [1,2,9], the method of characteristics [10,11], the method of stress characteristics [5], the limit analysis method [12,13,14], analytical upper-bound solution [15], the finite difference method [16,17] and the finite element method [4,18]. Different methods lead to different assumptions and procedures and consequently lead to discrepancies in their results.

Finite elements algorithm based elasto-viscoplastic solutions have been successfully employed to evaluate the bearing capacity [4,18]. The current note presents the computation of bearing capacity factor N_γ of strip footing, which is generally representative of the footings of buildings, using finite element algorithm based elasto-viscoplastic solution. The soil behavior is idealized as elastic perfectly plastic satisfying the Mohr Coulomb yield criterion with only non-associated flow rules because it tends to treat the actual behaviour of soil as the value of its frictional angle is always different from that of dilative angle. The total bearing resistance is computed using the superposition assumption in Terzaghi classical equation of the three basic components. The comparison of the authors' results with previous available published works, including experimental and numerical studies, have been presented and discussed. The results reveal a close agreement with existing published data.

2. Solution techniques

The solution technique of the solution is generally divided into two: analytical technique and numerical technique. Moreover, there are several methods are embedded inside these two techniques. In this paper, finite element algorithm based viscoplastic technique has proven to be an efficient way of solving the plasticity problem in geomechanics is employed. Likewise, this technique has been suggested by Zienkiewiczand Cormeau[19]; Zienkiewicz and Humpheson [20] and subsequently successfully used by Griffith[4] and Manoharan and Dasgupta [18]. This method is also known as the initial strain solution techniques. The plasticity problem is solved by iterating using the equivalent elastic solutions until any stresses which temporarily violated yield have returned back to the failure surface by obeying the fairly strict tolerances. To verify the change in body's forces from iteration to one another, the convergence criterion plays a vital role in doing so. As the stresses violate the yield, the balancing forces of the body (self-equilibrating body forces) are incremented at each iteration by the amount related to the magnitudes of violated stresses. As the increment in body forces diminished, the stresses also return to the yield surface. For this present work, convergence is allowed as long as the change in body forces with respect to the maximum absolute value, nowhere is more than 0.1%. Yet, Zienkiewicz et al. has also been reported a failure of the algorithm to converge at collapse [19,20]. Throughout this study, 500 is set to be the maximum restricted value for the number of steps of time integration. The displacement control finite element is employed satisfying the equilibrium and continuity. Eight-node quadrilateral isoparametric element with reduction of integration technique using a (2x2) Gaussian quadrature is used for all types of foundations to compute their stiffness. To define the bearing pressure, the average of the vertical stress component in the first row of integration points below the displaced nodes is calculated.

3. Definition of the problem

To complete the simulation in this paper, the plane strain solutions are raised to compute the bearing capacity factors for strip footing underlying the general cohesive-frictional soil. The ground surface is horizontal and besides the footing, the soil surface is subjected to the uniform surcharge pressure q, while the footing is subjected to vertical downward load without any shear stress (eccentricity). Smooth footing base is considered assuming there is no mobilized shear stress over the base, while rough footing base has a perfect bounding with the ground underneath. Table 1 shows the summary of the calculation terms for the bearing capacity of strip footings. The footing is placed on a general cohesive-frictional soil with thickness h. The ultimate bearing pressure (q_{ult}) is evaluated based on the equation (1). The notation Q_u, p, b, used in equation (1) represent the ultimate applied load, the vertical pressure at any point along the footing base, a half-width of strip footing. The mesh of the present analysis is illustrated in Figure 1. It has 1307 and 408 representing the node number and the element number, respectively.

Table 1. A summary of the calculation terms for strip footing.

Calculation terms	Strip footing
Plane strain	✓
Element type: 8 node quadrilateral element	✓
Ultimate pressure load	Equation (1) $q_{ult} = Q_u/b$, Where $Q_u = \int_0^b pdy$
Perfectly smooth and perfectly rough bases	✓
Soil friction angles ($\varphi = 5^o, 10^o, 15^o, 20^o, 25^o, 30^o, 35^o, 40^o, 45^o$)	✓

4. Soil constitutive relation

Soil behaviour satisfying the Mohr Coulomb yield criterion with only non-associated flow rules because it tends to treat the actual behaviour of soil as value of its frictional angle is always different from that of angle dilator. Regarding the determination of expansion angle, Vermeer and De Borst (1984) reported that soil dilation angle varies between 0° and 20° and is at least 20° lower than angle of soil friction [21]. Moreover, Cox (2008) utilized the light microscopy (LM) and scanning electron microscopy (SEM) imaging techniques for his experimental works to study the influence of grain shape on dilatancy using 4 different types of sand such as Silica 40 sand; Ottawa 20 sand; Rillito River sand and Santa Cruz River Sand. Melissa found that the limit of soil expansion angle is between 0° to 15° [22]. Hence, the present analysis is conducting by using the non-associated soil with the maximum dilation angle value of 15° if the internal friction angle of soil is greater than 35°.

The material properties of soil throughout the analysis in this paper are assumed such as Young's modulus E = $2\text{x}10^5$kN/m^2; Poisson's ratio ν=0.3, while cohesion c = 20kN/m^2; unit weight of soil γ=20kN/m^3 and surface surcharge q = 25kN/m^2. All these parameters are taken into account, if necessary. The friction angles of soil varied from 5° to 45° for the parametric study of bearing capacity factors.

The Mohr-Coulomb yield criterion of soil is described as a function of the three stress invariants $(\sigma_m, \bar{\sigma}, \theta)$ as [23]

$$F = \sigma_m \sin\varphi + \bar{\sigma}K(\theta) - c\cos\varphi \tag{2}$$

$$K(\theta) = \cos\theta - (\sin\varphi \sin\theta)/\sqrt{3} \tag{3}$$

is a function controlling the shape surface in the octahedral plane.

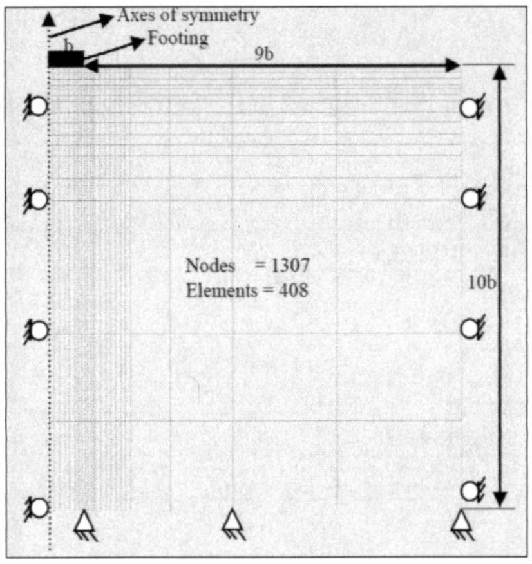

Figure 1. Mesh used in the present analysis.

5. Results and discussion

The bearing capacity of shallow strip footing based Terzaghi's superimpose assumption is expressed as:

$$q_{ult} = cN_c + q_s N_q + \gamma b N_\gamma / 2 \quad (4)$$

These three dimensionless bearing capacity factors N_c, N_q and N_γ are assumed to be a function of the soil friction angle. Finite element method has been utilized together with plasticity theory to predict bearing capacity factors. Griffiths used finite elements to compute each of the bearing capacity factors of strip footing in turn as a function of the soil friction angle by assuming (1) weightless soil with cohesion and friction to determine the N_c, (2) weightless, cohesion-less soil with friction receiving a uniform surface surcharge to evaluate N_q and (3) cohesion-less soil with friction and self-weight to compute N_γ [4]. The present analysis has been conducted for both perfectly smooth and rough footings with the soil friction angle ranging from $\varphi = 5^o$ to 45°. The vertical displacements applied to the footing are carried out continuously downwards with the constant rate of velocity until soil failure is reached. The soil resistance is based on parameters φ and c representing the angle of friction and cohesion of soil, respectively.

5.1. Computation of N_γ

Unlike the two previous soils bearing capacity factors N_c and N_q, there is no closed form solution for N_γ that contributes to bearing capacity because soil self-weight is included. To define solely the contribution of the N_γ term to the total bearing capacity of a footing, the assumption of soil to be weighted, cohesionless with no surcharge acting is made by vanishing the first two terms of the equation (2). In addition, the initial shear strength of soil prior to loading is constant for weightless cohesive-frictional soils and weightless cohesion-less soils with a uniform constant surcharge, while it varies for weighted cohesion-less frictional soil. Consequently, Footing-soil interface has a significantly effect on N_γ, however, it has a slight influence on N_c and N_q. To analyse N_γ, the weighted cohesion-less soil is taken into account and its initial vertical stresses are obtained by the product of the unit weight of soil and the distance of the element below the ground surface. The horizontal initial shear stresses in the others two horizontal directions are determined by the multiplication between its vertical stress and the coefficient of earth pressure at rest (k_0). k_0 is equal to unity in the present analysis. With the ultimate bearing capacity, N_γ value corresponding to the specified friction angle of soil calculated in the analysis is computed using [4] as:

$$N_\gamma = 2 \times q_{ult} \, / \, (\gamma \times \mathrm{b}) \quad (5)$$

where q_{ult} is computed from equation (1).

Figure 2 compares the calculation of bearing capacity factor N_γ for smooth strip footing from the current analysis with the available solution published in literature. At low and medium friction angles for $\varphi \leq 40^o$, the close agreement is found as compared the present solution with the available results in literature such as with the solutions obtained by Hansen and Christensen (1969) employing the method of stress characteristics[8], with results computed by Booker (1969) utilizing stress characteristics method[7], with solution calculated by Bolton and Lau (1993) [5]; Dongdong et al., (2016) [24] and Martin (2004) [25] using method of characteristics and with the mean value between lower and upper bound limit analysis produced by Ukritchon et al., (2003) [14]. However, their solutions are for high friction angles $\varphi > 40^o$noticeably greater than the current solution. Likewise, it also found out that the current solution is considerably smaller than the existing published solution in literature such as Hill (1950) mechanism [4];Sokolovskii (1965) [10] and Martin (2004) [25] used method of characteristics, Chen (1975) [12] employed upper bound analysis, Griffiths (1982) [4] solved by finite element method, Michalowski (1997) [6] ultilized upper bound analysis and Frydman and Burd (1997) [16] used finite different method.

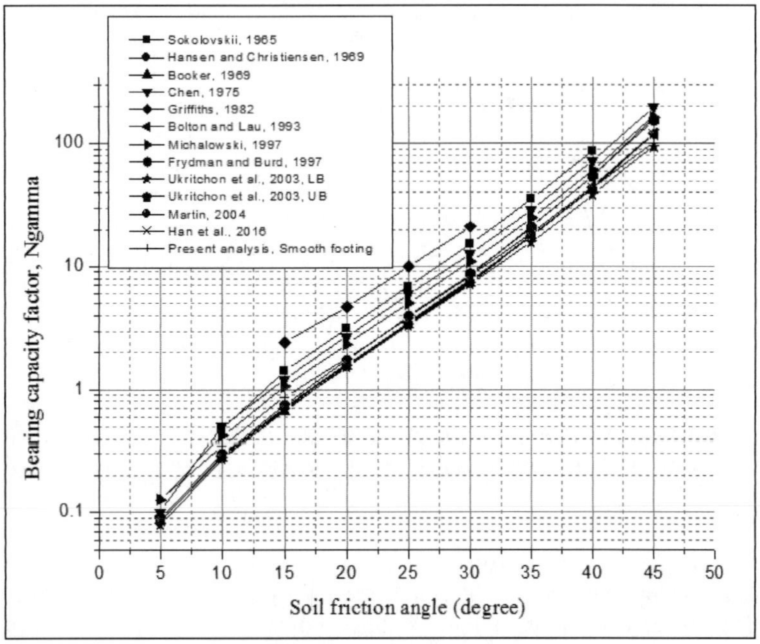

Figure 2. Comparison of bearing capacity factor N_γ of smooth strip footing from published solutions and present analysis.

The analogy of the computed bearing capacity factor N_γ for rough strip footing of the present analysis with previously available solution is plotted in Figure 3. The excellent agreement is discovered for low and medium friction angles for $\varphi \leq 40^o$, when the current result is compared to available results which are produced by (1) Meyerhof (1963) using Semi-empirical method [14], (2) Hansen and Christiensen (1969) employing the method of stress characteristics [8], (3) Booker (1969) utilizing the method of stress characteristics [7], (4) Griffiths (1982) applying finite element method [4], (5) Ukritchon et al., (2003) considering the mean value of N_γ from lower and upper bound limit analysis [14], (6) Kumar (2003) employing limit equilibrium [11], (7) Martin (2004) using method of stress characteristics [26], (8) Hjiaj (2005) utilizing upper bound analysis [26] and (9) Snodi (2011) employing the method of characteristics [27]. Nevertheless, their solutions are noticeably greater than the current solution for high friction angles $\varphi > 40^o$. On the other hand, at low and medium friction angles for $\varphi \leq 30^o$, the present result is slight lower as analogized to the existing solutions calculated by Terzaghi (1943) (used limit equilibrium) [1], Caquot and Kerisel (1953) (employed the method of stress characteristics) [14], Vesic (1973) (utilized limit equilibrium) [14], Michalowski (1997) (applied upper bound analysis) [6], Frydman and Burd (1997) (based on finite difference method) [16] and Soubra (1999) (employed upper bound analysis) [15]. Though, the noticeable discrepancies appear for $\varphi > 30^o$. Furthermore, the overestimation of N_γ over the current analysis for rough strip footing have presented such as Prandtl (1921) applying limit equilibrium [4], Chen (1975) using upper bound analysis [12] and Bolton and Lau (1993) [5] and Han et al., [24] employing the method of stress characteristics [5].

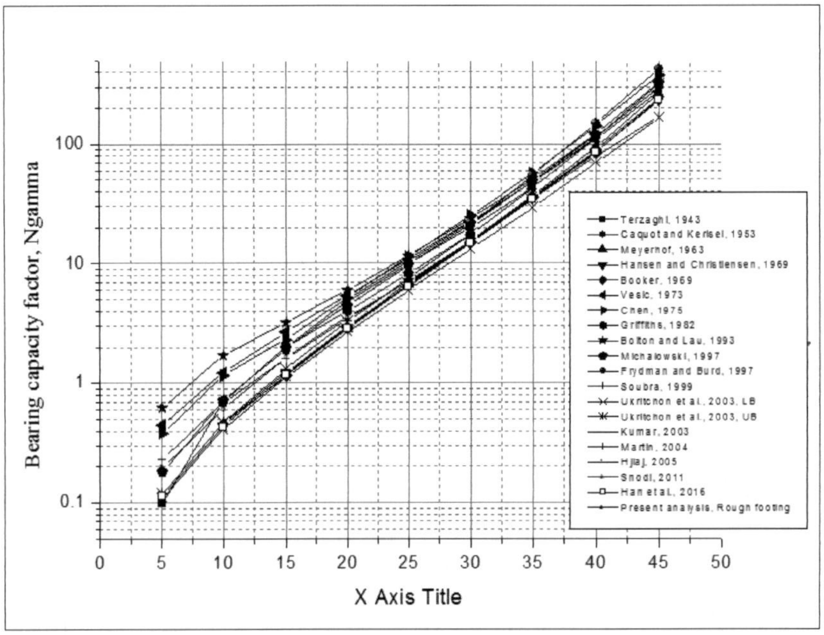

Figure 3. Comparison of bearing capacity factor N_γ of smooth strip footing from published solutions and present analysis.

6. Conclusion

Bearing capacity factor N_γ was computed for a strip footing using FEM. The analysis was executed by considering the actual properties of soil in which the dilation angle is generally less than 15°. The results reveal in close agreement with the existing published data. The assumption of additive contribution of soil cohesion, surcharge and weight was proposed by Terzaghi provides bearing capacity predictions sufficiently accurate. Bearing capacity factors depend on the angle of dilation, especially for higher values of the friction angle. The magnitudes of the factors for a rough footing base, especially with greater value of phi, were significantly higher than those for a smooth footing.

Acknowledgements

This work was supported by the research grant names Evaluation and Extension of current site specific assessment of jackup considering soil-structure interaction with the cost centre 0153AA-E88.

References

[1] Terzaghi K 1943. *Theoretical soil mechanics*. John Wiley & Sons, New York
[2] Meyerhof G G1951. The ultimate bearing capacity of foundations *Géotechnique.* **2** 301–332
[3] Meyerhof G G 1955. Influence of roughness of base and ground-water conditions on the ultimate bearing capacity of foundations *Géotechnique.* **5,** 227–242

[4] Griffiths D V 1982 Computations of bearing capacity factors using finite elements. *Géotechnique*32(3) 195–202

[5] Bolton M D, Lau C K 1993 Vertical bearing capacity factors for circular and strip footings on Mohr–Coulomb soil *Canadian Geotechnical Journal* **30,**1024–1033

[6] Michalowski R L 1997 An estimate of the influence of soil weight on bearing capacity using limit analysis. *Soils and Foundations.* **37**(4) 57–64

[7] Booker J R 1969. Australia. *Applications of theories of plasticity to cohesive frictional soils.* PhD thesis, Sydney University

[8] Hansen B, Christensen N H 1969 Discussion of 'Theoretical bearing capacity of very shallow footings', by A. L. Larkins. *J. Soil Mech. Found. Div., Am. Soc. Civ. Eng.* **95**(6) 1568–1572

[9] Zhu D Y, Lee C F and Jiang H D 2001 A numerical study of the bearing capacity factor N_γ *Canadian Geotechnical Journal.* **38** 1090–1096

[10] Sokolovskii V V 1965*Statics of granular media, Pergamon*, New York

[11] Kumar J 2003Nγ for rough strip footing using the method of characteristics. *Canadian Geotechnical Journal* **40** 669-674

[12] Chen W F 1975*Limit analysis and soil plasticity*. Elsevier, Amsterdam

[13] Sloan S W 1995 Limit analysis in geotechnical engineering *Proc., I.B. Donald Symposium on Modern Developments in Geomechanics*, C. M. Haberfield, ed., Monash Univ., Clayton, Melbourne, Victoria, 3168 Australia, **32**

[14] Ukritchon B, Whittle A J and Klangvijit C 2003 Calculation of bearing capacity factor Nγ using numerical limit analyses*Journal of Geotechnical and Geoenvironmental Engineering* **129**(6) 468-474

[15] Soubra A H 1999 Upper-bound solutions for bearing capacity of foundations. *J. Geotech. Geoenviron. Eng.* **125**(1) 59–68

[16] Frydman S and Burd H J 1997 Numerical studies of bearing capacity factor Nγ. *J. Geotech. Geoenviron. Eng.* **123**(1) 20–29

[17] Loukidis D and Salgado R 2009 Bearing capacity of strip and circular footings in sand using finite elements. *Comput.Geotech.* **36** 871–879

[18] Manoharan N and Dasgupta S P 1995 Bearing capacity of surface footings by finite elements. *Comput. Struct.* **54**(4) 563–586

[19] Zienkiewicz O C andCormeau I C 1974Viscoplasticity, plasticity and creep in elastic solids. A unified numerical solution approach*Int, J, Numer, Meth, Eng.* **8** 821-845

[20] Zienkiewicz O C, Humpheson C and Lewis R W 1975 Associated and non-associated viscoplasticityand plasticity in soil mechanics. *Geotechnique.* **25**(4) 671-689

[21] De Borst R and Vermeer P A 1984 Possibilities and limitation of finite elements for limit analysis. *Geotechnique.* **34**(2) 199-210

[22] Cox M R B 2008 *The influence of grain shape on dilation*. PhD Thesis, The University of Arizona

[23] Smith I M, Griffiths D V and Margetts, L 2013*Programming the finite element method*. John Wiley & Sons

[24] Dongdong H, XinyuXie, Lingwei Zheng, and Li Huang 2016 The bearing capacity factor Nγ of strip footings on c–ϕ–γ soil using the method of characteristics. *Springer Plus* **5**(1) 1482

[25] Martin C M 2004 *ABC—Analysis of bearing capacity*

[26] Hjiaj M, Lyamin A V and Sloan S W 2005 Numerical limit analysis solutions for the bearing capacity factor Nγ. *International Journal of Solids and Structures* **42** 1681-1704

[27] Snodi L N 2011 Estimation of N_γ for strip and circular footings using the method of characteristics *Diyala Journal of Engineering Sciences* **4**(2) 1-11

IOP Conf. Series: Materials Science and Engineering **513** (2019) 012012 doi:10.1088/1757-899X/513/1/012012

Information management for humanitarian aid distribution system in Malaysia

Suzila Mohd[1], Mohamad Syazli Fathi[2*] and Aizul Nahar Harun[1]

[1] Malaysia Japan International Institute of Technology (MJIIT), Universiti Teknologi Malaysia (UTM), Kuala Lumpur, Malaysia
[2] Razak Faculty of Technology and Informatics, Universiviti Teknologi Malaysia (UTM), Kuala Lumpur, Malaysia

[*]Corresponding author: syazli@utm.my

Abstract. Effective information management is one of the key success factors in managing the humanitarian aid distribution process during and post disasters and it depends on how much, how accurate, and how reliable the available information is. However, there are several key issues that need attention, such as redundancy of information about the disaster and fraudulent reports that could be retrieved by malicious persons which can have negative effects on the humanitarian aid distribution process. This paper aims to investigate the current practices of the information management system used by disaster management stakeholders in Malaysia during a natural disaster. Eighteen participants were interviewed to find out about their current strategies or plans for managing information during a natural disaster. Findings revealed that there is no information management process or protocol for non-government organizations (NGOs) requesting and receiving information on humanitarian aid. Conclusions are drawn about proposing a new information management process or protocol in the humanitarian aid distribution process.

1. Introduction

Since the Second World War, humanitarian aid has become an extensive activity throughout the world [1, 2]. Nowadays, the role of humanitarian aid has become a feature of the disaster management agenda during and post disasters of a natural disaster, which aims to lighten the burden of disaster victims. Humanitarian aid consists of several activities and several stakeholders. The main activities in humanitarian aid during and post disaster are search and rescue, welfare as well as health and medical [3-5]. The key drivers in managing the humanitarian aid activities consist of different stakeholders, which are Government agencies, non-government organizations (NGOs), volunteers, and the private sector [6-8].

Information management plays significant roles in managing humanitarian aid [9-11]. The effectiveness in managing disaster information during a natural disaster could avoid problems in the humanitarian aid distribution process. However, the involvement of several stakeholders in managing humanitarian aid has caused some problems in information management. Several papers have addressed the information management problems in managing the humanitarian aid process and have highlighted two (2) problems: the first one is redundancy of information on the disaster. This problem will lead the NGOs, volunteers or the private sector to respond to an individual request at the same time [12]. The second problem is fraudulent reports from malicious persons [12, 13]. The uncoordinated information management could lead to these fraudulent reports being retrieved by malicious persons for their own intentions [12, 13]. Consequently, these information management problems will affect the humanitarian aid distribution process and disaster victims in several ways, as follows:

- Evacuation centres become overloaded and clogged with non-critical items, which causes wastage [14];
- Chaotic processes arise in distributing goods and basic needs to disaster victims [14];
- Unequal distribution of goods and basic needs to the affected disaster area and disaster victims [14].

Therefore, this paper aims to investigate current practices of information management systems during a natural disaster occurrence. This is important to propose a new information management process in humanitarian aid distribution which will avoid information management problems and increase effectiveness and efficiency when managing the humanitarian aid distribution process.

2. Methodology

All the information in this paper is based on primary and secondary data. Multiple case studies were conducted to obtained data of this research at three (3) natural disaster risk areas in Malaysia, which are Kelantan, Terengganu and Pahang [15, 16]. First, primary data was collected through semi-structured interviews with 9 government officers (GO) and 9 local non-government organizations (NGOs). The GO participants are consisting of GO from Department of Social Welfare (JKM) (3 district officers, 3 state officers and 1 central officer), 1 GO from Malaysian Civil Defence Force (APM) (central level) and 1 GO from National Disaster Management Agency (NADMA). All the participants were experienced in managing the humanitarian aid distribution process (more specifically in foods and basic necessities distribution) at the selected case studies area.

Second, secondary sources were gained through a literature review on humanitarian aid distribution, which covered its concepts, processes, disaster management teams, enabling technologies, and issues. Various sources were reviewed, such as books, journal articles, international conference papers, government reports, and material available on the internet. Content analysis techniques were used to analyse both primary and secondary sources. All the findings presented in this paper are through text, images and expressions.

3. Disaster management stakeholders

The disaster management stakeholders according to Asproth, Amcoff Nyström [6] comprise government agencies [17], non-government organizations (NGOs) [17, 18], volunteers [19] and the private sector [9, 19]. The roles and responsibilities of disaster management stakeholders in disaster management is to provide humanitarian aid to the disaster victims. Findings from the interviews revealed that, in Malaysia, government agencies form the focal agency in disaster management, while NGOs, volunteers and the private sector are classified as supporting stakeholders in disaster management. All the disaster stakeholders were involved in disaster management by providing humanitarian aid distribution to the disaster victims during and post disaster. In Malaysia, the government agencies are divided into three (3) command and control levels as shown in Figure 1. The three (3) Command and Control levels of the government agencies in disaster management are shown in Figure 1, they are level I: District disaster management and Relief committee, level II: State disaster management and Relief committee and level III: Central disaster management and Relief committee. Each level consists of several government agencies.

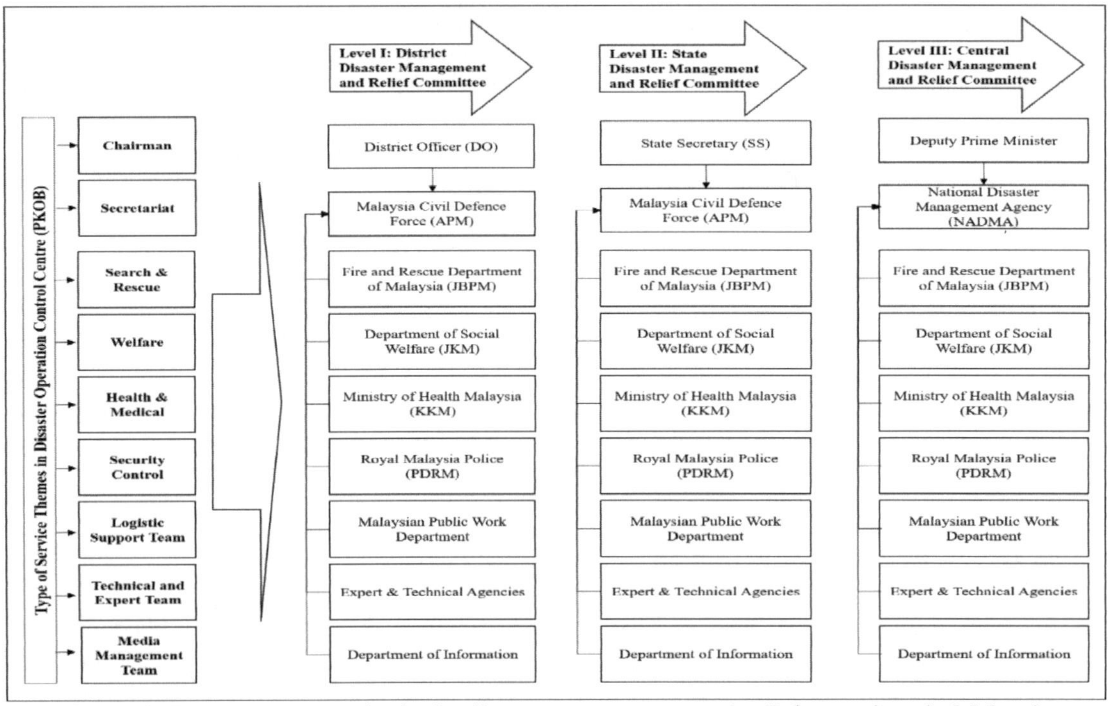

Figure 1. Government agencies in the disaster management and relief committee in Malaysia.

Finding from both the document review process (current legislation: Directive No.20) and interviews with GO participants revealed that the roles and responsibilities of Government agencies are well-defined in Directive No.20 compared to those of NGOs, volunteers and the private sector as supporting stakeholders in disaster management which are loosely defined. This is agreed by all the participants in this research, which highlighted the need to develop a standard operating procedure (SOP) to define the roles and responsibilities of NGOs, volunteers and the private sector as supporting stakeholders as well as a procedure or process for disaster management. This is in line with Dahari, Yusof [20], which explained the importance of a SOP as a guidance in planning and monitoring the NGOs activities during all phases of disaster management by government agencies. More important, the existing of the SOP is believed could reduce problems in managing humanitarian aid distribution process during and post disaster.

4. Current information management flow in humanitarian aid

It is very important to understand on how is the disaster information management is been manage by disaster management stakeholders. Figure 2 presents the current information management flows conducted by one of the disaster management stakeholders during a disaster, which is the Social and Welfare Department (JKM). According to all GO participants, there are four (4) main roles and responsibilities of the JKM in disaster management, as follows:

- Prepare and maintain evacuation centres for disaster victims;
- Prepare and distribute foods, clothes, and others basic needs to the disaster victims;
- Carry out the registration of disaster victims for recovery purposes;
- Provide guidance, advice/counselling to disaster victims.

The implementation of the roles and responsibilities of JKM is divided into three (3) stages, which are: (1) pre-disaster, (2) during a disaster and (3) post-disaster. The current information management flow for humanitarian aid done by JKM is discussed in this paper based on all the findings gained from GO participants, as follows:

4.1. Level I: District disaster management and relief committee

The main roles and responsibilities of the Social and Welfare Department at the district level (JKMD) of the disaster management and relief committee are to recover and give help to the disaster victims. JKMD will open the evacuation centres that have been declared safe to become a shelter for disaster victims. Then, the JKMD will manage the registration of the disaster victims to track the record of their information. The disaster victims will be provided with several basic necessities throughout the disaster. The registration process will continue throughout the disaster. At the same time, the JKMD will also need to identify and record the information on disaster victims that have returned home. All the information will be updated to the district officer (DO), the Malaysia Civil Defence Force (APM) officer (the secretariat at district level) and the state Social and Welfare Department (JKMN) 4 times per a day. Moreover, the JKMD will provide food and water to the disaster victims. Each evacuation centre will be provided with sufficient stocks of food and water for 7 days.

4.2. Level II: State disaster management and relief committee

The information updated by the JKMD will be analysed and used by JKMN. The information will be used to update the Social and Welfare Department portal, called the 'Info Banjir JKM V2.0'. All the information will also be used during disaster management meetings at state level. Besides that, the JKMN is also responsible for providing food and other necessities to the disaster victims as well as being responsible for managing the volunteers and other government officers mobilized to the district or area affected by the disaster, if necessary. Besides that, JKMN also needs to update all the disaster information to the federal JKM 4 times per a day. Other roles and responsibilities of the JKMN explained by GO participants are listed as follows:

- Managing logistic facilities for land, air and water transportation for the purpose of supplies delivery.
- Update the latest disaster information to the Prime Minister, other ministers, the chairman of the disaster management committee at state level; post-scene control (PKTK), and the JKM control centre.
- Coordinate humanitarian aid requests from the district.
- Coordinate the contribution of goods and make distributions.
- Record types and the total amount of humanitarian aid that have been channelled to the disaster victims.
- Coordinate and monitor additional officers who have been placed in their respective states.
- Provide lists of volunteers' teams to assist disaster operations in other states if needed.

4.3. Level III: Central disaster management and relief committee

At central level, JKM needs to open a disaster operation centre after receiving a command from the KPKM. JKM at national level will also receive a disaster report from JKMN 4 times per a day. At the same time, JKM national must use the latest disaster information report from JKMN and update the information to the:

- Minister of Women, Family and Community Development (KPWKM);
- Deputy Minister of KPWKM;
- Chief Secretary of KPWKM;
- Highest Management of KPWKM;
- Director of Social and Welfare (KPKM);
- Highest Management of JKM;

IOP Conf. Series: Materials Science and Engineering **513** (2019) 012012 doi:10.1088/1757-899X/513/1/012012

- National Disaster Control Centre (NDCC).

Other than that, the roles and responsibilities of JKM federal are listed as follows:

- Coordinate humanitarian aid requests and the needs of disaster victims from the state.
- Coordinate the purchase of disaster-relief necessities according to their suitability and transfer to disaster relief storage facilities.
- Coordinate the needs of the additional volunteers needed by the state.
- Provide the latest disaster-related information to the ministers, KPWKM, KPKM, the media and others.
- Coordinate the contribution of goods and manage humanitarian aid distribution.

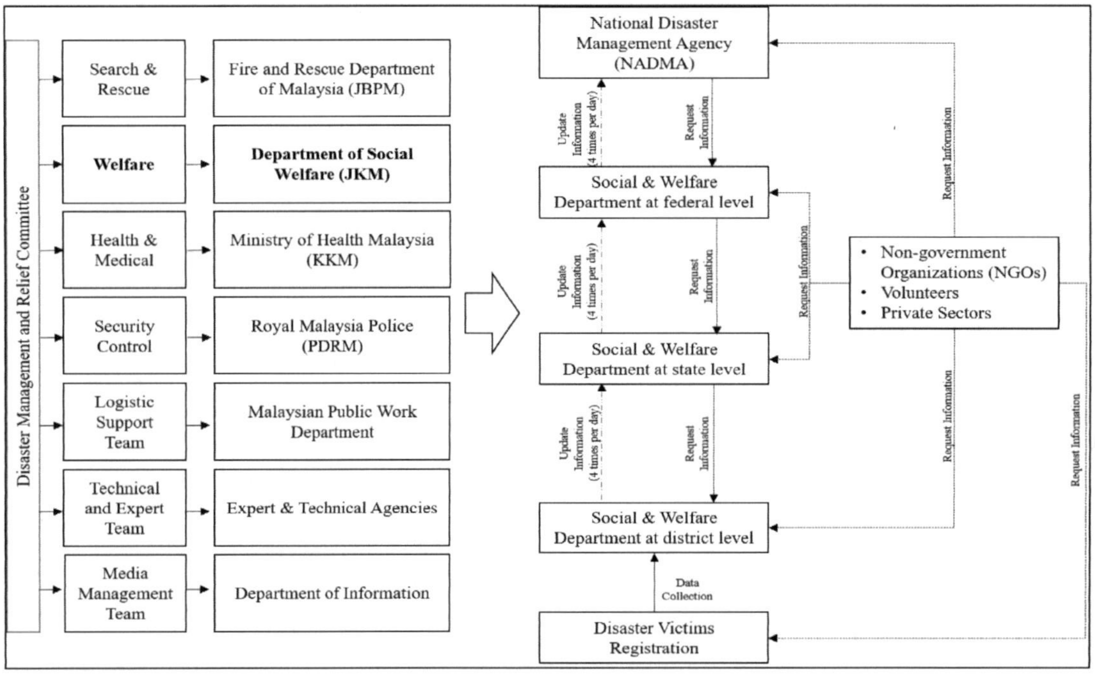

Figure 2. Current information management flows of the social and welfare department during a disaster.

Meanwhile, three (3) of the NGO participants explained that their organizations will gather information on a disaster by referring to one of the disaster management stakeholders: The National Disaster Management Agency (NADMA), JKM federal, JKMN or JKMD. Five (5) of the NGO participants explained that their organizations gather information on disaster victims based on their own surveys. All the NGO participants stated that the information is important to their organizations to make decisions based on their capabilities in managing humanitarian aid.

5. Current disaster information management tools

According to all GO participants, current disaster information management tools used to disseminate disaster information among the disaster management stakeholders are through WhatsApp application, Short Message Service (SMS), Fax, Telephone, Media and warning system. This is in line with research conducted by Zakaria, Said [21], where all the tools were mainly used in disseminating the disaster information among disaster management stakeholders. All GO participants agreed that the used of WhatsApp application as a tool is the faster way in disseminating disaster information among the Government agencies however, the information could not be fully controlled by the Government agencies. This is explained by the GO participants that, most of the disaster information spread through WhatsApp group created by them is too overwhelms and sometime are redundant. In the chaos situation, overwhelms and redundant information can affect the humanitarian aid activities especially in search and rescue task as well as welfare task.

Moreover, all GO participants explained that overwhelms WhatsApp and SMS from the NGOs, private sectors as well as volunteers that request for disaster information's during disaster were also disturbing their task. All GO participants explained that during disaster they could not respond to NGOs, private sectors and volunteers to their request immediately due to chaotic situation. Consequently, all NGO participants explained that, most of NGOs, private sectors and volunteers tend to request information from others Government agencies that involved during disaster. This is one of the factors contribute to redundancy of disaster information that lead to problem in managing humanitarian aid distribution process. Besides that, all NGO participants were explained that, due to limited information gained from the government agencies has encouraged them to use their own initiative to obtain disaster information. Without a proper guideline from the Government agencies will lead to several problems in humanitarian aid management and it could affect the disaster areas as well as disaster victims.

6. Discussion

Regarding management of the humanitarian aid distribution information, it should be noted that, commonly, disaster management stakeholders should collaborate in managing and distributing such information. This will enhance the communication with better understanding between the disaster management stakeholders. This research has considered the main disaster management stakeholders providing humanitarian aid that focus on providing food and the distribution of basic necessities to the disaster victims. To improve humanitarian aid information management between disaster management stakeholders, therefore we would like to propose a new humanitarian aid information management system as shown in Figure 3. The proposed humanitarian aid information management system was developed based on information from all the participants in this research.

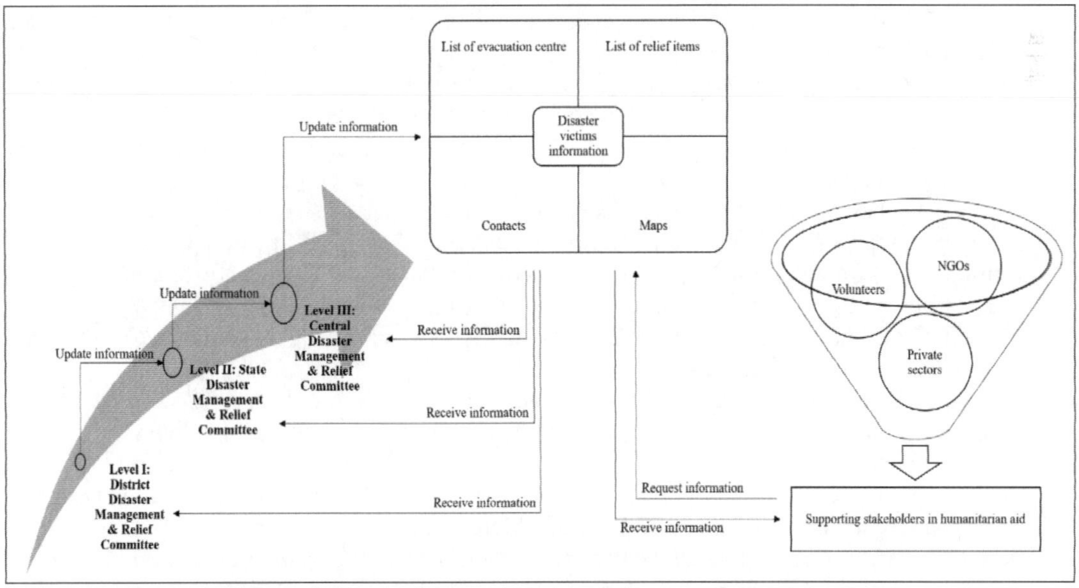

Figure 3. Humanitarian aid information management system.

The new humanitarian aid information management system is very useful in order to improve and control the information management in a disaster. The system will also have the potential to avoid known issues with information management, which are fraudulent reports and redundancy of disaster information. Due to the absence of a standard operating procedure (SOP) that explains the roles and responsibilities of NGOs in managing humanitarian aid distribution, this system is one of the potential platforms to allow Government agencies and NGOs to control and disseminate disaster management information.

7. Conclusion

Information management is the main important factor to manage the effectiveness and efficiency of the humanitarian aid distribution process. The different stakeholders within the humanitarian aid distribution process require strong collaboration and coordination to control and disseminate disaster management information. The major concern within the distribution process is how to facilitate, collaborate, and integrate the various stakeholders within humanitarian aid so that their mutual goal can be attained. The disaster management stakeholders should accomplish the humanitarian aid distribution process by using the humanitarian aid information management system to improve effectiveness and ensure efficient delivery of disaster management information. Consequently, to enhance and improve the humanitarian aid information management system, it could be integrated with new technology such as mobile applications. Therefore, further research will be conducted to integrate the humanitarian aid information management system with mobile applications in the hope of significantly mitigating the major issues in the humanitarian aid distribution process.

Acknowledgment

This work was financially supported by a Seed Grant of "Advanced Institute on Knowledge-Based Actions for Disaster Risk Reduction (AI-KBA)" from The International Council for Science - Regional Office for Asia and The Pacific (ICSU-ROAP) with UTM Registration no R.K130000.7317.4B323.

References

[1] Oloruntoba, R. and R. Gray, *Humanitarian aid: an agile supply chain?* Supply Chain Management: an international journal, 2006. **11**(2): p. 115-120.

[2] Thérien, J.-P. and C. Lloyd, *Development assistance on the brink.* Third World Quarterly, 2000. **21**(1): p. 21-38.

[3] The Sphere Project, *Humanitarian Charter and Minimum Standards in Humanitarian Response*, ed. P. Greaney, S. Pfiffner, and D. Wilson. 2011: The Sphere Project.

[4] National Security Council, N., *Directive No.20.*, J.P. Menteri., Editor. 2012, Majlis Keselamatan Negara: Malaysia.

[5] Careem, M., et al. *Sahana: Overview of a disaster management system.* in *Information and Automation, 2006. ICIA 2006. International Conference on.* 2006. IEEE.

[6] Asproth, V., C. Amcoff Nyström, and D.M. Dubois. *Preparing for Emergency Situations.* in *AIP Conference Proceedings.* 2010.

[7] Abiew, F.K., *Humanitarian Action under Fire: Reflections on the Role of NGOs in Conflict and Post-Conflict Situations.* International Peacekeeping, 2012. **19**(2): p. 203-216.

[8] de Oliveira, P.M., T.C. Fontainha, and A. Leiras, *The Role of Private Stakeholders in Disaster and Humanitarian Operations.* JOSCM: Journal of Operations and Supply Chain Management, 2016. **9**(1): p. 77.

[9] van der Laan, E., et al., *Demand forecasting and order planning for humanitarian logistics: An empirical assessment.* Journal of Operations Management, 2016. **45**: p. 114-122.

[10] Yilmaz, H. and Ö. Kabak, *A Multiple Objective Mathematical Program to Determine Locations of Disaster Response Distribution Centers.* IFAC-PapersOnLine, 2016. **49**(12): p. 520-525.

[11] Sankaranarayanan, K., J.A. Castañeda, and S. Villa, *Future Research in Humanitarian Operations: A Behavioral Operations Perspective*, in *The Palgrave Handbook of Humanitarian Logistics and Supply Chain Management*. 2018, Springer. p. 71-117.

[12] Gao, H., et al., *Harnessing the crowdsourcing power of social media for disaster relief.* 2011, DTIC Document: Arizona State Univ Tempe. p. 10-14.

[13] Ha, K.-M., *Disasters can happen to anybody: The case of Korea.* Environmental Impact Assessment Review, 2016. **57**: p. 1-9.

[14] Husain, W., et al., *Enhancing Supply Chain Management System to Support Effective Flood Disaster Relief Operation*, in *Laporan Akhir Persidangan Kajian Bencana Abanjir 2014-Bahagian 2: Environmental and Physical Impact (EPI), Governance and Capability Building (GCB), Health and Clinical Science (HCS)*, Z. Yusop and N. Mahpof, Editors. 2014, Universiti Teknologi Malaysia (UTM): Kementerian Pendidikan Tinggi (KPT), Universiti Teknologi Malaysia (UTM). p. 229.

[15] PLANMalaysia, *Lampiran: Dasar Perancangan Fizikal Negara, DPF Desa Negara 2030*, J.P.B.D.D. Malaysia, Editor. 2016, Jabatan Perancangan Bandar Dan Desa Malaysia: Malaysia.

[16] Center for Excellence in Disaster Management and Humanitarian Assistance, C.-D., *Malaysia - Disaster Management Reference Handbook.* 2016, Malaysia.

[17] Kapucu, N. and Q. Hu, *Understanding multiplexity of collaborative emergency management networks.* The American Review of Public Administration, 2016. **46**(4): p. 399-417.

[18] Lynn, M.L., *Congregational Aid: North American Protestant Engagement in International Relief and Development.* The Journal of Development Studies, 2016. **52**(7): p. 965-985.

[19] Martin, E., I. Nolte, and E. Vitolo, *The Four Cs of disaster partnering: communication, cooperation, coordination and collaboration.* Disasters, 2016.

[20] Dahari, J.R., et al., *Consolidating Policy and Legal Framework For National Disaster Response*, in *Laporan Akhir Persidangan Kajian Bencana Abanjir 2014-Bahagian 2: Environmental and Physical Impact (EPI), Governance and Capability Building (GCB), Health and Clinical Science (HCS)*, Z. Yusop and N. Mahpof, Editors. 2014, Universiti Teknologi Malaysia (UTM): Kementerian Pendidikan Tinggi (KPT), Universiti Teknologi Malaysia (UTM). p. 122.

[21] Zakaria, N.B., et al., *Knowledge Base Framework to Support Decision Making for Flood Disaster Relief*, in *Laporan Akhir Persidangan Kajian Bencana Abanjir 2014-Bahagian 2: Environmental and Physical Impact (EPI), Governance and Capability Building (GCB), Health and Clinical Science (HCS)*, Z. Yusop and N. Mahpof, Editors. 2014, Universiti Teknologi Malaysia (UTM): Kementerian Pendidikan Tinggi (KPT), Universiti Teknologi Malaysia (UTM).

Rainfall induced slope failure detection using infiltration type slope stability method applying non-linear failure envelope

M J Md Noor[1*], M A Muda[1,2] and A S Abdul Rahman[1]

[1] Faculty of civil engineering, Universiti Teknologi Mara, Shah Alam, Selangor, Malaysis
[2] Public Works Department, Kuala Lumpur, Malaysia

[*]Corresponding author: mohdjamaludinmdnoor@yahoo.com

Abstract. Rainfall induced shallow slope failures in hilly terrain in the tropics is a complex soil mechanical behavior when it is difficult to be back analysed using the conventional slope stability method. The basis of the problem is in the applications of the linear soil shear strength failure envelope with cohesion intercept that over-estimate shear strength on the low stress levels i.e. for effective stresses less than 100 kPa or for failure depth less than 5 m. At this low stress levels the genuine shear strength behaviour is in fact non-linear with zero cohesion. Besides, the failure is not related to the groundwater table (GWT) when it is located at greater depth and has no influence on the failure. Theoretically the shallow slip failure surface is very difficult to be replicated with factor of safety (FOS) less than unity and that pose difficulties in understanding the failure. A slope stability method that deployed mechanics of unsaturated soils and calculate stability base on the advancement of the wetting front has been applied to characterize the shallow slips. The stability method uses the soil true non-linear soil shear strength behavior with respect to effective stress and suction. The method has been validated against a shallow slips triggered by rainwater infiltration that occurred along a road side in Cameron Highland, Malaysia.

1. Introduction

Pre-detection of potential landslide is very important to avoid the risk of lives and properties. However this is the problem faced by the geotechnical engineers. It is the rainfall induced slope failure that is difficult to be quantified theoretically. The developments of the slope stability methods started since the nineteen seventies when geotechnical researchers in Hong Kong find difficulties to back analyse the shallow land slips using the conventional slope stability methods. When the plane envelope soil shear strength model of Fredlund *et al.* [10] was introduce then engineers began to understand the role of rain in triggering landslide. When the soil is in the state of unsaturation there exists suction that pulls the soil particles together and produce extra shear strength known as apparent shear strength. The infiltration of rain water into the slope resulted in the reduction of suction in the ground which in turn reduces the apparent shear strength and may end up with failure. Therefore the application of unsaturated soil mechanics is inevitable in order to understand rainfall induced slope failures.

Then Rahardjo and Fredlund [30] introduced a slope stability method that incorporates the contribution of suction on the stability. This is the first time the role of rainfall infiltration has been incorporated in the slope stability mechanics to quantify slope instability. This is the real understanding on how rainfall actually influences stability of slopes. It is the advancement of the

IOP Conf. Series: Materials Science and Engineering **513** (2019) 012013 doi:10.1088/1757-899X/513/1/012013

wetted band into the slope that triggers the slope failure in the hillsides and not the rise of the GWT as commonly understood. Nevertheless, the mechanics apply the plane envelope soil shear strength model of Fredlund *et al.* [10] which over-estimate the shear strength for shallow slips as shown in Figure 1. Then the developments of this soil mechanics continued with the introduction of the curved surface envelope soil shear strength model of Md Noor and Anderson [21]) and its application in the state-of-the art infiltration type slope stability method introduced by Md Noor [20] and Md Noor *et al*, [22]. This method considers the advancement of the wetting front into the slope forming a band of wetted zone underlying the slope surface parallel to the contour of the slope face. As infiltration continues, this band becomes thicker and the wetted front progress deeper into the slope. This wetted band is considered to have zero suction. The slope would fail when the critical thickness of this wetted band is reached. By this method, slope instability influence by rainfall infiltration can be accurately detected. The reliability of this method will be validated in Section 4.0.

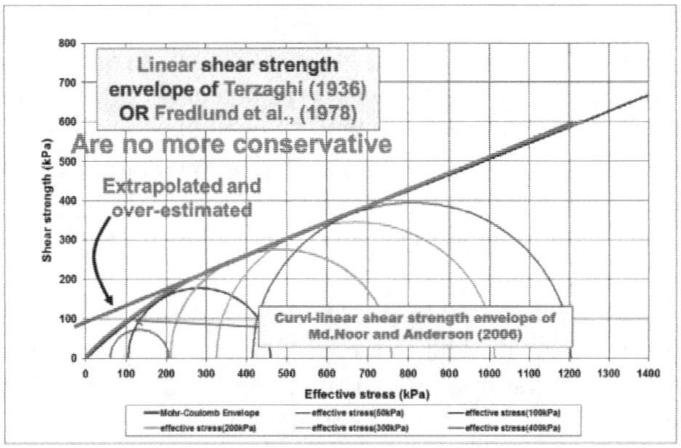

Figure 1. Over-estimation of shear strength at low stress levels when using linear soil shear strength model with cohesion intercept.

2. Problems in the conventional slope stability methods to quantify rainfall induced slope failure

The incidence of slope failure in the tropics has already puzzled geotechnical engineers. This is because they faced problem to quantify the failures theoretically. This dilemma need to be resolved and the right mechanics of failure need to be developed and validated. There was a study conducted by Othman [24] to check the stability of 118 slopes along Karak Highway, Malaysia and his finding showed that 90% of the slope has FOS less than unity but the slopes are still standing. This is giving a strong indication that there is something wrong with the current slope stability methods and there is a direct requirement for those methods to be revamped. In the contrary, engineers have difficulties to achieve FOS less than unity for failed slopes. Slopes that have failed would produce FOS less than unity when being back analysed. In addition, back analysis should also replicate the actual failure surface whether it is shallow or deep seated slip. Especially the shallow slip surface is very difficult to be modeled.

Brand [4] quoted his experience dealing with slopes in Hong Kong that in highland area the groundwater is too far down to have influence on the failure at the top. This means that the shallow rainfall induced shallow failure is not associated with the rise of the GWT and instead it is very much related to the infiltration of the surface water [14]. Despite that, the technique of modelling rainfall induced slope failure by elevating the groundwater table (GWT) has become a common practice among geotechnical engineers despite not replicating the actual occurrence in the field. This practice is still insisted because that is the only available method to quantify. When the GWT rises then the effect of buoyancy will reduce the effective stress between the soil grains and thereby reduce the shearing resistance or the shear strength of the soil. This shearing resistance is actually resisting the sliding failure and if it is being reduced then the resisting variable become less and when it has become

IOP Conf. Series: Materials Science and Engineering **513** (2019) 012013 doi:10.1088/1757-899X/513/1/012013

smaller and being overcome by the disturbing variable derived from the weight of the soil or slope external loading then failure will be triggered. This dilemma was earlier quoted by Lumb [14] as follows;

"*The association of failures with heavy rain is clear, but this qualitative association must be quantified on a physical basis, if at all possible, before reliable design methods can be established. Some advances can be made in this direction if infiltration is postulated as the dominating factor.*"

In the nineteen seventies there is still no engineering mechanics for stability of slopes that relates to rainfall except the practice of elevating the GWT. Therefore the effect of rainfall on slope failure is still not understood. Then Brand [3] has quoted a controversial statement in the 11th International Conference on Soil Mechanics and Foundation Engineering, San Francisco regarding predicting the performance of residual soil slopes in Hong Kong as follows;

"*It is concluded that our soil mechanics predictive tools are far from adequate for analysing residual soil slopes, largely because of the difficulties of predicting pore pressures and of modeling geological detail which often controls the mode of slope failure.*"

In Hong Kong, residual soil slopes are located in elevated hilly terrain where the GWT is far below the ground surface and the failure is inevitably due to surface water infiltration. In other words, Brand [3] is nullifying the applicability of all the slope stability equations that were developed before 1985 for predicting rainfall induced slope failures. This includes the slope stability equation of Fellenius [8,9], Simplified Bishop Method [2], Morgenstern and Price [23], Spencer's method [27,28] and Sharma method [25]. These are the common methods used in the slope stability software and the effect of rainfall is modeled by elevating the GWT. These methods would produce various values of FOS and the validity of those FOS is questionable. All those slope stability equations applied soil shear strength equation of Terzaghi [2] and this equation is a linear type of equation with cohesion intercept and it does not replicate the true soil shear strength behaviour. Then the FOS produced by those slope stability equations will be dubious. Therefore it is very important to use the legitimate non-linear soil shear strength equation that can replicate the true non-linear shear strength behavior. There are many non-linear soil shear strength equations that have been published like De Mellow [7], Hoek and Brown [12], Maksimovic [16], Md Noor and Anderson [21] and Lade [1]. This is already giving the indication that many have realized that the true soil shear strength behavior with respect to. effective stress is in fact non-linear. Its application in analyzing slope stability is inevitable especially for shallow rainfall induced failure. Then, it is very important to choose the one that can make excellent replication of the shear strength envelope as obtained from laboratory tests.

The over-estimation of the FOS when the shear strength equation of Terzaghi [29] with cohesion intercept is applied in slope stability analysis is further emphasized by Lade [13] with his quotation as follows;

"*It is not safe to employ the classical infinite slope failure analysis procedure in which the Coulomb failure criterion is used, because a very large portion of the factor of safety is assigned to the effective cohesion which is not present in the soil.*"

Besides, Md Noor and Hadi [17] also quoted that;

"*The practice of extrapolating linear shear strength envelope at low stress levels to intercept the τ axis is dangerous since it can produce a misleading high slope stability factor.*"

Therefore it is very important to deploy the right non-linear soil shear strength equation with zero cohesion intercept in slope stability analysis for shallow rainfall induced slope failure in order to avoid getting a misleading over-estimated FOS. The paradox of zero cohesion intercept for all soil under drained condition has been thoroughly discussed by Md Noor [19]. This is a fundamental concept which is not realized by many. Even if Weald clay [11], London clay and Avonmouth clay [1] have zero cohesion intercept then why would other soils which has only a portion of clay would need to be assigned a small value of cohesion intercept.

The current slope stability software available in the market applied either the linear model of Terzaghi [29] or the incline plane type of Fredlund *et al.* [10] soil shear strength model with cohesion intercept. The true soil shear strength behavior is in fact non-linear with zero cohesion intercept for drained type of analysis or also known as effective stress analysis unless the soil is cemented like cement in rock mass. This true aspect of shear strength behavior makes a lot of difference at low stress levels and is very important for analysis of shallow landslide. Applying the linear or the plane type of soil shear strength model would over-estimate the shear strength at the lower stress range as shown in Figure 2 and its application in analysis would produce a higher FOS [26,13,19]. Therefore the application of the linear soil shear strength model with cohesion intercept is no more conservative as initially thought by geotechnical engineers. In other words, slope is thought to be safe with a fake high FOS and this is why Lade [13] referred this as being "not safe". When this method is applied in slope failure investigation the geotechnical engineer would has the difficulties to achieve the theoretical failure condition with FOS less than unity. Thus the failure is hard to explain. Therefore, it is very important to apply the true non-linear soil shear strength behavior where the shear strength is neither over-estimated nor under-estimated and the resulted FOS in the analysis would be a realistic value [19].

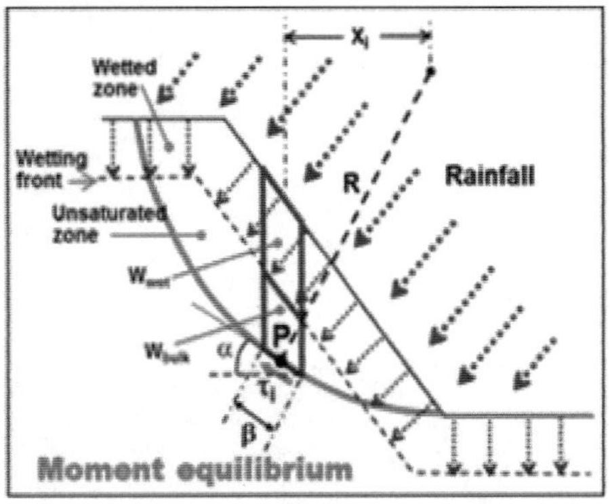

Figure 2. Typical slice considered in the infiltration type slope stability method.

3. The state-of–the art infiltration type slope stability method

Slope failure due to rainfall is triggered by the advancement of the wetting front into the slope which is termed as infiltration [5,6]. The infiltration type slope stability method replicates exactly the mechanics of slope failure due to rainfall. This is in contrary to the conventional method which assumes the failure due to rainfall is triggered by the rise of the GWT as resulted from the continuous rainfall.

Most importantly, this state-of-the art method make use of the latest advancement of soil shear strength model which is the shear strength model of Md Noor and Anderson [21] which replicates very well the non-linear shear strength behavior with respect to effective stress and suction for tropical residual soil as shown in Figure 3. Essentially the soil shear strength behaviour is non-linear with respect to effective stress and suction whereas the earlier model assume linear in both respect. The uniqueness of these equations is that they matched the non-linear soil shear strength behavior base on the geometrical dimensions that representing the shear strength parameters. Unlike the rest of the proposed non-linear equations that make use of the fitting parameters in the exponential type of equations. Even though the equations of Md Noor and Anderson [21] look slightly complex but most importantly it is able to make excellent agreement with the laboratory data for both shear strength variation with respect to net stress/ effective stress and suction. This makes it superior than the existing equations. The algorithm for the slope stability FOS calculation in the infiltration type slope stability method using the strength variation with respect to net stress/ effective stress and suction as shown in figure 4.

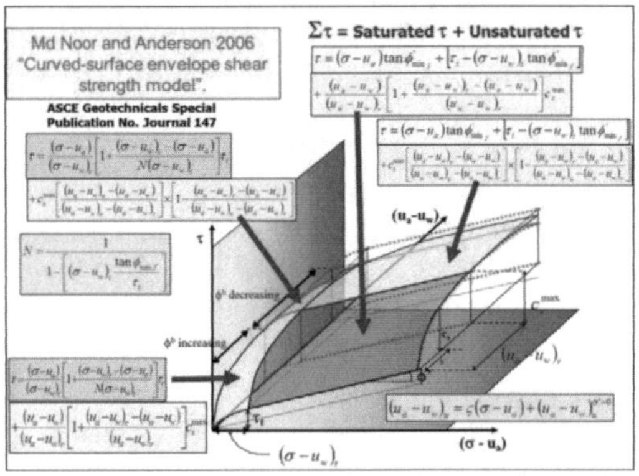

Figure 3. Curved surface envelope of Md Noor and Anderson [21] able to replicate the true soil shear strength behaviour and being applied in the state-of-the art slope stability software.

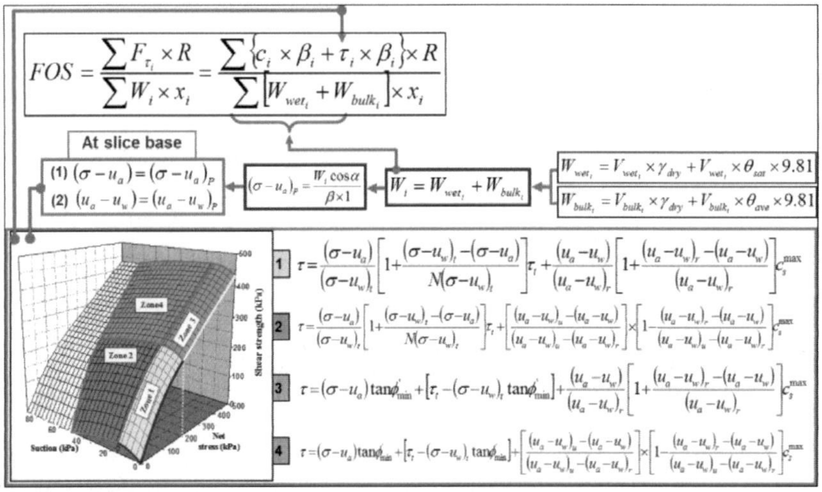

Figure 4. The algorithm for the slope stability FOS calculation in the infiltration type slope stability method.

4. Validation of the infiltration type slope stability method against shallow rainfall induced failure in Malaysia soil

Any new introduction of slope stability method needs to be validated against a failure case. It should be able to replicate the actual failure surface whether it is shallow or deep type of failure with FOS less than unity. This is the basic requirement in order to verify its applicability. If this cannot be proved then the method is not valid to be applied. It is to be noted that geotechnical engineers faced difficulties to back calculate rainfall induced slope failures with FOS less than unity.

A road side up slope, land slip has occurred in Cameron Highland. The incidence happened during a rainy season. The failure is a shallow slip of 2 m deep as shown in a survey cross sectional drawing in Figure 5. There is no groundwater table (GWT) encountered in the boreholes. Therefore the failure is not associated with the elevation of GWT. CIU triaxial tests have been conducted on three undisturbed soil specimens to determine shear strength failure envelope of in-situ soil. A test at low stress level has been specifically conducted at effective stress less than 50 kPa in order to substantiate that the non-linear failure envelope converges to zero cohesion intercept. The Mohr circles obtained are plotted in Figure 6 and the interpreted linear and non-linear failure envelopes are also shown. Figure 7 shows the Mohr circles and the envelopes as defined in the software SLOPE-RAIN applied to

IOP Conf. Series: Materials Science and Engineering **513** (2019) 012013 doi:10.1088/1757-899X/513/1/012013

conduct stability analysis. The inset shows the Mohr circles data as obtained from the CIU triaxial tests. The result of the analysis is shown in Figure 8. Essentially the software produced a FOS of 0.99 at 2 m depth of infiltration with wetted suction in the wetted band of 5 kPa. This means that the failure was triggered even before the wetted band is fully saturated. The failure slip circle is confined within the wetted band and this is in accordance to the findings by Md. Noor and Hadi [18] and Md.Noor *et al.* [15]. This proves exactly how the failure has occurred at the site.

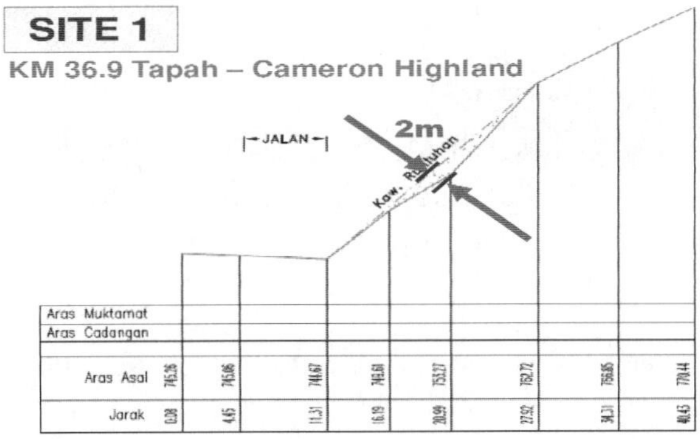

Figure 5. Shallow slope failure at Tapah, Cameron Highland: Site 1.

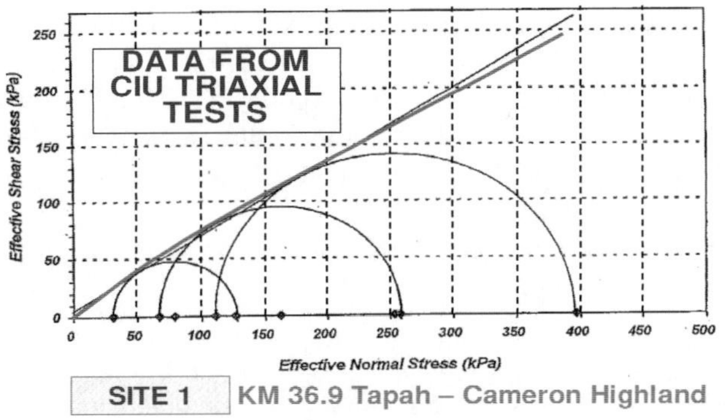

Figure 6 .Mohr circles as obtained from conducting CIU triaxial tests on undisturbed soil samples and the deduced linear and non-linear failure envelopes.

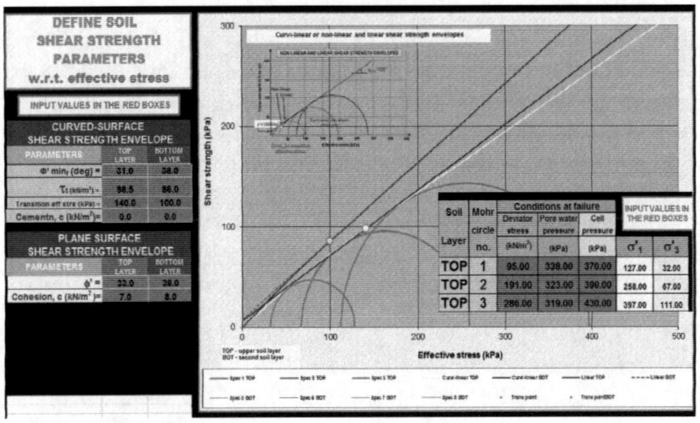

Figure 7. Definition of the shear strength envelopes according to linear and non-linear failure envelopes in the software.

IOP Conf. Series: Materials Science and Engineering **513** (2019) 012013 doi:10.1088/1757-899X/513/1/012013

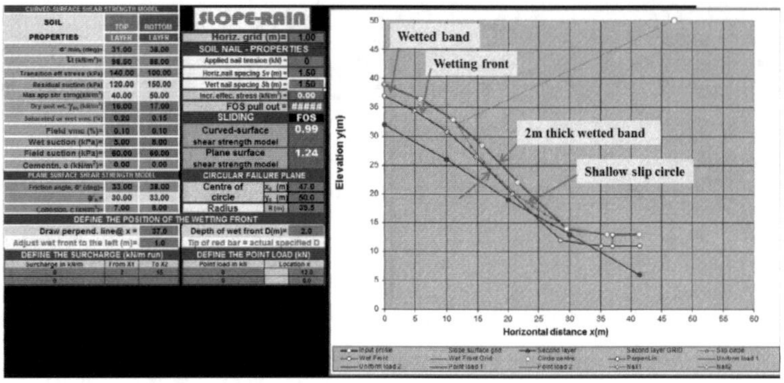

Figure 8. Essentially the infiltration type slope stability method produced a FOS of 0.99 at 2 m depth of infiltration to prove exactly as failure has occurred.

5. Conclusions

The conclusions that can be drawn from this technical paper are that the infiltration type slope stability method:-

1. Can back calculate failure by replicating the actual slip plane with FOS less than unity and thus it is very reliable to be applied in slope failure investigation triggered by infiltration of surface water.
2. Can be applied be applied to check the genuine state of stability of a slope in effect of rainfall.
3. Can be applied to reliably check for potential slope instability due to rainfall.
4. Can be used by the government departments, geotechnical consultants and the state Municipal Council to check the real state of stability of their slopes and with this potential mishap can be pre-determined and the potential risk can be identified earlier and precautions can be taken to avoid loss of lives and damaging buildings.
5. Can be effectively applied in slope risk mapping projects to produce a reliable risk map against rainfall induced failures.

References

[1] Bishop, A.W., 1955. "*The use of the slip circle in the stability analysis of slope.*" Geotechnique, Vol.5, No.1, March, p. 7-17

[2] Bishop, A W 1966 "*The strength of soils as engineering materials.*" Geotechnique, **16(2)**, p. 91-130.Bishop, A.W., 1955. "*The use of the slip circle in the stability analysis of slope.*" geotechnique, Vol.**5**, No.1, March, p. 7-17

[3] Brand, EW 1985 *Predicting the performance of residual soil slopes* Proceedings of the 11th international conference on soil mechanics and foundation engineering, San Francisco, august 12-16, 1985. Volumes 1-5 and golden jubilee volume

[4] Brand, E.W 1981 *Some thoughts on rain-induced slope failures* Proc. of the 10th Int. Conf. on Soil Mechanics and Foundation Eng. Stockholm **3** p. 373 – 376

[5] Corominas, J 2001 Landslides and Climate, *Keynote Lectures from the 8th International Symposium on Landslides* No **4** pp 1–33

[6] Collins, B. D. and Znidarcic 2004 D.: *Stability analyses of rainfall induced landslides, Geotechnical Geoenvironmental Eng* **130** 362–372

[7] De Mello, V F B 1977 *Reflections on design decisions of practical significance to embankment dams* Geotechnique **27** (3) 281354

[8] Fellenius, W 1927 *Erdstatische Berechnungen mit Reibung and Kohaesion* Berlin: Ernst

[9] Fellenius, W 1936 *Calculation of stability of earth dams* Transactions 2nd Congress Large Dams, Washington, D.C. **4** p. 445 – 462

[10] Fredlund, D G, Morgenstern, N R, and Widger, R A 1978 Shear Strength of Unsaturated Soils *Canadian Geotechnical Journal* **15**(3), 313-321

[11] Henkel, D J 1958 *The correlation between deformation, pore-water pressure and strength characteristics of saturated clay* Ph.D. Thesis, University of London

[12] Hoek, E and Brown, ET 1980 Empirical strength criterion for rock masses J. Geotech. Engng Div ASCE 106(GT9)1013-1035

[13] Lade P V 2010 *The mechanics of surficial failure in soil slopes* Engineering Geology Volume **114** Issues 1–2 23 June 2010 57-64

[14] Lumb, P 1975 *Slope failures in Hong Kong* Quarterly Journal of Engineering Geology and Hydrogeology **8** pp. 31-65

[15] M. J. Md.Noor A. S. A Rahman and B. Hadi 2014 *Application of non-linear failure envelope in back analysis for shallow rainfall induced slope failure* World Landslide Forum 3, China 2-6 June 2014

[16] Maksimovic, M. 1996 *A Family of Nonlinear Failure Envelopes for Non-Cemented Soils and Rock Discontinuities* Electronic Journal Geotechnical Engineering, Premiere Volume **1**

[17] M. J. Md.Noor and B. Hadi 2011 Effect of soaking on stability of cut slope in granitic residual soil *5TH Asia-Pacific Conference On Unsaturated Soils, 29 February, 1-2 March 2012, Pattaya, Thailand* ISBN 978-616-7522-77-7, pp 735-740

[18] Md. Noor M J and B.A. Hadi 2010 *The significance of curved shear strength envelope in modeling rainfall induced shallow failure of slopes* Unsaturated Soils **2** pp 1261-1265, ISBN 978-0-415-60430-7, eBook ISBN 978-0-203-83401-5, Taylor and Francis Group, London, UK , Indexed by SCOPUS, 5TH International Conference Unsaturated Soils, 6-8 September 2010, Barcelona, Spain

[19] Md. Noor M.J 2016 *Landslide and the hidden roles of shear strength.*" © UiTM Press, UiTM Malaysia ISBN 978-967-363-385-2

[20] Md.Noor, M J 2007 Slope stability method incorporating the curved surface envelope shear strength model *16th South East Asian Getechnical Conference, 8-11 May 2007, Kuala Lumpur*

[21] Md.Noor, M J and Anderson, W F 2006 *A comprehensive shear strength model for saturated and unsaturated soils*" Proc. 4th Int. Conf. on Unsaturated Soils, ASCE Geotechnical Special Publication No. 147, Carefree, Arizona, Vol. **2**, pp 1992-2003 ISBN 0-7844-0802-5

[22] Md.Noor, M J,Hadi B A and Hafez M A 2009 *Modeling of shallow slope failure using curved-surface envelope extended Mohr-Coulomb model.*" 4th Asia Pacific Conference on Unsaturated Soils, Newcastle, Australia, 23th-25th November 2009. Theoretical and Numerical Advances in Unsaturated Soil Mechanics, pp 909-914, CRC Press/ Balkema, Taylor and Francis Group, London, UK; ISBN 978-0-415-87716-9, eBook ISBN 978-0-203-85609-3, Indexed by ISI THOMPSON

[23] Morgenstern, N R and Price, V E 1965 The Analysis of the Stability of General Slip Surfaces. Géotechnique, 15, 79-93

[24] Othman, M A 1989 *Highway cut slope instability problems in West Malaysia* Ph.D. Thesis, Department of Geography, University of Bristol, United Kingdom

[25] Sharma, S K 1973. Stability analysis of embankments and slopes. Géotechnique, 23(3): 423-433

[26] Seely, D D B and Trandafir, A C 2008 Nonlinear failure envelope of a nonplastic compacted silty-sand." Proceedings of the 10th International Symposium on Landslides and Engineered Slopes, Xi'an, China, CRC Press, ISBN 978-0-415-41196-7

[27] Spencer, E. 1967 A method of analysis of the stability of embankments assuming parallel interslices forces. Geotechnique 17(1), 11-26

[28] Spencer, E. 1973 Thrust line criterion in embankment stability analysis. Geotechnique, Volume 23 Issue 1 March 1973, pp. 85-100

[29] Terzaghi, K., 1936. "*The shear resistance of saturated soils.*" Proceedings for the 1st. International Conference on Soil Mechanics and Foundation Engineering (Cambridge, MA), 1, p. 54 - 56.

[30] Rahardjo, H. and D.G. Fredlund 1991 Calculation Procedures for Slope Stability Analyses Involving Negative Pore-Water Pressures *Proceedings of the International Conference on Slope Stability Engineering- Developments and Applications, Shanklin, Isle of Wight, England, April 15-19, The Institution of Civil Engineers U.K., pp 43-49*

IOP Conf. Series: Materials Science and Engineering **513** (2019) 012014 doi:10.1088/1757-899X/513/1/012014

Breakdown properties of aged low voltage cross-linked polyethylene insulated cable

H N Faezah[1*], M A M Azreen[1], K Y Lau[2] and Han-Seung Lee[3]

[1]School of Civil Engineering, Faculty of Engineering, Universiti Teknologi Malaysia, Johor Bahru, Johor, Malaysia
[2]School of Electrical Engineering, Faculty of Engineering, Universiti Teknologi Malaysia, Johor Bahru, Johor, Malaysia
[3]School of Architecture and Architectural Engineering, Hanyang University, South Korea

[*]Corresponding author: nfaezah@gmail.com

Abstract. In this study, the electrical breakdown strength tests were performed on low voltage insulation material, crosslinked polyethylene (XLPE) insulated cable rated at 0.6/1 (1.2) kV. In Malaysia, this type of cable was found to have premature degradation within three to five years of service. The objective of this experiment is to study the characteristic of XLPE used as the insulation material of the aged cable. The allowable maximum operating temperature of this cable is at 90 °C. New cables of different manufacturers were tested and a degraded cable taken from field was used as a reference sample. The thermal consequences on each cable were compared. XLPE sheets sliced from the cable insulations were prepared as the test samples. New samples were heated to a temperature of 90 °C to accelerate the ageing process in an oven. An electrical breakdown tests were carried out to assist the evaluation of the degradation on the unaged and aged cables. It has been shown by the experiment that the electrical properties of the aged XLPE of low voltage cable for both manufacturers were greatly affected when exposed to the heat. This can be considered as defect to the manufacturing process because the insulation layers could possibly be made of a substandard material, that fails to perform according to the specification.

1. Introduction

Cross-linked polyethylene (XLPE) is manufactured from polyethylene (PE) plastic with a three dimensional-molecular bond that is created within the structure of the plastic. XLPE is now commonly used in electrical cable construction for a wide range of voltages. Silane cross-linking is the most common practice for manufacturing low voltage XLPE cable (Vandbakk, 2012). Low voltage is defined as a supply system voltage in the range of 50 to 1000 V AC as defined by International Electrotehnical Commision (IEC).

XLPE is favoured over other insulation material such as polyvinyl chloride (PVC), ethylene propylene rubber (EPR) and silicone rubber due to its toughness against temperature. Excellent dielectric strength, high insulation resistance and a low dissipation factor at all frequencies are among other benefits of using XLPE as insulation material.

In Malaysia, the low voltage XLPE insulated cable are available in both single-core and multi-core form, designated for general use and for underground cabling, where lightness and convenience of terminating are major considerations. XLPE insulated cable also withstands smaller radius bending, allowing for easy and reliable installation. The design of low voltage XLPE insulated cable must conform to IEC 60502-1 and rated at 0.6/1 (1.2) kV. It has plain circular conductors made of compacted stranded copper or stranded aluminium. The XLPE insulation layer is rated at 90°C.

Previous studies and researches showed that XLPE cable has great performance over high temperature and could last 40 years or more (Anwar ul-Hamid et al, 2015). However, there were increments in the number of complaints for insulation failure for less than five years. During installation, cables are subjected to repeated bending or random flexing causing its copper conductors to be stressed. Mechanical stresses during installation are generally more severe than those encountered while in service. Physical damage caused by mishandling of the cables may cause injury to the insulation layer.

One of the issues with XLPE insulation is degradation under service conditions. As much as the advantages of using XLPE is well known, their long-term behaviour is still undetermined. The XLPE cable is constantly exposed to the heat under its operational condition, causing damage and affects the physical properties of the insulation. Being under temperature stress over certain period of time may cause changes to the chemical composition of XLPE. Consequently, the degradation may cause limitations on the effective service life of the cable (Nobrega et al, 2013).

2. Materials and preparation

The materials employed in this study is an XLPE insulated cable with 25 mm^2 conductor area and 2 mm insulation thickness. Cables are supplied by two different manufacturers, Cable A and Cable B. Both cables comply with the cable requirement according to IEC 60502-1 standard. The cables are rated at 0.6/1 (1.2) kV with XLPE insulated maximum operating temperatures at 90 ^{O}C. A degraded cable taken from site is also used in this study. The degraded cable is an XLPE insulated, with 25 mm^2 conductor area and 2 mm insulation thickness and complied with the IEC 60502-1 standard.

2.1 Thermal Ageing Procedure

The XLPE insulation layer of Cable A and Cable B were heated for this procedure. The heating process was conducted in an air ventilated oven at a temperature of 90 ^{O}C for 7 hours a day. This process was repeated for another 7 days and the total heating duration was approximately 50 hours. After the heating process, the samples were left to cool at room temperature.

3. Experimental techniques

3.1 Breakdown test of XLPE insulation

This test was done for the purpose of determining the dielectric breakdown strength of the XLPE insulation. Samples for this experiment were fresh and aged XLPE insulation of Cable A, fresh and aged XLPE insulation of Cable B and degraded sample. Sample were cut into smaller size with length of 2 mm.

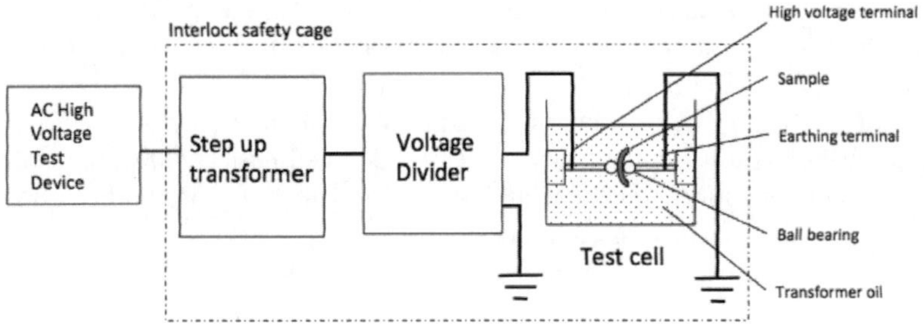

Figure 1. Breakdown test configuration.

IOP Conf. Series: Materials Science and Engineering **513** (2019) 012014 doi:10.1088/1757-899X/513/1/012014

Figure 1 shows the main component used in the test. Sample was placed between two steel ball-bearing electrodes before being immersed in transformer oil. The electrode on the left side was connected to high voltage supply while the electrode on the right side was earthed. Voltage of 50 Hz was injected starting from 1 kV and the applied voltage was steadily increased at manually controlled voltage rate until the sample experienced breakdown. Crackling sound could be heard when the XLPE layer began to fail. After the breakdown, the voltage amplitude was reduced immediately by turning the voltage regulator to '0' position. Figure 2 shows the condition of the samples after the test.

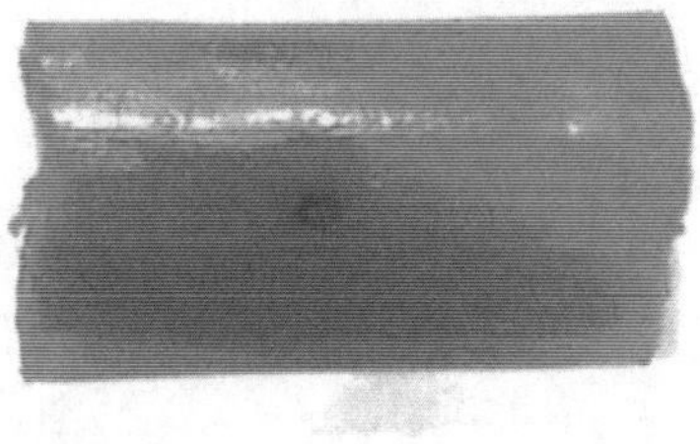

Figure 2. The condition of XLPE materials after being injected with high voltage and experienced breakdown.

Test was completed within 5 minutes for each sample. The test was repeated for 5 times for each type of samples (fresh, aged and degraded) and a total number of 25 breakdown test had been performed. The transformer oil was replaced when the oil colour had changed to dark brown with a smell. The readings for each breakdown test were recorded manually and the breakdown field was calculated by dividing the breakdown voltage by the sample thickness. These data were then analysed using two-parameter Weibull distribution (Kwan Yiew Lau, 2013).

3.2 Result and analysis

Figure 3 compare the Weibull plots of the breakdown strength of fresh, aged XLPE insulated layer of Cable A and degraded XLPE insulation. The breakdown strengh in Cable A for fresh sample was 174 ± 6 kV/mm and aged sample was 144 ± 8 kV/mm. Figure 4 compare the Weibull plots of the breakdown strength of fresh, aged XLPE insulated layer of Cable B and degraded XLPE insulation. The breakdown strengh in Cable B for fresh sample was 166 ± 5 kV/mm and aged sample was 151 ± 10 kV/mm. The breakdown strength in degraded sample was 128 ± 2 kV/mm.

The breakdown field for each cable is tabulated in Table 1.

Table 1. Breakdown strength data.

Sample		**Alpha (kV/mm)**	**Beta**
Cable A	Fresh	174 ± 6	24 ± 20
	Aged	144 ± 8	14 ± 11
Cable B	Fresh	166 ± 5	28 ± 19
	Aged	151 ± 10	11 ± 9
Cable Degraded		128 ± 2	58 ± 39

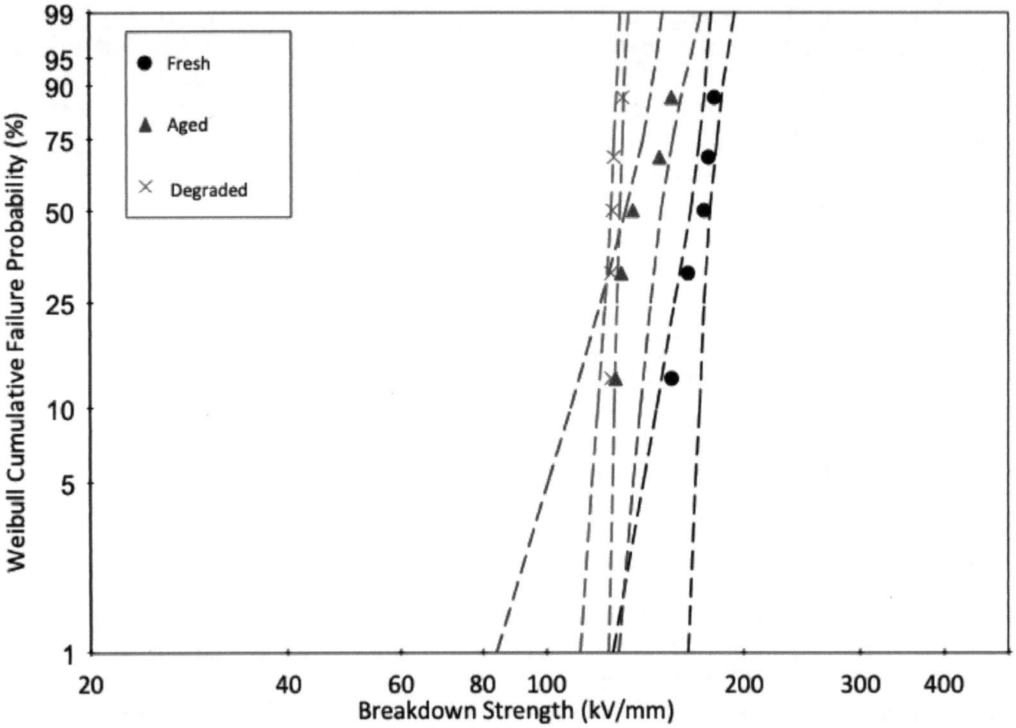

Figure 3. Weibull plots comparing the AC breakdown strength of XLPE insulation layer of fresh and aged Cable A, and degraded cable.

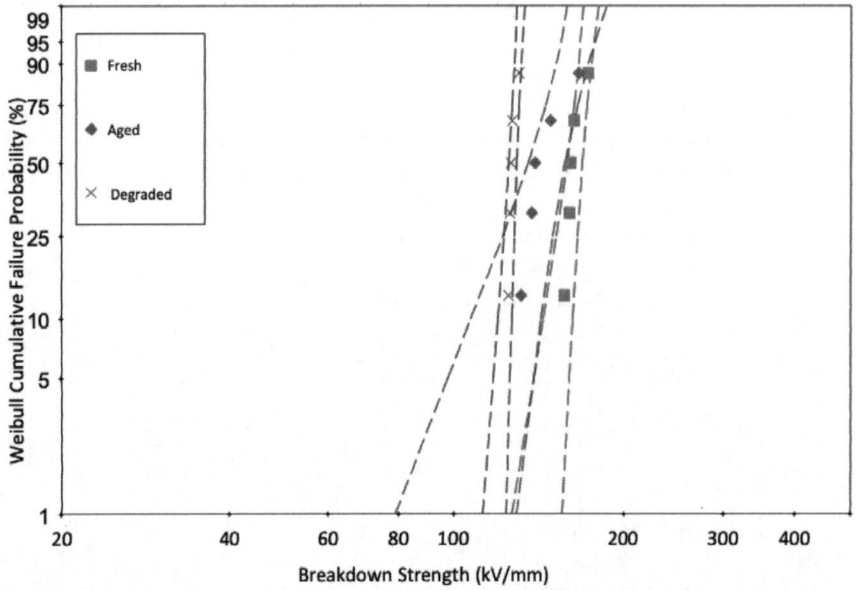

Figure 4. Weibull plots comparing the AC breakdown strength of XLPE insulation layer of fresh and aged Cable B, and degraded cable.

4. Discussion

The fresh sample of Cable A and B were kept at room temperature and the aged sample of Cable A and Cable B had been exposed to the heat at 90 °C for 50 hours. It can be observed that the breakdown voltage in both aged samples were slightly decreasing, where it fell from 174±6 kV/mm to 144±8 kV/mm for Cable A with a decrease in beta value. Similarly, samples of Cable B also show a fall in breakdown strength, from 166 ± 5 kV/mm to 151 ± 10 kV/mm with a decrease in beta value. Thermal ageing appears to affect the breakdown voltage of the XLPE materials for both Cable A and B.

The effect of accelerated ageing is related to the effect of the degraded sample used as the reference. Although in this study, we cannot presume that the thermal ageing is the only factor that cause the degradation in the degraded cable. The lower breakdown strength due to thermal ageing can be considered a significant reliability issue for the insulating material. Through this experiment, the correlation between AC breakdown strength and thermal ageing is in good agreement with most results pubished (Mecheri et al., 2013; Jong-Il Weon., 2014; and Geng et al. 2018).

5. Conclusion

In this work. we investigated the XLPE performance for low voltage XLPE insulated cable. It was observed that the thermal ageing leads to an early degradation of the XLPE materials. Temperature variations had caused a decrease in the electrical properties and affected the breakdown strength. The results confirm the expected correlation between thermal ageing and breakdown strength and clearly demonstrate that thermal ageing lowers the breakdown strength of the XLPE insulation layer. The degraded sample used as reference shows that there is a general trend that the breakdown strength decreases as the cable aged.

Acknowledgements

The authors would like to thank the School of Civil Engineering and the School of Electrical Engineering of Faculty of Engineering, Universiti Teknologi Malaysia, Johor Bahru through financial grant RJ130000.7351.4B350 for the financial support of this research.

References

[1] Alan Melo Nóbrega, Manuel Luis Barreira Martinez and Alvaro Antonio Alencar de Queiroz (2013). Investigation and analysis of electrical ageing of XLPE insulation for medium voltage covered conductors manufactured in Brazil. *IEEE Transactions on Dielectrics and Electrical Insulation*, 20(2), 628-640. https://doi.org/10.1109/TDEI.2013.6508767

[2] Anwar Ul-Hamid, Khaled Y. Soufi, Luai M. Al-Hadrami, Ahsan M. Shemsi (2015). Failure investigation of an underground low voltage XLPE insulated cable. *Anti-Corrossion Method and Materials*, Vol 62 Issue:5, pp.281-287 https://doi.org/10.1108/ACMM-02-2014-1352

[3] International Electrotechnical Commission. *Power cables with extruded insulation and their accessories for rated voltages from 1 kV(U_m=1,2 kV) up to 30 kV (U_m = 36 kV) - Part 1: Cables for rated voltages of 1 kV (U_m=1,2kV) and 3 kV (U_m=3,6 kV).* Switzerland, IEC 60502-1. 2004

[4] Kwan Yiew Lau (2013). Structure and electrical properties of silica-based polyethylene nanocomposites, Universiti of Southampton, Faculty of Physical Science and Engineering, PhD Thesis, 94-97.

[5] Pulong Geng, Jiancheng Song, Muqin Tian, Zhipeng Lei and Yakun Du (018). Influence of thermal ageing on AC leakage current in XLPE insulation. *AIP Advance* 8(2). https://doi.org/10.1063/1.5017297

[6] Vandbakk, M.(2012). Organic Contamination in Sub-Marine AC and DC High Voltage Cables, Norwegian University of Science and Technology, Department of Physics, Master Thesis, 36-40.

[7] Weon, Jong-Il (2011). Effects of thermal ageing on mechanical and thermal behaviors of linear low density polyethylene ppipe. *Polymer Degradation and Stability*. 95. 14-20.https://doi.org/10.1016/j.polymdegradstab.2009.10.016

[8] Yacine Mecheri, L. Boukezzi, A. Boubakeur and M. Lallouani (2000). Dielectric and mechanical behaviour of cross-linked polyethylene under thermal ageing. *Conference oon Electrical Insulation and Dielectric Phenomena (CEIDP)*, Annual Report, 2 (May 2016) 560-563. https://doi.org/10.1109/CEIDP.2000.884022

Damage identification in plate using wavelet transform and artificial neural network

Muyideen Abdulkareem[1*], Abideen Ganiyu[2] and Muhd Z. Abd Majid[1]

[1]Construction Research Centre, Universiti Teknologi Malaysia, Johor, Malaysia
[2]Department of Civil Engineering and Quantity Surveying, Military Technological College, Muscat, Oman

[*]Corresponding author: muyikareem@gmail.com

Abstract. Detection of damage at early stage is necessary to prevent catastrophic collapse of civil structure. The quantification of small damage has however, not received the much needed attention from researchers. Amongst various non-destructive methods studied by researchers to identify structural defects are Wavelet Transform (WT) and Artificial Neural Network (ANN). In this paper, a technique is proposed to quantify damage by combining WT and ANN. The study is divided into two phases. The first phase involves detection and location of damage in a plate numerical model by WT decomposition of the mode shape difference. Due to difficulties in obtaining higher mode shapes in practice, the difference of the first mode shapes of the damaged and undamaged plates are applied. After obtaining the damage location, the coordinates of the damage location and the mean values of the obtained WT moduli are applied as input to the designed neural network. The output of the ANN is the severity of damage in the plate model. This method is demonstrated by using a numerical square steel model with all the four sides fixed. The results indicate the ease of the method and that reliable damage identification can be obtained by combining ANN and WT.

1. Introduction

Civil structures are the major assets of any nation. The health status of these structures are paramount to safety of lives and properties, as well as rising cost of maintenance. The health status of structures can be improved by early detection of damage. Such detection has gained attention of researchers over the past two decades. Researchers have applied different vibration-based damage detection methods such as change in frequency [1], mode shape [2], frequency response function (FRF) [3], modal curvature [4] and artificial neural network (ANN) [5] to identify and assess defects in different civil structures.

In applying WT to detecting damage, Garstecki et al. [6] applied dynamic and static numerical models to detect defects in structures by using discrete wavelet. Spanos et al. [7] identified crack damage in beams by establishing a WT modulus map. The WT modulus map had the boundary distortion eliminated and the local maxima due to damage were easily identified. Their technique was verified by employing pseudo-experimental and numerical test of beams with cracks up to three. Wu and Wang [8] applied Gabor wavelet to detect cracks of different depths in an aluminum cantilever beam. Static displacement was imposed on the structure and laser sensors were used to measure the deflection profile that was then decomposed. Rucka [9] examined the effects of higher modes in damage detection. The WT technique considered Gaussian wavelets Gaus4, Gaus6, and Gaus8 and a cantilever beam by imposing varying depth rectangular notches at different locations. Cao et al. [10] employed a WT method to deal with measurement noise in modal curvature from mode shapes. A Teager energy operator was applied to analyse the modal curvature to produce Teager energy operator-WT modal curvature that has distinct suppressing noise capabilities, thus boosting the main feature caused by damage. The technique was validated using an experimental test consisting mode shapes acquired from a damaged aluminum beam obtained by a scanning laser vibrometer. Montanari et al. [11] proposed a spatial Continuous Wavelet Transform (CWT) damage identification method to determine the minimum number of sampling intervals required for optimal damage detection. In achieving this, the first three free vibration modes of a cantilever and a simple supported beam with an edge open crack were numerically simulated. In addition, the authors carried out a parametric analysis by considering the main parameters governing the problem.

More studies were done applying WT when Douka et al. [12] applied continuous wavelet transform (CWT) to detect crack damage in a plate by analyzing the plate's mode shapes. Law et al. [13] proposed a WT technique to for assessing noise by analyzing the structure's acceleration response. Even though the technique was able to detect local damage by taking measurements at few points, the technique showed flaws due to its sensitivity to modelling error and measurement noise. In the works of Sarrafi and Mao [14], a probabilistic WT method was applied to quantify uncertainty by using WT coefficients, and using outlier analysis to obtain the optimal detection threshold. Recently, Abdulkareem et al. [15] proposed a non-probabilistic WT technique to detect damage in plate structures. The WT coefficient values of the decomposed damaged and undamaged mode shape structures are obtained. Thereafter, interval analysis method was utilised to identify the position of damage.

A review of previous studies reveals that existing WT-based damage detection techniques are unable to detect and quantify of damage simultaneously. Although a few WT-based techniques tried to quantify damage [16], they have only been successful in estimating the damage to the overall structure [17]. On the other hand, ANN has the ability to recognize linear and non-linear relationships between two variables. It is the most explored artificial intelligent approach in vibration-based damage detection [18]. Another advantage is that it possesses noise filtering ability and can be trained with incomplete data.

One of the pioneers ANN applications in structural damage detection was done by Wu et al. [19]. The authors applied the FRF data of a numerical three-storey structure to identify damage that was initiated by reduction of stiffness of the columns. Although damage was accurately identified when the ANN input data were from the third floor, the results were flawed when the ANN was fed with measurements taken from the second floor and third floor. Sahin and Shenoi [20] applied different structural parameters as input to an ANN. The first three frequencies changes, maximum absolute difference in mode shape curvature, absolute difference in mode shape curvature and their corresponding damage locations as input vector of the neural network. The results indicated that the absolute difference in mode shape curvature provided the best result for damage identification. However, change in frequencies neither predicted damage location nor damage severity. Yam et al. [21] proposed a neuro-wavelet method to detect damage in composite structures. The energy variations of the structural responses were decomposed and damage features extracted were applied to train the neural network. The proposed method accurately located all the damage cases, however, few incorrect predictions were observed in the experimental study.

In addition, Lee et al. [22] utilised a simple beam and a multi-span bridge to compare results obtained by training an ANN with inputs of mode shapes, their ratios and their differences. They concluded that mode shape differences and mode shape ratios provided good results. Rucka and Wilde [23] applied ANN to identify damage in plate and shell structures by using the decomposed mode shape signals. Bagchi et al. [24] successfully obtained damage data of a bridge structure by employing ANN and modal parameters. The location of damage was identified first by using the natural frequencies and mode shapes, and subsequently using ANN to obtain the damage severity. Li et al.

[25] presented an ANN technique by using details of the residual FRF as input. The FRF was reduced by using principal component analysis (PCA) to obtain damage features that were applied to train different neural networks. They further applied an ensemble neural network to fuse the outcome of each neural network. It was shown that the network ensemble provided better results than the individual neural networks. Hakim et al. [26] applied ANN to detect damage in an I-beam by using natural frequencies and mode shapes as inputs. Recently, Khoshnoudian et al. [27] fed damage features from FRF extracted by using 2D PCA into a neural network to identify damage in truss bridge. In addition to using these methods, researchers have combined them with other methods to identify damage in structures. The combination of methods allows complementing of drawbacks of methods. Dackermann et al. [28] utilized ANN and FRF methods to identify defects on a frame structure. The efficiency of the technique showed that the bearable limit of uncertainty was 10%. This work was furthered by Fallahian et al. [29] when they added PCA and replaced the ANN with deep neural network (DNN). The effects of the uncertainties caused by temperature change on the undamaged structures were reduced by DNN. The PCA reduced the FRF to a defined pattern, while using temperature as the input to the DNN. These studies have shown that identification of structural damage is more feasible by combining other methods with ANN.

This study aims at proposing a damage detection method to detect, locate and quantify damage by combining WT and ANN. The proposed neuro-wavelet technique is to identify small damage in a steel plate structure. The plate structure is inflicted with damage at different locations. The proposed method involves applying WT to decompose the mode shape difference to detect and locate damage in the plate structure. Lee et al. [22] showed that the sensitivity of mode shape differences to modelling error is less compared to mode shapes due to the baseline finite element model and the accuracy of the finite element model greatly influences the effectiveness of damage detection algorithms. In addition, the training data applied in ANNs are mostly obtained from finite element models, this property makes mode shape differences a perfect input vector for training neural networks. Upon damage localization, the mean value of the WT moduli of the decomposed signal and the coordinates of the located damage are applied to train the neural network to predict damage severity.

2. Procedure of the neuro-wavelet technique

In the proposed method, data (mode shape) from the modal analysis of plate model in the damaged and undamaged states are used. This is facilitated by numerical data generated from a finite element model that enables collection of sufficient data needed to train the neural network. The plate structure is square and made of steel, and all the side boundaries are fixed. The proposed method is designed to be able to provide information on the three levels of damage identification, namely, detect presence of damage, detect location of damage and detect severity of damage. The 1st two levels are accomplished by using WT, while the third was achieved by using features extracted from the first WT and damage located coordinates.

Continuous wavelet transform (CWT) is the WT applied to decompose the mode shape difference. Unlike Fourier transform that is only localized in frequency, WF is localized in both frequency and time [30]. WT examines local data (e.g., displacement), thus it is able to detect and locate local damage; it also performs excellently in the presence of noise. The major problem of WT is selection of the most appropriate mother wavelet. The best option usually requires trial and error simulations. Decomposition of the mode shape difference provides information on damage detection and location. The coordinates of detected damage and mean value of the WT moduli serve as input to the neural network to predict damage severity. The procedure of damage severity prediction in the plate structure consists of the following stages:

1. Find the mode shapes of the monitored structures.
2. Evaluate the mode shape difference of the first mode.
3. Calculate the WT moduli of the first mode shape difference to locate damage.
4. Feed the mean value of WT moduli and detected damage coordinates to the ANN system.
5. Evaluate the output regarding damage severity (severity ranges from 0 to 1).

3. Continuous wavelet transform

A mother wavelet function $\psi(x)$ creates a wavelet family $\psi_{u,s}(x)$.

$$\psi_{us}(x) = \frac{1}{\sqrt{s}}\psi\frac{(x-u)}{s} \tag{1}$$

where s and u are scale and position, respectively.
The deflection of the plate as a one-dimensional signal f(x), the CWT defined by Mallat [31] is:

$$Wf(u,s) = \frac{1}{\sqrt{s}}\int_{-\infty}^{+\infty} f(x)\,\psi\left(\frac{x-u}{s}\right)\mathrm{dx} \tag{2}$$

where $Wf(u,s)$ is wavelet coefficient of the wavelet $\psi_{u,s}(x)$. The n vanishing moment of a wavelet is vital in detecting singularity of signals, which is based on the equation:

$$\int_{-\infty}^{+\infty} fx^{k}\ \psi(x)\mathrm{dx} = 0, \ k = 0,1,2,\ldots\ldots,n-1 \tag{3}$$

Rewriting the above equation gives:

$$Wf(u,s) = \frac{1}{\sqrt{s}}\int_{-\infty}^{+\infty} f(x)\,\psi\left(\frac{-(u-x)}{s}\right)\mathrm{dx} = \frac{1}{\sqrt{s}}f * \psi\left(\frac{-u}{s}\right) = f * \bar{\psi}_s(u) \tag{4}$$

and

$$\bar{\psi}_s(u) = \frac{1}{\sqrt{s}}\psi\left(\frac{-x}{s}\right) \tag{5}$$

Re-writing as the nth order derivation of a smooth function $\theta(x)$ [25]:

$$Wf(u,s) = \frac{s^n}{\sqrt{s}}\int_{-\infty}^{+\infty} f(x)\frac{d^n}{\mathrm{dx}^n}\theta\left(\frac{x-u}{s}\right)\mathrm{dx} = \frac{s^n}{\sqrt{s}}\frac{d^n}{\mathrm{dx}^n}\int_{-\infty}^{+\infty} f(x)\,\theta\left(\frac{-(u-x)}{s}\right)\mathrm{dx}$$

$$\bar{\theta}_s(x) = \frac{1}{\sqrt{s}}\theta\left(\frac{-x}{s}\right) \tag{6}$$

where $f * \bar{\theta}_s$ is convolution. In two-dimensional CWT, the function $f(x, y)$ is given by Rucka and Wilde [32] as:

$$W^i f(u,v,s) = \frac{1}{s}f * \psi^i\left(\frac{-u}{s},\frac{-v}{s}\right) = f * \bar{\psi}_s(u,v), \ i = 1,2 \tag{7}$$

The horizontal $\psi^1(x,y)$ and vertical $\psi^2(x,y)$ wavelets are constructed with separable products of scaling $\emptyset$ and wavelet function ψ :

$$\psi^1(x,y) = \emptyset(x)\psi(y) \ \ \psi^2(x,y) = \psi(x)\emptyset(y) \tag{8}$$

4. Numerical plate model

The plate model applied in this study is a square steel plate. The dimensions of the steel plate model are: 560mm length, 560mm width and 2mm thickness, and the material properties are: Young Modulus E = 200Gpa, Poisson's ratio ν = 0.3 and steel density ρ = 7850kg/m^3. The four sides of the plate are fixed. The plate is modelled by using Finite Element Analysis (FEA) program SAP2000 into 672 plate elements of size 20mm x 20mm elements, and the modal parameters (mode shape and frequencies) are obtained. To demonstrate damage detection of WT, three locations (middle, side and top of the plate) of damage are applied in this section. These locations are shown in Figure 1. The three damage locations are simulated by reducing the plate's thickness at the specified area. The damages are in rectangular shape with the dimension of 40mm x 40mm which represents approximately 0.005% of the plate's area. Figure 2 shows a cross-section of the plate with a middle damage.

The mode shape difference of the damaged and undamaged states is generated by subtracting the displacement value at each node in the first mode shape. Figure 3 shows the first mode shape of the undamaged plate model. The obtained mode shape differences are analyzed with CWT to obtain the damage locations. To accurately determine the damage location, appropriate mother wavelet needs to be chosen in decomposing the mode shape differences. Paul mother wavelet is chosen after several trials and errors [33]. Figure 4 shows the locations of the detected damage for the three damage locations. The coordinates of the damage are included in Figure 4, and these coordinates and the mean wavelet coefficients serve as input into the neural network.

IOP Conf. Series: Materials Science and Engineering **513** (2019) 012015 doi:10.1088/1757-899X/513/1/012015

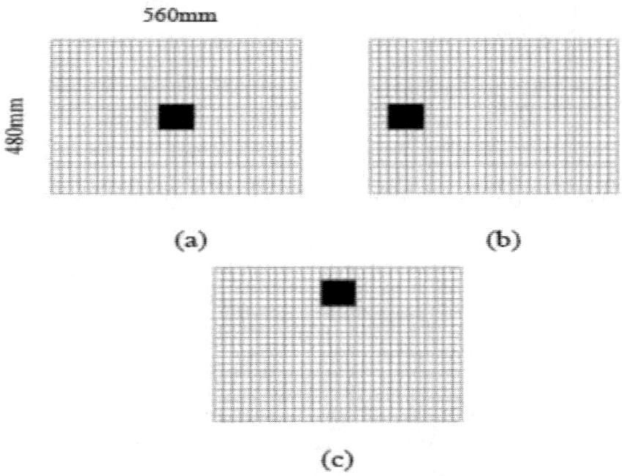

Figure 1. Damage locations: (a) middle; (b) side; (c) corner.

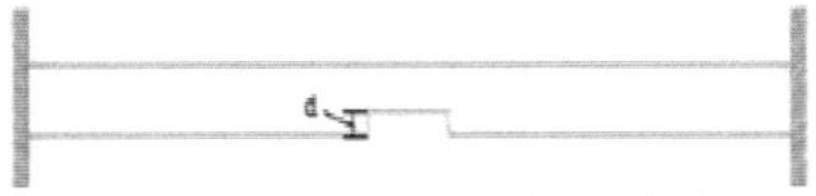

Figure 2. Damage section of plate.

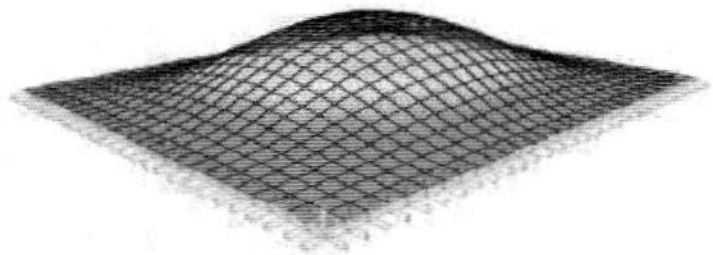

Figure 3. First mode of the plate model.

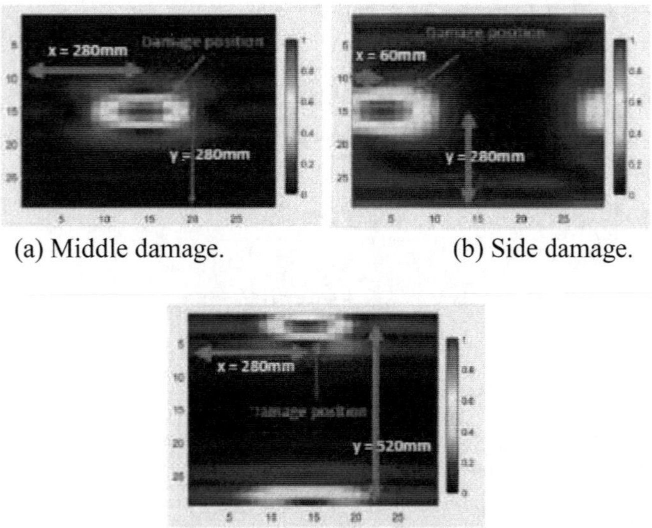

(a) Middle damage. (b) Side damage.

(c) Top damage.

Figure 4. Wavelet coefficient showing detected damage location in the plate model.

5. Artificial neural network

5.1. Architecture of neural network

The architecture of the ANN consists of two hidden and one output layers. Each of the hidden layers consists of 20 neurons. Figure 5 shows a typical architecture of a neural network model. The optimum number of hidden layers and the required neurons were estimated by applying trial and error. The input layer has 3 nodes that correspond to the mean value of WT moduli and the x- and y-axes of the detected damage. Levenberg–Marquardt (LM) backpropagation algorithm was applied for training the ANN. The neurons in the ANN are taken to be fully connected and sigmoid function was selected as the activation function for all hidden layers. The output layer was assigned the linear activation function. To prevent the overcrowding of the ANN, the input and output values were trimmed to between -1 and +1.

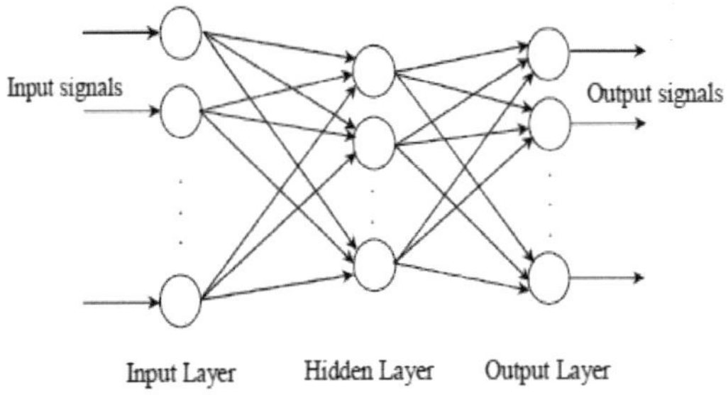

Figure 5. Typical architecture of an ANN model.

5.2. Training of neural network
The training of the neural network is done by using noise-free numerical data. The damage locations and damage orientations are shown in Figure 1 and Figure 2. The damage severity is defined by the thickness loss in the plate and taken as b, as shown in Figure 2. The damage locations are characterized by their coordinates (x- and y-axes), and 25 different damage locations and 50 different damage severities were randomly selected on the plate model. The numerical deflection lines were computed by SAP2000. The Paul mother wavelet was selected for the CWT after several trials and errors.

The total number of input pattern was 672, and the inputs were the x- and y-axes coordinates of the detected damage and the WT moduli mean value. Training of the neural network stopped at 700 epochs or when the *msereg reaches* 10^{-3}. The performance of the neural network for best prediction is shown in Figure 6(a). The predictions of the neural network are shown on the y-axis, while the actual values are on the x-axis. From Figure 6(a), all points lie very near the line called the best linear fit. The correlation coefficient is 0.9904 and is very close to 1, this indicates very good compatibility between the outputs and targets.

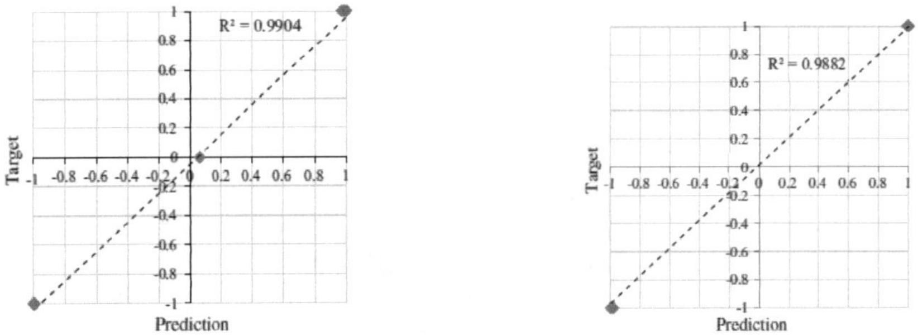

(a) Training mode (b) Testing mode

Figure 6. Damage severity results.

5.3. Testing of neural network
Noise-free numerical data of the steel plate was applied for testing of the trained neural network. A total of 10 different damage locations and 15 different damage severities were randomly used to test the neural network. The damage set comprised of 100 patterns. These damage patterns were totally different from the damage patterns applied in training the neural network. However, the neural network was able to provide good predictions that were in good agreement with the actual values. The correlation coefficient reaches 0.9882. The results of the testing mode are shown in Figure 6(b). It can

be concluded that neural network can accurately predict damage severity when the appropriate damage WT parameters are applied as input.

6. Conclusion

This paper presented a neuro-wavelet damage detection method for damage identification in plate structures. A numerical example of a square steel model was utilised. Damage identification was done in two phases. The first phase involved WT decomposition of mode shape differences to detect and locate damage. The second phase utilised neural network to predict the damage severity. A backpropagation neural network was trained using damage parameters obtained from the WT moduli and detected damage coordinates. This study showed that WT can successfully detect and locate small damage in plate structure by utilizing the mode shape difference. In addition, neural network can predict damage severity in plate structures with a high degree of precision. The architecture of neural network to damage detection plays a role in obtaining accurate result and reducing computational time.

References

[1] Laory I, Trinh T N, Smith I F C and Brownjohn J M W 2001 Methodologies for predicting natural frequency variation of a suspension bridge. *Eng. Struct.* **80** 211-221.

[2] Cao M, Radzieński M, Xu W and Ostachowicz W 2014 Identification of multiple damage in beams based on robust curvature mode shapes. *Mech. Sys. Sig. Pro.* **46** 468-80.

[3] Dackerman U, Li J and Samali B 2013 Identification of member connectivity and mass changes on a two-storey framed structure using frequency response functions and artificial neural networks. *J. Sound and Vib.* **332** 3636-3653.

[4] Dawari V B and Vesmawala G R 2013 Modal curvature and modal flexibility methods for honeycomb damage identification in reinforced concrete beams. *Pro. Eng.* **51** 119-124.

[5] Bakhary N, Hao H and Deeks A J 2010 Substructuring technique for damage detection using statistical multi-stage artificial neural network. *Adv. in Struct. Eng.* **13** 619-39.

[6] Garstecki A, Knitter-Piatkowska A, Pozorski Z and Ziopaja K 2004 Damage detection using parameter dependent dynamic experiments and wavelet transformation. *J. of Civil Eng. and Mgt.* **10** 191-197.

[7] Spanos P D, Failla G, Santini A and Pappatico M 2006. Damage detection in Euler–Bernoulli beams via spatial wavelet analysis. *Struct. Cont. and Health Moni.* **1** 472-87.

[8] Wu N and Wang Q 2011 Experimental studies on damage detection of beam structures with wavelet transform. *Int. J. of Eng. Sci.* **49** 253-61.

[9] Rucka M 2011 Damage detection in beams using wavelet transform on higher vibration modes. *J of Theoretical and Appl. Mech.* **49** 399-417.

[10] Cao M, Xu W, Ostachowicz W and Su Z 2014 Damage identification for beams in noisy conditions based on Teager energy operator-wavelet transform modal curvature. *J. Sound and Vib.* **333** 1543-53.

[11] Montanari L, Spagnoli A, Basu B and Broderick B 2015 On the effect of spatial sampling in damage detection of cracked beams by continuous wavelet transform. *J. Sound and Vib.* **345** 233-49.

[12] Douka E, Loutridis S and Trochidis A 2004 Crack identification in plates using wavelets analysis. *J. Sound and Vib.* **270** 279-95.

[13] Law S S, Li X Y, Zhu X Q and Chan S L 2005 Structural damage detection from wavelet packet sensitivity. Eng. Struct. **27** 1339-48.

[14] Sarrafi A and Mao Z 2016 Probabilistic uncertainty quantification of wavelet-transform-based structural health monitoring features in *Health Moni of Struct. and Bio. Sys.* pp. 9805.

[15] Abdulkareem M, Bakhary N, Vafaei M, Noor N M and Padil K H 2018 Non-probabilistic wavelet method to consider uncertainties in structural damage detection. *J. Sound and Vib.* **433** 77-98.

[16] Reda Taha M M 2010 A neural-wavelet technique for damage identification in the ASCE benchmark structure using phase II experimental data. *Adv. in Civil Eng.*

[17] Vafaei M, Adnan A B, Abd. Rahman A B 2014 A neuro-wavelet technique for seismic damage identification of cantilever structures. *Struct. and Infra Eng.* **10** 1666-1684.

[18] Padil K H, Bakhary N and Hao H 2017 The use of a non-probabilistic artificial neural network to consider uncertainties in vibration-based-damage detection. *Mech. Sys. and Sig. Pro.* **83** 194-209.

[19] Wu X, Ghaboussi J and Garrett Jr J 192 Use of neural networks in detection of structural damage. *Comp. & Struct.* **42** 649-659.

[20] Sahin M and Shenoi R 2003 Quantification and localisation of damage in beam-like structures by using artificial neural networks with experimental validation. *Eng. Struct.* **25** 1785-1802.

[21] Yam L, Yan Y and Jiang J 2003 Vibration-based damage detection for composite structures using wavelet transform and neural network identification. *Compo. Struct.* **60** 403-412.

[22] Lee J, Lee J W, Yi J H, Yun C B and Jung H Y 2005 Neural networks-based damage detection for bridges considering errors in baseline finite element models. *J. Sound and Vib.* **280** 555-578.

[23] Rucka M and Wilde K 2010 Neuro-wavelet damage detection technique in beam, plate and shell structures with experimental validation. *J. of theoretical and appl. mech.* **48** 579-604.

[24] Bagchi A, Humar J, Xu H and Noman A S 2010 Model-based damage identification in a continuous bridge using vibration data. *J. of Perf. of Constr.* Fac. **24** 148-158.

[25] Li J, Dackermann U, Xu Y L and Samali B 2011 Damage identification in civil engineering structures utilizing PCA compressed residual frequency response functions and neural network ensembles. *Struct. Contr. Health Moni.* **18** 207-226.

[26] Hakim S, Razak H A and Ravanfar, S 2015 Fault diagnosis on beam-like structures from modal parameters using artificial neural networks. *Measurement* **76** 45-61.

[27] Khoshnoudian F, Talaei S and Fallahian, M 2016 Structural Damage Detection Using FRF Data, 2D-PCA, Artificial Neural Networks and Imperialist Competitive Algorithm Simultaneously. *Int'l J of Struct. Stab. and Dyn.* pp. 1750073.

[28] Dackermann U, Li J, Samali B 2013 Identification of member connectivity and mass changes on a two-storey framed structure using frequency response functions and artificial neural networks. *J. Sound and Vib.* **332** 3636-53.

[29] Fallahian M, Khoshnoudian F and Meruane V 2017 Ensemble classification method for structural damage assessment under varying temperature. *Struct. Health Moni.* 1475921717717311.

[30] Reda Taha M M, Noureldin A, Lucero J L and Baca T J 2006 Wavelet Transform for Structural Health Monitoring A Compendium of Uses and Features. *Struct. Health Moni.* **5** 267–295.

[31] Mallat S 1998 A Wavelet Tour of Signal Processing. Academic Press, New York.

[32] Rucka M and Wilde K 2006 Application of continuous wavelet transform in vibration based damage detection method for beams and plates. *J. Sound and Vib.* **297** 536-550.

[33] Abdulkareem M, Bakhary N and MV 2015 Application of Wavelet Transform to Damage Detection in Plates using Response-only Measurements. *16th Asia Pacific Vibration Conference.*

IOP Conf. Series: Materials Science and Engineering **513** (2019) 012016 doi:10.1088/1757-899X/513/1/012016

2D finite element simulation of air-coupled impact echo testing on concrete slab

A R Norinah[1*], F N Adnan[1] and R Hamid[1]

Smart and Sustainable Township Research Centre, Faculty of Engineering and Built Environment, Universiti Kebangsaan Malaysia, 43600 UKM Bangi, Selangor, Malaysia.

[*]Corresponding author: norinah@ukm.edu.my

Abstract Impact-echo is an effective method in identifying and detecting delamination in concrete structures. Recently, the contact sensor is replaced with air-coupled sensor due to the interest in rapid testing method. This paper presents a study of the optimum test configuration of air-coupled impact echo and conventional concrete-coupled impact echo tests by means of numerical simulation. Two-dimensional finite element analyses were performed on concrete slab to investigate the effect of impact-to-sensor distance and sensor height from concrete surface. Two concrete slabs, one without delamination and one with delamination, were modelled. For intact concrete, both sensors were effective in detecting concrete thickness at any point within the specified location range. However, for air-coupled IE performed on concrete with 40 mm delamination depth, optimum sensor location is 60 to 100 mm away from impact point at height of 10 mm.

1. Introduction

Non-destructive evaluation techniques have become important for the maintenance of existing structures as well as for the construction quality control. Impact-echo is a widely used non-destructive testing method that has been demonstrated to be effective in identifying and detect delamination in concrete structures. Impact-echo is one of the techniques which work on the principle of stress wave propagation that has been introduced since 1980s. It can be used to determine the location of flaw such as cracks, delaminations, estimating the depth of delamination and slab thickness. Impact-echo is a mechanical wave-based NDE technique which transient point load is applied at specimen surface to initiate waves propagation. This impact generates body waves (P-waves and S-waves) and surfaces waves (Rayleigh waves) that simultaneously propagate in the concrete. The resulting transient surface motion which is set up by a vibration resonance through the thickness of the element is detected by a sensor coupled to specimen surface. The transient time response of the solid structure is commonly measured with a contact sensor coupled to the surface at some distance to the impact source. The time domain signal is transformed to the frequency domain where the frequency at the maximum spectral amplitude is obtained. The thickness or depth of delamination, d is related to P-wave velocity V_p and peak frequency, f_{max} of the frequency spectrum by

IOP Conf. Series: Materials Science and Engineering **513** (2019) 012016 doi:10.1088/1757-899X/513/1/012016

$$d = \frac{V_p}{2f_{max}} \tag{1}$$

This is the basic equation for determining the thickness of solid concrete structures using the impact-echo method. However, previous research consisting of experiment and numerical simulations has demonstrated there should be a shape factor, β, which is function of the geometry of the tested concrete structure, to convert the P-wave speed from the Equation (1) or the P-wave speed test to the apparent P-wave speed measured in the impact-echo test. For plate-like structure, the shape factor is equal to 0.96 and Equation (1) can be written as

$$d = \frac{\beta V_p}{2f_{max}} \tag{2}$$

The use of an air-coupled sensor was demonstrated to be effective in improving the IE testing speed [1, 2, 3]. In an air-coupled IE testing system, the contact sensor is replaced with a contactless sensor. The basic idea of this test is that waves propagates into the air initiated by the wave motion of the surface can be captured by a microphone, which is an air-coupled sensor. Numerous investigations have demonstrated that air-coupled IE data are consistent, reliable and are equivalent to that collected using conventional contact sensor [3].

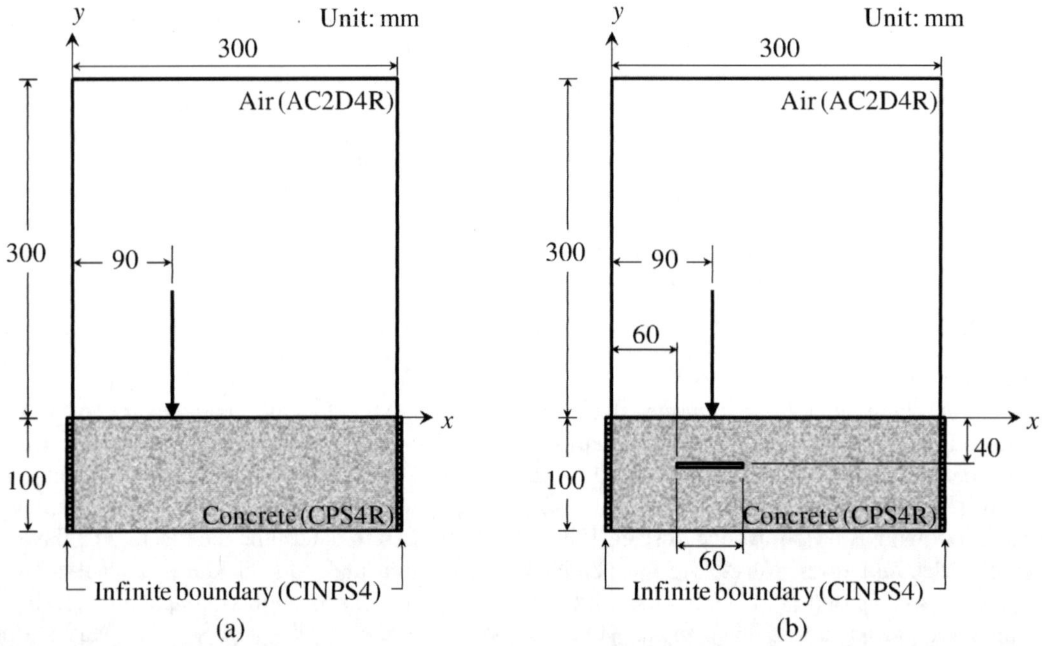

Figure 1. 2D model for an air-concrete systems (a) concrete slab without delamination and (b) concrete slab with 60mm-long delamination at depth of 40mm.

In IE test, the distance between impact point and sensor is important. However, not much consideration was made in deciding the location of sensor relative to source location. If the distance is too large, the response will not dominated by the reflected P-wave, and the simple relationships expressed by Equations. (1) and (2) are not applicable. If the distance is too small, the response is dominated by the effect of the surface wave [4]. [5] suggest that spacing from 0.2 to 0.5 of delamnation depth or concrete thickness was found to give acceptable results based on a series of analytical studies. [6] recommend a spacing of less than 0.4 of the concrete thickness. When the source is impacted to concrete surface, stress waves generated in the air and concrete, and later, waves in the concrete radiated into air nearby air-concrete interface. For air-coupled IE test, [7] setup the microphone at 20 mm above concrete surface and at 0.5 of concrete thickness distance from impact point. The test results agree well with numerical simulation. [8] performed air-coupled IE with MEM microphone array as a transducer. The transducer was placed 30 mm above concrete surface and concluded that zero mode resonance causing a surface displacement that is coherent in an area around the impact location with the radius approximately as the concrete thickness. Therefore, [8] suggested that transducer should be located not less than the concrete thickness from impact point. Microphone height was set closer to concrete surface by [9] at 10 mm. Summary of previous study/suggestion on impact echo test configurations are shown in Table 1.

Table 1. Previous studies on IE test configurations.

	Impact-to-sensor distance, *x*	**Sensor height, y (mm)**
Groschup (2015)	$x > d$	*30*
Shin et al (2012)	-	*10*
Tsai & Zhu (2012)	$x = 0.5d$	*20*
Sansalone & Street (1997)	$x > 0.4d$	-
Carino (1986)	$0.2d < x < 0.5d$	-

**d = concrete thickness/flaw depth*

2. Finite element simulation

Finite element models were modelled to represent the air-solid system using ABAQUS software [10]. 2D planar models were employed to stimulate an impact point source applied on the surface of a solid specimen. Model parameters should be selected properly so that the wave propagation in the air-solid system could be accurately simulated.

Table 2. Material Properties.

Material	**Acoustic-wave Velocity (m/s)**	**S-wave Velocity (m/s)**	**P-wave Velocity (m/s)**	**Density (kg/m3)**	**Poisson's Ratio**
Concrete	-	2510	4100	2500	0.2
Air	343	-	-	1.204	-
Void	-	-	-	-	-

2.1. Model parameters

A 2D model consisting of a solid and air sections was built to represent air-concrete slab system as presented in Figure 1. A 300 mm by 100 mm solid section representing concrete slab is seeded with 4-node bilinear plane stress quadrilateral element (CPS4R). The air part is represented by a 300 mm by 300 mm section seeded with 4-node linear 2-D acoustic quadrilateral elements (AC2D4R). To diminish reflections of stress waves from the left and right boundaries, low-reflecting boundaries were specified on both left and right sides of the concrete plate. A layer of infinite elements (CINPS4) were utilized along the both edge of solid section. Material properties for the solid and air are listed in the Table 2. A maximum impact force of 1 N was used, with contact duration of 30 μs. This impact is similar to that produced by steel sphere 7 mm in diameter, which is a common size use in impact-echo testing. Meshing the model geometry is the last step to complete prior to computing. For transient analysis, a sufficiently fine mesh should be used to resolve the waveforms of the smallest wavelength

IOP Conf. Series: Materials Science and Engineering **513** (2019) 012016 doi:10.1088/1757-899X/513/1/012016

of interest. Element size of 2 mm is used uniformly throughout the model and an incremental time step of 1 μs is used for internal stability of the finite element analysis.

2.2. Test configurations

Ten contact sensors were coupled to the concrete surface at every 10 mm away from impact point. While air-coupled sensors located at three different lift-offs line parallel to concrete surface which are 10, 20 and 30 mm. At every height, ten contactless sensors placed at every 10 mm starting from impact location. Impact point is aligned with the centre of concrete delamination. The location of impact are maintained for both model. Location for sensors is the same form both sound and defected models. In total, 40 sensors involved in each simulation. Therefore, analyzing each frequency spectra automatically would be beneficial to speed up data reduction process.

2.3. Maximum frequency search

Analyzing many data sets manually requires much time and effort. An algorithm for rapid search of peak frequency in frequency domain is developed as shown in Figure 2. The algorithm is implemented in MATLAB software. All data collected in the simulation were processed using this algorithm.

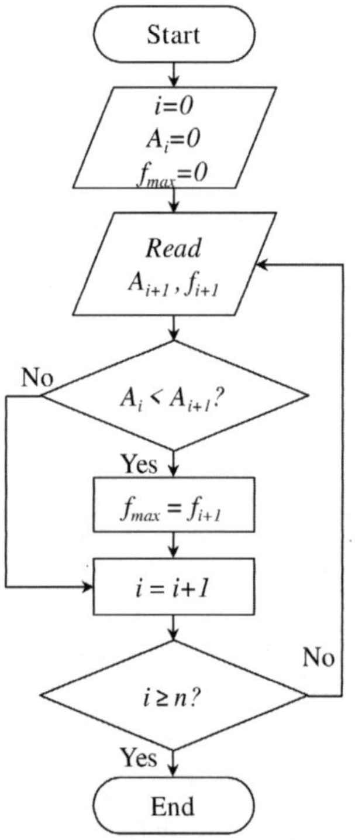

Figure 2. Global peak frequency search algorithm.

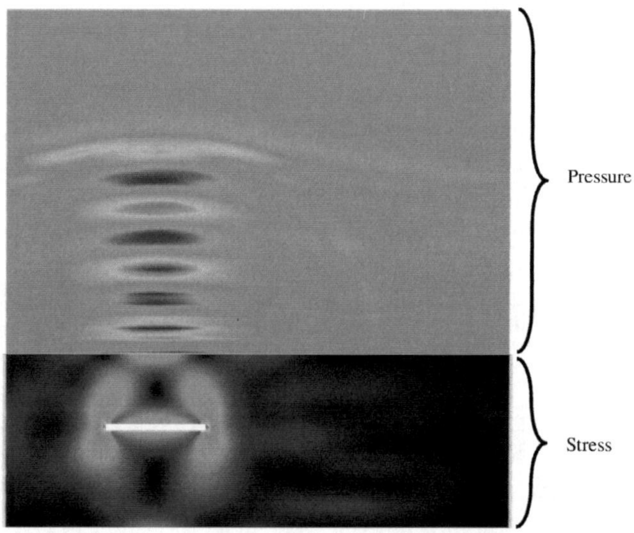

Figure 3. Snapshot of stress and pressure field at time, t = 0.4 ms after impact.

3. Result and analysis

Figure 3 shows the pressure field in the air and stress in solid at time, t = 0.4 ms after impact force is applied on the concrete slab. Three types of wave are observed which are leaky Rayleigh waves, direct acoustic waves and airborne (due to flexural mode in concrete) waves. Leaky Rayleigh waves radiates in a conical shape wavefront. Direct acoustic waves radiating with a velocity of 343 m/s creates perfect spherical wavefront. A series of airborne periodic plane waves above the impact location are also observed in Figure 3. These waves indicate local resonant mode were built around impact location. Hollow zone is observed in concrete slab on both delamination edge, on the top of delamination surface and at the air-concrete boundary is caused by a flexural mode vibration of the entire delaminated region. The resulting pressure and air fields show good agreement with the result form previous study [7]; thus, validity of the model was verified.

Table 3 summarized peak frequencies obtained at different impact-to-sensor distance, x and sensor height from concrete surface, y for conventional impact echo (IE) and air-coupled impact echo (ACIE) tests performed at concrete slab model without delamination. Based on Eq. 1, expected frequency peak for sound concrete with 100 mm thickness and 0.96 shape factor is 19.68 kHz. A consistence peaks are observed throughout entire testing locations for ACIE test at 19.53 kHz. A slight difference occurs starting at 80 mm and above is cause by frequency scale where sensitivity of the scale is 0.12 kHz. Consistency of IE test result is observed until impact-to-sensor distance reaching 70 mm, where 36.74 and 59.94 kHz peaks are collected after 60 mm. The ratios of these peaks with expected resonance frequency are 1.88 and 3.07, corresponding to second and third resonance modes. Therefore, three resonance modes are successfully captured by this numerical simulation. Normalized amplitude of these peaks are plotted in Figure 4. Variation of peak frequency captured is based on search algorithm explained in section 2.3. By analyzing plots in Figure 4 manually, the expected frequency is always exists and significant, however maximum frequency captured by the algorithm is a global maximum without considering any expected value. Therefore, IE and ACIE are actually able to detect concrete thickness (100 mm) at any sensor distance up to 100 mm from impact point. This is also true for microphone with height from concrete surface up to 30 mm. The result implied that all previous studies summarized in Table 1 are acceptable. The differences in preferred configurations are due to factors affecting during tests conducted. In this study, the FEM model is in ideal condition that is not affected by any external factors.

Table 3. Variation of peak frequency (kHz) obtained at different impact-to-sensor distance and sensor height for concrete slab model without delamination. (Expected frequency = 19.68 kHz)

Test	Height, y (mm)	Distance from impact source, x (mm) 10	20	30	40	50	60	70	80	90	100
IE	-	19.53	19.53	19.53	19.53	19.53	19.53	59.81	59.94	36.74	36.74
ACIE	10	19.53	19.53	19.53	19.53	19.53	19.53	19.53	19.41	19.41	19.41
	20	19.53	19.53	19.53	19.53	19.53	19.53	19.53	19.41	19.41	19.41
	30	19.53	19.53	19.53	19.53	19.53	19.53	19.53	19.41	19.41	19.41

Table 4. Variation of peak frequency (kHz) obtained with different impact-to-sensor distance and sensor height for concrete slab with delamination. (Expected frequency = 49.20 kHz)

Test	Distance, y (mm)	Distance from impact source, x (mm) 10	20	30	40	50	60	70	80	90	100
IE	-	48.58	48.71	9.89	10.01	48.58	48.71	50.29	50.29	51.03	50.05
ACIE	10	9.89	9.89	9.89	10.01	10.01	48.58	48.71	50.29	50.78	9.89
	20	9.89	9.89	9.89	10.01	10.01	9.89	1.46	9.89	10.01	10.01
	30	9.89	9.89	9.89	9.89	10.01	1.46	1.59	9.89	9.89	9.89

Table 4 summarized peak frequencies obtained at different impact-to-sensor distance, x and sensor height from concrete surface, y for conventional impact echo (IE) and air-coupled impact echo (ACIE) tests performed at concrete slab model with delamination. Based on Equation 1, expected frequency peak for concrete with 40 mm delamination depth and 0.96 shape factor is 49.20 kHz. Result for IE test is considerably consistence by ignoring scale sensitivity effect, except for distance 30 and 40 mm. Based on Figure 5, frequency spectrum at distance 30 and 40 mm showing a clear peak at expected resonance frequency. As previously discussed, the algorithm is picking up global maximum. Therefore, IE test is again able to detect concrete delamination depth (40 mm) at any sensor distance up to 100 mm from impact point. However for ACIE measurement, only sensors at height of 10 mm are capturing the right resonance frequency at distance between 60 to 90 mm. The presence of hollow stress contour around delamination may contribute to this issue. As can be seen in Figure 6 (b) to (d), frequency domain plot is dominated by low frequency energy corresponding to flexural mode peak response. Therefore, it can be concluded that ACIE magnifying flexural modes more compared to thickness-types mode. Similar finding are also found in previous study by [1].

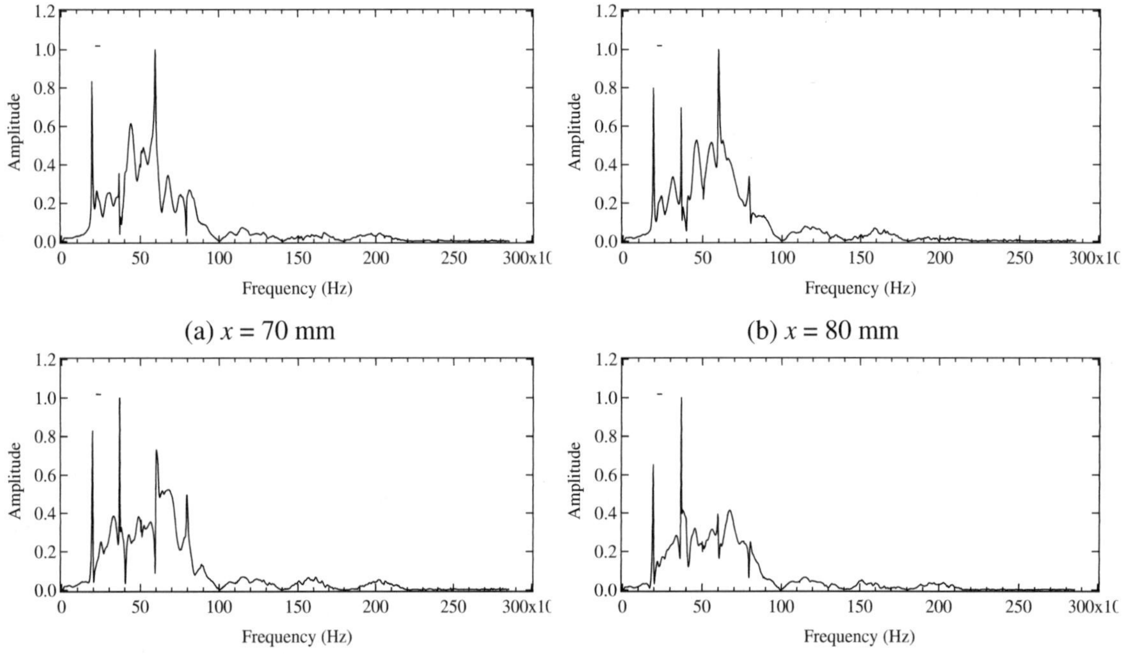

(a) x = 70 mm (b) x = 80 mm

(c) x = 90 mm (d) x = 100 mm

Figure 4. Frequency spectrum of acceleration for concrete model without delamination. Sensor-to-impact distance ranging from 70 mm to 100 mm.

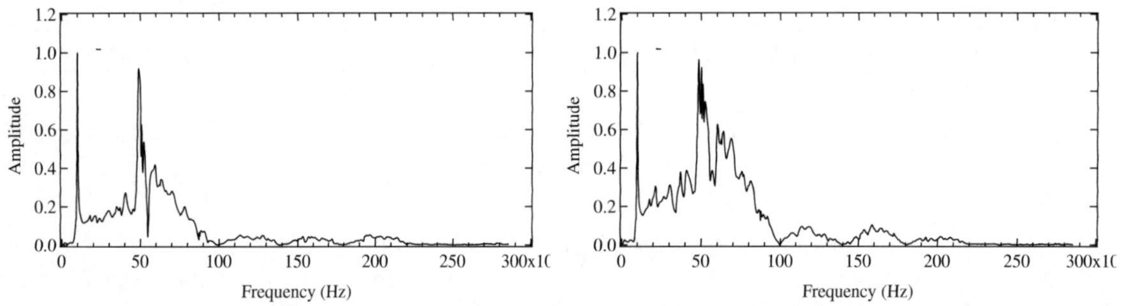

(a) x = 30 mm (b) x = 40 mm

Figure 5. Frequency spectrum of acceleration for concrete model with delamination. Sensor-to-impact distance is at 30 mm and 40 mm.

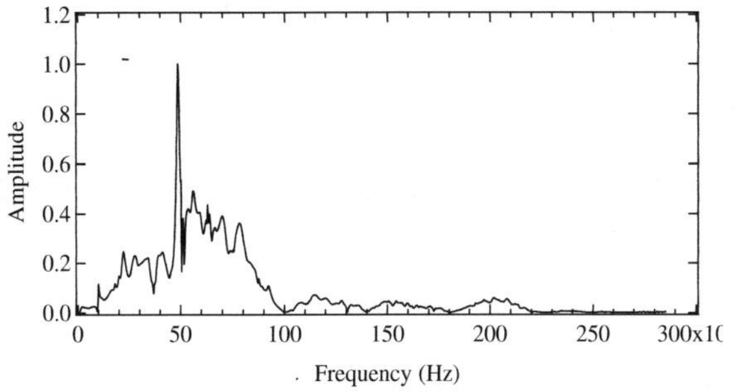

(a) Acceleration

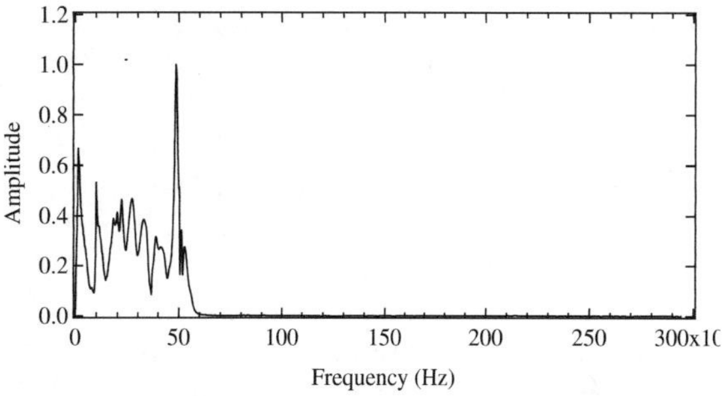

(b) Pressure at $y = 10$ mm

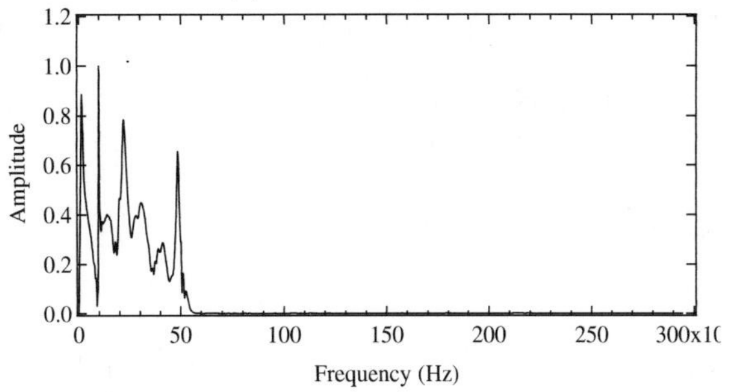

(c) Pressure at $y = 20$ mm

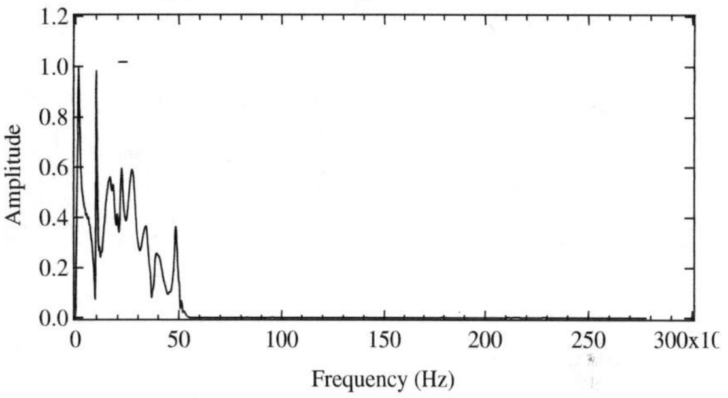

(d) Pressure at $y = 30$ mm

Figure 6. Frequency spectrum of acceleration and pressure for concrete model with delamination. Sensor-to-impact distance, x at 60 mm and sensor height form concrete surface is ranging from 10 mm up to 30 mm.

4. Summary and Conclusions

Numerical simulations performed to investigate the effectiveness of IE test with contactless sensor compared to contact sensor in term of sensor-to-source distance. Effects of contactless sensor lift-off from concrete surface were also investigated. Two concrete models were built, one without delamination and one with delamination. 40 sensors points with 10 mm distance apart distributed at 100 mm by 30 mm area above the concrete surface were considered in each simulation. Time signal data were transformed into frequency spectra by fast Fourier transform. The validity of FEM model was verified against previous study. An algorithm dedicated for peak frequency search was developed. The algorithm sped up data reduction process. However, in some cases it fails to extract actual peak frequency and therefore, manual consideration were involved. For intact concrete, both IE and ACIE tests were able to detect concrete thickness at any sensor location within the specified location range in this study. Effectiveness of both sensor types was considerably equivalent. However, for defected concrete with 40 mm depth delamination, only 10 mm sensor height was effective in detecting delamination depth starting from 60 mm away from the impact point.

Acknowledgments

The authors acknowledged Universiti Kebangsaan Malaysia for their financial support via the allocation grant under project GGP-2017-035.

References

[1] Kee, S. and Gucunski, N. 2016. Interpretation of flexural vibration modes from Impact-echo testing. *Journal of Infrastuctures Systems*. **22**(3) 1-10.

[2] Oh, T., Kee, S., Arndt, R. W., Poupinet, G. and Zhu, J. 2013. Comparison of NDT methods for assessement of a concrete bridge deck. *Journal of Engineering Mechanics* 139 (March) 305-314.

IOP Conf. Series: Materials Science and Engineering **513** (2019) 012016 doi:10.1088/1757-899X/513/1/012016

[3] Zhu, J. and Popovics, J. S. 2007. Imaging concrete structures using air-coupled Impact-echo. *Journal of Engineering Mechanics.* **133**(6) 628-640.
[4] Carino, N. J. 2004. The Impact-echo method: An Overview. *Proceedings of the 2001 Structures Congress & Exposition.* **89**(3) 1-18.
[5] Carino, N.J., Sansalone, M., and Hsu, N.N., 1986b, Flaw Detection in Concrete by Frequency Spectrum Analysis of Impact-Echo Waveforms *in International Advances in Nondestructive* Testing, Ed. W.J. McGonnagle, Gordon & Breach Science Publishers, New York, pp. 117-146.
[6] Sansalone, M., and Streett, W. B., 1997, Impact-Echo: Nondestructive Testing of Concrete and Masonry, Bullbrier Press, Jersey Shore, PA.
[7] Zhu, Y. T. J. 2012. Simulation and Experiments of Airborne Zero-Group-Velocity Lamb Waves in Concrete Plate **133**(4) 373–382.
[8] Groschup, R. and Grosse, C. U. 2015. Enhancing air-coupled Impact-echo with microphone arrays 2. *The Impact-Echo Method.* **8**(1) 1-12.
[9] Shin, S. W., Popovics, J. S., and Oh, T. 2012. Cost Effective Air-Coupled Impact-Echo Sensing for Rapid Detection of Delamination Damage in Concrete Structures. *Advances in Structural Engineering,* **15**(6) 887–895.
[10] ABAQUS v. 6.10. Dassault Systems Simulia.

IOP Conf. Series: Materials Science and Engineering **513** (2019) 012017 doi:10.1088/1757-899X/513/1/012017

Dynamic load test of full-scale pile for the construction and rehabilitation of bridges

S T Noor[1*], M S Islam[2], M Mumtarin[2] and N Chakraborty[2]

[1] Department of Civil Engineering, University of Asia Pacific, 74/A Green Road, Dhaka-1205, Bangladesh
[2] Prosoil Foundation Consultant, 88/B Indira Road, Dhaka-1215, Bangladesh

[*]Corresponding author: sarah@uap-bd.edu

Abstract. Seven dynamic load tests were performed on six large diameter piles, installed near the Gumti river Bangladesh. The diameter of five cast in situ piles was 1.5 m, and that of the steel pile was 1 m. The steel pile was tested twice: under both drive and re-drive conditions. The pile lengths varied from 45 m to 71.8 m. The distribution of pile bearing resistance under static load were evaluated using the program DLTWAVE by signal matching. In addition, standard penetration tests were conducted in the vicinity of all the six piles. The results of dynamic load test were calibrated by comparing the static capacity obtained from static and dynamic load tests, conducted on two identical cast in place piles. Therefore, the skin frictional resistance and pile tip resistance are calculated using the conventional formulae of static analysis utilizing the developed correlation for the clay layers. The consistency of clay soil is observed to play an important role in the development of shaft resistance. For very stiff to hard clay, DLTWAVE indicates the development of shaft resistance smaller than that obtained from the formula (static analysis). Greater skin frictional resistance was obtained in DLT re-drive, as compared to that during initial drive.

1. Introduction

Determination of pile capacity from load test is recognized important, as the soil condition surrounding the pile may be significantly modified during the construction/installation [1]. At present, dynamic load test (DLT) on pile is practiced in more than 40 countries to evaluate the static capacity and structural integrity of pile using the measurement of both force and velocity. The test method of DLT has been standardized by different relevant codes, such as ASTM 4945. In DLT, dynamic impact is applied, signals are captured, and then static capacity, skin frictional resistance and tip resistance of pile are obtained after post-processing of signals. In addition to the assessment of pile's static capacity, DLT provides information about pile driveability [2], pile integrity and changes in cross-section (if any). DLT is significantly faster and economical than static test, as at least two bored piles and even more driven piles can be tested in one day with proper arrangements of DLT. In determining pile's capacity, DLT has several advantages over static load test in the following cases:

- Large diameter piles
- Offshore and onshore piling
- If there is space constraint to conduct static load test, especially for jetties, flyover, bridge and other construction projects in crowded area

Moreover, the use of the formulae of static analysis is limited in bridge construction projects, as it requires large number of borelog so that the variability in soil deposition at the river-bed can be addressed. In order to measure skin frictional resistance and pile tip resistance from static load test, pile needs to be instrumented with strain gauges at different sections of the pile, while in DLT these can be measured without any such instrumentation. Moreover, time after EOD (end of driving) influences the measurements of pile capacity and skin frictional and tip resistances [5]. DLT makes it possible to investigate "set up' time that has a significant effect on the pile capacity measurement, as it can be executed easily [6, 7]. Pile resistances can be reasonably obtained from DLT in 50% shorter period after EOD, as compared to static load test. [5]. Ultimate pile capacity of a square pile (30 cm x 30 cm) from a DLT was about 75% of that obtained from static load tests [8]. DLT is found as a useful method for predicting pile capacity for helical pile in cohesive soil [9]. DLT predicts pile resistances reasonably well but correlations and further analysis based on the DLT results. In recent years researchers paid attention in numerical modelling of pile for predicting pile resistances [10]. The objectives of the present study are to evaluate DLT in the determination of pile capacity, shaft resistance developed on the pile surface and pile tip resistance, when the pile is embedded in layered soils. A program DLTWAVE of Profound BV is used to obtain the static load settlement curve of the pile by signal matching.

2. Experimental program

In this study, seven dynamic load tests on six single vertical driven piles, along with DLTWAVE signal matching, were conducted. DLT on steel pile was repeated by re-driving for comparing the pile soil interaction during driving and re-driving. The sub-surface soil exploration, near each test pile location, was carried out by conducting standard penetration tests. In addition, undisturbed specimens were collected from cohesive layers to determine UU triaxial tests. Table 1 presents the detail of the experimental program carried out in-situ and in the laboratory as well.

Table 1. Experimental Program.

	Name of the test	No. of tests	Outcome
Field Tests	Dynamic load test	7	skin frictional resistance, end-bearing resistance and ultimate pile capacity
	Standard penetration test	6	borelog
Laboratory Tests	UU Triaxial	5	undrained shear strength
	Direct Shear Test	1	angle of internal friction

2.1. Test site

The test site was located near the Gumti river in Bangladesh. The Gumti river is a tributary of the Ganges. During monsoon, the hilly river has strong and rapid current, and its breadth is about 100 m (on an average). The Gumti river causes floods as a common and regular phenomena. Bangladesh water development board (BWDB) took several measures to train the river. At present, there is a bridge across the Gumti river and the construction of the second is in progress. Moreover, the rehabilitation projects of the existing bridges across other rivers with strong current are currently on-going. Therefore, the site has been chosen for the investigation of pile resistance by employing advanced technology of dynamic load test.

2.2. *Soil condition*

During the sub-surface exploration, standard penetration tests were conducted at 1.5 m interval up to 80 m depth below the ground level. The borelogs, as given in table 2 and table 3, demonstrate that different layers of SP-SC, SC, ML and CL soil layers exist at the test sites, according to unified soil classification system. SC and CL types of soils are most commonly encountered at the test site. Deep stratum with SPT N of 50 was found at 65 m or greater depths. According to SPT resistances, CL soil layers at the depth up to 45 m were found stiff to very stiff, while those at greater depths were very stiff to hard. The consistency of SC soil layers was identified as medium dense type.

In the laboratory, UU triaxial tests were carried out on undisturbed cohesive specimens, and the undisturbed shear strength varies from 45 to 129 kPa. The water level was found to vary from 0-5 m above the river bed, as of June 2017. The specimen collected from SC soil layer was tested in direct shear test apparatus, and the angle of internal friction was found between 32° and 37°. During SPT, N_{60} of SC soil layer gave a value ranging from 12 to 14.

2.3. *Pile*

In this study, dynamic load tests were conducted on 5 concrete piles and 1 steel pile. The dimensions and types of piles are given in table 4. The locations of test piles are also given in table 4.

Table 2. Soil profile at the locations of piers : P5, P15 and P16.

P5			P15			P16		
Depth	Soil Type	N_{avg}	Depth	Soil Type	N_{avg}	Depth	Soil Type	N_{avg}
0 - 8.5	SC	10	0 - 5	ML	2	0 - 4	SP-SM	16
8.5 - 15.5	SC	30	5 - 10	SM	5	4.1 - 11.1	ML	2
15.5 - 17.5	CL	20	10 - 22	SM	17	11.1 - 14.1	SC	6
17.5 - 22.5	SC	21	22 - 26	SM	32	14.1 - 16.1	CL	19
22.5 - 27.5	CL	18	26 - 29	CL	15	16.1 - 30.1	SC	21
27.5 - 30	CL	17	29 - 31	SM	24	30.1 - 40.1	CL	23
30 - 31	SC	17	31 - 44	CL	25	40.1 - 49.1	SC	27
31 - 44	CL	22	44 - 56.5	CL	34	49.1 - 64.1	CL	25
44 - 56.5	CL	35	56.5 - 62	SM	50	64.1 - 70.1	CH	30
56.5 - 72	SC	50	62 - 67	ML	32	70.1 - 83.1	SC	50
-	-	-	67 - 78.2	SM	50	-	-	-

IOP Conf. Series: Materials Science and Engineering **513** (2019) 012017 doi:10.1088/1757-899X/513/1/012017

Table 3. Soil profile at the locations of piers : P1, A1 and A2.

P1			A1			A2		
Depth (m)	Soil Type	N_{avg}	Depth	Soil Type	N_{avg}	Depth	Soil Type	N_{avg}
0 - 8.2	SC	5	0 - 3.2	SC	14	0 - 6.2	SC	12
8.2 - 15.2	SP-SC	24	3.2 - 21.2	SC	32	6.2 - 13.2	CL	14
15.2 - 30.7	SP-SC	29	21.2 -35.2	SC	37	13.2 - 24.2	SC	33
30.7 - 36.2	CL	29	35.2 - 39.2	CL	25	24.2 - 30.2	CL	23
36.2 - 50.2	CL	21	39.2 - 43.2	CL	43	30.2 - 57	CL	35
50.2 - 55.2	CL	32	43.2 - 52.2	CL	33	-	-	-
55.2 - 59.2	CL	30	52.2 - 57.5	ML	40	-	-	-
59.2 - 66.2	CL	20	-	-	-	-	-	-
66.2 - 71.8	CL	50	-	-	-	-	-	-

Table 4. Piles tested in this study.

Location	Co-ordinate	Pile ID	Diameter (m)	Length (m)	Pile type
Pier 1	N-2604059.006m E-264739.936 m	P1	1.5	71.8	cast in place
Pier 15	N-2604119.910m E-265956.400 m	P15	1.5	70.1	cast in place
Pier 16	N-2604120.696m E-266038.646 m	P16	1.5	71.4	cast in place
Abutment 1	N-2604023.240m E-264701.230 m	A1	1.5	57.5	cast in place
Abutment 2	N-2604103.123m E-266104.697 m	A2	1.5	46.5	cast in place
Pier 5	N-2604076.790m E-265088.470 m	P5	1	45	Driven, steel

3. Dynamic pile load test

Dynamic load test was carried out with two pairs bolt-on strain and acceleration transducers (sensors) attached to diagonally opposite sides of pile at some distance (1.5 times pile diameter) below the pile head (figure 1). This high strain dynamic pile test conforms ASTM D4945 standard. For connecting sensors to the pile, anchor bolts and welded mounting block were used for concrete and steel piles, respectively. The compressive stress wave that was generated by the strike of driving hammer, travels down the piles and reflects upward from the pile toe. The reflected stress waves are picked up by the transducers and the signals are stored in the computer. The overall arrangement of dynamic load test is shown in figure 2.

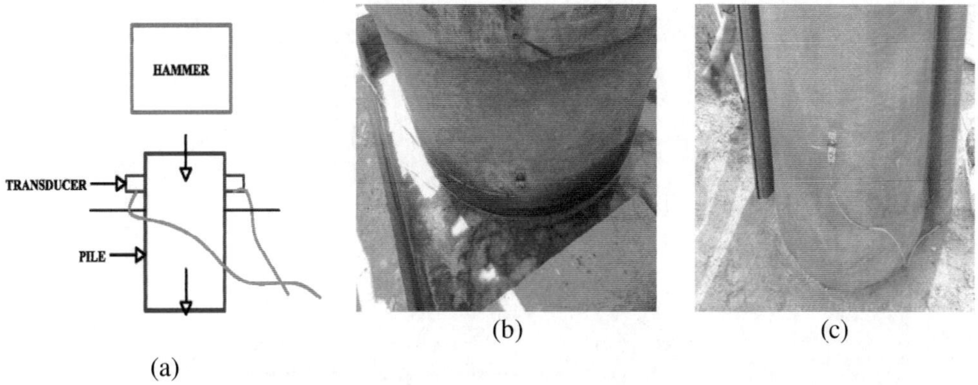

Figure 1. Transducers mounted on pile surface: (a) schematic diagram, (b) cast-in-place pile and (c) steel pile.

While striking pile by falling a heavy hammer from a pre-determined height, the strain transducers attached to the pile measure induced strains, whereas accelerometers record the accelerations generated in the pile. The Pile Driving Analyzer converts strain into force, and acceleration records into velocities. Both force and velocity x impedence signals with time were recorded. The resistance developed by the pile is then a function of force and velocity, and it includes few factors such as the quake and damping parameters as inputs based on the soil type. The maximum pile top compression is obtained by integrating the pile top velocity. A more accurate value of these parameters is then obtained from software analysis conducted on field data.

Figure 2. Overall arrangement of dynamic load test.

Further the analysis is carried out in DLTWAVE (version 8.173), a signal matching program of Profound BV. One-dimensional wave equation theory is applied to model the pile-soil interaction. The signal matching process utilizes an iterative method, in which the results of each analysis are compared to the actual measured pile behaviour (figure 3) employing wave equation theory. The downward and upward forces are expressed by eq 1 and eq 2, respectively. At a pile section, force is the unbalanced portion of upward and downward forces.

$$F^{\downarrow} = (F + Z.v)/2 \quad (1)$$
$$F^{\uparrow} = (F - Z.v)/2 \quad (2)$$

where, Z = impedence and v = velocity

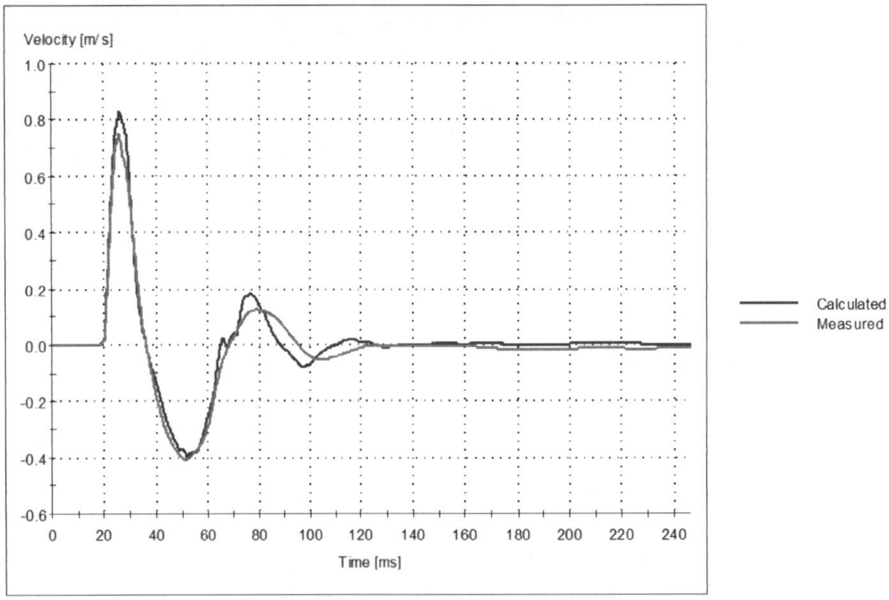

Figure 3. Signal matching in modelling pile-soil interaction using DLTWAVE.

Reliability of DLT results in determining static capacity is checked by field calibration and comparing the value of static capacity obtained from DLT with that from static load test. Customarily, both static pile load test and DLT are carried out on the same test pile, while DLT is conducted on the same pile after a setting time after conducting static load test; 7 days and 15 days for pile in sand and clay, respectively. This study finds that static capacity obtained from DLT is found less than that from static load test, when both the tests are conducted on the same pile. Therefore, for field calibration, these two tests were conducted on two identical piles at the same pier location.

4. Results and analysis

The pile tips of four (out of five piles) cast in place concrete piles, tested in this study, rest on cohesive soil bed. The undrained cohesion of clay soil, existing below the pile tip, was determined in the laboratory and the correlation between c_u and N_{60} was observed, as shown in figure 4. It can be noted that undrained cohesion is found greater for lean clay with significant portion of fine sand than that obtained for lean clay only, though the N_{60} is the same for both of them.

In this study, the results of dynamic load test were calibrated by comparing the static capacity obtained from static and dynamic load tests, conducted on two identical cast in place piles (1.5 m in diameter and 45.1 m in length) at location of Abutment 1. Both the tests gave static capacity of pile within acceptable range: 1336 ton from static load test and 1006 ton from DLT.

In the prediction of pile tip resistance in low plastic clay, both DLT and undrained cohesion (estimated from SPT N_{60}) in the formula of static analysis follow the same trend-line with respect to N_{60} (shown in figure 5) for the test site near the Gumti river. While analyzing the influence of soil below the pile tip, clay soil was found to produce greater tip resistance than clayey sand though the value of N_{60} remained the same (figure 6).

The skin frictional resistance predicted from the formula of static analysis and DLTWAVE signal matching, is compared in table 5. It can be noted that skin friction during re-drive test is found greater that that obtained from initial drive. When the greater portion of pile length is in contact with clay soil (e.g., CL in present investigation) and its consistency is 'very stiff to hard', skin frictional resistance is found significantly greater than those obtained from the formula of static analysis. This indicates that much less skin frictional resistance was developed at the contact between 'very stiff to hard' clay and the pile surface than expected from the formula of static analysis.

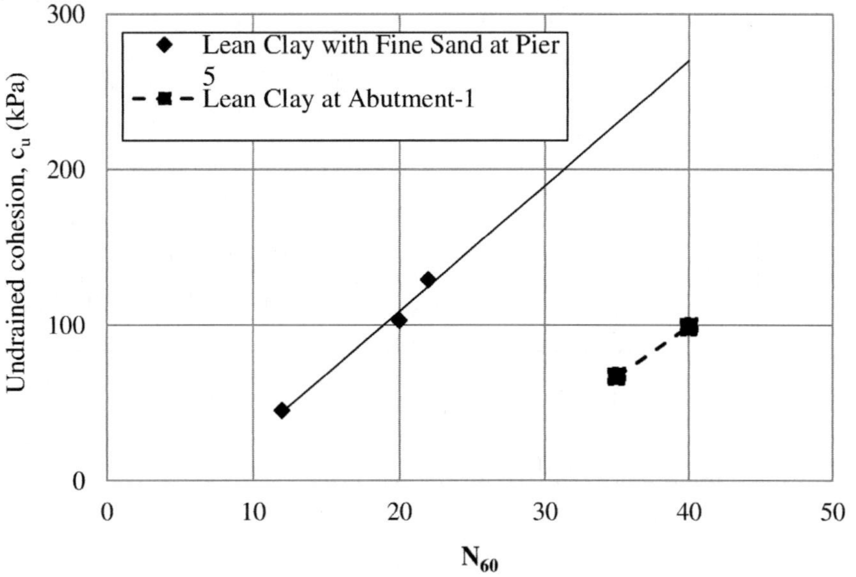

Figure 4. Undrained cohesion versus N_{60}.

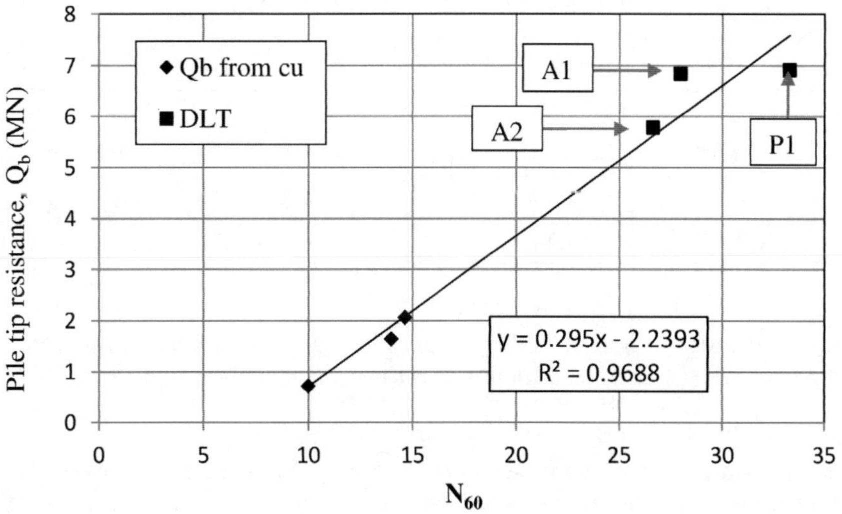

Figure 5. Pile tip resistance versus N60 [Tip of concrete piles on clay; D = 1.5 m].

IOP Conf. Series: Materials Science and Engineering **513** (2019) 012017 doi:10.1088/1757-899X/513/1/012017

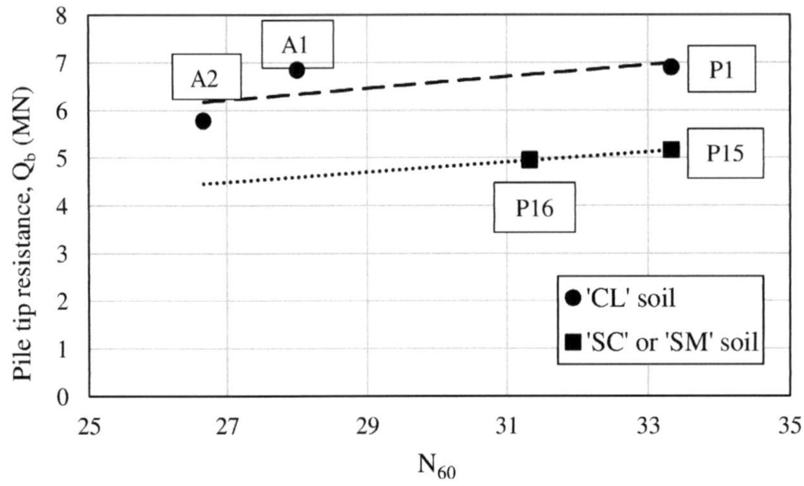

Figure 6. Pile tip resistance versus N_{60} [D = 1.5 m].

Table 5: Comparison of results: Skin frictional resistances.

Pile ID	Q_s from formula of static analysis	Q_s from DLT (Initial Drive))	Q_s from DLT (Re- Drive))	D	Pile length in contact with clay		Consistency of 'CL' soil
	MN	MN	MN	m	m	%	
P5	1.93	1.2	2.5	1	24	53.3	Stiff to very stiff
A2	7.5	7.7	-	1.5	13	27.9	
P1	7.3	4.3	-	1.5	41	57.1	Very stiff to hard
A1	5.9	3.2	-		22	38.3	

For the piles in contact with 'stiff to very stiff' clay (P5 and A2), skin frictional resistances from static analysis and DLT are in good agreement. For the piles in contact with 'very stiff to hard' clay (P1 and A1), static analysis derives greater skin frictional resistance than that obtained from DLT.

5. Conclusion

Based on the results of geotechnical investigation (such as SPT), undrained cohesion determined from laboratory investigation on undisturbed specimens and Dynamic Load Test (DLT), the following conclusion may be drawn:

1. Both DLT and the formula of static analysis gave a unique trend-line with respect to N_{60} (shown in Figure 3) for the test site near the Gumti river.
2. Clay soil was found to produce greater tip resistance than clayey sand though the value of N_{60} remained the same.
3. The greater the pile surface is in contact with very stiff to hard clay, the smaller the skin friction developed according DLT signal matching. This observation is found less pronounced for 'stiff to very stiff' consistency of clay soil than that for 'very stiff to hard' consistency.
4. Skin frictional resistance in re-drive test is found greater than that obtained from DLT during initial drive.

More DLTs in re-drive are needed to be carried out to investigate the variation in pile skin frictional resistance when significant portion of pile length is in contact with 'very stiff to hard' clay.

References

[1] Rausche, F., Likins, G., & Hussein, M. H. (2008). Analysis of post-installation dynamic load test data for capacity evaluation of deep foundations. *From Research to Practice in Geotechnical Engineering* pp 312-30

[2] Hussein, M. H., Woerner, II, W. A., Sharp, M., & Hwang, C. (2006). Pile driveability and bearing capacity in high-rebound soils. *GeoCongress 2006: Geotechnical Engineering in the Information Technology Age* pp 1-4

[3] Rausche, F., Goble, G. G., & Likins Jr, G. E. (1985). Dynamic determination of pile capacity. *Journal of Geotechnical Engineering*, **111**(3), 367-383

[4] Rausche, F., Likins, G., Liang, L., & Hussein, M. (2010). Static and dynamic models for CAPWAP signal matching. *Art of foundation engineering practice* pp 534-53

[5] Abu-Farsakh, M. Y., Haque, M. N., & Tsai, C. (2017). A full-scale field study for performance evaluation of axially loaded large-diameter cylinder piles with pipe piles and PSC piles. *Acta Geotechnica*, **12**(4), pp 753-72

[6] Haque, M. N., Chen, Q., Abu-Farsakh, M., & Tsai, C. (2014). Effects of pile size on set-up behavior of cohesive soils. *Geo-Congress 2014: Geo-characterization and Modeling for Sustainability* pp 1743-49

[7] Haque, M. N., Abu-Farsakh, M. Y., Tsai, C., & Zhang, Z. (2016). Load-Testing Program to Evaluate Pile-Setup Behavior for Individual Soil Layers and Correlation of Setup with Soil Properties. *Journal of Geotechnical and Geoenvironmental Engineering*, **143**(4), 04016109

[8] Zhussupbekov, A.Z., Syrlybaev, M.K., Lukpanov, R.E., & Omarov, A.R. (2016). The applications of dynamic and static piling tests of Astana. The 15th Asian Regional Conference on Soil Mechanics and Geotechnical Engineering, *Japanese Geotechnical Society Special Publication*, **2**(79), pp 2726-29

[9] Beim, J., & Luna, S. C. (2012). Results of dynamic and static load tests on helical piles in the varved clay of Massachusetts. *The Journal of the Deep Foundations Institute*, **6**(1), pp 58-67

[10] Noor, S.T., Hanna, A.M. and Mashour, I. (2013). Numerical Modeling of Pile in Collapsible Soil subjected to Inundation. *Journal of Geomechanics*, **13**(4) pp. 514-26

IOP Conf. Series: Materials Science and Engineering **513** (2019) 012018 doi:10.1088/1757-899X/513/1/012018

Optimal sensor distance for damage detection considering wavelet sensitivity and uncertainties

Muyideen Abdulkareem[1*], Mujedu Kasali[2,3], Esan Martins[2,4], Olukotun Nathaniel[2,5] and Muhd Z. Abd Majid[1]

[1] Construction Research Centre, Universiti Teknologi Malaysia, Johor, Malaysia
[2] School of Civil Engineering, Universiti Teknologi Malaysia, Johor, Malaysia
[3] Department of Civil Engineering Federal Polytechnic, Ede, Osun State, Nigeria
[4] Department of Building Technology, Federal Polytechnic, Ede, Osun State, Nigeria
[5] Department of Building Technology, Kogi State Polytechnic, Kogi State, Nigeria

[*]Corresponding author: muyikareem@gmail.com

Abstract. Wavelet Transform (WT) is an efficient signal processing tool used extensively to detect damage in various types of structures. It is capable of describing signals in both time and frequency domains. However, improper sensor distance may result to false detection of damage, thus affecting the reliability of the method. A small sensor distance leads to large number of sensors, thus increases the financial cost and causes higher computational time, while large sensor distance may provide inadequate data for accurate damage detection. In this paper, the optimum sensor placement for damage detection is investigated by considering the effect of noise and sensitivity of WT. This involves using Continuous Wavelet Transform (CWT) to decompose the mode shape difference of a plate numerical model. Different measurement distances and different levels of uncertainties are added to the mode shape data to evaluate the optimum sensor distance that provide the best damage detection result. A numerical plate model with all four sides fixed is used as an example. The results indicate that increase of noise reduced the detectability of damage. It is also observed that excessive sensor distance increment significantly effects damage detectability.

1. Introduction

Vibration-based damage detection techniques that are applied to assess the structural condition are referred to as a global method of damage identification. These techniques are cheaper, faster and can be easily applied to large structures. The basic principle of these techniques in detecting damage in structures is based on the alteration of the physical properties (mass, stiffness and damping), dynamic properties (natural frequencies, mode shapes, damping ratios and frequency response function) [1] and dynamic responses (acceleration, velocity and displacement) [2]. Several methods have been developed to analyse these data for accurate detection, location and severity of damage; which includes modal data [3], frequency response function (FRF) [4] and time history [5].

In addition to the above mentioned techniques, Wavelet Transform (WT), an efficient signal processing tool has been extensively applied to detect damage in various types of structures. It is capable of describing signals in both time and frequency domains. Many studies have proven that WT is reliable in detecting damage. For example, Rucka and Wilde [6] analysed the estimated mode

IOP Conf. Series: Materials Science and Engineering **513** (2019) 012018 doi:10.1088/1757-899X/513/1/012018

shapes of a cantilever beam and steel plate with 1-D and 2-D CWT respectively for damage detection in these structures. Bagheri et al. [7] proposed an effective technique for detecting linear flaws in plate structures using 2-D Discrete Wavelet Transform (DWT) to decompose the mode shape signals. Ovanesova and Suarez [8] applied wavelet transform to detect cracks in plane frame structures using response signals from static and dynamic loading of the frames. Recently, Katunin and Przystalka [9] assessed damage in a composite structure using fractional wavelet transform. Also, Vafaei et al. [10] localised damage in a beam by decomposition of the mode shape using DWT. These authors have shown that with measurement data, damage can be identified in structures with WT.

In practice, there are two types of noise which may lead to false damage detection results. Modelling errors from inaccurate Finite Element (FE) parameters may result into the initial FE model not accurately representing the actual condition of the real structure. This type of noise is always the major issue in many methods that requires initial model to construct the relationship between the vibration parameters and damage detectability, such as artificial neural network (ANN) [11], Genetic Algorithm (GA) [12] and model updating [13]. On the other hand, the existence of measurement noise in the measured data due to environmental effects, electronic devices, measurement errors and etc., may also lead to false damage detection. Several studies have investigated the effect of noise in damage detection based on vibration parameters. For example, Chandrashekhar and Ganguli [14] investigated the effect of noise in assessing structural damage with geometric and measurement uncertainties by mode shape curvatures. The authors employed a Fuzzy Logic System (FLS) to analyse a cantilever beam with single and multiple damage. The FLS identified multiple damages with 94% average accuracy when the measurement noise was 10%. Garcia-Gonzalez et al. [15] checked the capacity of ANN for noise filtering using natural frequencies as inputs in damage localisation in a beam structure. Bandara et al. [16] proposed a damage identification method to identify damage using noise polluted Frequency Response Function (FRF) which involved the calculation of FRF and damage index, development of an ANN and introduction of damage indices into the ANN method for damage detection. Aydin and Kisi [17] studied the effect of noise with ANN using frequencies, mode shape rotation and other beam properties as input to detect cracks in a beam. Most of the studies that investigated the effect of noise in damage detection have been using modal data as the main parameter to detect damage. However, with wide application of WT for damage detection, studies on noise effects are quite limited. Thus, it is very important to study the effect of noise in measured data of structures to wavelet ability to detect damage.

The distance between the sensors may also influence the degree of accuracy of damage identification. When the number of sensors is small, the distance between sensor are larger, thus lead to less accurate estimation of the structural damage, while a large number increases the financial cost of the monitoring system and causes higher computational time. Researchers have come up with several proposed methods for sensor placement using different damage detection methods. Li et al. [18] determined the optimal sensor location for structural vibration measurements using the uniform design method. This method appears to be simple, however, its application is limited to simple structures and not so accurate for complex structure. Xie and Xue [19] applied a hybrid algorithm for optimal number of sensors and placement by using improved reduced system and singular value decomposition. Yi et al. [20] proposed an optimal sensor location by reducing the maximum value of the off-diagonal elements contained in the modal assurance criterion matrix and presented a hybrid method for sensor placement based on multiple optimization methods respectively. In both studies, the methods presented are too complex and requires much computational efforts. Recently, Alem and Benazzouz [21] presented a structural adjacent matrix method of optimal sensor placement for fault detection by finding the vertex-disjoint paths. A detailed review on sensor placement can be found in Yi and Li [22], Li et al. [23] and Li et al. [24]. However, study on optimum sensor distance focusing on WT method is limited.

Therefore, this study investigates the optimum sensor placement for damage detection by considering the sensitivity of CWT and uncertainties (noise) in a plate-like structure. For the purpose of damage detection, the mode shape difference before and after damage served as the input rather than conventional mode shape of the damaged structure [25]. The vibration data obtained from the numerical analysis of the plate structure is used in computing the mode shape difference. To detect and locate the damage, the mode shape differences are then decomposed using CWT. The effect of noise and sensors' distances to CWT in detection of different degrees of damage severity in a plate structure is analysed in this study. To study the effect of noise to damage detectability, various levels of noise are added to the values of the mode shape differences. The effect of the sensor distance is investigated by varying the distance between the measurement nodes. The effects of noise and sensor distance are evaluated based on damage detectability by applying a damage index and pictorial representation of the decomposed signal.

2. Methodology

2.1. Continuous wavelet transform

A wavelet is an oscillation with amplitude which originates at zero, increases progressively, and then decreases back to zero. The function $\psi(x)$, which is the mother wavelet, is localised in space and frequency domain, creating a wavelet family $\psi_{u,s}(x)$.

$$\psi_{u,s}(x) = \frac{1}{\sqrt{s}}\psi\frac{(x-u)}{s} \tag{1}$$

where *s* and *u* are scale and position respectively.

The continuous wavelet transform (CWT) of a 1-Dimensional signal *f(x)* is the integral of the product of the signal function with the complex conjugate $\psi(x)$ of the wavelet function. Taking the deflection of the structure as a 1-dimensional signal *f(x)*, the CWT by Mallat [26] is

$$Wf(u,s) = \frac{1}{\sqrt{s}}\int_{-\infty}^{+\infty} f(x)\psi\left(\frac{x-u}{s}\right)dx \tag{2}$$

$Wf(u,s)$ is the wavelet coefficient of the wavelet $\psi_{u,s}(x)$.

The *n* vanishing moment of a wavelet is vital in detecting singularity of signals, which is based on the equation

$$\int_{-\infty}^{+\infty} x^k\, \psi(x)dx = 0,\ k = 0,1,2,\ldots\ldots,n-1. \tag{3}$$

Rewriting equation (2) as,

$$Wf(u,s) = \frac{1}{\sqrt{s}} f * \psi\left(\frac{-u}{s}\right) = f * \bar{\psi}_s(u) \tag{4}$$

$$\bar{\psi}_s(u) = \frac{1}{\sqrt{s}}\psi\left(\frac{-x}{s}\right) \tag{5}$$

The wavelet with *n* vanishing moment can be re-written as the *n*th order derivation of a smooth function $\theta(x)$ [6], which can be expressed as a multiscale differential operator:

$$Wf(u,s) = \frac{s^n}{\sqrt{s}}\frac{d^n}{du^n} f * \theta\left(\frac{-u}{s}\right) = s^n \frac{d^n}{du^n} f\bar{\theta}_s(u) \tag{6}$$

$f * \bar{\theta}_s$ denotes convolution of functions which is the average of *f*(x) over a domain proportional to the scale *s*.

2.2. Noise

To analyse the effect of noise, noise levels of 1%, 2%, 5% and 10% (gaussian white noise) were added to the obtained mode shape difference data. The mode shape difference data were imported from the Finite Element Analysis (FEA) program SAP2000 to Matlab software. For a particular noise level *n*%, the corresponding noise value *e* is obtained by:

$$e = 20log_{10}\left({}^{1}/_{n}\right) \tag{7}$$

where *n* is noise level. This noise level *n* is added to the data *m* through the noise value *e*. The noise value *e* for 10% noise level is 20. The values for 1%, 2% and 5% are 40, 33.9 and 26 respectively.

2.3. Sensor locations

The sensor locations are taken as the nodes in the models. The distance between these nodes are 80 mm, 40 mm, 20 mm, 10 mm and 5 mm, and this translates to distance ratio r (total distance l / measurement distance l_e) of 7, 14, 28, 56 and 112, and are represented by i_{80}, i_{40}, i_{20}, i_{10} and i_5 respectively. This is tabulated in Table 1 and Figure 2 shows the distance between the nodes.

3. Numerical model

The mode shape of the plate is computed using the commercial Finite Element Analysis (FEA) program SAP2000 and the wavelet analysis was done with Wavelet Toolbox on Matlab software. The steel plate is square with dimensions: 560mm, 560mm and 2mm for length l, width b and thickness h. Material properties are: Poisson's ratio ν = 0.3, Young Modulus E = 200Gpa and steel density ρ = 7850kg/m^3. Boundary condition at all four sides of the plate is fixed ends. Damage is assumed to be caused by rust effect and this is simulated by reducing the thickness of the plate at the specified damage area. Two damage locations are considered: middle and side damages of the plate. These points represent near and far from support damage conditions. Figure 1 shows a diagram of the depiction the positions of the damages. The damages are rectangular shaped with dimension of 80mm x 80mm which is approximately 2.041% of the plate's area. The damages cases are simulated by reducing the thickness of the plate at the damage location, the reduced thickness ranged from 1.90mm to 0.25mm (see Figure 2). These cases are 190mm, 180mm, 175mm, 150mm, 125mm, 100mm, 75mm, 50mm and 25mm which are equivalent to 5%, 10%, 12.5%, 25%, 37.5%, 50%, 62.5%, 75% and 87% thickness reduction respectively. These damage cases are tabulated in Table 2. A total of 360 models were used for the accuracy and noise sensitivity analyses.

Table1. Plate distance ratio

Measurement distance (mm) i	80	40	20	10	5
Distance ratio (MDR) r	7	14	28	56	112

Table 2. Thickness reduction of plate

Case	1	2	3	4	5	6	7	8	9
Thickness reduction (%)	5	10	12.5	25	37.5	50	62.5	75	87.5
Plate thickness (mm)	1.90	1.80	1.75	1.50	1.25	1.00	0.75	0.50	0.25

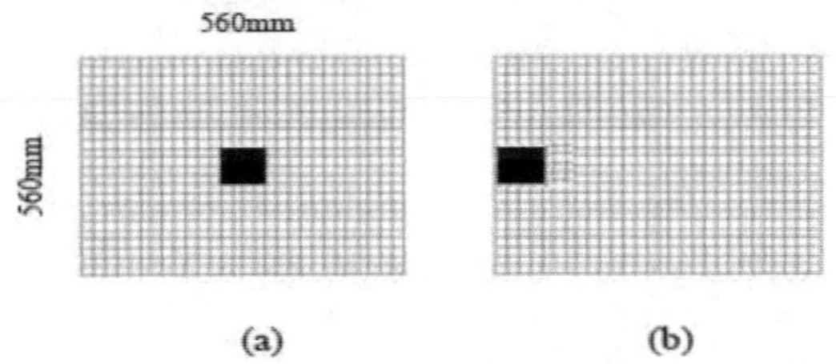

Figure 1. Damage location: (a) middle; (b) side

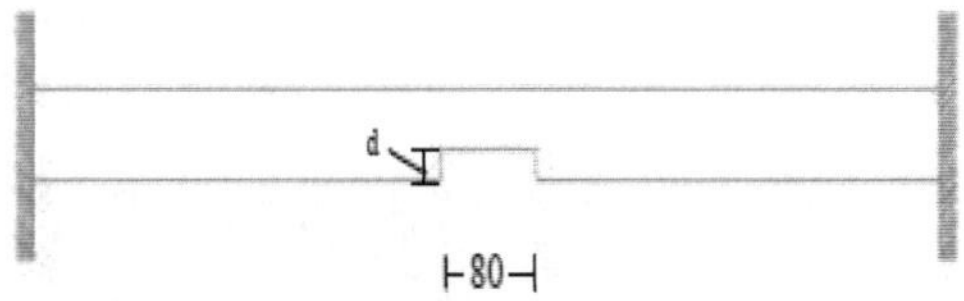

Figure 2. Thickness reduction in plate

4. Results

The results presented in this section are noise free. Damages are identified from the pictorial representation of the decomposed signal. This is based on the sensitivity of WT to sudden and abrupt change in the decomposed mode shape differences. Damages were not detectable at i_{80}, that is, measurements at 80mm. This is caused by the inability to capture adequate data for proper estimation of the state of the plate due to the large measurement distance. Figure 3 is a display of the decomposed signal of the plate with case 9 damage (middle and side) at i_{80}. Typical examples of detected damages are shown in Figure 4. The bar charts in figures 5 to 8 show the accuracy of the side and middle damages detected in the plate. The dark and lighter bars represent the accuracies of the x and y coordinates of the detected damage location, respectively.

(a) (b)

Figure 3. Undetected Case 9 Damage (i_{80}): (a) middle damage; (b) side damage.

(a) (b)

Figure 4. Detected Damages: (a) middle damage; (b) side damage.

In Figure 5, the results show damage detectability at i_{80}. Figure 5(a) shows the accuracy of middle detected damages. Damage of 5% thickness loss was not detected in this case. The damage was not captured in the decomposed signal to effect the detection of the damage because the severity was small (5%). Damages which were above 10% thickness loss were accurately detected, almost 100% accurate. Figure 5(b) shows the side damage accuracy. In this scenario, damages of 5% and above thickness loss were accurately detected. The detection of damage (side) of 5% thickness loss may be a result of its close proximity to the support. Considering the bar charts (Figures 5 to 8), damage is either detected with a high degree of accuracy or it is undetected. The detected middle damages are of higher accuracy. The detectability of middle damage is reduced as the measurement distance is increased i.e. for 10% thickness loss, middle damage is undetectable at i_{40}. This can be attributed to low damage severity and inability to capture the damage signal due to increase of the measurement distance. This is also noticed in the side damage as well, as the side damage accuracy decreases as i increases. From the above results, by using this method, both side and middle damages can be accurately detected in a noise free signal.

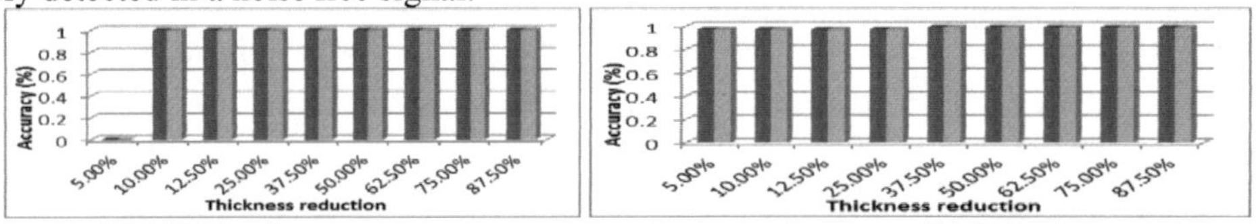

(a) Middle damage (b) Side damage

Figure 5. Measurement at 5mm (i_5)

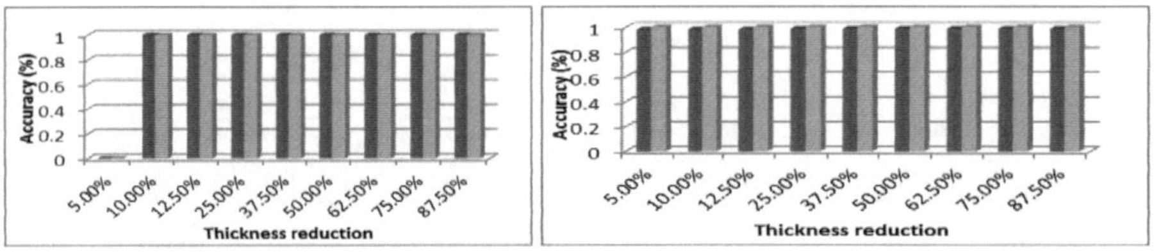

(a) Middle damage (b) Side damage

Figure 6. Measurement at 10mm (i_{10})

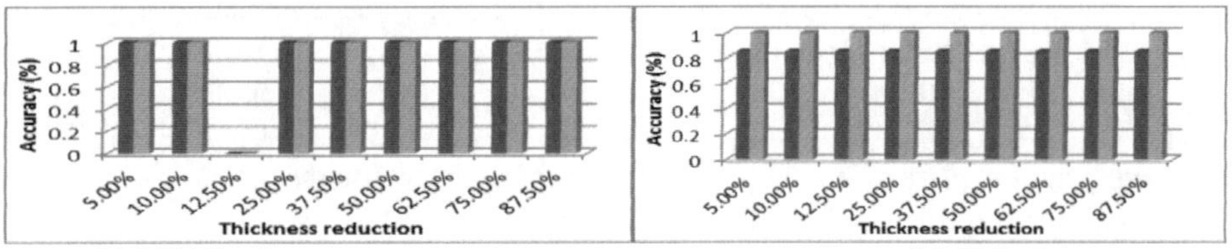

(a) Middle damage (b) Side damage

Figure 7. Measurement at 20mm (i_{20})

5. Sensitivity to noise effect

The effects of noise in experimental and field experiments cannot be neglected. This is modelling error which is ascribed as the imperfection of numerical models to simulate real structures. In order to put these effects into consideration, various levels of noise were added to the numerical results i.e. 1%, 2%, 5% and 10% noise levels as described in the methodology section above. The Figures 9 – 12 represent the noise effects (sensitivity) for i_{40}, i_{20}, i_{10} and i_5. The figures show the damage detection sensitivity to noise, i_x and severity of damage

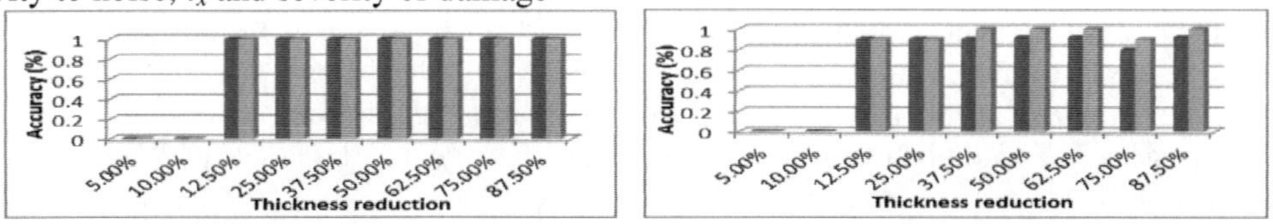

(a) Middle damage (b) Side damage

Figure 8. Measurement at 40mm (i_{40}).

Figure 9 summarises the result (middle and side damages) for i_5 with noise level varying from 1% to 10%, and damage severity from 5% to 87.5%. When the noise level increases (and damage severity decreases), it is observed that the sensitivity to damage decreases. Take Figure 9(a) for example (middle damage), when noise is increased (to 2%, 5% or 10%), damage is not detectable until the severity is up to 25%. Also, when the damage severity is 12.5%, damage is detectable when the noise is reduced to 1%. The reason for this is that the sensitive damage features in the data have been submerged by noise. It is also observed that the accuracy of the detectability is high, close to 100%, and on the other hand, the non-detected damages are completely undetectable. Figure 9(b) shows the sensitivity of side damage for i_5. The side damage can be detected at 5% thickness loss, which was undetected in the middle damage. This could be as a result of the boundary effects. Figures 10 – 12 summarises the sensitivity i_{10}, i_{20} and i_{40}, respectively. It is observed that for i_5 and i_{20}, the low intensity side damages are detectable even in the presence of 15% noise. Except for i_{20}, light damages imposed at the middle of plate are not detectable even when the noise level was minimal. Results also indicate that even severe damages may not be detected when the noise level is high and measurements are performed with larger distances. In short, it can be concluded that, measurement distances and noise levels have a great impact on the detectability of imposed damages. However, a direct relationship cannot be found between change in the noise level or measurement distance and detectability of damages. This may be related to the pattern at which the noise is being added to the data. Since the noise is added randomly to the mode shape differences data by considering the average, it is possible to have a distortion in the signal produced as a result of the addition of the noise. This distortion may aid or hinder damage detection.

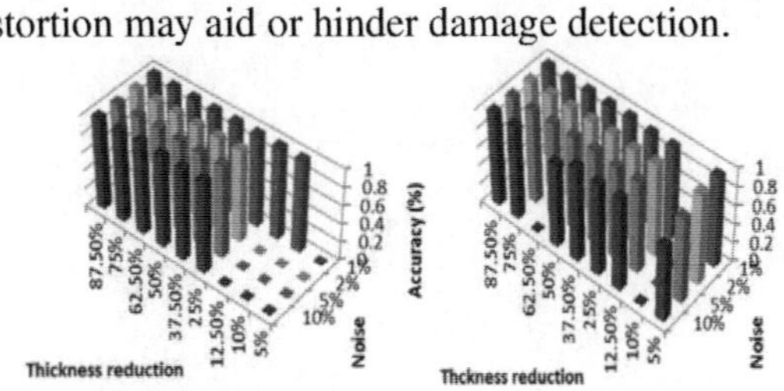

(a) Middle damage (b) Side damage

Figure 9. Measurement at 5mm (i_5).

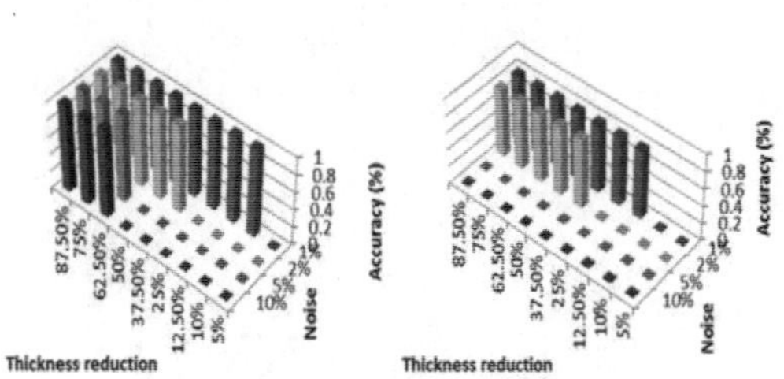

(a) Middle damage (b) Side damage

Figure 10. Measurement at 10mm (i_{10}).

6. Parametric study

This section provides a more detailed sensitivity investigation of damage detection using WT. This analysis determines the sensitivity of this method to different values of i_x, damage severity (thickness loss) and noise (noise levels considered are 1%, 2%, 5% and 10%). Figures 13 to 16 summarize the damage detectability level of the plate in this study. The vertical lines correspond to the detectability limit for different levels of damage severities. The numbers in the rectangular boxes represent the i_x value of each limit. The area on the left side of each line represents the area where damage is undetectable under those conditions. The area on the right side of each of these vertical lines represents the area where damage is detectable under those conditions. Figure 13(a) shows the middle damage detectability level at 1% noise. This figure shows that at i_5 and noise level at 1%, the minimum severity of damage (thickness loss) that can be detected is 10%. This implies that middle damage which is less than 10% thickness reduction (say 5%) cannot be detected with i_5 in the presence of 1% noise. On the other hand, when middle damage is 10% thickness loss (or more), an i_x value of i_5 is capable of detecting this damage provided the noise level is not beyond 1%. This is because the damage signals captured overshadows the effect of the noise. Figure 13(b) represents the side (near support) damage case, and Figures 14 – 16 show the damage detectability at 2%, 5% and 10% respectively.

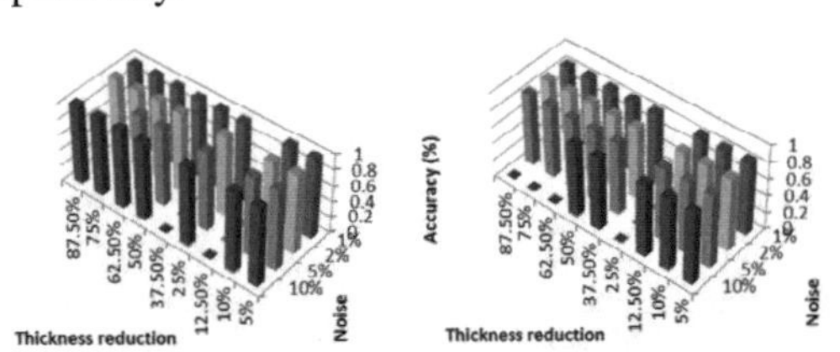

(a) Middle damage (b) Side damage

Figure 11. Measurement at 20mm (i_{20}).

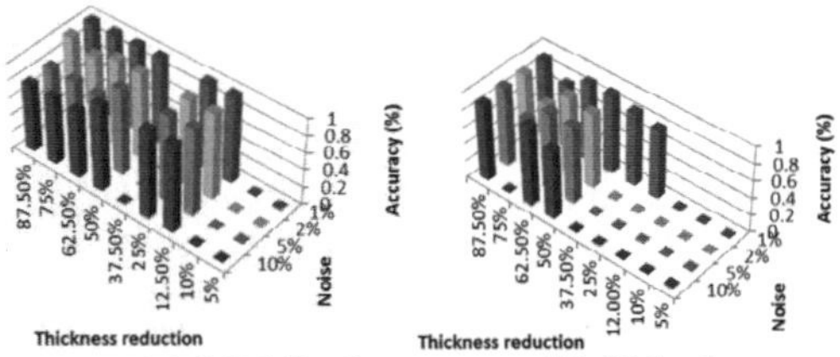

(a) Middle damage (b) Side damage

Figure 12. Measurement at 40mm (i_{40}).

It is observed in the figures that in most cases, the detectability level decreases as the noise increases, this can be seen in both damage locations (far and near support). For example, at 1% noise level, the detectable damage for i_{10} is 10% thickness reduction, whereas for 2% and 5% noise levels are 25% thickness reduction. A closer observation reveals that at low noise level (1%), middle damage can be detected with lower severity (10% thickness reduction) compared to damages near support (12.5%). This middle damage detectability decreases (25% thickness reduction) as the noise increases (25, 5%, and 10%) while the detectability of side damage remains at 12.5% even as the noise is increased to 10%. These results show that damage detectability depends on the damage severity, noise level and i_x value.

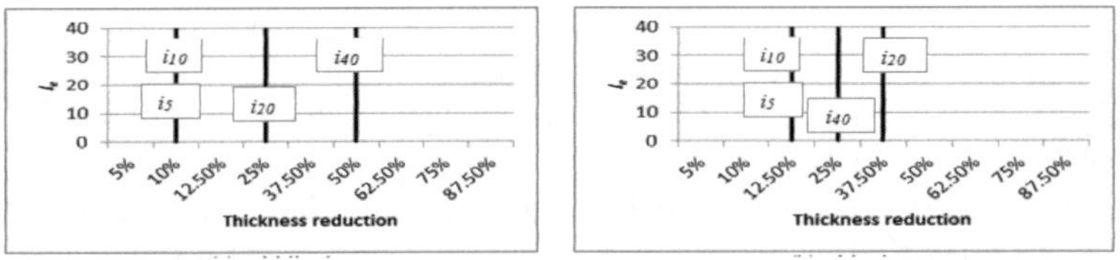

(a) Middle damage (b) Side damage

Figure 13. Damage detectability at 1% noise.

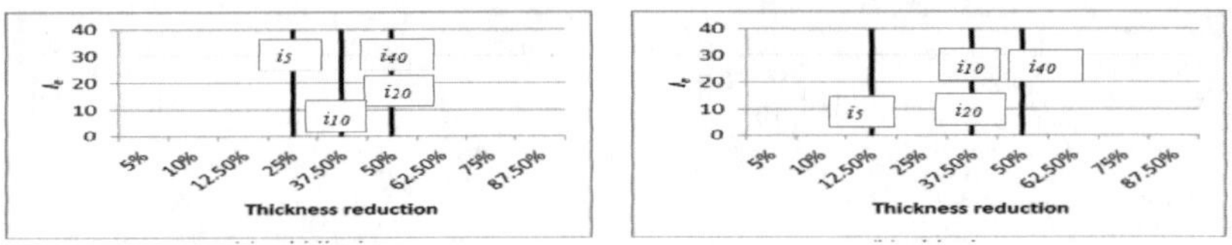

(a) Middle damage (b) Side damage

Figure 14. Damage detectability at 2% noise.

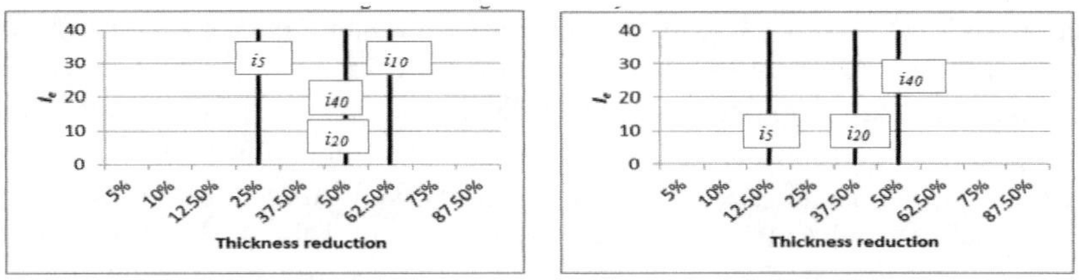

(a) Middle damage (b) Side damage

Figure 15. Damage detectability at 5% noise.

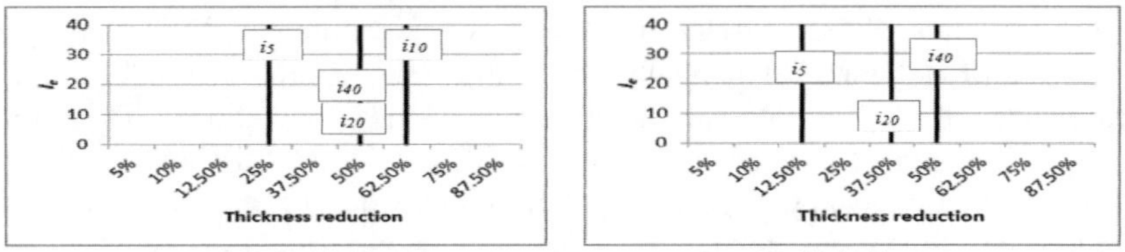

(a) Middle damage (b) Side damage

Figure 16. Damage detectability at 10% noise.

7. Conclusion

The study presents an investigation on the Wavelet Transform (WT) sensitivity and attributed uncertainties for damage detection of plate-type structures using mode shape difference. The results have shown the sensitivity of WT to damage and its ability to accurately detect damage in the presence of noise. It is observed that damage is either detected accurately or it's not detected at all. For all studied cases, damage was not detectable when the measurement distance was 80mm. In most cases, the presence of noise, increment of measurement distance and decrease of damage severity reduced the detectability of damage. Measurements at i_5, i_{10}, i_{20} and i_{40} are adequate for mode shape difference analysing with WT for damage detection.

References

[1] Sampaio R P , Maia N M and Silva J M 2003 The frequency domain assurance criterion as a tool for damage detection. *Pro of Key Eng Materials: Damage Assess. of Struc.* **245** pp. 69–76.

[2] Feng D and Aymerich F 2014 Finite element modelling of damage induced by low-velocity impact on composite laminates. *Compo. Struct.* **108** pp. 161–171.

[3] Ong Z C, Rahman A G and Ismail Z 2014 Determination of damage severity on rotor shaft due to crack using damage index derive from experimental modal data. *Experi. Mech.* **38** pp. 18-30.

[4] Bandara R P, Chan T H and Thambiratnam D P 2014 Structural damage detection method using frequency response functions. *Struct. Health Moni.* **13** pp. 418–429.

[5] Zhu X Q and Law S S 2006 Wavelet-based crack identification of bridge beam from operational deflection time history. *Int'l J. of Solids and Struct.* 2006. **43** pp. 2299-17.

[6] Rucka . and Wilde K 2006 Application of continuous wavelet transform in vibration based damage detection method for beams and plates. *J. of Sound and Vib.* **297** pp. 536-550.

[7] Bagheri A, Ghodrati Amiri G, Khorasani M, Bakhshi H 2011 Structural damage identification of plate base on modal data using 2D discrete wavelet tranform. *Struct Eng Mech* **40** pp. 13-28.

[8] Ovanesova A V and Suárez L E 2004 Applications of wavelet transforms to damage detection in frame structures. *Eng. Struct.* **26** pp. 39-49.

[9] Katunin A and Przystałka P 2014 Damage assessment in composite plates using fractional wavelets transform of modal shapes with optimized selection of spatial wavelets. *Engrg Appli. of Arti. Intell.* **30** pp. 73–85.

[10] Vafaei M, Alih S C, Rahman A B and Adnan A B 2014 A wavelet-based technique for damage quantification via mode shape decomposition. *Struct. and Infra. Eng.* **11** pp. 869-883.

[11] Meruane V and Mahu J 2014 Real-time structural damage assessment using artificial neural networks and antiresonant frequencies. *Shock and Vib.*

[12] Malekzehtaba H. and Golafshani A A 2013 Damage detection in an offshore jacket platform using genetic algorithm based finite element model updating with noisy modal data. *The 2nd International Conference on Rehabilitation and Maintenance in Civil Engineering.*

[13] Aghagholizadeh M and Catbas F N 2015 A Review of Model Updating Methods for Civil Infrastructure Systems. Comp. Techni. for Civil and Struct. Engr. pp. 83-99

[14] Chandrashekhar M and Ganguli R 2009 Damage assessment of structures with uncertainty by using mode-shape and fuzzy logic. *J. of Sound and Vib.* **326** pp. 939–957.

[15] Garcia-Gonzalez A, Gonzalez-Herrera A and Garcia-Cerezo A 2010 Damage Localization based on Modal Parameters using the Finite Element Method and Neural Networks. Proceedings of the 10th International Conference on Computational Structures Technology.

[16] Bandara R P, Chan T H and Thambiratnam D P 2011 Structural Damage Identification with Noise polutted frequency response function. Proceedings of the 14th Asia-Pacific Vibration Conference, Hong Kong Polytechnic University, Hong Kong Polytechnic University.

[17] Aydin K and Kisi O 2014 Damage detection in Timoshenko beam structures by multilayer perceptron and radial basis function networks. *Neural Comput & Applic.* **24** pp. 583–597.

[18] Li Z N, Tang J and Li Q S 2004 Optimal sensor locations for structural vibration measurements. *Appl. Acoust.* **65** pp. 807-818.

[19] Xie Q and Xue S T 2006 An optimal sensor placement algorithm for structural health monitoring. Pro. of the 2nd Int'l Conference on Struc. Health Moni. Intelligent Infrastructure, China.

[20] Yi T H, Li H N and Gu M 2011 Optimal sensor placement for structural health monitoring based on multiple optimization strategies. *Struct. Des. of Tall & Special Bdgs* **20** pp. 881-900.

[21] Saiuml A L and Djamel B E 2013 Optimal sensor placement for fault detection and isolation by the structural adjacency matrix. *Int'l J of Phy. Sci.* **8** pp. 225-230.

[22] Yi T H and Li H N 2012 Methodology developments in sensor placement for health monitoring of civil infrastructures. *Int'l J of Distributed Sensor Networks* **8** pp. 612726..

[23] Li D S, Li H N and Fritzen C P 2007 Representative least squares method for sensor placement. Proceedings of the 3rd Int'l Conf on Struc Health Moni and Intelli Infra, Vancouver, Canada.

[24] Li D S, Li H N and Fritzen C P. 2007 The connection between effective independence and modal kinetic energy methods for sensor placement. *J. of Sound and Vib.* **305** pp. 945–955.

[25] Lee J, Lee W, Yi H, Yun B and Jung Y 2005 Neural networks-based damage detection for bridge considering error in baseline finite element model. *J. of Sound & Vib.* **280** pp. 555-78.

[26] Mallat S 1998 A Wavelet Tour of Signal Processing. Academic Press, New York.

IOP Conf. Series: Materials Science and Engineering **513** (2019) 012019 doi:10.1088/1757-899X/513/1/012019

Seismic fragility of concrete box girder bridges in Malaysia

Aqilah Ghazali[1], Hasan Al-Haris Alaydrus[1], Sophia C. Alih[2] and Mohammadreza Vafaei[1*]

[1]Faculty of Engineering, School of Civil Engineering, Universiti Teknologi Malaysia, Johor, Malaysia
[2]Faculty of Engineering, School of Civil Engineering, Institute of Noise and Vibration, Universiti Teknologi Malayisa, Johor Malaysia

[*]Corresponding author: vafaei@utm.my

Abstract. In Malaysia, the majority of structures and infrastructures have not been designed for seismic actions. Recent earthquakes in the country, however, has raised the attention of authorities toward the seismic vulnerability of these structures. Bridges are among the crucial infrastructures that their functionality during ground motions is of great importance. In this study, three bridges with different pier heights of 10 m, 20 m, and 30 m were selected and their nonlinear response was determined through pushover and incremental dynamic analysis. By using 15 far-field earthquake records the seismic fragility curves of the bridges were derived. Peak ground acceleration was considered as the ground motion intensity measure and the drift ratio was selected as the engineering demand parameter. The obtained fragility curves revealed different patterns for the probability of exceeding light to severe damage for the studied bridges. While the difference between the probabilities of exceeding light and severe damage in the shortest bridge was significant, in the tallest bridge it was negligible. The bridge with the pier height of 20 m had the largest probability of exceeding light to severe damage when compared to other bridges. The tallest bridge, on the other hand, showed the lowest seismic vulnerability when compared with other bridges. Comparison between Malaysia national annex to Eurocode 8 and the obtained fragility curves revealed that for the estimated PGA in Kuala Lumpur, all studied bridges could remain functional.

1. Introduction

Seismic performance of bridges can be affected by various uncertain parameters, therefore, a probabilistic assessment approach is required to address their vulnerability against seismic actions [1]. Seismic fragility curves have been widely employed as an efficient tool for probabilistic seismic risk assessment of structures [2]. Fragility curves can be obtained by empirical, analytical and hybrid methods[3]. However, a review of literature shows that the analytically obtained fragility curves have been frequently used for the seismic vulnerability assessment of bridges and other type of structures[4-12]. The fragility curves enable national authorities to prepare retrofitting scheme, seismic risk map and emergency measures. It is worth noting that experience from past earthquakes have shown that many bridges have suffered substantial damages during ground motions[13,14]. It should be also mentioned that seismic induce damage to bridges can impede emergency response operations during earthquakes. Moreover, the repair cost of damaged bridges are often high and time consuming.

IOP Conf. Series: Materials Science and Engineering **513** (2019) 012019 doi:10.1088/1757-899X/513/1/012019

Most of the bridges constructed in Malaysia have not been designed for seismic actions because seismic design has not been compulsory in the country and no local seismic code has been established for seismic design of structures. During past decades, seismic activity in Malaysia has been considered to be very low mostly because of being far from strong active faults and the lack of historical data that shows seismic damage to structures. However, in 2015, an earthquake with the magnitude of 6.0 hit Ranau in Sabah and imposed significant damage to the structures. The Ranau earthquake raised the attention of authorities toward the vulnerability of existing structures and infrastructures. However, seismic vulnerability studies on buildings and bridges are limited in Malaysia and existing studies have mostly concentrated on buildings[15].

In this study, three concrete box girder bridges were selected from the current practice in Malaysia for the seismic vulnerability study. These three bridges have different pier heights of 10 m, 20 m, and 30 m but they share a similar concert deck. The finite element models of the concrete box girder bridges were established using CSI Bridge software [16]. Pushover analysis which has been widely used for seismic assessment of structures [17,18]was employed to estimate the drift capacities of the bridges. For the development of seismic fragility curves, incremental dynamic analysis (IDA) was applied on 15 far-field earthquake records. Three structural performance levels of immediate occupancy (IO), life safety (LS), and collapse prevention (CP) recommended by FEMA 356 [19] were employed as the limit states in the development of fragility curves.

2. Selected structures

The selected bridges in this study are similar to those used in the line 2 of the Mass Rapid Transit (MRT) project in Kuala Lumpur, Malaysia. Figure 1 shows a sample of the 3D finite element model of bridges which is established in CSI Bridge software[16]. All simulated bridges have five similar spans with the length of 39.8 m. As it is shown in figure 2, the width of the bridge deck in all investigated bridges is 9.8 m and its thickness is 0.24 m. The deck of bridges is not continuous and sits on two rubber bearings at both ends of each span. The deck is post-tensioned by 16 tendons each of which having the jacking force of 4082 kN. The yield stress, ultimate stress, and the modulus of elasticity of tendons are 169 MPa, 186 MPa and 196 MPa, respectively. The yield and ultimate stress of reinforcing bars in piers and the bridge deck are 485 MPa and 600 MPa, respectively. The compressive strength of concrete for all of the structural element in the bridges is 40 MPa. As it is shown in figure 3 the cross section of piers is tapered along their height such that the bottom and upper parts are larger than the middle part. In the finite element models, piers are divide into six sections along their height. Table 1 shows the size of cross section and reinforcing bars of each section for the pier with the total height of 30 m. In all piers a reinforcement ratio of 2.5% was considered. The effective stiffness of structural elements were calculated based on the recommendation of ASCE 41-13 [20]as shown in Table 2. It should be mentioned that, in the nonlinear analysis, in order to consider the inelastic behavior of piers the lumped plasticity model that has been widely used by other researchers [21,22]is adopted. Plastic hinges are assigned to the both end of each section of piers. The modeling parameters of plastic hinges and their corresponding acceptance criteria are determined based on the recommendation of ASCE 41-13[20]. The deck of the bridge was simulated using shell elements. The deck was meshed using the auto mesh option of the software. The deck to pier connection was simulated using link elements suggested by the software. The piers connection to the ground was assumed to be fixed. In order to validate finite element models the fundamental natural frequencies of the simulated bridges were compared with those obtained from analytical calculations. As can be seen from table 3 the obtained values from both methods are close to each other indicating a good agreement between the finite element models and analytical results.

IOP Conf. Series: Materials Science and Engineering **513** (2019) 012019 doi:10.1088/1757-899X/513/1/012019

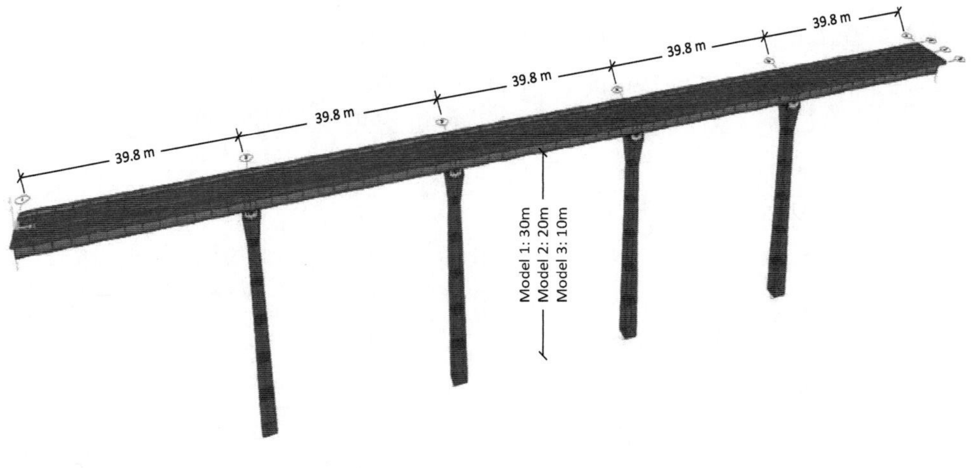

Figure 1. Finite element model of a concrete box girder bridge.

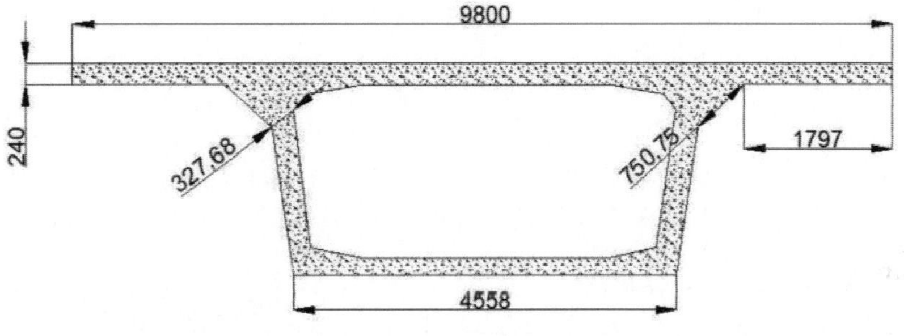

Figure 2. Cross section of the bridges' deck (all dimension in mm).

Figure 3. Pier with six sections and the height of 30 m simulated in the software.

Table 1. Dimension and reinforcing bars inside the 30 m height pier.

Section	Length (mm)	Width (mm)	No. of 50 mm bar in 3-dir.	No. of 50 mm bar in 2-dir.
Section 1	2800	3000	26	28
Section 2	2640	2840	23	25
Section 3	2480	2680	21	22
Section 4	2320	2520	18	20
Section 5	2160	2360	16	17
Section 6	2000	2200	14	15

Table 2. Effective stiffness considered for the bridges.

Component	Flexural rigidity	Shear rigidity
Column with compression caused by design gravity load $\leq \mathbf{0.5A_g f_c{}'}$	$0.3E_c I_g$	$0.4E_c A_w$
Deck (pre-stressed)	$E_c I_g$	$0.4E_c A_w$

Table 3. Comparison of natural frequency.

	Pier height 10 m		Pier height 20 m		Pier height 30 m	
Direction	Analytical	CSI Bridge	Analytical	CSI Bridge	Analytical	CSI Bridge
Longitudinal	0.30 sec	0.39 sec	0.90 sec	1.01 sec	1.70 sec	1.80 sec
Transverse	0.28 sec	0.44 sec	0.83 sec	0.94 sec	1.54 sec	1.57 sec

3. Derivation of fragility curves

The probability of exceeding the damage state (i.e. IO, LS and CP) was calculated by using equation (1)[23]:

$$P(DS/SI)=1-\Phi\left(\frac{(\lambda_C-\lambda_{D/SI})}{\sqrt{\beta_{D/SI}^2+\beta_C^2+\beta_\mu^2}}\right) \quad (1)$$

$$\beta_{D/SI} = \sqrt{Ln(1+S_E^2)} \quad (2)$$

where, S_E^2 is the standard error of demand drift; DS is damage state; SI is seismic intensity; Φ is standard normal distribution; $\lambda_{D/SI}$ is natural logarithm of the median demand drift given the seismic intensity from the best fit power law; λ_C is natural logarithm of the median of drift capacities for particular damage state; β_C and β_μ are uncertainties related to capacity and modelling, respectively. The uncertainties of capacities were calculated from the results of Incremental Dynamic Analysis (IDA) and that of the modelling was taken as 0.3[23]. The drift capacities of bridges were calculated by using both the pushover analysis and IDA. However, for the calculation of fragility curves conservative values were selected for the drift capacities by comparing the results of pushover analysis and IDA. Table 4 shows the employed drift capacities for the derivation of fragility curves. It is evident from this table that the drift capacities related to the LS and CP performance levels of the bridge with the 30 m height pier are close to each other. This indicates a limited reserved displacement ductility in the bridge. It can also be seen that the taller piers have larger drift capacities when compared with the shorter piers. In this study, the drift ratio was selected as the engineering demand parameter and Peak Ground Acceleration (PGA) was considered as the seismic intensity measure. The fragility curves were derived for PGAs varying from 0.1g to 0.5g as shown in figure 4, figure 5, and figure 6 for 10 m, 20 m, and 30 m height piers, respectively. Figure 4 shows that the probability of a light damage (i.e. IO) to the bridge with the 10 m height pier is significantly larger than severe damage (CP). It can also be seen that the probability of exceeding a moderate damage (LS) is close to the probability of exceeding a severe damage. On the other hand, the bridge with the 20 m height pier shows a significant difference between the probabilities of exceeding a moderate and a severe damage. For this bridge, the probability of exceeding a light damage compared to a moderate damage is not as much as it is for the bridge with the 10 m height pier. Similar to the bridge with the 10 m height pier, increase in the PGA significantly raises the probability of exceeding the light damage. However, unlike the first bridge, increase in the PGAs also significantly raises the probability of moderate and severe damage. This implies that the bridge with the 20 m height pier has a larger chance of collapse during strong earthquakes when compared with the bridge having the 10 m height pier. Figure 6 shows that in the bridge with the 30 m height pier the probability of exceeding light damage is slightly larger than moderate and severe damage indicating a limited reserved ductility in the bridge to sustain large deformations. The probability of exciting light, moderate and severe damage in the bridge with the 30 m height pier is smaller than the other two bridges. This implies that, for an identical intensity of PGA, this bridge will experience less damage when compared with other bridges. The main reason for such observation relies on the fundamental natural period of the bridges and the frequency content of the selected records. In addition, the cross-sectional size of piers in the tallest bridge is larger than the other two bridges.

In accordance to the seismic hazard map given in the Malaysian national annex to Eurocode 8[24], the peak acceleration on the bed rock in Kuala Lumpur is 0.09g. Considering a maximum soil amplification of 1.4, as given in the national annex, the value of PGA is 0.13g for this city. Therefore, as can be seen from figure 4, figure 5, and figure 6, the probability of exceeding light damage for bridges with the 10 m, 20 m, and 30 m height piers are all below 10%. This implies that the studied bridges will remain almost intact when the intensity of PGA is 0.13g. It can also be seen that, for a strong earthquake with the PGA of 0.5g, only the bridge with 20 m height pier is expected to experience a light damage.

Table 4. Drift Capacities of the studied bridges.

Pier height	IO (%)	LS (%)	CP (%)
10 m	1.10	1.96	2.62
20 m	2.22	2.82	3.38
30 m	2.88	3.46	3.89

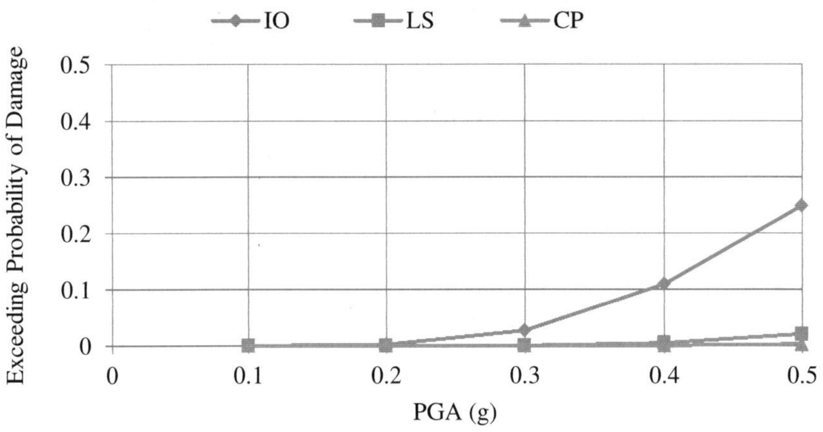

Figure 4. Seismic fragility curve of the bridge with the 10 m height pier.

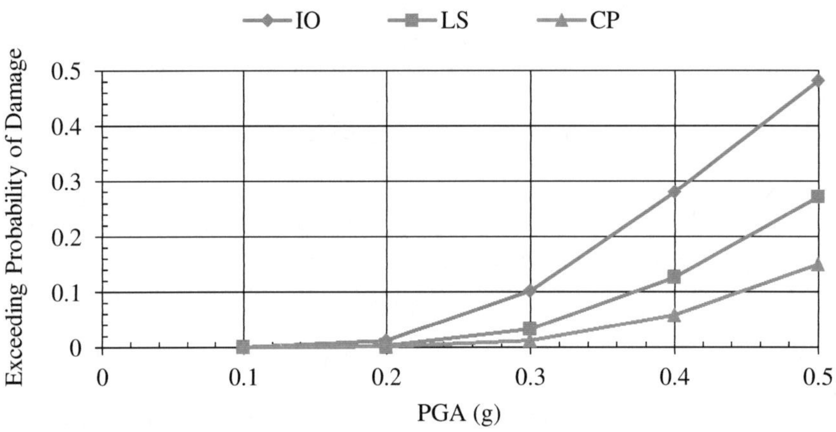

Figure 5. Seismic fragility curve of the bridge with the 20 m height pier.

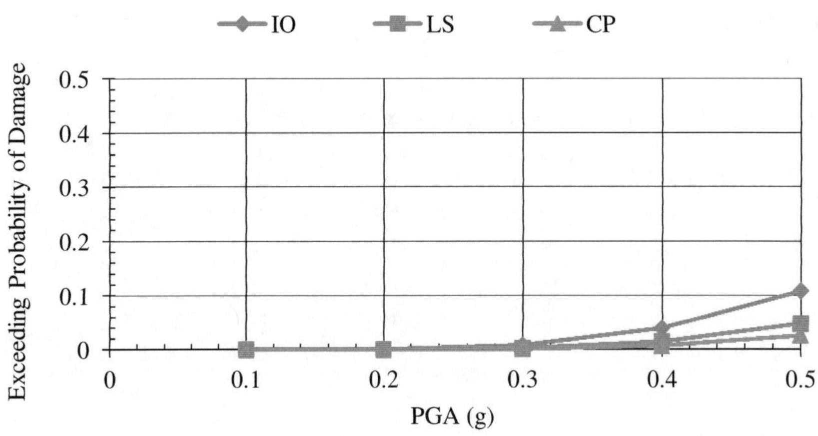

Figure 6. Seismic fragility curve of the bridge with the 30 m height pier.

4. Conclusions

This study investigated the seismic vulnerability of concrete box girder bridges in Malaysia through the framework of fragility curves. Three bridges with different pier heights of 10 m, 20 m, and 30 m were simulated in the CSI Bridge software. A 2.5% reinforcement ratio was considered for all piers. Nonlinear static and nonlinear incremental dynamic analyses were applied on the simulated bridges. Totally, 15 far-field natural earthquake records which were scaled to the PGAs of 0.1g to 0.5g were employed for the derivation of seismic fragility curves. Results indicated that the bridge with the pier height of 20 m is more vulnerable compared to the other studied bridges. However, for a PGA of 0.13g, which is the estimated peak ground acceleration in Kuala Lumpur, all studied bridges were able to maintain their functionality during low to medium intensity earthquakes.

Acknowledgement

The authors would like to acknowledge the financial supports from Universiti Teknologi Malaysia under the grant no. of 17H80 and 19H36.

References

[1] Moehle J and Eberhard M 2003 Earthquake Damage to Bridges *Bridge Engineering Seismic Design* ed W F Chen and L Duan (CRC Press) pp 52–84
[2] Martínez A, Hube M A and Rollins K M 2017 Analytical fragility curves for non-skewed highway bridges in Chile *Eng. Struct.* **141** 530–42
[3] Calvi G M, Pinho R, Magenes G, Bommer J J, Restrepo-Vélez L F and Crowley H 2007 Development of Seismic Vulnerability Assessment Methodologies over the past 30 Years *ISET J. Earthq. Technol.* **43** 75–104
[4] Karim K R and Yamazaki F 2001 Effect of earthquake ground motions on fragility curves of highway bridge piers based on numerical simulation *Earthq. Eng. Struct. Dyn.* **30** 1839–56
[5] Nateghi F and Shahsavar V 2004 Development of fragility and reliability curves for seismic evaluation of a major prestressed concrete bridge 13th World Conf. on Earthquake Engineering *13th World Conf. on Earthquake Engineering* (Vancouver, BC Canada) pp 1–6
[6] Choi E, DesRoches R and Nielson B 2004 Seismic fragility of typical bridges in moderate seismic zones *Eng. Struct.* **26** 187–99
[7] Banerjee S and Shinozuka M 2007 Nonlinear Static Procedure for Seismic Vulnerability Assessment of Bridges *Comput. Civ. Infrastruct. Eng.* **22** 293–305
[8] Lee S M, Kim T J and Kang S L 2007 Development of fragility curves for bridges in Korea *KSCE J. Civ. Eng.* **11** 165–74
[9] Av Ö 2012 Development of Analytical Seismic Fragility Curves for Ordinary Highway Bridges in Turkey *15th World Conference on Earthquake Engineering* (Lisbon, Portugal)
[10] Parool N and Rai D C 2015 Seismic Fragility of Multispan Simply Supported Bridge with Drop Spans and Steel Bearings *J. Bridg. Eng.* **20** 04015021
[11] Simon J and Vigh L G 2016 Seismic fragility assessment of integral precast multi-span bridges in areas of moderate seismicity *Bull. Earthq. Eng.* **14** 3125–50
[12] Vafaei M and Alih S C 2018 Seismic vulnerability of air traffic control towers *Nat. Hazards* **90** 803–22
[13] Ghasemi H, Otsuka H, Cooper J D and Nakajima H 1996 Aftermath of The Kobe Earthquake *Fed. Highw. Adm. Res. Technol.* **60**
[14] DeBlasio A J, Zamora A, Mottley F, Brodesky R, Zirker M E and Crowde M 2002 *Bridge Engineering Handbook*
[15] Vafaei M, Alih S C and Abdul Rahman Q 2016 Drift demands of low-ductile moment resistance frames (MRF) under far field earthquake excitations *J. Teknol.* **78** 82–92
[16] Computers and Structures 2015 CSI BRIDGE Berkeley, California.

[17] Vafaei M, Adnan A bin and Rahman A B A 2014 Seismic performance evaluation of an airport traffic control tower through linear and nonlinear analysis *Struct. Infrastruct. Eng.* **10** 963–75
[18] Vafaei M, Adnan A B and Yadollahi M 2011 *Seismic damage detection using pushover analysis* vol 255–260
[19] American Society of Civil Engineers (ASCE) 2000 FEMA 356 Prestandard and Commentary for the Seismic Rehabilitation of Building *Rehabilitation* 221
[20] Anon 2013 ASCE/SEI 41-13,Seismic Evaluation And Retrofit Of Existing Buildings
[21] Vafaei M, Adnan A bin and Abd. Rahman A B 2014 A neuro-wavelet technique for seismic damage identification of cantilever structures *Struct. Infrastruct. Eng.* **10** 1666–84
[22] Moravej H, Vafaei M and Bakar S A 2016 Seismic performance of a wall-frame air traffic control tower *Earthq. Struct.* **10** 463–82
[23] Mwafy A 2012 Analytically derived fragility relationships for the modern high-rise buildings in the UAE *Struct. Des. Tall Spec. Build.* **21** 824–43
[24] MS EN 1998-1:2015 2017 Malaysia National Annex To Eurocode 8 - Design Of Structures For Earthquake Resistance, Part 1: General Rules, Seismic Actions And Rules For Buildings

IOP Conf. Series: Materials Science and Engineering **513** (2019) 012020 doi:10.1088/1757-899X/513/1/012020

Damage Assessment of High-Rise Reinforced Concrete Buildings in Peninsular Malaysia Subjected to Ranau Earthquake

R.Ismail*[1,2], A.Ibrahim[2], A.Adnan[3]
[1] Institute for Infrastructure Engineering and Sustainable Management (IIESM)
Universiti Teknologi Mara (UiTM) Shah Alam, Selangor, MALAYSIA.
[2] Department of Civil Engineering
Universiti Teknologi Mara (UiTM) Shah Alam, Selangor, MALAYSIA
[3] Department of Civil Engineering
Universiti Teknologi Malaysia (UTM) Skudai, Johor, MALAYSIA

Abstract. This paper shows the analysis of high reinforced concrete buildings in Ranau that have experienced low intensity intensity earthquakes. This study refers to the performance of five (5) high rise reinforced concrete frame when subjected to a variation of low earthquake intensity analysis with 5% damping ground motion measurements. The IDARC software is used to study structures through nonlinear dynamic analysis. Beam-column points are examined to determine damage to the index and to build damage levels subject to varying seismic load. Result shows the first yielding occurred at 4.82 seconds for the beam element at when 0.05g load applied. Based on the result, all buildings cannot withstand seismic load when goes up to 0.2 ground acceleration. It may have arisen that the building is categorized to the extent of the major damage where there is structural affect and structural affect up to 0.15g intensity.

1. Introduction

Estimation of damage the structure is an essential demand for performance structural analysis. To better understand comprehensive, comprehensive performance criteria that are capable of predicting damage levels structures subject to earthquake excitation shall be developed. Earthquake excitations often cause a certain state of damage to structures, the extent of which is generally expressed by using a Damage Index (DI). DI becomes more frequent accepted as a powerful tool for quantitative assessment damage to the structure caused by the earthquake. In seismic area, the damage index has a major role in decide on retrofit. Some researchers have attempted study the performance of structural due earthquake effect in the last decade [1-8]. According to Li [1], in the repercussions of seismic ground motion, buildings must be properly evaluated for structural safety due building damage and collapse that might occur.

Moreno and Bairan [2] demonstrated the performance of buildings in the low earthquake area. The results show that the building is expected to cause damage due to unexpected performance.

IOP Conf. Series: Materials Science and Engineering **513** (2019) 012020 doi:10.1088/1757-899X/513/1/012020

Ditommaso et al, [9] describes the period of reinforced concrete building to initiate structural damage and also identify non-structural elements that are expected to damage the effects of earthquakes.

Joel and Mary [10] found that flexible models produce more displacement and interstorey drift than solid frames using IDARC 2-Dimensional software for reinforced concrete buildings. Achievements usually involve decent measurement processes. Due to the reflection and expectation of injury, the values and parameters must be quantitatively made because of the performance level. Table 1 shows the degree of damage ATC-13 [12] to justify the degree of damage that has been taken to be considered in this study. The degree of damage was also used by Ismail et al., [13] in her research.

Table 1. Description of Damage levels. [12]

Damage Level	Damage Indices
Minor Level	< 1%
Moderate Level	1% - 3%
Major Level	> 3% < 1.0
Collapse	1.0

2. Description of Structure and Modeling

This study focuses on the performance of seismic public buildings held by the government. Random site location selected from peninsular Malaysia. To examine high performance seismic performance, one key frame is selected from each building for a model with finite element analysis. The IDARC software is used to study structures through nonlinear dynamic analysis. There were six (6) districts around peninsular Malaysia including Johor Bahru, Negeri Sembilan, Kuala Lumpu, Kuala Terengganu, Ipoh and Pulau Pinang. Table 2 shows the details of buildings analysed in this study. They are from 11 up to 17 number of storey which being categorised as high rise building with the height range from 36.20m to 73.0m . Table 3 shows the building configurations. The materials of properties are 2500N/mm^2 (E_c), 460N/mm^2 (f_y), 27.6N/mm^2 (f_c) and 25mm for concrete cover (c). Loadings applied in the analysis was determined according to British Standard 8110: Part 1 [14].

The height of the buildings will play an important part in prediction of damage indices as we know that high rise buildings will be likely affected to ground motion when earthquake incidents occurred. Beam elements and columns are modeled as nonlinear frame elements with susceptible sensitivity by specifying the plastic hinges at both ends of beams and columns. In the beam, the axial force is assumed to be zero because all floors are considered to be hard on the plan to explain the action of the concrete slab diaphragm so that the nodes in each Floor level are deformed with the same value later. The contribution of floor slabs to strength and immobility is neglected and columns are assumed to be in underground stages. The damping coefficient value of 5% RC frame structure is used. The predicted structure may not be the same as performance monitoring on the site after the earthquake incident because the model used in this study ignores the flexibility of foundations and other components helping with structural strength such as shear wall.

Table 2. Information of building

No.	Building	Floor level	Total Height (m)
1.	Wisma Persekutuan Kota Bharu	13	48.54
2.	Kuarters Polis Ipoh	11	36.20
3.	Jabatan Kerja Raya KL	17	64.20
4.	Majlis Perbandaran Ampang Jaya	17	73.00
5.	Wisma Dewan Perniagaan Seremban	12	51.00

Table 3. Configuration of Modeling concept

Building Name	Section Element		Material Properties
	Beam	Column	
Wisma Persekutuan Kota Bharu	Ground: 230x150mm (top:3T20), (bottom:3T20) 1st-12th: 230x150mm (top:3T20), (bottom:3T20) 13th: 897x308mm (top:2T32), (bottom:2T25)	610x108mm (10T25)	
Kuarters Polis Ipoh	Ground: 900x350mm (top:3T32), (bottom:2T25) 1st-11th: 750x350mm (top:4T20), (bottom:3T20)	Ground: 500x1000mm (24T25) 1st-11th: 550x300mm (10T25)	
Jabatan Kerja Raya KL	Ground: 750x350mm (top:4T20), (bottom:3T20) 1st- 3rd: 600x300mm (top:3T20), (bottom:3T20) 3rd-17th: 520x250mm (top:2T25), (bottom:2T25)	Ground: 600x450mm (16T32) 1st-17th: 550x300mm (10T25)	Ec = 2500N/mm^2 fy = 460N/mm^2 fc = 27.6N/mm^2 fys = 276N/mm^2 c = 25mm
Majlis Perbandaran Ampang Jaya	Ground: 750x350mm (top:4T20), (bottom:3T20) 1st-17th: 450x300mm (top:5T25), (bottom:3T20)	Ground: 500x1000mm (24T25) 1st-17th: 550x300mm (10T25)	
Wisma Dewan Perniagaan Seremban	Ground: 230x150mm (top:3T20), (bottom:3T20) 1st-13th: 768x308mm (top:3T20), (bottom:3T20)	610x108mm (10T25)	

3. Earthquake Data

This study used the intensity recorded by the Malaysian Meteorological Department. The intensity recorded on May 6, 2015 in Kota Kinabalu with a magnitude of 5.9 on the Richter level is shown in Figure 1 with 3538 output numbers in 0.02 steps at the time with the highest acceleration amongst others' different intensities. Ground motion data are classified into four levels of peak ground acceleration (PGA) of 0.05g, 0.10g, 0.15g and 0.20g respectively. This study uses a slower 5% spectrum acceleration.

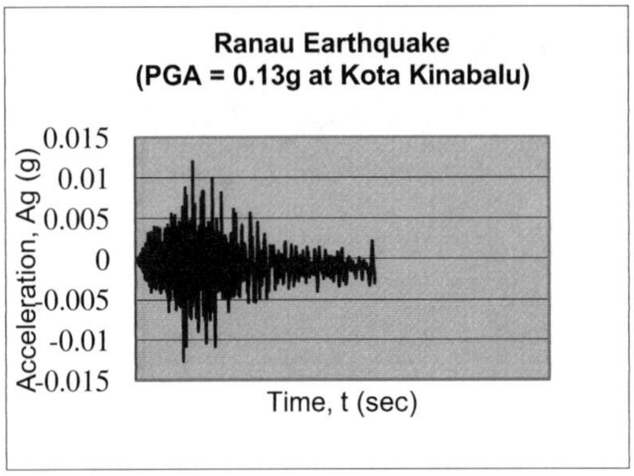

Figure 1. Data of ground acceleration for Ranau Earthquake.

4. Result

The term plastic hinges are used to describe the deformation of the frame section comprising columns and columns where elastic plastics occur. Typically, plastic hinges are currently resisting structural building types due to local structural failures occurring in beams and columns. Figure 3 illustrates the formation of damage location conditions and plastic hinges at the time of the resistance frame with four intensities of different ground acceleration. This represents the sequence of damage to the state of the frame under the loading of the earthquake.

4.1 Plastic Hinge Formation

The term plastic hinges are used to describe the deformation of the frame section comprising beams and columns where elastic plastics occur. Typically, plastic hinges are currently resisting structural building types due to local structural failures occurring in rays and columns. This study focuses on the extent of building damage and knows when yielding begins. Based on the result, for the intensity of 0.05g, no plastic hinges are formed.

Figure 4 illustrates the formation of a state of damage and plastic hinges at the time of the resistance frame at four different intensities of ground acceleration. The frame shows the location of the elemental structure where it begins to yield, crack or collapse. This represents the sequence of damage to the state of the frame under the loading of the earthquake. As a result, the concrete frame has cracked and yielded elements. The 'x' symbol indicates the appearance of cracking for concrete frames and 'o' symbols showing the results of plastic hinges constructed. For the intensity of 0.05g, it shows that the frame does not experience anything that degrades or collapse, but for the intensity of 0.10g to 0.20g, the frame has yielded and cracked. For example, at 0.05g intensity, it shows that the beams on the left begin to damage at the second floor. For 0.15g intensity, it is described as the resulting frame. However, the concrete frame undergoes infusion and cracking in the first floor when subjected to the intensity of the earthquake 0.20g (Figure 4). It shows that the structure element (beam no-1) starts to produce at 5.46 seconds when the intensity of 0.10g is used and the overall index of the recorded damage is 0.35g. All the buildings have experienced plastic hinges on the same floor level (floor 1). Table 4 shows the location of plastic hinges and the first yielding point of each building with respect to PGA.

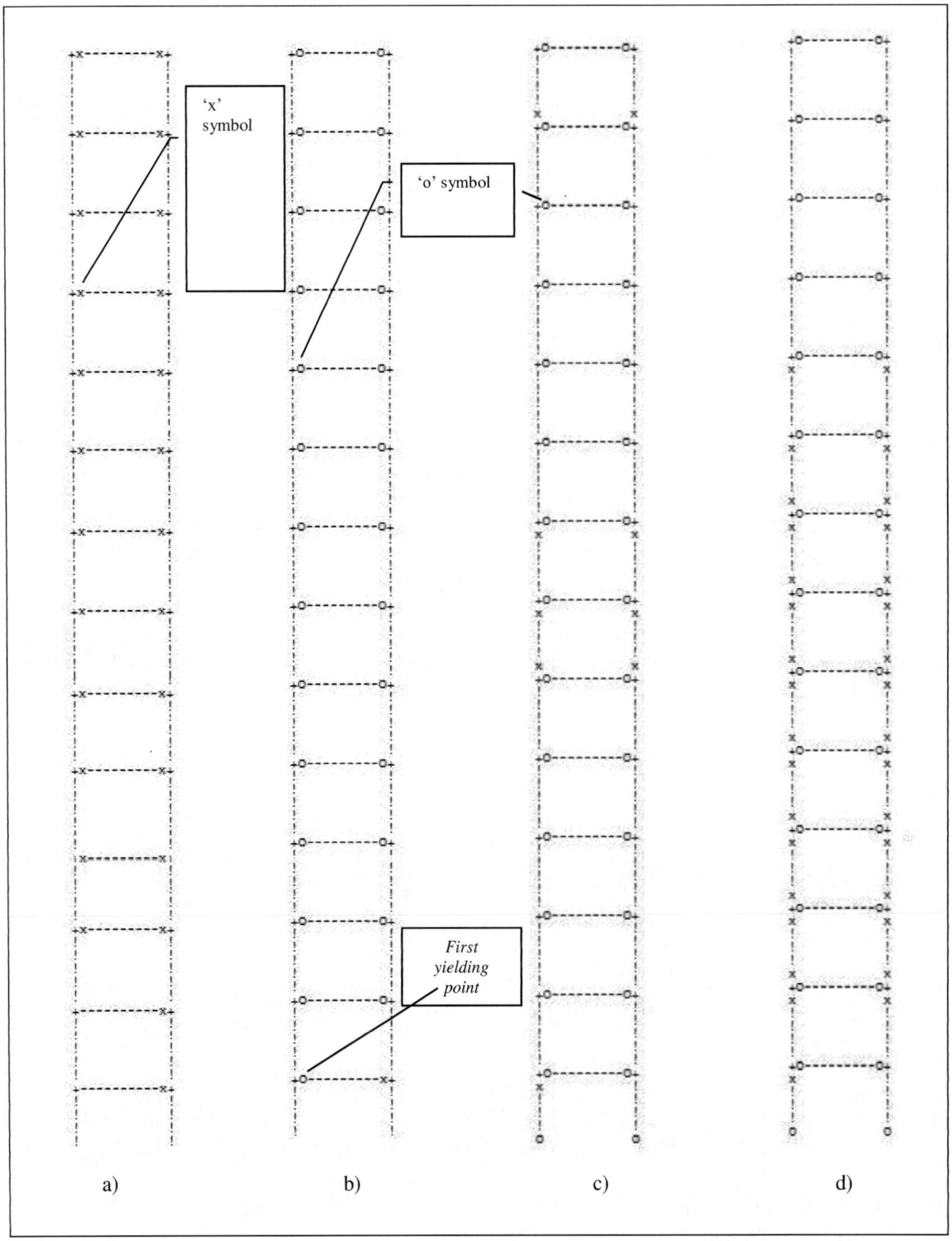

Figure 4. Illustration and the formation of a state of damage and plastic hinges for Wisma Persekutuan Kota Bahru building.

IOP Conf. Series: Materials Science and Engineering **513** (2019) 012020 doi:10.1088/1757-899X/513/1/012020

Table 4. Location and time for first yielding point.

Building Name	No. of Story	Location of Plastic Hinge	Floor level	Intensitie s (g)	Time of First Yielding (sec)
Wisma Persekutuan Kota Bharu	13	Column	1	0.10	5.46
Kuarters Polis Ipoh	11	Beam	1	0.05	4.82
Jabatan Kerja Raya KL	17	Column	1	0.05	6.98
Majlis Perbandaran Ampang Jaya	17	Beam	1	0.10	5.22
Wisma Dewan Perniagaan Seremban	12	Column	1	0.05	7.62

4.2 Time History Analysis

Based on the results of such an analysis, IDARC [15] defines the overall damage index for the entire structure. The overall damage index for buildings can be determined at the SEAOC Level Damage outlined in Table 1. The results are shown in Table 5. The Wisma Dewan Perniagaan Seremban has no damage when the intensity of 0.05g is enforced. The result of structural damage is referred to as the weighted index of the story and every index of the story is referred to as the weighted index. Total damage index when 0.10g intensity is used is also categorized as moderate. This means no structural damage occurs. However, some structural impairments occur but in this intensity, no elemental structure produces. The damage level has become the major damege level for intensity up to 0.15g. Local failures are often built into beam joints followed by column extensions. All buildings cannot withstand 0.20g intensity.

Table 5. Overall finding on damage index for different intensities.

No.	Building	5%	10%	15%	2%
1.	Wisma Persekutuan Kota Bharu	0.01	0.35	0.74	1.00
2.	Kuarters Polis Ipoh	0.02	0.27	0.68	1.00
3.	Jabatan Kerja Raya KL	0.01	0.26	0.72	1.00
4.	Majlis Perbandaran Ampang Jaya	0.02	0.38	0.85	1.00
5.	Wisma Dewan Perniagaan Seremban	0.004	0.21	0.63	1.00

5. Conclusion

Based on the results of the study, it can be concluded that:

- The results indicate that the potential for serious damage to the building may be indicated by the injury index.
- The highest PGA value for certain types of buildings may be appropriate to the possible damage to the structure
- IDARC results indicate that there is an approximate linear relationship between damage index and PGA
- Different plastic hinge formation patterns for different models, with better approval for larger PGAs.

Acknowledgments
Acknowledgement goes to Universiti Teknologi MARA (UiTM) and support from Faculty of Civil

Engineering, Universiti Teknologi MARA (UiTM) as part of a project Bestari Perdana Research Fund with File No 600-IRMI/PERDANA 5/3 BESTARI (081/2018).

References

[1] Li, G., Zhang, Y., and Li, H 2013 Seismic Damage Analysis of Reinforced Concrete Frame Using the Force Analogy Method." *J. Eng. Mech.*, 139(12), 1780–1789.

[2] Moreno-Gonzalez R. & J. M. Bairan 2013 Seismic Damage Assessment for Waffled-Slabs Reinforced Concrete (RC) Buildings in Barcelona, *International Journal of Architectural Heritage: Conservation, Analysis, and Restoration*, 7(1).

[3] Ismail, R., & Ismail, M. I 2017 Dynamic analysis of Concrete Faced Rockfill Dam using Finite Element Method. Journal of Engineering and Applied Sciences, 12(7), 1772-1776. DOI: 10.3923/jeasci.2017.1772.1776.

[4] R. Ismail, A. Ibrahim and A. Adnan 2018 Damage Assessment of Medium-Rise Reinforced Concrete Buildings In Peninsular Malaysia Subjected To Ranau Earthquake, *International Journal of Civil Engineering and Technology, 9(7)*, pp.881–888.

[5] Ismail, R., & Abdul Karim, M. R 2017 Concrete bridge pier performance under earthquake loading. *Journal of Engineering and Applied Sciences*, *12*(9), 2254-2258. DOI: 10.3923/jeasci.2017.2254.2258.

[6] Anudai S, Hamid N.H.A, Hashim M.H.M 2016 Experimental Study on Seismic Behavior of Repaired Single and Double Unit Tunnel Form Building Under In-Plane Cyclic Loading Malaysian *Construction Research Journal*, 19 (2), 19-28.

[7] R. Ismail, A.Ibrahim, H.Hamid, M.Rusop, A. Adnan 2018 Experimental study on mechanical properties of elastomer containing carbon nanotubes. *Journal of Engineering Science and Technology*, Vol. 13, No. 3 (2018) 656 – 664.

[8] R. Ismail, A. Ibrahim, M. Rusop and A. Adnan 2018 Determination of Mechanical Properties Natural Rubber Compounds Using Double Shear Test Pieces. *International Journal of Civil Engineering and Technology*, 9(8), pp. 37-43.

[9] Ditommaso R., M. Vona1, M. R. Gallipoli, and M. Mucciarell 2013 Evaluation and considerations about fundamental periods of damaged reinforced concrete buildings, *Nat. Hazards Earth Syst. Sci*., 13, 1903–1912.

[10] Joel M. Barron and Mary Beth D. Hueste 2004 Diaphragm Effects in Rectangular Reinforced Concrete Buildings, *ACI Structural Journal: Technical Paper*, September-October, Vol. 3, 89-98.

[11] Kunnath, S. K., Reinhorn, A. M., and Lobo, R. F 1992 IDARC Version 3.0: A Program for the Inelastic Damage Analysis of Reinforced Concrete Structures, Report No. NCEER-92-0022, *National Center for Earthquake Engineering Research*, State University of New York at Buffalo.

[12] Surya Kumar V.Gunturi 1992 *Building Specific Earthquake Damage Estimation*, Doctor of Philosophy Dissertation, Stanford University, San Francisco.

[13] Ismail R, Adnan A and Ibrahim A. (2011) Vulnerability of Public Buildings in Sabah Subjected to Earthquake by Finite Element Modeling , *The 2nd International Building Control Conference 2011*, Procedia Engineering 20 (54-60).

[14] British Standard 8110: Part 1 (1997) Structural use of concrete. Code of practice for design and construction. BSI.

[15] Park, Y. J., Ang, A. H.-S., and Wen, Y. K. (1984) Seismic Damage Analysis and Damage-Limiting design of R/C Buildings, Civil Engineering Studies, *Technical Report* No. SRS 516, University of Illinois, Urbana.

IOP Conf. Series: Materials Science and Engineering **513** (2019) 012021 doi:10.1088/1757-899X/513/1/012021

Comparison between cyclic response of RC columns transversely reinforced with FRP strips and carbon steel

Nur Hajarul Falahi Abdul Halim[1], Sophia C. Alih[2] and Mohammadreza Vafaei[3*]

[1]School of Civil Engineering, Faculty of Engineering, Universiti Teknologi Malaysia, 81310 Johor Bahru, Johor, Malaysia
[2]School of Civil Engineering, Faculty of Engineering, Institute of Noise and Vibration, Universiti Teknologi Malaysia, 81310 Skudai, Johor, Malaysia
[3]School of Civil Engineering, Faculty of Engineering, Forensic Engineering Center, Universiti Teknologi Malaysia, 81310 Skudai, Johor, Malaysia

[*]Corresponding author: vafaei@utm.my

Abstract. This study investigates the effects of different confinement methods for longitudinal reinforcement bars in reinforced concrete, RC columns. Comparison is made between three RC columns; two were transversely reinforced with carbon steel in single tie and continuous spiral form, and the third sample were transversely reinforced with Fibre Reinforced Polymer, FRP strips in continuous spiral form. Numerical simulation is conducted using validated numerical models. All three RC column models were subjected to a constant axial load and cyclic loading. Hysteresis curves and backbone curves were used to analyse cyclic behaviour of each samples. Comparisons between the RC columns were made based on the ultimate load capacity and energy dissipation capacity ratio. Results show that, RC column transversely reinforced with FRP strips exhibits the highest ultimate load and energy dissipation capacity. The ultimate load capacity is 15% and 10% higher than the RC column transversely reinforced with carbon steel in single tie and continuous spiral form respectively. The FRP strips also increases the energy dissipation capacity up to 35% and 22% compared to the one transversely reinforced with carbon steel in single tie and spiral form respectively. It can be concluded that the usage of FRP strips as transverse reinforcement in RC column is able to improve the cyclic responses of the conventional RC columns.

1. Introduction

The ability of reinforced concrete (RC) columns to dissipate energy well when subjected to lateral loads such as earthquake plays a vital role in determining seismic behaviour and structural stability[1]. Provisions of providing stringent confining reinforcement in plastic hinge regions can enhance the seismic behaviour of concrete elements. However, most of the older buildings were designing to carry out gravity loads alone without seismic consideration [2]. Figure 1 shows the failure of columns due to poor construction and inadequate shear reinforcement during earthquakes. Inadequate length of anchorage with 90° end hooks and excessive spaced between stirrups contributing to ties opened, excessive concrete crushing and spalling, followed by buckling of longitudinal steel bars, result to a very brittle or sudden failure of the structure [3].

Incorporation of continuous spiral reinforcement in concrete elements such as columns and beams can improve the seismic performance of the structure [4–6]. Higher ultimate load capacity, energy dissipation and ductility ratio were reported for continuous spiral as internal shear reinforcements compared with conventional stirrup shear reinforcement. For well-confined concrete columns with spiral steel, the increase in concrete strength can be in range 2.1 to 4.0 times the lateral load [5]. However, due to the corrosion behaviour of the steel bar, robust materials were demanded.

Fibre reinforced polymer (FRP) has been widely used for making reinforcement and confining materials of concrete components to enhance the performance of reinforced concrete element [7–9], [10]. They exist in two form; bars and sheet, which consist of glass fibre, carbon fibre, aramid and basalt fibre [11]. FRP bars usually used to substitute steel reinforcement, whereas FRP sheets usually applied externally bonded RC structures for strengthening or repairing as shown in Figure 2. High strength and stiffness to weight ratio, high corrosion resistance, and lightweight of FRP can be used to increase the strength and the ductility of RC columns under severe loadings [12]. However, the application of FRP strips as transverse reinforcement in RC columns is very limited.

Therefore, this study is conducted to investigate the cyclic responses of RC column transversely reinforced with FRP strips compared to carbon steel through numerical simulation. This method has been proven effective in determining structural response under dynamic loadings [13,14]. Two types of confining methods were applied for the transverse reinforcement; the conventional single tie and continuous spiral form. Numerical simulation is conducted using validated numerical modelling methods and non-linear material models [15]. Cyclic responses of all models were analysed based on the hysteresis curve and backbone curves obtained from the numerical simulation.

Figure 1. Column failure during the earthquakes; (a) Northridge earthquake, (b) Abruzzo earthquake, (c) Kobe Earthquake (d) The Great East Japan Earthquake [16–18].

IOP Conf. Series: Materials Science and Engineering **513** (2019) 012021 doi:10.1088/1757-899X/513/1/012021

(a) (b)

Figure 2. FRP as confining reinforcements; (a) FRP bar and (b) FRP sheets [11][12].

2. Non-linear numerical modelling

Three-dimensional (3D) non-linear finite element, FE models of cantilever columns were developed using FE software ABAQUS 6.14 version [21]. In this section, a brief description of models, modelling methods and procedures were presented.

2.1. Model description

Three RC columns transversely reinforced with different methods as listed in Table 1 were used in this study. Two columns were transversely reinforced with carbon steel in single tie form (CSC) and continuous spiral form (CSS). The third model is transversely reinforced using FRP strips in continuous spiral form (SFRP). The dimensions of all samples were similar; 1500mm in length, with cross section of 200mm x 200mm. All columns are supported with identical footing designed according to Eurocode 2 [22] as depicted in Figure 3. Figure 4 (a-b) shows the dimension and reinforcement details of RC columns using carbon steel for transverse reinforcements, CSC and CSS. Figure 4(a) display design details for the column with transverse reinforcement in single tie form (CSC), and Figure 4(b) shows the details for the column with transverse reinforcement in continuous spiral form (CSS). Carbon steel bars with diameter of 12mm were used as longitudinal reinforcement bars and 6mm diameter carbon steel bar was used as the transverse reinforcement at 100 mm distance.

Figure 4(c) shows the dimension and reinforcement detailing for the sample with FRP strips (SFRP). The equivalent dimension of FRP strip with the 6mm diameter carbon steel bar was calculated according to Eurocode 2 [22] as in reference [15]. Based on this calculation, FRP strips with 15 mm width were used as stated in Table 2. Three links size 6 mm diameter was placed at the bottom, middle and top of the longitudinal reinforcement as support to the longitudinal reinforcement bars. It is worth mentioning that the modeling of reinforcement bars for this sample is according to the construction method proposed for RC column transversely reinforced with FRP strips.

Table 1. Types of RC column models simulated in FEM.

Model	Material for transverse reinforcement	Confinement method	Model name
1	Carbon steel	Single tie	CSC
2	Carbon steel	Continuous spiral	CSS
3	FRP strips	Continuous spiral	SFRP

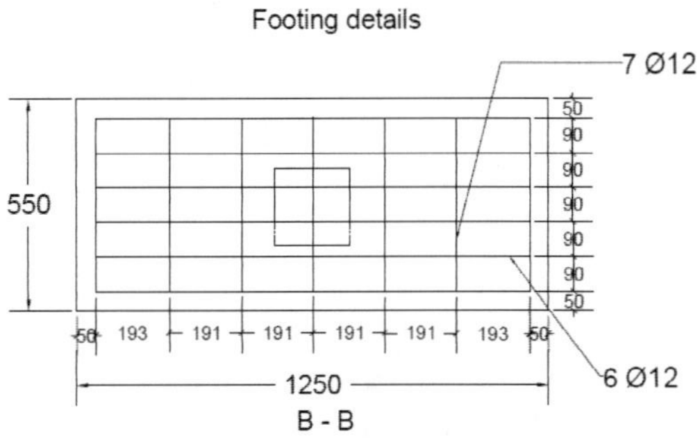

Figure 3. Dimension and reinforcement bars detailing of footing.

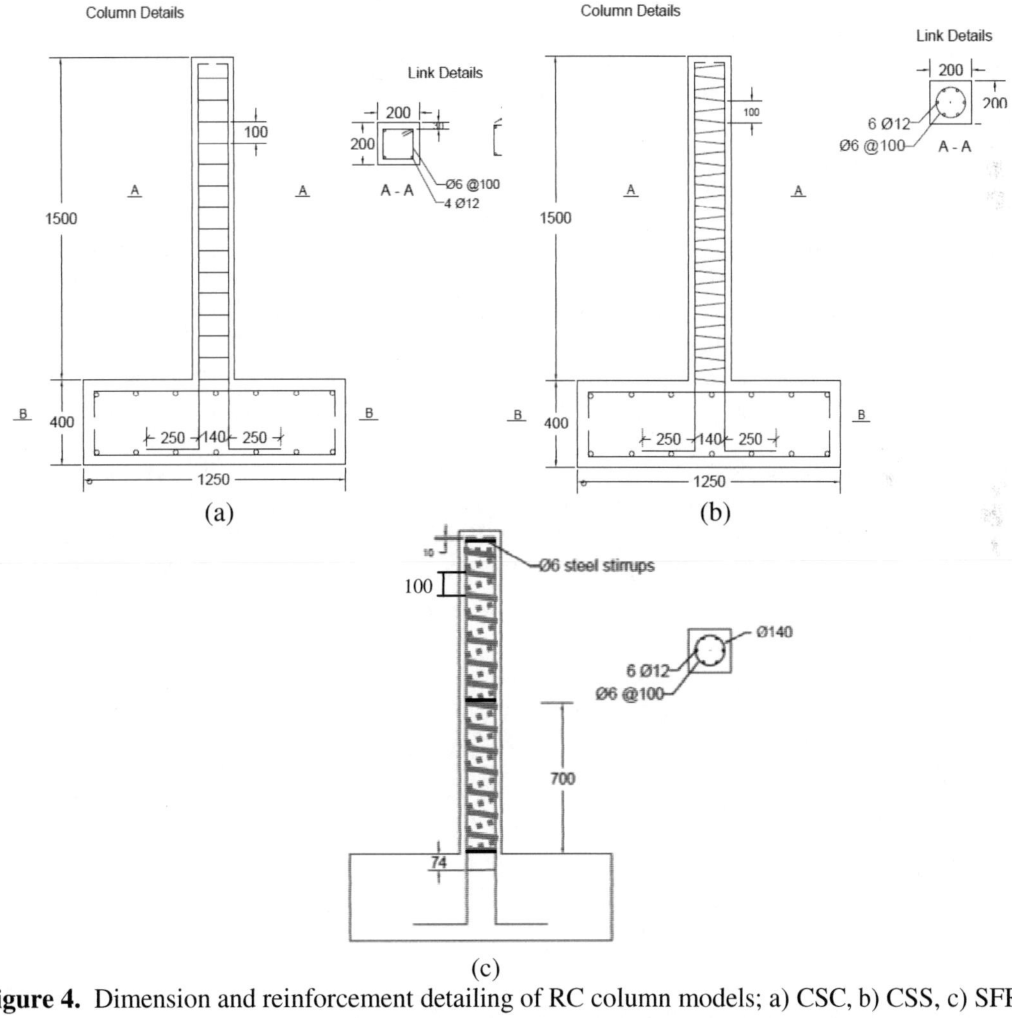

Figure 4. Dimension and reinforcement detailing of RC column models; a) CSC, b) CSS, c) SFRP.

Table 2. Equivalent geometrical design of FRP strips with carbon steel bar [23].

Bar dia. (mm)	Area of bar (mm^2)	Width of FRP strips (mm)	Thickness of FRP strips (mm^2)
6	28.27	15	0.164

2.2. Finite element model

All concrete elements of the columns were modelled using eight-node 3D brick elements (C3D8R). Concrete damage plasticity (CDP) model usually applied for cyclic loadings were used to represent the properties of concrete [8,21,24]. Details of the concrete material model is available in the reference [15]. For reinforcement bars, a bilinear elastic-plastic model was developed using 2-node linear 3-D truss (T3D2). Figure 5 shows the stress-strain curves of carbon steel bars. Next, for carbon FRP composites, a 4-node doubly curved thin or thick shell, reduced integration, hourglass control, finite membrane strains (S4R) were used to model the FRP. Table 3 shows the mechanical properties of FRP given by the manufacturer [25]. The bond condition between FRP strips and steel reinforcements were model using 'surface to surface' contact option as modelled in the reference. This option is used to determine surface-based cohesive behaviour in the analysis of mechanical contact. All reinforcements were embedded in the concrete elements. It is worth mentioning that all material models have been verified in the previous study [15].

The models then meshed into small elements to make sure that each of the concrete elements contains rebar. Different numbers of elements and nodes have been selected for each type of model. This is because, the interaction between the rebars and the concrete tends to reduce the mesh sensitivity. For CSC, total number of elements and nodes are 6050 and 7398 respectively, while, for CSS, 6276 and 7627. Since the model with FRP strips spiral (SFRP) has more complicated surface contact, smaller elements were used in order to avoid convergence problems. The total number of elements and nodes are 6465 and 7684, respectively.

Figure 6 shows the boundary condition of the FE models. The columns were assigned fixed at the bottom of the footing. F_y is a constant axial load, which is 90kN, whereas F_x is the direction for cyclic lateral loadings based on FEMA 461 [26]. At least 10 cycles of loading should be applied for each sample during the test. Each cycle repeated twice with increasing of amplitude from previous cycle by 40%. Displacement controlled cyclic loading is applied to the RC models using loading protocol depicted in Figure 7.

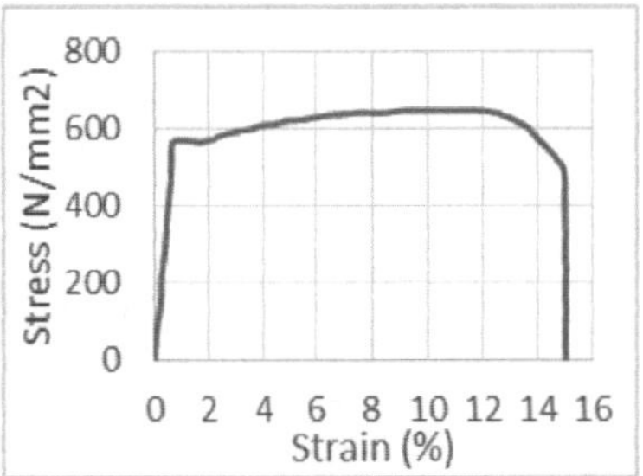

(a)

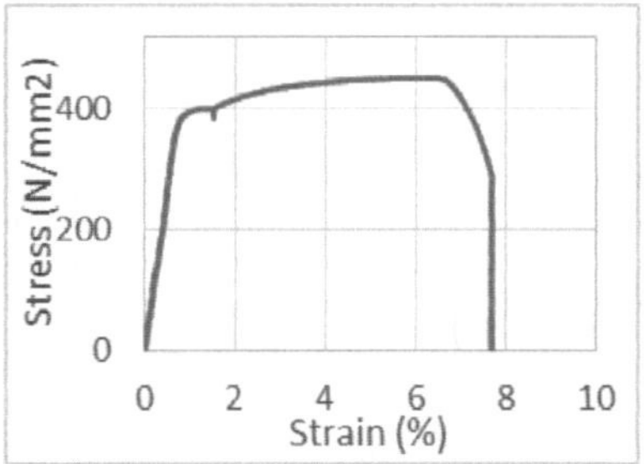

(b)

Figure 5. Stress-strain curves of steel reinforcement bars; (a) ribbed rebar 12 mm diameter and (b) plain rebar 6mm diameter.

Table 3. Mechanical properties of CFRP sheets [25].

Type of fibre	Weight (g/m^2)	Density (kg/m^3):	Thickness (mm)	Tensile strength (N/mm^2)	Modulus of elasticity (N/mm^2)	Elongation at breakage (%)
Carbon fibre	300	1800	0.164	≥ 4900	$252{,}000 \pm 2\%$	≥ 2

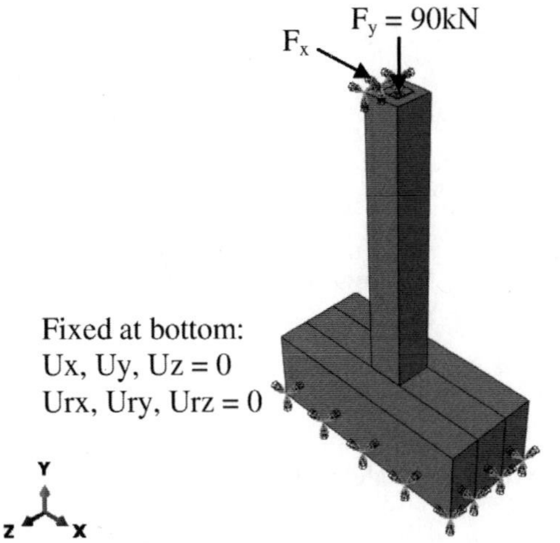

Figure 6. Boundary condition of columns.

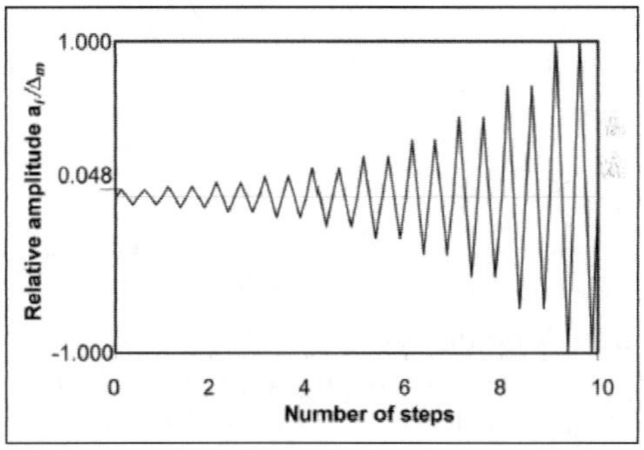

Figure 7. Loading protocol according to FEMA 2007.

3. Results of numerical modelling

Based on the numerical modelling, hysteresis curves, backbone curves and energy dissipation capacity were plotted for each column models. These cyclic responses are discussed in the following sections.

3.1. Hysteresis curves and backbone curves

Hysteresis curves are one of the important properties to determine cyclic response of a structure. From this curve, backbone curves can be obtained, and energy dissipation capacity ratio can be calculated. Figure 8 and 9 present the hysteresis and backbone curves of the RC column models. It can be observed that the initial stiffness is similar in all models. However, upon reaching the ultimate load capacity, the RC column transversely reinforced with carbon steel in single tie form (CSC) failed immediately. This indicates brittle behaviour in CSC model when subjected to cyclic loading. Compared to the RC columns with spiral form transverse reinforcement (CSS and SFRP), the columns can withstand certain displacement upon reaching their ultimate load capacity. Based on the backbone curves (Figure 9), it can be observed that the RC column transversely reinforced with FRP strips (SFRP) has the highest ultimate load capacity. The ultimate load capacity in SFRP model is 15% and 10% higher than CSC and CSS model respectively.

IOP Conf. Series: Materials Science and Engineering **513** (2019) 012021 doi:10.1088/1757-899X/513/1/012021

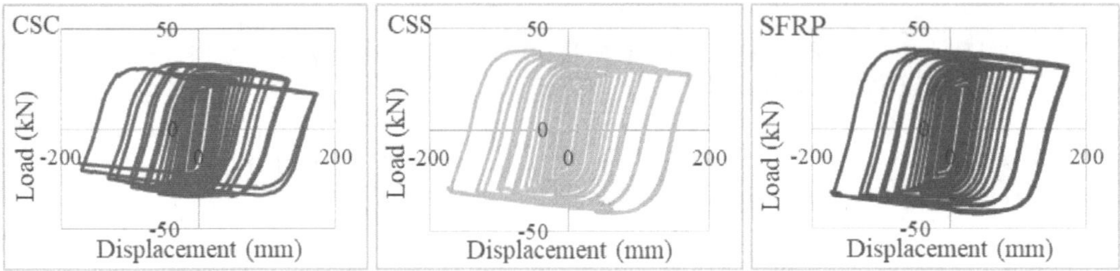

(a) (b) (c)

Figure 8. Hysteresis curves for each model; (a) CSC, (b) CSS and (c) SFRP.

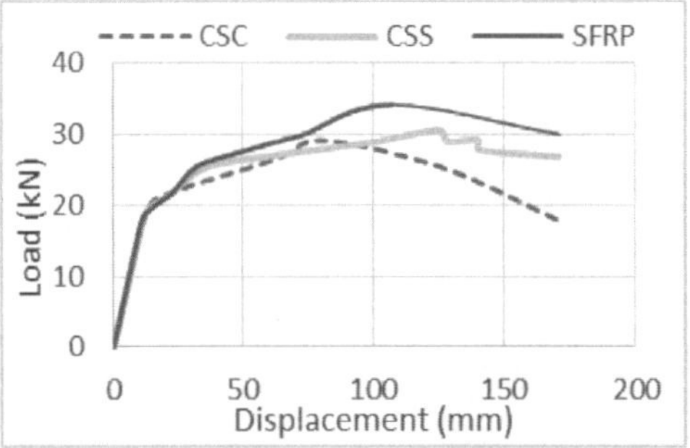

Figure 9. Backbone curves of FE models.

3.2. Energy dissipation capacity

Energy dissipation capacity indicates the amount of energy that can be dissipated when a structure is subjected to cyclic loadings [27]. This property is important especially when the structure is imposed with earthquake loadings [28]. Figure 10 shows the energy dissipation capacity of all column models. It is evident from the graph that RC column transversely reinforced with FRP strips (SFRP) dissipated more energy when subjected to cyclic loading compared to the RC column models transversely reinforced with carbon steel. The energy dissipation capacity of SFRP model is 35% and 22% higher than the CSC and CSS model respectively.

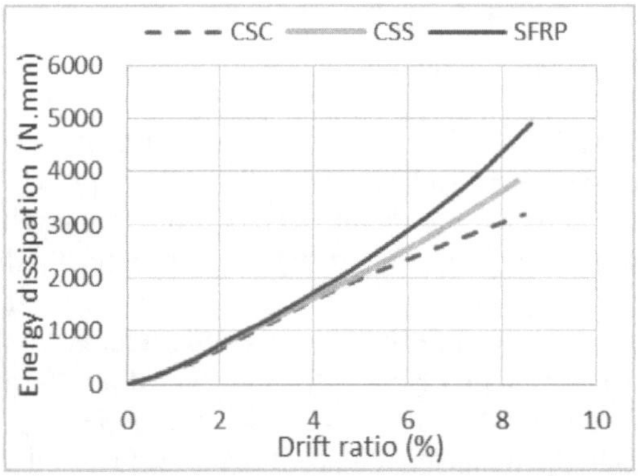

Figure 10. Cumulative energy dissipation capacity of all RC column models.

4. Conclusions

This study investigated the cyclic responses of RC column transversely reinforced with FRP strips and carbon steel bars. Two types of confinement method are selected; single tie form and continuous spiral form. Using validated numerical model, all RC column models were simulated with a constant axial load and cyclic load. Based on the obtained hysteresis curves and backbone curves, ultimate load capacity and energy dissipation capacity are compared between the studied models. It is observed that the RC column transversely reinforced with FRP strips, SFRP has the highest ultimate load capacity and energy dissipation capacity followed by the one transversely reinforced with carbon steel in spiral form, CSS. The RC column with conventional confinement method; using carbon steel in single tie form, CSC has the lowest values among all models. The ultimate load capacity of SFRP model is 15% and 10% higher than the CSC and CSS respectively. While the energy dissipation capacity of the SFRP is 35% and 22% higher than the CSC and CSS model respectively. This study demonstrated the potential application of FRP strips as transverse reinforcement in RC column to increase their cyclic responses.

Acknowledgments
The authors would like to acknowledge supports from Universiti Teknologi Malaysia and financial support from the Ministry of Higher Education of Malaysia through the RUG Vot. No. of 17H80 and 19H36, and FRGS Vot. 4F716.

References

[1] A.S. Rajput, U.K. Sharma 2018 *Structures* **13** 26–35
[2] M. Vafaei, S.C. Alih, Q. Abdul Rahman 2016 *J. Teknol.* **78** 82–92
[3] K. Sharma, L. Deng, C.C. Noguez 2016 *Eng. Struct.* **121** 61–74
[4] K.H. Yang, G.H. Kim, H.S. Yang 2011 *Constr. Build. Mater.* **25** 911–18
[5] S.A. Sheikh, T.T. Murat 1993 *ACI Struct. J.* **90** 542–53
[6] M. Azimi, A. Bagherpourhamedani, M.M. Tahir, A.R.B.M. Sam, C.K. Ma 2016 *Adv. Struct. Eng.* **19** 730–45
[7] G.B. Maranan, A.C. Manalo, B. Benmokrane, W. Karunasena, P. Mendis, T.Q. Nguyen 2018 *Compos. Struct.* **187** 454–65
[8] J. Kim, M. Kwon, W. Jung, S. Limkatanyu 2013 *Constr. Build. Mater.* **43** 563–74
[9] T.C. Rousakis, A.I. Karabinis 2012 *Mater. Struct.* **45** 957–75
[10] F.R. Mansour, S.A. Bakar, M. Vafaei, S.C. Alih 2017 *PCI J.* **62** 78–89
[11] U. Tamon 2005 *Jt. Semin. Concr. Eng.* 54–68
[12] N.H.F.A. Halim, S.C. Alih, M. Vafaei, M. Baniahmadi, A. Fallah 2017 *Int. J. Appl. Eng. Res.* **12** 12519–33
[13] M. Vafaei, S. C. Alih 2017 *Neural Comput. Appl.* **30** 2509-18
[14] M. Vafaei, S. C. Alih 2016 *Bull. Earthq. Eng.* **14** 3441-61
[15] N.H.F. Abdul Halim, S.C. Alih, M. Vafaei 2018 *Int. J. Civ. Eng. Technol.* **9**
[16] L. Braile 2003 *Earth Atmospheric Planetary Sciences* (Indiana: Purdue University) p 19
[17] C. Beschi, P. Riva, G. Metelli, A. Meda 2015 *J. Earthq. Eng.* **19** 25–47
[18] S. Koshimura, N. Shuto 2015 *Philos. Trans. R. Soc. London A Math. Phys. Eng. Sci.* **373**
[19] H.M. Mohamed, Z.A. Mohammad, B. Benmokrane 2014 *J. Bridge Eng.* 1–12
[20] T. Alkhrdaji 2015 *Building Blocks* 18–20
[21] Simulia DS. Abaqus 6.14 Documentation 2014
[22] Eurocode 2 2008 *Design of Concrete Structures: British Standard* (London: BSi)
[23] N.H.F.B.A. Halim, Sophia C. Alih, M. Vafaei 2018 *Proc. 7th Int. Grad. Conf. Eng. Sci. Humanit.* pp. 142–4
[24] C. Li, H. Hao, K. Bi 2017 *Eng. Struct.* **148** 373–86
[25] MapeWrapC UNI-AX 2018 (Malaysia: MAPEI Malaysia)
[26] FEMA 461 2007 *Interim Testing Protocols for Determining the Seismic Performance Characteristics of Structural and Nonstructural Components* (Washington, D. C.: FEMA) p 113
[27] H. Shad, A. Adnan, H. Behbahani, M. Vafaei 2016 *Struct. Eng. Mech.* **60** 131-48
[28] Vafaei M and Alih S C 2015 *Earthq. Struct.* **8** 541-53

IOP Conf. Series: Materials Science and Engineering **513** (2019) 012022 doi:10.1088/1757-899X/513/1/012022

Buckling behavior of cold-formed stub channels under compression

R Ismail[1,2*], N I Bisnan[2], Y C Bon[2] and I S Ishak[2]

[1]Institute for Infrastructure Engineering and Sustainable Management (IIESM) Universiti Teknologi Mara (UiTM) Shah Alam, Selangor, Malaysia
[2]Department of Civil Engineering Universiti Teknologi Mara (UiTM) Shah Alam, Selangor, Malaysia

[*]Corresponding author: rozaina_fka_uitm@yahoo.com

Abstract. Channel section is one of the most common cold-formed steel sections uses today. Cold formed steel members are widely used in roof trusses. Yielding and crushing are the modes of failure for short compression members or stub column. Design equations can be used to determine local buckling load capacity of stub column. In building construction, the stub column is found in the roof trusses. The purpose of this study is to understand buckling behaviour of the cold formed stub channels under compression. Seven (7) specimens were tested in this study. The results of local buckling, maximum load capacity and axial displacement of seven (7) cold-formed stub channels specimens were obtained by conducting the stub column test (compression test) and tensile test. Five (5) specimens are conservative to mean ratio and two (2) specimens are unconservative. Load-displacement graphs were plotted and two (2) specimens had achieved the expected curve. The failure patterns of the stub channels showed that the buckling occurred at the top edge and bottom edge of stub channels.

1. Introduction

Cold formed steel is manufactured from steel sheet, strips, plates or flat bars in roll-forming machines or by press brake or bending brake operations. Cold-formed steel (CFS) members are being used extensively in building construction [1]. Cold formed steel products usually use in structure member, sheeting and sandwich panel. The AISI Standard provides integrity formulas for cold formed steel members. Given that the web and the flange of cold formed steel members are light and thin, the behaviour of member components become more complex. There is variety of failure modes such as the following: Flexural buckling, torsion buckling, flexural-torsion buckling and local buckling [2].

Cold formed steel structural members delivered many benefits in building construction including high strength-to-weight ratio, non-shrinking and non-creeping behaviour at ambient temperature and ease of prefabrication and mass production. Another major advantage of cold-formed steel construction is the sections allowing for compact packaging and shipping and easily for transportation. Cold formed channels are normally used to construct walls and partitions called stud and trusses. Trusses made of cold formed members may be found in industrial, storage buildings and roof truss. The main chords are usually channel sections joined back to back. The web members are normally single channels. Stub column is a short vertical structural member that are bolted between two other members such as a rafter and ceiling beam. Currently, cold-formed steel (CFS) is regularly employed in low-to-mid-rise building construction both for non-structural and structural walls [3].

In building construction, the stub column is found in the roof truss. Cold formed channel is commonly used for interior wall such as studs and trusses. In trusses the compression force is acting to the cold formed channels members. Under axial loads the compression members can fail by yielding, local buckling and overall buckling. Yielding and local buckling are controlled by the slenderness of their section only, whereas overall buckling is dependent on the slenderness of the section as well as the slenderness of the column length. For short compression members or stub column, only the first two failure modes occur, that are yielding and buckling. Local buckling capacity of stub column can be determined using design equations. However, the channel could exhibit post buckling behaviour which gives higher load capacity. Therefore, there is a need to study the bulking behaviour of cold-formed stub channels under compression.

2. Method

Laboratory test was conducted for the analysis of the buckling behaviour of stub channel. Seven (7) stub channel column specimens were tested for compression test and yield strength. The laboratory tests to be carried out were stud column test and tensile test.

2.1. Stud column test

Stub column tests were conducted to know the relationship between load and axial displacement that occur during the compression by continuous load applied to the Specimen. Local buckling pattern can also be determined. From the stub column test, stress-strain curves were obtained. The stress and strain are obtained by dividing the applied load against the initial section area and the shortening divided by the initial length, respectively [4].

2.1.1. Stud column specimen

Seven specimens of stub channel were prepared as below:

i. The stub column specimens were cold-sawed from the stock at a distance at least equal to the shape depth away from a flame-cut end.
ii. The length of each stub column was less than three times the largest dimension of the cross section and not more 20 times the least radius-of-gyration.
iii. The end of the column was milled plane and perpendicular to the longitudinal axis of the column.
iv. The thickness of the flange and webs and the length and cross sectional area of the stub column were measured and recorded.
v. Descriptions of specimens are shown in Table 1.

Table 1. Geometry of stub channel.

Specimen No.	Specimen (LC x B x b x t)	Thickness, t (mm)	Web Width, B (mm)	Flange Width, b (mm)	Length, L (mm)
1	75Cx25x20x1	1.0	25	20	75
2	150Cx50x25x1		50	25	150
3	225Cx75x25x1		75	25	225
4	300Cx100x25x1		100	25	300
5	375Cx125x25x1		125	25	375
6	450Cx150x30x1		150	30	450
7	600Cx200x40x1		200	40	600

Linear Variable Differential Transducers (LVDT) were used to determine the axial displacement or vertical displacement during test. Transducer provides linear output for small displacements where the core remains within the primary coils. The position of LVDT are placed at top bearing plate to measure the necessary displacements as shown Figure 1.

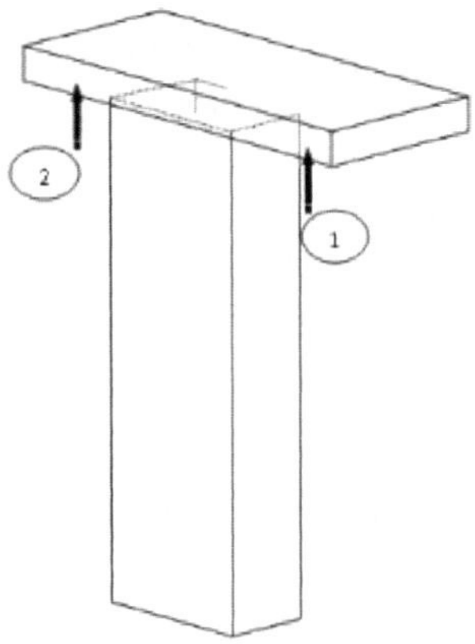

(a) (b)

Figure 1. (a) The position of transducer (b) LVDT.

2.1.2. Test set-up

The stub column must be uniformly strained in compression over the entire section during test. To do so, a spherical base is adjusted before loading such that end faces, which are finishing flat, of the specimen keep full contact with the bearing plates, after which the rotation of the spherical base restrained by steel wedges. Linear Variable Differential Transducers (LVDT) was attached at top of bearing plate to determine the axial displacement when the load is applied to compress. Test setup of a stub column is shown in Figure 2.

Figure 2. Test setup of stub column.

The load increments in the elastic region should be less than 10% of the expected yield load [5]. In this experiment, the chosen load rate was 20N/sec.

2.2. Tensile test

Tensile tests were carried to determine the yield strength of the specimen. Yield strength of the material is defined as the stress applied to the material at which deformation start to occur while the material loaded. The determination of the yield strength is necessary because this value is to determine the effective width of stub column then applied into capacities calculation.

2.2.1. Test specimen

In this study, the specimen of test is in accordance with the BS EN 10002-1: 2001. Types of test piece to be used for thin products: sheets, strips and flat between 0, 1mm and 3mm thick. The thickness of specimen used in this study 1mm. The dimension specimen is shown in the Figure 3.

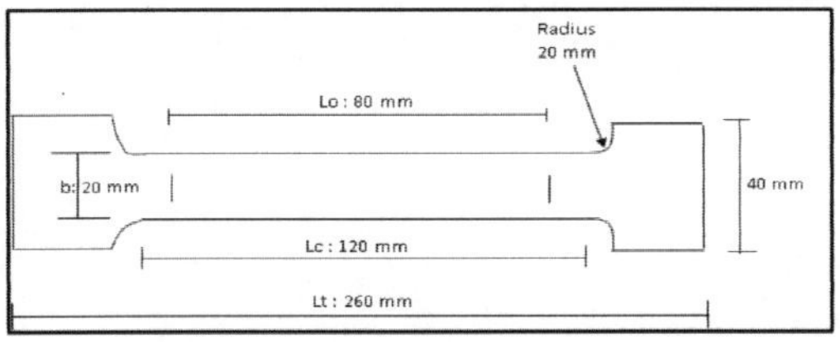

Figure 3. Dimension of test specimen.

2.2.2. Tensile test procedure

A test specimen was grasped at each end by a tensile testing machine and subjected to a tensile axial load. A standard specimen was used for tensile testing for calibration purposes. The specimen is generally a cylindrical rod with a precise diameter, although other specimen geometries can be used to evaluate certain situations. Gauge marks were scribed onto the surface of the specimen at specified distance. The specimen was loaded to a universal testing machine and load was applied to the specimen at a specific rate. As the specimen is axially loaded in tension, the distance between the gauge marks was monitored. This measurement can be made mechanically with an extensometer or optically with digital video.

The stress can be calculated based on the applied load. The change in the distance between the gauge marks can be converted into strain. These two factors can be used in order to plot a graph against each other to form a stress-strain diagram. A stress-strain curve illustrates the relationships between the applied load, the geometry of the part, the material properties of the part, and the reaction of the part to the applied load.

3. Result and discussion

3.1. Tensile stress

Table 2 shows the experimental results of tensile tests, for this study the yield stress used was 345 N/mm^2.

Table 2. Yield stress of cold-formed plate.

	Specimen 1	Specimen 2	Specimen 3	Average
Area (mm^2)	20	20	20	20
Yield load (kN)	7.15	6.78	6.78	6.9
Yield Stress (N/mm^2)	357.5	339	339	345

3.2. Load vs displacement relationship stub channels under compression

In investigating the performance of the stub channels section, the test results were analyzed in the context of load-displacement profiles, shortening and mode of buckling. At the end of the experiment, it was found that the results of each stub channels section that can sustain are shown in the Table 3.

Table 3. Experimental results of the cold-formed stub channel.

Specimen No.	Specimen	Maximum load (kN)	Axial Displacement (mm) at Maximum load
1	75Cx25x20x1	20.96	0.57
2	150Cx50x25x1	20.31	0.99
3	225Cx75x25x1	20.59	0.46
4	300Cx100x25x1	20.79	1.16
5	375Cx125x25x1	25.41	0.78
6	450Cx150x30x1	27.46	1.11
7	600Cx200x40x1	21.51	0.91

Figure 4 shows the load-axial displacement response of Specimen 1. At the 0.57 mm axial displacement, the stub channel could sustain the maximum load of 20.96 kN. After the specimen reached the maximum strength, the load decreased and axial displacement increase. The maximum axial displacement of the specimen was 2.10 mm shorter from the original length.

Figure 5 shows at the 0.99 mm axial displacement of the stub channel the maximum load can sustain was 20.31 kN. After the specimen reached the maximum strength, the load decreased and axial displacement increase. The maximum axial displacement of the specimen is 2.37 mm from the original length.

Figure 6 shows at the 0.46 mm axial displacement of the stub channel the maximum load can sustain was 20.59 kN. After the specimen reached the maximum strength, the load decreased and axial displacement increase. The maximum axial displacement of the specimen is 1.73 mm from the original length. Figure 7 shows at the 1.16 mm axial displacement of the stub channel the maximum load can sustain was 20.79 kN. After the specimen reached the maximum strength, the load decreased and axial displacement increase. The maximum axial displacement of the specimen is 2.17 mm from the original length.

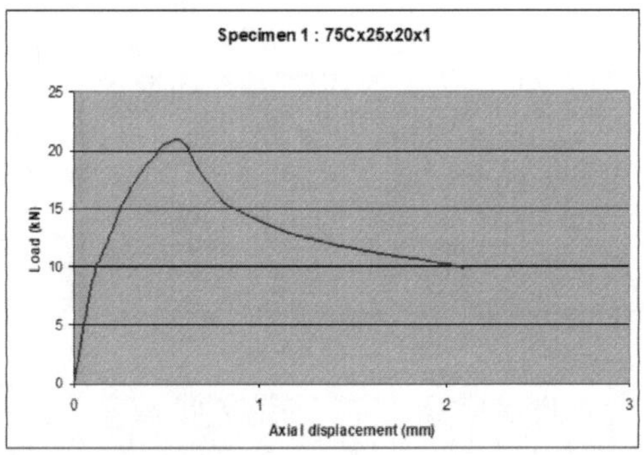

Figure 4. Graph of load versus axial displacement for specimen 1:75Cx25x20x1.

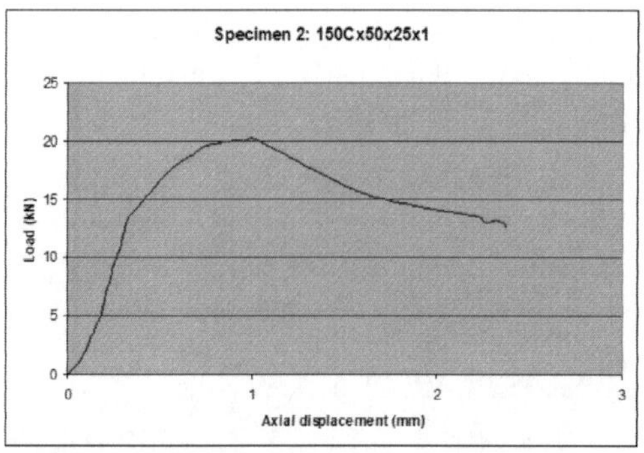

Figure 5. Graph of load versus axial displacement for specimen 2: 150Cx50x25x1.

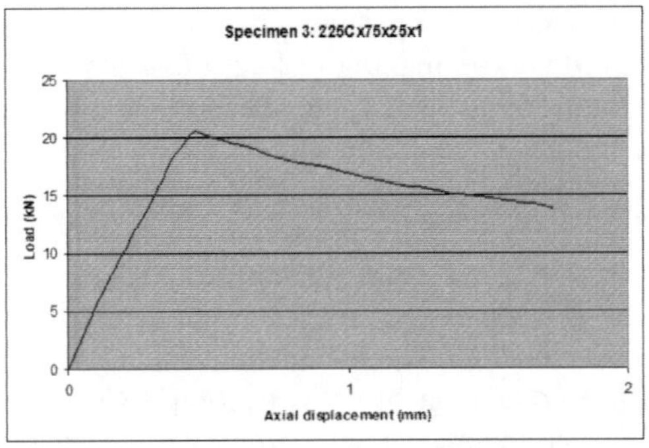

Figure 6. Graph of load versus axial displacement for specimen 3: 225Cx75x25x1.

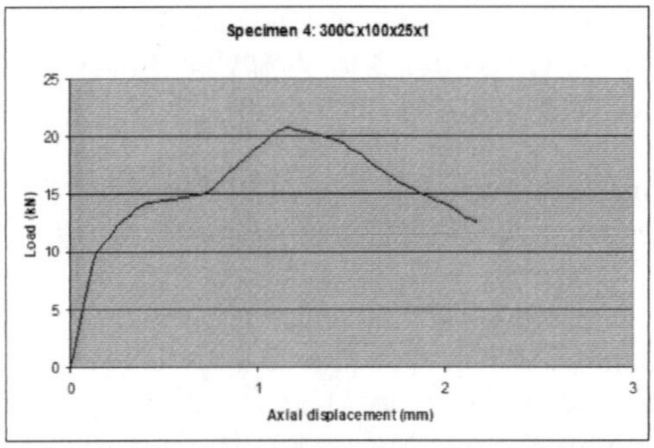

Figure 7. Graph of load versus axial displacement for specimen 4: 300Cx100x25x1.

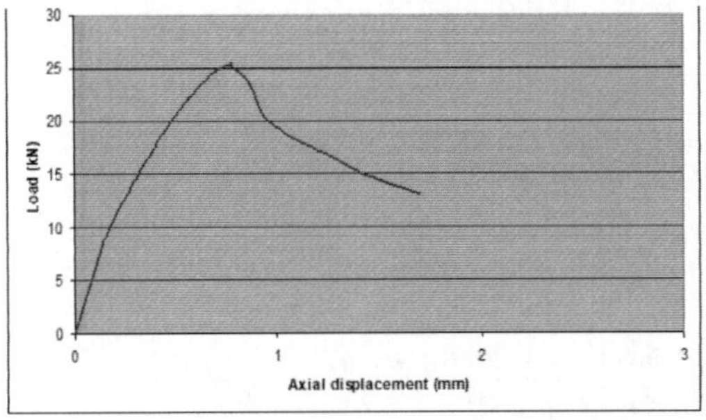

Figure 8. Graph of load versus axial displacement for specimen 5: 375Cx125x25x1.

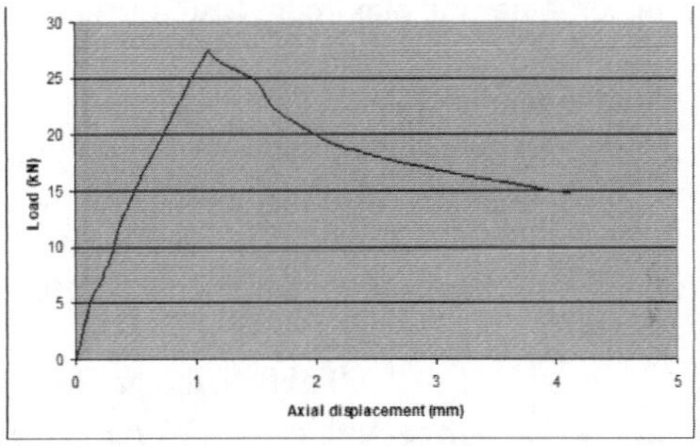

Figure 9. Graph of load versus axial displacement for specimen 6: 450Cx150x30x1.

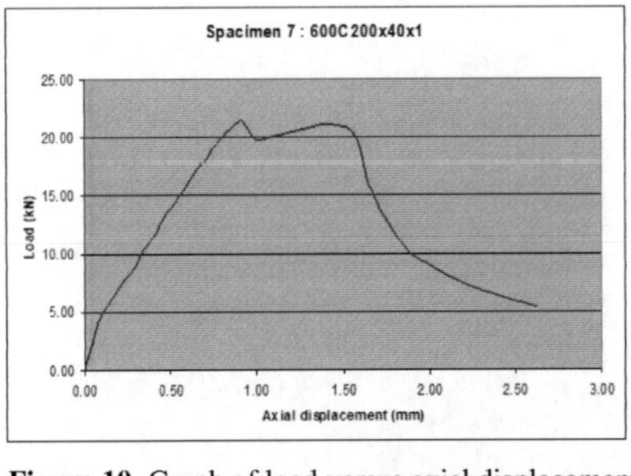

Figure 10. Graph of load versus axial displacement for specimen 7: 600Cx200x40x1.

From Figure 8, it is shown at the 0.78 mm axial displacement of the stub channel the maximum load can sustain was 25.41 kN. After the specimen reached the maximum strength, the load decreased and axial displacement increase. The maximum axial displacement of the specimen is 1.70 mm from the original length. Figure 9 shown at the 1.11 mm axial displacement of the stub channel the maximum load can sustain was 27.46 kN. After the specimen reached the maximum strength, the load decreased and axial displacement increase. The maximum axial displacement of the specimen is 4.12 mm from the original length. From Figure 10 shown at the 0.91 mm axial displacement of the stub channel the maximum load can sustain was 21.51 kN. After the specimen reached the maximum strength, the load decreased and axial displacement increase. The maximum axial displacement of the specimen is 2.62 mm from the original length.

The relationship of load and axial displacement is presented to know the behavior of mode buckling when the load is applied to specimens and to determine the maximum load (ultimate load) that the specimens can sustain before local buckling occurred. The expected load-displacement curve should comprise the post-buckling stage, in which after reaching the critical load, there were still further increased of applied loads until to the maximum load where local buckling take place and the specimens could no longer sustain the load.

From the experimental result, two specimens (150Cx50x25x1 and 300Cx125x25x1) were considered conservative as they achieved the expected curve. For other specimens, the curves showed immediate increased of load until they achieved the maximum load. All the specimens could carry the maximum load up to 20 kN and the range of axial displacement when reaching the maximum load was between 0.5 mm to 1.5 mm. The specimen 375Cx125x25x1 had the maximum load carrying capacity compared to other specimens. When the load increases, the non-linear behavior of load-axial displacement can be seen, most probably due to the on set of local buckling that occurred on the specimens. After the maximum load, the specimens start to unloading condition.

3.3. Load capacity

BS 5950: Part 5: 1987 provide the prediction of load capacity of the specimens. The calculation is based on the effective width approach where the yield stress is assumed to act on the reduced effective area. Then, the classification of element is used to determine the stiffened element and unstiffened element. From there, the load capacity is determined by using the stress formula.

Table 4. Comparison of compression load capacity.

Specimen No.	Specimen	Load capacity (Experiment) kN	Load capacity (Theoretical) kN	The Ratio (Experimental/ Theoretical Load Capacity)
1	**75Cx25x20x1**	20.96	20.25	1.03
2	**150Cx50x25x1**	20.31	23.63	0.85
3	**225Cx75x25x1**	20.59	21.74	0.95
4	**300Cx100x25x1**	20.79	22.39	0.93
5	**375Cx125x25x1**	25.41	22.77	1.12
6	**450Cx150x30x1**	27.46	23.39	1.17
7	**600Cx200x40x1**	21.51	24.74	0.87
			Mean	**0.99**

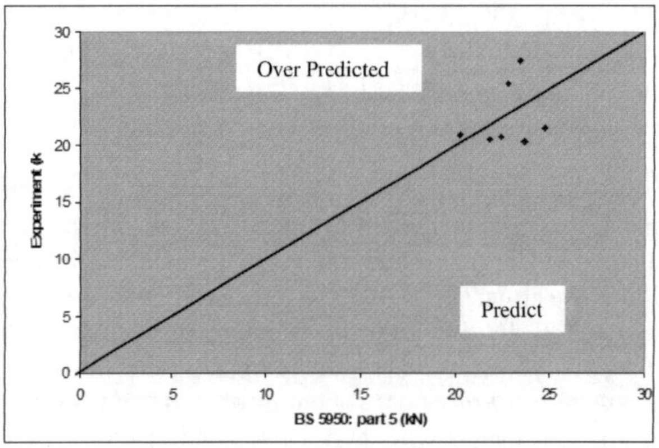

Figure 11. Experiment vs BS 5950:Part 5.

Table 4 and Figure 11 show the comparison between the experimental result and the predicted theoretical using the BS 5950: Part 5: 1987. The mean for the ratio of experimental result to the predicted theoretical were also calculated base ~~to~~ on the geometry of the stub channels. Mean value for all specimens is 0.99, where the ratios of the predicted to theoretical result are close to one. The BS 5950: Part 5: 1987 predictions are more conservative. From Figure 11, it can be concluded that four specimens (150Cx50x25x1, 225Cx75x25x1, 300Cx100x25x1 and 600Cx200x25x1) is conservative and close to 1.0. However the prediction for specimen 75Cx25x20x1 is more accurate, close to 1.0 and the experimental result, and associated errors are less than 5% for most specimens. Specimen 375Cx150c25x1 and 450Cx150x25x1 are not conservative, the experimental loads are respectively 12% and 17% higher than predicted result.

3.4. Failure pattern of stub channels column under compression

The load was applied on the specimens incrementally until it failed. Buckling was observed on the specimen as they occurred at every increment. The load corresponding to the appearance of first buckling and the failure load were recorded. Buckling was observed at various positions on the specimens. Mostly the specimens found to be failed by buckling at the top edge and bottom edge of specimens. Figure 12 shows how the specimens buckled for all specimens. Almost in many cases, the buckling occurred at the web which at the stiffened element. Three (3) specimens buckled at the bottom edge and remaining four (4) specimens occurred at top edge. The buckling location can be concluded to occur near the support of the stub channel. The patterns of the buckling show that the unstiffened element of stub channel transferred the load to the stiffened element to cater the load and the occur early, at edge of stub channels.

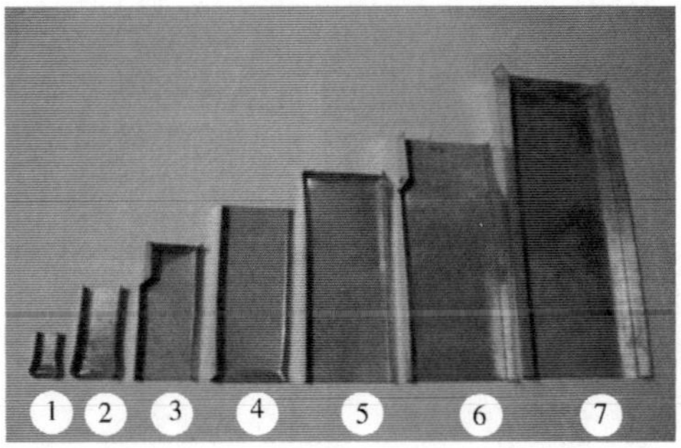

Figure 12. Failure patterns of stub channels.

4. Conclusion

Buckling behaviour of cold-formed stub channels under compression was studied. The axial load and deformation characteristics of the specimens were measured experimentally. Load capacity predictions using BS 5950: Part5: 1987 are comparable with the experimental values. Five (5) specimens were conservative whereas two (2) specimens unconservative to the mean ratio.

Specimens 150Cx50x25x1 and 300Cx100x2x1 achieved the objective of study. This specimen complied with the expected load-displacement curve and predicted theoretical calculation with experimental result. The pattern of buckling is similar where buckling occurred at the bottom edge of stub channels respectively.

From the engineering principles, the cold-formed stub channel could be considered as an ideal stub column. Meaning that, the specimen is perfectly straight before loading and the load is to be applied through the centroid of the cross section. It is shown that cross section affects the ability of the specimen to sustain applied load.

Acknowledgement
The authors would like to acknowledge that this research has been carried out as part of a project Bestari Perdana Research Grant with File No 600-IRMI/PERDANA 5/3 BESTARI (081/2018) internal fund, funded by Universiti Teknologi MARA (UiTM) and support from Faculty of Civil Engineering, Universiti Teknologi MARA (UiTM).

References
[1] D. Bondok, H. Salim, A. Saucier, and R. Dinan 2015 "Static resistance function of cold-formed steel stud walls," *Thin Walled Struct.* **94** 177–187.
[2] D. M. Lue, P. T. Chung, J. L. Liu, and C. L. Pan 2009 "The Compression Strength of Slender C-Shape Cold-formed Steel Member with Web Openings 2009 " *Int. J. Steel Struct.* **9**(3) 231–240.
[3] K. D. Peterman, N. Nakata, and B. W. Schafer 2014 "Hysteretic characterization of cold-formed steel stud-to-sheathing connections," *J. Constr. Steel Res.* **101** 254–264.
[4] H. Kuwamura 2003 "Local Buckling of Thin-Walled Stainless Steel Members TT -," *Int. J. Steel Struct. TA* **3**(3) 191.
[5] T. V. Galambos 1998 *Guide to Stability Design Criteria for Metal Structures*, 5th ed. John Wiley & Sons, 1998.
[6] BS EN 10002-1, *Metallic materials-Tensile Testing, Part 1: Method of test at ambient temperature*. 2001.
[7] BS 5950: Part 5, *Structural use for steelwork in building, Part 5 : Code of practice for design of cold formed section*, 5th ed. London: British Standard Institution, 1987.

IOP Conf. Series: Materials Science and Engineering **513** (2019) 012023 doi:10.1088/1757-899X/513/1/012023

Influence of steel fibre, electrical waste copper wire fibre and electrical waste glass fibre on mechanical properties of concrete

A Sofi[1*] and Ganesh Naidu Gopu[2]

[1] Associate Professor, Department of Structural and Geotechnical Engineering, School of Civil Engineering, VIT, Vellore, Tamilnadu, India
[2]Research Scholar, School of Civil Engineering, VIT, Vellore, Tamil Nadu,India

[*]Corresponding author: sofime@gmail.com

Abstract. Fibre-reinforced concrete (FRC) contains fibrous material which increases its structural integrity. The application of irregular arrangement of fibres to concrete altogether upgrades its essential characteristics, for instance, static flexural strength, influence quality, elasticity and flexural stiffness. Filaments are added to cement to control breaking because of plastic shrinkage and to drying shrinkage. This paper exhibits the use of steel fibre, electrical waste copper wire fibre and electrical waste glass fibre of various percentages of volume fractions such as 0.25%, 0.5%,0.75%,1.0% and 1.25% incorporated in concrete. Mechanical properties such as compressive strength, splitting tensile strength and flexural strength tests were conducted for normal concrete and fibre reinforced concrete for a curing period of 7 days, 14 days and 28 days. In addition to that modulus of elasticity of concrete for various percentages of fibre content was investigated for a curing period of 28 days. The test results of fibre reinforced concrete were compared with normal concrete to determine the influence of fibres used.

1. Introduction

In the recent years, the various fibres were developed and used in the construction industries. Steel fibre is mainly used in concrete for various applications[1]. Other types of fibres such as polythene fibres, carbon fibres, polyamide fibres, sisal fibre are now developed and also used in construction, industrial and infrastructure development. Some electronic waste fibres can also be used in concrete[2]. Electronic waste (E-Waste) is considered more toxic than municipal waste. When E-waste used in concrete, it reduces the environmental pollution and reduces the solid waste problem[3]. E-waste in concrete is the new revolutionary concept of sustainable concrete. When E-waste plastic type fibre are used at small size the results obtained are good when compared with the larger size [4]. In this paper, the extracted outer casting of electronic wire was used for making E-waste fibre [5]. Fibre reinforced concrete helps to reduce the crack propagation and increase the mechanical properties compared to the normal concrete[6]. In hybrid concrete, the micro fibres delayed the development of macro cracks and shows greater strength and crack resistance [7]. It also improves the mechanical properties of concrete[8]. The strength and the corresponding deflection was increase linearly with increasing steel fibre content in concrete[9].This paper exhibits influenceof mechanical properties of concrete by use of steel fibre, electrical waste copper wire fibre and electrical waste glass fibre of various percentages of volume fractions in concrete.

2. Experimental investigation

2.1. Test materials

2.1.1. Cement
Ordinary Portland cement (OPC) of 53 Grade was used in entire program. Specific gravity was found to be 3.15 [10].Initial and final setting times of cement is found to be 200 and 280 minutes respectively. Properties of the cement as show in Table-1

2.1.2. Fine aggregates
Good quality locally available river sand was used as a fine aggregate. The sand was conforming to zone II [11]. Aggregate passing through 4.75 mm sieve and retained on the 75 µm sieve. Specific gravity of the fine aggregates is found to be 2.66. Properties of fine aggregates was show in Table -2

2.1.3. Coarse aggregates
The Crushed granite stone aggregate was used from the local quarry. In this experiment the aggregate was used of 20mm down and Specific gravity of the fine aggregates is found to be 2.58. Properties of coarse aggregates was show in Table -2

Table 1. Properties of cement.

Test	Results
Fineness	227 m^2/kg
Soundness	10 mm
Setting time	
1.initial	32 min
2.final	581min
Compressive strength	
1. 7days	27 N/mm^2
2. 14 days	37 N/mm^2
3. 28 days	53 N/mm^2

Table 2. Properties of aggregates.

Physical tests	Coarse	Fine
Specific gravity	2.58	2.66
Fineness modulus	4.32	2.32
Bulk density (kg/m^3)	1540	1780

2.1.4. Steel fibres (SF)
Steel fibres confirming to ASTM A 820/ A 820M – 04were used[12].Steel fibres improve ductility, flexural strength and toughness (shown in Figure 1)The properties of the steel fibres show in Table 3.

IOP Conf. Series: Materials Science and Engineering **513** (2019) 012023 doi:10.1088/1757-899X/513/1/012023

2.1.5. *Electrical waste copper wire fibre*

E-waste consists of all waste from electronic and electrical appliances. It includes computer and its accessories monitors, printers, internet cables, power supply cables, electrical machines windings, central processing units, mobile phones and chargers, remotes, compact discs, headphones, batteries, LCD/Plasma TVs, air conditioners, refrigerators and other household appliances. The length of the fibre used was 30mm(shown in Figure 2). Properties of the electrical waste copper wire fibre as shown in Table 3.

2.1.6. *Electrical waste glass fibres*

Electrical waste glass fibres confirming to ASTM D3517-14 were used [13]. The dried fibres were then cut to the required lengths to attain the desired aspect ratio of 30(shown in Figure 3). Properties of electrical waste glass fibres were shown in Table 3.

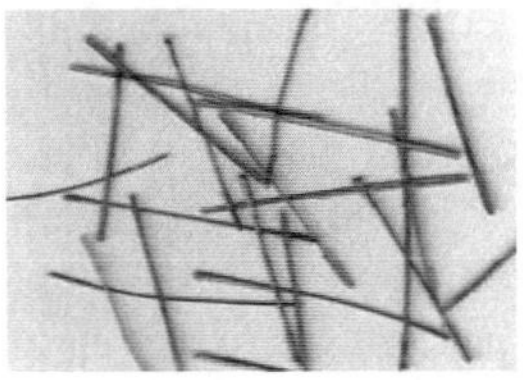

Figure 1. Steel fibres.

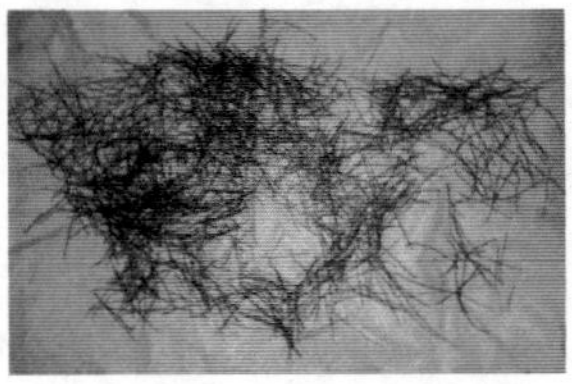

Figure 2. Electrical waste copper wire fibres.

Figure 3. Electrical waste glass fibres.

Table 3. Properties of fibres.

Fibre category	Steel fibres	Electrical waste copper wire fibre	Electrical waste glass fibre
Fibre shape	Straight	Straight	Straight
Fibre length [mm]	40	30	50
Fibre diameter [mm]	0.8	0.6	1
Aspect ratio	50	50	50
Tensile strength of fibres [N/mm^2]	1220	966	1950

2.2. *Workability*

The slump test was done for all the concrete mix to check the workability of concrete and is shown in Table 4. No admixture is used in the concrete mix to improve its workability [14].

Table 4. Slump values.

Fibres used (%) by volume of concrete	Steel fibre reinforced concrete (mm)	Electrical waste copper wire fibre reinforced concrete (mm)	Electrical waste glass fibre reinforced concrete (mm)
0	61	61	61
0.25	59	57	60
0.5	58	55	59
0.75	55	54	56
1.00	54	52	55
1.25	53	51	53

2.3. *Testing of specimens*

Casting and testing of concrete cubes, cylinders, beams were done as per IS code recommendations. The concrete mix adopted was M20 concrete [15] with varying percentage of fibres ranging from 0%, 0.25%, 0.5% and 0.75%. Nominal concrete cubes (15 cm x 15 cm x 15 cm), concrete cylinders (15 cm diameter and 30 cm long)and prism (10 cm×10cm × 50cm)were used. A blend of fibres were mixed with the aggregate while casting the specimens, it was ensured that fibres were uniformly distributed throughout the mix. Tests were conducted on concrete cubes using different percentage of fibres to check for variations in compressive, splitting tensile and flexural strength. Three sets of nine cubes, nine cylinders and nine prisms of M20 mix were cast without fibres. Later, different sets of cubes, cylinders and prisms were cast with fibre content ratio as 0.25%, 0.5%, 0.75%, 1.0% and 1.25%. The specimens were then transferred to curing tank for the required period of curing and tested.

2.3.1. *Compressive strength of concrete*

Compressive strength of concrete cubes(15cm×15cm×15cm) were made according to IS516 [15]. No packing material was placed in between the cube and loading frame. The load was applied gradually without any shock. Compressive strength of cubes were determined at a curing age of 7,14 and 28 days. The results were tabulated in Table 5 and the graphical representation in Figure 4.

2.3.2. *Splitting tensile test of concrete*

The Splitting tensile strength of cylinders (15cm dia and 30cmlength) were made according to IS 5816 [16]. Diametric lines on each of cylinder were drawn to ensure that they were in the same plane. The splitting tensile strength was determined at curing periods of 7,14 and 28 days. The results were tabulated in Table 6 and the graphical representation in Figure 5.
The Splitting tensile strength (T) was determined from the equation-1.

$$T = 2P/\pi \mathrm{LD} \tag{1}$$

2.3.3. *Flexural strength of concrete*

Flexural strength of prisms (10cm×10cm×50cm) casted are determined at a curing age of 7,14 and 28 days respectively as per IS516 [15].The load is applied gradually without any shock or vibration. Flexural strength of prisms were determined at a curing age of 7,14 and 28 days [17]. The results were tabulated in Table7 and the graphical representation in Figure 6. The flexural strength of the specimen was determined from equation-2.

$$\mathrm{F_b} = \frac{\mathrm{Pl}}{\mathrm{Bd^2}} \tag{2}$$

Table 5. 28 days compressive strength of concrete cube with different fibres in N/mm^2.

% of mixing proportion by volume of concrete	Steel fibre reinforced concrete	Electrical waste copper wire fibre reinforced concrete	Electrical waste glass fibre reinforced
0.00	27.36	27.36	27.36
0.25	34.23	30.29	30.43
0.50	35.11	32.51	31.10
0.75	37.42	33.86	31.56
1.00	37.46	32.63	29.53
1.25	39.14	31.13	28.19

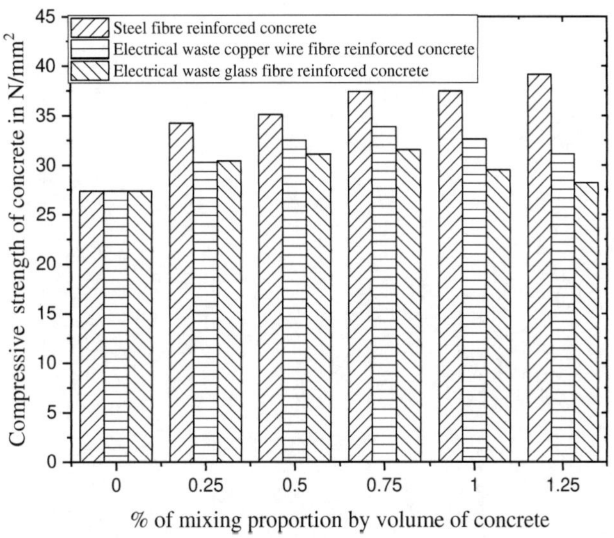

Figure 4. Compressive strength of concrete cube with different fibres.

Table 6. 28 days splitting tensile strength of cylinder with different fibres in N/mm^2.

% of mixing proportion by volume of concrete	Steel fibre reinforced concrete	Electrical waste copper wire fibre reinforced concrete	Electrical waste glass fibre reinforced concrete
0.00	2.67	2.67	2.67
0.25	3.10	2.79	3.25
0.50	3.34	3.12	4.05
0.75	3.83	3.41	4.78
1.00	4.20	3.64	5.03
1.25	4.85	3.91	5.10

IOP Conf. Series: Materials Science and Engineering **513** (2019) 012023 doi:10.1088/1757-899X/513/1/012023

Table 7. 28 days flexural strength of prism with different fibres in N/mm^2.

% of mixing proportion by volume of concrete	Steel fibre reinforced concrete	Electrical waste copper wire fibre reinforced concrete	Electrical waste glass fibre reinforced concrete
0.00	4.44	4.44	4.44
0.25	4.62	4.51	4.99
0.50	4.93	5.23	5.18
0.75	5.26	5.41	5.52
1.00	5.65	5.62	5.60
1.25	6.54	6.17	5.85

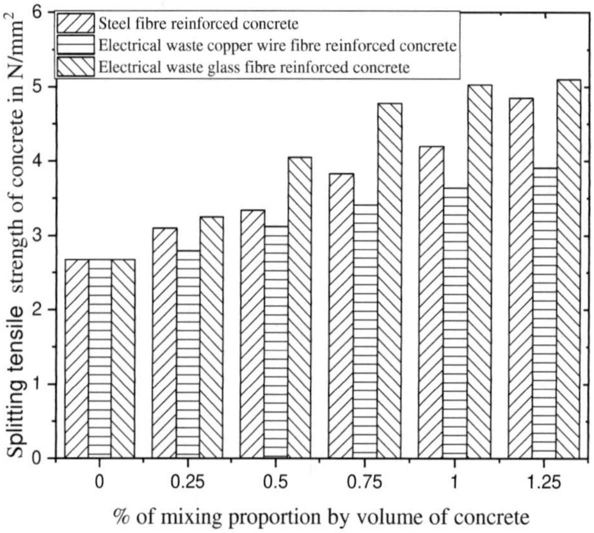

Figure 5. Splitting tensile strength of concrete with different fibres in N/mm^2.

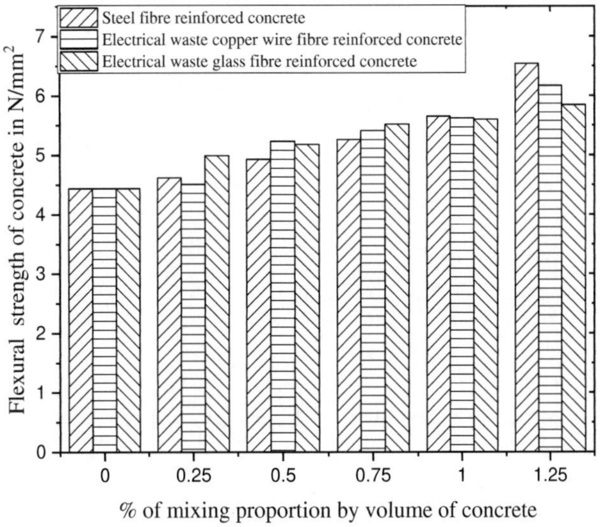

Figure 6. Flexural strengths of prism with different fibres in N/mm^2.

2.3.4. Elastic modulus of concrete

Elastic modulus was determined at 28 days of curing period on cylindrical specimen (150mm dia and 300mm height) under compression load as per ASTM C469-02 [18] and the deformation measured by means of dial gauge fixed between certain gauge lengths.The elastic modulus of the concrete incorporate with steel fibre and electrical waste copper wire fibres at various volumes is mentioned in Table 8. Figure 7 and 8 shows the stress-strain curve for the modulus of elasticity which has taken as the slope of the chord from the origin to some arbitrary point, the secant modulus calculated was for 40% of maximum stress.

Table 8. Elastic Modulus of concrete with different fibres.

% of mixing proportion by volume of concrete	Steel fibre reinforced concrete	Modulus of elasticity (GPa)	Electrical waste copper wire fibre reinforced concrete	Modulus of elasticity (GPa)
0.00	27.36	28.91	27.36	28.91
0.25	34.23	29.25	30.29	29.28
0.50	35.11	31.44	32.51	31.21
0.75	37.42	32.44	33.86	31.84
1.00	37.46	33.38	32.63	32.03
1.25	39.14	35.32	31.13	32.68

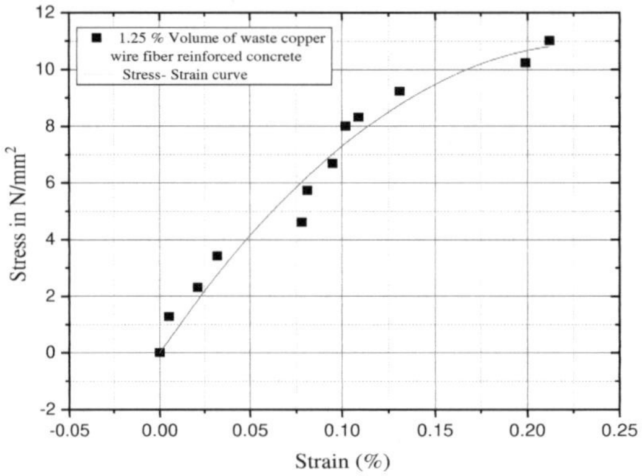

Figure 7. Stress –strain behavior of the electrical waste copper wire fibre reinforced concrete.

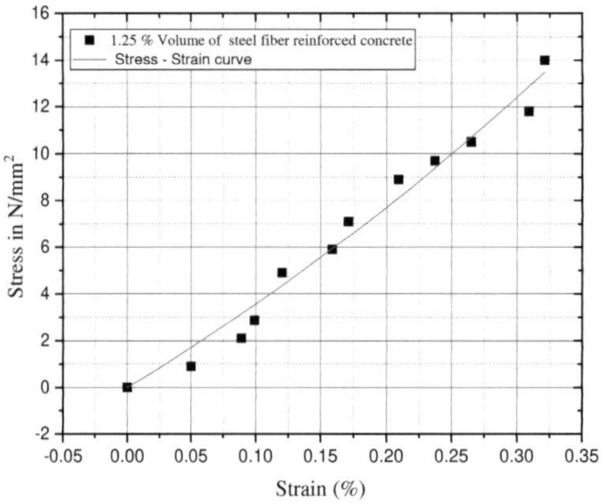

Figure 8. Stress –strain behavior of the steel fibre reinforced concrete specimen.

3. Conclusion

The following conclusions could be drawn from the present investigation.

- Compressive strength of concrete by using steel fibre, waste copper fibre and Electrical waste glass fibres were increased by 42.6%, 23.76 and 15.35% when compared to normal concrete.
- Splitting tensile strength of concrete by using steel fibre, waste copper fibre and Electrical waste glass were increased by 81.6%, 46.4% and 90.1% when compared to normal concrete.
- Flexural strength of concrete by using steel fibre, waste copper fibre and Electrical waste glass fibres were increased at 46.1%, 38.8% and 31.7% when compared to normal concrete.
- Elastic modulus of concrete by using steel fibre and waste copper fibre were increased at 21.1%and 13.04% when compared to normal concrete.

Hence this paper can be concluded that the addition of steel fibre, copper fibre and electrical waste glass fibre incorporated at 0.25,0.5%, 0.75 %, 1.0% and 1.25% by volume of concrete improve the compressive strength, splitting tensile strength, flexural strength and modulus of elasticity of the concrete and it also reduces the crack width under different loading conditions that was observed in concrete testing conditions.

References

[1] Sofi A. and Phanikumar, B.R., Effect of pond ash and steel fibre on engineering properties of concrete, *Ain shams Engineering Journal*, **7**, (2016), pp.89-99.

[2] Arjun, R.K. and Senthil Kumar, K., Novel fibrous concrete mixture made from recycled PVC fibres from electronic waste, *J. Hazard Toxic Waste* **21** (2016).

[3] Ashwini Manjunath, B.T, Partial replacement of E-plastic waste as coarse aggregate in concrete, *Procedia Environmental Sciences* **35** (2016), 731-739.

[4] Byung-wan, Jo., Young-Hyun, S., Young-Jin, K., The Evaluation of elastic modulus for steel fibre reinforced concrete, *Russian Journal of Non-destructive Testing* **37** (2001), 152-161.

[5] Doo-Yeol., Yoo, H.O, Young-Soo, Y, Material and bond properties of ultra high performance fibre reinforced concrete with micro steel fibre, *Composites of Engineering* **28** (2014),122-133.

[6] Balanji, E. K. Z., Sheikh, M. Neaz. and Hadi, M. N. S. Performance of high strength concrete columns reinforced with hybrid steel fibre under different loading conditions. *Proceedings of the First European and Mediterranean Structural Engineering and Construction Conference.*

[7] Senthil Kumar, K., Baskar, K., Development of Eco-friendly concrete incorporating recycled high impact polystyrene from hazardous Electronic waste, *J. Hazard. Toxic Waste* **19** (2015), 1-11.

[8] Iftekar Gull, Balasubramanian, M., A new paradigm on experimental investigation of concrete for E-plastic waste management, *J. Eng Trend and Tech* **10** (2014), 180-186.

[9] Sofi A., Phanikumar, B.R., Durability properties of fibre-reinforced pond ash-modified concrete, *Journal of Engineering Science and Technology*, **11** (2016), 1385 – 1402.

[10] IS 12269, Indian Standard code for Specification of 53 grade Ordinary Portland Cement, 1987.

[11] IS 383, Indian Standard code for Specification of Coarse and fine aggregates, 1987.

[12] ASTM A 820/ A 820M, Standard Specification for Steel fibres.

[13] ASTM D3517-14, Code for Glass fibre specifications and testing.

[14] IS 10262, Indian Standard code for Mix Design of concrete mixes, 2009

[15] IS 516, Indian Standard Methods of test for strength of concrete, 1959.

[16] IS 5816, Indian standard code for measurement of splitting tensile strength of concrete, 1959.

[17] Sofi A and Phanikumar, B.R., An experimental investigation on flexural behaviour of fibre reinforced pond ash-modified concrete, *Ain shams Engineering Journal*, **6** (2015), pp. 1133-1142.

[18] ASTM C469-02 Standard Test Method for Static Modulus of Elasticity and Poisson's ratio of concrete in compression. Annul Book of ASTM Standards (2005).

IOP Conf. Series: Materials Science and Engineering **513** (2019) 012024 doi:10.1088/1757-899X/513/1/012024

Effect on Salak Tinggi residual soil mixed Bentonite as compacted clay liner

Norazlan Khalid[1*], Mazidah Mukri[1], Mohd Fadzil Arshad[1], Norbaya Sidek[1] and Faizah Kamarudin[2]

[1]Faculty of Civil Engineering, UiTM Shah Alam, Malaysia
[2]Faculty of Civil Engineering, UiTM Pulau Pinang, Malaysia

[*]Corresponding author: aln_kh82@yahoo.com

Abstract. This paper present the result from experimental studies on using bentonite mixed with residual soil. The entire testings were performed on Salak Tinggi residual soil mixed with various percentage of bentonite content (5%, 10% and 15%). A series of laboratories were performed to investigate the effectiveness of mixed samples associated to physical properties testing such as: particle size distribution, *pH*, plastic index (*PL*), liquid limit (*LL*), plasticity index (*PI*) and linear shrinkage (*LS*). Meanwhile the standard proctor compaction test was conducted on mixed soil-benonite samples to determine the maximum dry density (MDD) and optimum moisture content (OMC). In addition the permeability test to determine the hydraulic conductivity was conducted at MDD condition at effective stress of 100 kPa. It was found that the addition of bentonite to the Salak Tinggi residual soil changed the physical properties result. Meanwhile, compaction results show a decrement in the MDD with increment of bentonite percentage mixed. The result also showed the bentonite mixed decreases the hydraulic conductivity value ($k < 1x10^{-9}$ m/s) for Salak Tinggi soil significantly. This finding shows that by using the small amount of 5% bentonite significantly can be used as compacted of soil liner materials.

1. Introduction

Residual soils are widely distributed over the globe covering and it's covered especially in tropical climate country. Malaysia is a tropical climate and humid country had residual soil broadly distributed and its covers more than quarters of the land area of Peninsular Malaysia [1]. Thus, residual soils have been used extensively in various earthwork projects, as a construction material and it suggests as a municipal landfill area are mainly constructed on residual soil.

Therefore, major concern issues for residual soils in tropics area are not feasible as compacted clay liner (CCL) application due to varies in characteristics. In current practices, there is different kind of clay liners used during the construction of landfill areas, such as inorganic clays or clayey soils due to have low hydraulic conductivity. In Malaysia, Compacted Clay Liners (*CCL*) and Geo-synthetic Clay Liners (*GCL*) are commonly used as a protective layer on residual soil surface in order to prevent groundwater contamination [2]. However, these liners are subjected to desiccation effect during long hours of exposure to the atmosphere thus causes shrinkage and eventually cracks failure [3].

Hydraulic conductivity is the most significant factor effecting soil liner performance. Liner are commonly composed of compacted soils and compaction characteristics consider as important role in determining the hydraulic conductivity [4]. This good compaction characteristic encourages the use of tropical residual soil as compacted clay liners (CCL) as a barrier not to only the economic application using local materials as CCL. Normally the hydraulic conductivity value for compacted liner must be less than or equal to 1 x 10^{-9} m/s [4-5]. Thus, more study is needed on the suitability use of bentonite (clay mineral) mixed with residual soil in order to determine the effectiveness of bentonite mixture on sedimentary residual soil. Therefore, it is useful for designing these types of engineered fill systems and to understand these mixtures as a purpose for compaction process and compaction energy that is used.

Hence the aim of this study was to evaluate the effectiveness of various percentage of bentonite content mixed with Salak Tinggi residual soil on the physical properties and engineering properties in hydraulic conductivity. This paper particularly reports the result finding on physical properties and results of hydraulic investigation on compacted Salak Tinggi residual soil mixed with bentonite from the several testing.

2. Materials

2.1. Salak Tinggi residual soil

Disturbed soil samples of yellowish residual soil Grade IV used in this study were derived from sedimentary residual soil and was collected in Salak Tinggi, Selangor, Malaysia. This soil samples were named as Salak Tinggi residual soil. To minimize the organic matter content, the soil was taken at 1.5 m to 2.5 m below from ground surface. Based on the geological map obtained from Minerals and Geoscience Department Malaysia [6], the Salak Tinggi area classified as sedimentary soil derived from phyllite, schist and slate: limestone and sandstone prominent. The yellowish-brown colours of Salak Tinggi soil were due to the changes on the mineral composition from the erosion of the top layer of parent rock. Seepage of the soluble minerals in the eroded materials deep underneath surface layer making the insoluble Fe and Al ions remains on the top. These metallic oxides create a new soil and gave the brownish colour. The physical properties were investigated guided from BS 1377: Part 2: 1990 and the result shown in Table 1. From the result, Salak Tinggi soil considered as strong acidic soil due to the average of *pH* value is 3.5 which are common in tropical climates [7]. It shows the Salak Tinggi soil sample considered as coarse soil with the major size of sand 69.2% and silt size 20.9% and clay size 9.9%. Meanwhile the plastic limit and liquid limit are 16% and 29% respectively. Based from the properties result, Salak Tinggi residual soil classified as Silty SAND (Low Plasticity, ML).

Table 1. Formatting sections, subsections and subsubsections.

Properties	Values
Depth (m)	1.5 – 2.5
Colour	Yellowish
Natural Moisture Content (%)	20.2
pH	3.5
Specific Gravity, (*Gs*)	2.67
Liquid Limit, *LL* (%)	29
Plastic Limit, *PL* (%)	16
Plasticity Index, *PI* (%)	13
Hydraulic Conductivity, *k* (m/s)	1.005 x 10^{-9}
Gravel (%)	0
Sand (%)	69.2
Silt (%)	20.9

Clay (%)	9.9
Classification	Very Silty SAND of Low Plasticity, ML

2.2. Bentonite

The sodium bentonite used in this study was in grey powder provided from the main supplier. Scientifically, bentonite is a term used to name smectic clay in its sedimentary form and has high content of clay mineral (usually montmorillonite). This clay mineral capable of swelling during uptake of water, has large specific surface area, excellent plasticity, and has low hydraulic conductivity [8]. Due to these characteristics, bentonite has a major role in industrial, environmental, and civil engineering activities that require sealants, absorbents, and hydraulic barriers. The laboratories testing result shown in Table 2 for the bentonite properties conducted based on BS 1377: Part 2: 1990. The results show the bentonite considered as a strong alkaline due to the high of *pH* value about 9.6 and has extremely of high plasticity index.

Table 2. Physical properties for bentonite [9].

Properties	Values
Colour	Grey powder
pH	9.62
Specific Gravity, (*Gs*)	2.14
Liquid Limit, *LL* (%)	419
Plastic Limit, *PL* (%)	190
Plasticity Index, *PI* (%)	229

3. Samples preparation and laboratories testing

The Salak Tinggi residual soil sample was mixed with bentonite and was prepared manually. Four (4) different samples were prepared based on the different percentage of bentonite as shown in Table 3. The samples of natural Salak Tinggi soil without mixed of bentonite consider as control samples. The air-dried soil samples passing 20 mm was mixed with different percentage of bentonite at 5%, 10% and 15% by dry weight of soil. After that, the mixture samples were moistened with distilled water to allow all mixture blended for overall mixed sample to get the homogenous mixed and the samples were allowed to air dried at least for 24 hour prior for properties testing and compaction testing [9].

Table 3. Details of mixed sample.

Sample	Bentonite (%)	Symbol
Soil	0	S
Soil + 5B	5	S+5B
Soil + 10B	10	S+10B
Soil + 15B	15	S+15B

All entire prepared mixed samples were used to determine the physical properties such as *pH,* Atterberg limit, size distribution and linear shrinkage. All physical testing is based from BS 1377: Part 2: 1990. Meanwhile, standard proctor compaction testing was done on soil samples to determine the optimum moisture content (OMC) and maximum dry density (MDD) based from BS1377: Part 4:1990:3.3.

The mixed samples for permeability test were prepared and mixed thoroughly with distilled water at OMC value in order to get the homogenous mixture. After that samples were compacted using standard compaction proctor test. Basically, the test specimens for permeability test proposed by BS 1377: Part 6: 1990 is about 100 mm diameter and 100 mm high, although specimens of other dimensions from 38 mm diameter upwards may be used. Therefore, the approaches taken in this study were using the sample about 75 mm in diameter and 35 mm in height. It was obtained that the longer sample was used, the longer it takes to stabilize. This approach was considered from previous researchers [10-11].

The permeability testing was conducted on all entire mixed soil samples to determine the hydraulic conductivity value; *k*. Samples for permeability test was compacted at MDD condition using standard compaction test. Permeability test was carried out using Triaxial consolidation test in a hydraulic consolidation cell as stated in BS1377: Part 6:1990. This method is suitable for soils with low and intermediate permeability and produced hydraulic conductivity at fast measurement rates for soil containing fine grain soils. This method covered the measurement of the coefficient of permeability at laterally confined specimen of soil under a known vertical effective stress, and under the application of a back pressure. The hydraulic conductivity value *k* is the rate of discharge of water under laminar flow conditions through a unit cross sectional area of porous medium under a unit hydraulic gradient and standard temperature conditions. The effective stress of 100 kPa were applied to samples during the permeability test and the volume of water passing through the mixed soil samples in a known time, and under a constant hydraulic gradient were measured.

4. Results and discussion

The soil engineering characteristics are depending on the particle size distribution. Table 4 shows the summarize tabulation grading of particles size distribution result for Salak Tinggi residual soil and mixed samples of Salak Tinggi Residual with bentonite. From the results, it shows an increment of fine grain particles along with reduction on the coarse grain size after addition of bentonite from 5% to 15%. It can be seen, addition of bentonite was increased clay size content and alter to reduce the silt size and sand size content. This due to the bentonite has high content of clay mineral of montmorillonite. This shows mixed samples considered as good liner because a good soil liner should have at least 30% of fine grain content and 15% of clay content [12].

Table 4. Summarize of the tabulation of grading size results for Salak Tinggi residual mixed bentonite.

Sample	Coarse grain		Fine grain		Classification (from BSCS)
	Gravel (%)	Sand (%)	Silt (%)	Clay (%)	
Soil	0	69.14	20.92	9.94	Very Silty SAND
Soil + 5B	0	63.63	17.93	18.44	Sandy CLAY
Soil + 10B	0	61.42	15.59	22.99	Sandy CLAY
Soil + 15B	0	59.33	16.11	24.56	Sandy CLAY

Figure 1 shows that a mixture of bentonite with Salak Tinggi residual soil were affected to alter the pH value from acidic to alkaline condition after addition of 5%, 10% and 15% bentonite from pH of 8.21, 8.64 to 8.96 respectively. In acid condition (pH $<$ 7) the soil particle tend to flock particle and flock each other in a random orientation produced a flocculated structure. These flocculants structure possesses a high void ratio and will result a high on hydraulic conductivity value [12-14]. Consequence on soil with pH $>$ 7 produced the low of hydraulic conductivity value.

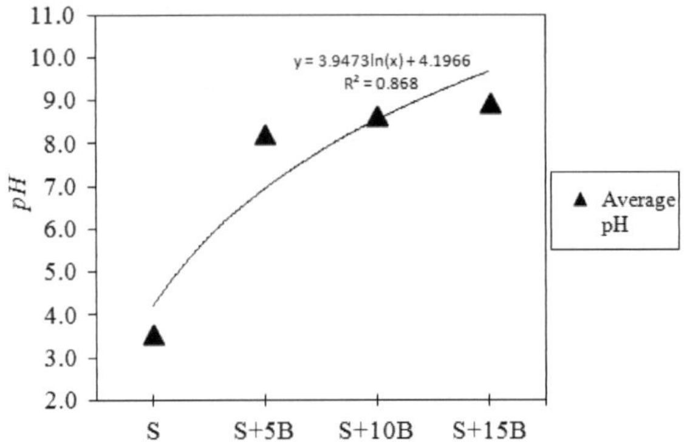

Figure 1. The *pH* result of Salak Tinggi Residual mixed bentonite.

Figure 2 shows the consequence of addition various percentage bentonite with Salak Tinggi residual soil on the liquid limit (*LL*), plastic limit (*PL*) and plastic index (*PI*). It was found that, with increase of bentonite percentage used in soil was increased the *LL* and as well as the *PL*. These also were resulting in increment of *PI* value of mixed samples. This was proved that bentonite action was changed the characteristics of soil properties. From the results, it shows that the mixed samples classification was changed the soil classification to low plasticity after addition 5% bentonite and change to Intermediate plasticity after addition of 10% bentonite and change to high of plasticity after addition of 15% bentonite. The increasing trend of *PI* index for mixed soil samples indicates the increase of clay portion in mixture soil and having a higher of surface activity. The plasticity index of soil is directly related to the mineralogy of the soil. Soil that produces high *LL* and *PI* generally contains more clay minerals and typically presents a low hydraulic conductivity [4]. It indicates that the *LL* and *PI* of mixed soil are primarily controlled by its clay content from bentonite. The suitable requirement for liner materials when the LL>20 and PI>7[12-14], meanwhile the low of shrinkage can be expected if the PI<35 [12-14]. Therefore, the results shows the addition of bentonite 5%, 10% and 15% were fulfil the requirement as liner materials.

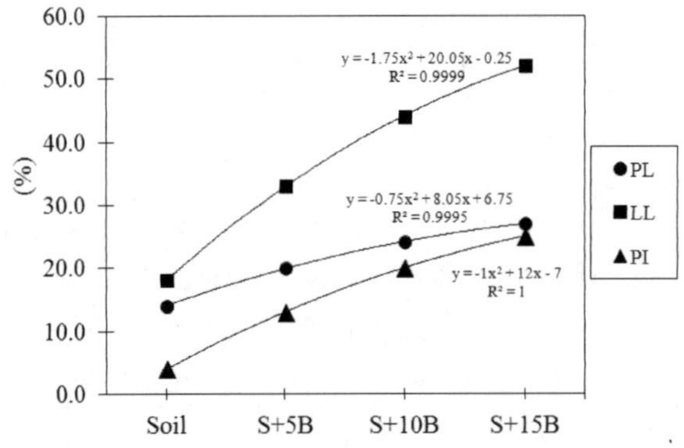

Figure 2. The consistency limit result of Salak Tinggi residual soil mixed bentonite.

Figure 3 shows the linear shrinkage result for mixed soil samples with bentonite. It shows that the linear shrinkage of the mixed soil sample increase as the percentage of bentonite increase and it shows a significant of increment on linear shrinkage increasing with the increasing percentage of bentonite. This indicates that the soil mixed bentonite has a higher change in volume. However, soil with a limit of linear shrinkage of less than 12% indicates the ability of a small change in volume [12-14]. Shrinkage of the soil increase due to high plasticity index of bentonite and the significant increase of plasticity index and linear shrinkage limit over 15% bentonite is due to the high water absorption capacity of bentonite. Besides that, shrinkage behaviour significantly affected by fine particles consisted in the soil distribution rather than by the plasticity characteristics and fine particles in the soil were a better indicator of shrinkage than plasticity characteristics. The high percentage of fine particles that consist in bentonite and the ability of soil to absorb water increased as the clay content (bentonite) were increased.

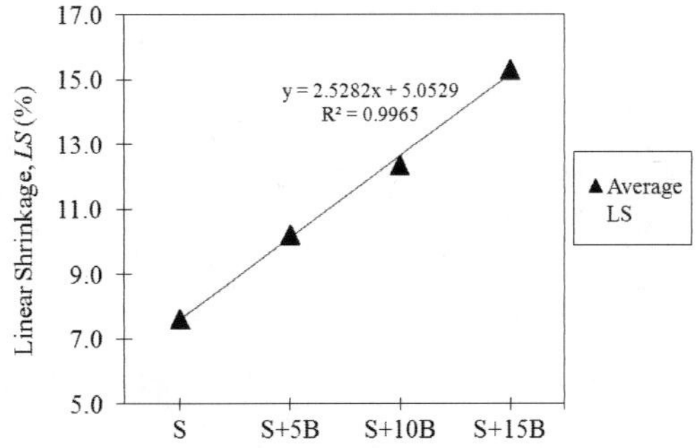

Figure 3. The linear shrinkage result of Salak Tinggi residual soil mixed bentonite.

Figure 4 show the compaction curve result for Salak Tinggi residual soil mixed with different percentage of bentonite at standard compaction. The different percentage of bentonite used was affecting the maximum dry density (MDD) and optimum moisture content (OMC) result.

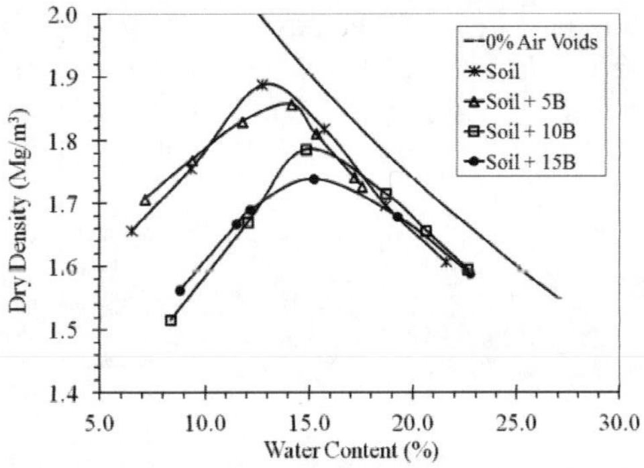

Figure 4. Compaction curves for Salak Tinggi residual soil mixed with different percentage of bentonite at standard compaction.

Figure 5 and Figure 6 shows the result for MDD and OMC from standard proctor compaction at different percentage of bentonite. It shows an increment OMC value and at once decrement of MDD after addition of bentonite. It can be seen the MDD value for natural Salak Tinggi soil is 1.89 kN/m^3 and dropped down to 1.86 kN/m^3, 1.79 kN/m^3 and 1.74 kN/m^3 after the addition of 5%, 10% and 15% bentonite respectively due to the increment of bentonite content. Otherwise the OMC for natural Salak Tinggi soil is 13% and increased to 14%, 15% and 15.5% as the bentonite content were increased from 5%, 10% and 15% respectively. The reduction in MDD with increasing of bentonite is due to the high swelling of bentonite characteristics that forms a gel around the soil particles. These causes increment in effective size of particle and at once increase void volume and thus decreased dry unit weight. Meanwhile the increase of OMC was related to require more moisture for hydration reaction comes from the increment of fine grains size [12-15].

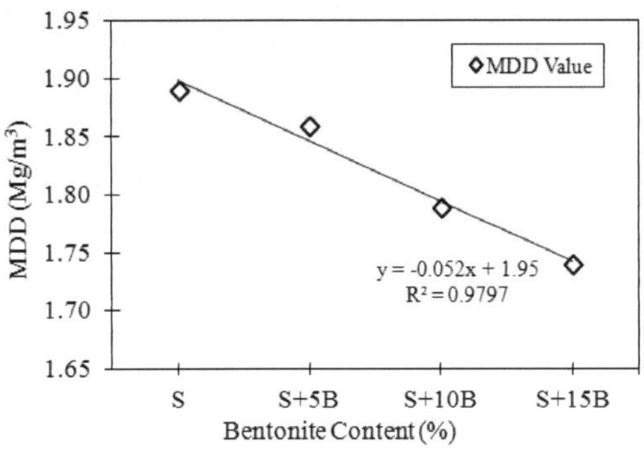

Figure 5. The MDD value at different percentage of bentonite mixture at standard compaction.

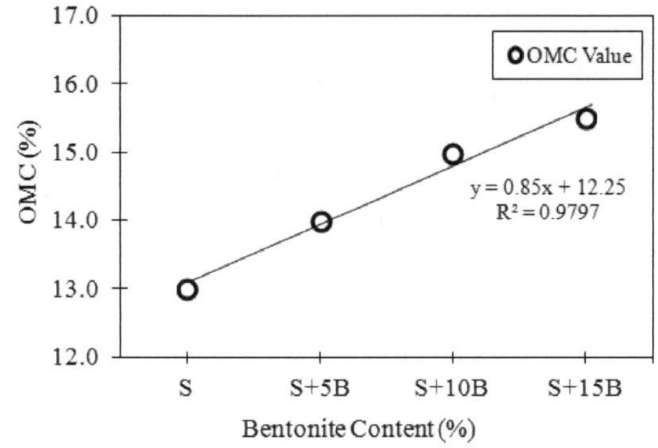

Figure 6. The OMC value at different percentage of bentonite mixture at standard compaction.

Figure 7 shows the result of hydraulic conductivity value for Salak Tinggi residual soil mixed with different percentage of bentonite at standard compaction. It is clearly demonstrate that the increase in bentonite content considerably reduced the hydraulic conductivity value at the highest confining stress of 100 kPa. It shows the hydraulic conductivity value for natural Salak Tinggi soil is 1.005 x 10^{-9} m/s and dropped down to 2.430 x 10^{-10} m/s, 9.342 x 10^{-11} m/s and 5.111 x 10^{-12} m/s after the addition of 5%, 10% and 15% bentonite respectively. It shows that soils compacted at OMC tend to have a lower hydraulic conductivity due to the increasing water content may resulted increased ability to break down clay aggregate and eliminate inter-aggregates pores [16-17]. In addition, increasing water content resulted in orientation of clay particles and producing a large water film around particles which causes swelling of clay and at once reducing the inter particles pores together resulting in low of hydraulic conductivity [16-18].

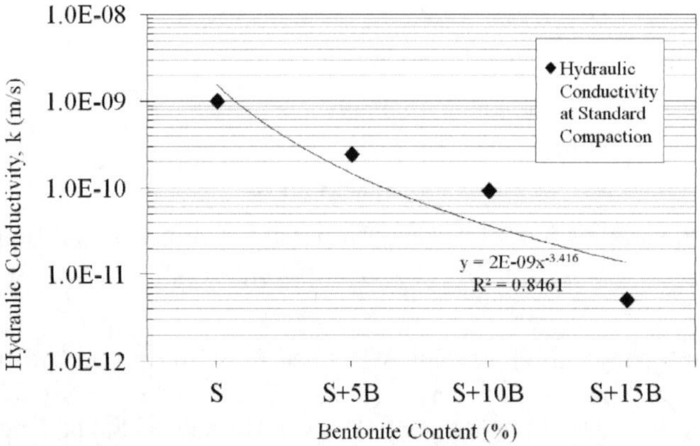

Figure 7. The effect of bentonite percentages on the hydraulic conductivity value, k at MDD conditions for standard proctor compaction.

5. Conclusion

Some basic characteristics of Salak Tinggi residual soil mixed bentonite have been studied from this study for the purposes as potential use as liner material. The physical properties, compaction and hydraulic conductivity were measured. It can be concluded that the addition of bentonite was resulted in changes in physical properties. It shows and increment of *pH*, plastic index, liquid limit, plasticity index, linear shrinkage and clay fraction content. It also changed the soil classification from clay low plasticity to clay high plasticity for addition bentonite 5% to 15% respectively. The compaction result shows the reduction of MDD and at the same time increase the OMC from addition of bentonite in line with increment of LL and PI. From this study, the suitable percentage of bentonite were determined in aout 5% to be used as mixture to the Salak Tinggi residual soil to produce the suitable physical properties for soil liner with suitable hydraulic conductivity value at standard compaction. However, it shows hydraulic conductivity value reduced by mixing with higher content of bentonite at standard compaction energy.

Acknowledgments
Authors would like to express deepest thanks to the Faculty of Civil Engineering, UiTM Shah Alam, Malaysia for providing a budget to complete this study and to geotechnical laboratory and advanced geotechnical laboratory to accomplish this study.

References
[1] Taha M R and Kabir M H 2005 Tropical residual soil as compacted soil liners Environmental Geology **47**(3) 375–381.
[2] Faizal M, Tajul B, Ir B, Rashid A, Othman B, Fairus M and Aman B. 2006 Assessment on Physico-Chemical Properties for Granite Residual Soil as Landfill Liner Material at Batu Pahat, Johor, Malaysia. Technology and Innovation for Sustainable Development Conference, (January), 282–285
[3] Yesiller N, Miller C J, Inci G, and Yaldo K 2000. Desiccation and cracking behavior of three compacted landfill liner soils. Engineering Geology, **57**(1), 105-121
[4] Benson C H, Daniel D E and Boutwell G P 1999 Field performance of compacted clay liners. Journal of Geotechnical and Geoenvironmental Engineering, ASCE, Vol. No. **5**, pp. 390-403.
[5] Benson C H, Daniel D E and Boutwell G P 1999 Field performance of compacted clay liners. Journal of Geotechnical and Geoenvironmental Engineering, ASCE, Vol. No. **5**, pp. 390-403.
[6] Minerals and Geoscience Department Malaysia Updated: 2012 Geological Map of Peninsular Malaysia based on the 8th edition Geological Map in 1985.
[7] Singh H, Huat B K, Sew G and Ali F 2004. Origin, formation and occurrence of tropical residual soils. In S. Gue and F. Ali (Eds.) Tropical Residual Soils Engineering. London: Taylor and Francis Group.
[8] Liu Y, Gates WP and Bouazza A 2013 Acid induced degradation of the bentonite component used in geosynthetic clay liners Geotextiles and Geomembranes Volume **36** (2013), pg 71 – 80.
[9] Norazlan Khalid and Mazidah Mukri 2017 Compaction Characteristics on Dengkil Residual Soil Mixed Bentonite Electronic Journal of Geotechnical Engineering, Vol. **22** (Bund. 2): 605-612
[10] Daniel D E and Benson C H 1990 Water content-density criteria for compacted soil liners Journal Geotech.Eng. ASCE 116 (GT12): 1811-1830
[11] Daniel D E and Liljestrand H M 1984 Effect of landfill leachates on natural liner systems. Report GR83-6, Geotechnical Engineering Centre, University of Texas, Austin.
[12] Benson C H, Zhai H and Wang X 1994 Estimating hydraulic conductivity of compacted clay liners. J Geotech Engrg ASCE 120(2):366–387.
[13] Daniel D E 1993a Clay Liners In Geotechnical Practice for Waste Disposal, (ed. David E. Daniel) Chapman and Hall, London, UK, pp 137-163.
[14] Daniel D E and Benson C H 1990 Water content density criteria for compacted soil liners.Journal Geotech. Eng. ASCE. 116 (12): 1811-1830.
[15] Mitchell J K and Soga K 2005 Fundamentals of Soil Behaviour. 3rd Edition. John Wiley and Sons Inc.
[16] Garcia-Bengochea I, Lovell C and Altschaeffl A 1979 Pore distribution and permeability of silty clays. J Geotech Eng ASCE 105(7):839–856.
[17] Osinubi K J and Nwaiwu C M O 2006 Hydraulic Conductivity of Compacted Lateritic Soil J. Geotechnical and Geoenvironmental Engineering, ASCE, Vol. **131**, No. 8, 1034 – 1041.

IOP Conf. Series: Materials Science and Engineering **513** (2019) 012024 doi:10.1088/1757-899X/513/1/012024

[18] Amadi A A, Afolayan J O and Osinubi K J 2014 Sensitivity Analysis of the Hydraulic Conductivity of Compacted Laterirtic Soil and Bentonite Composite. Geo-Congress 2014 Tehcnical Paper, GSP 234, ASCE, 921 – 930.

IOP Conf. Series: Materials Science and Engineering **513** (2019) 012025 doi:10.1088/1757-899X/513/1/012025

Preliminary study on influence of silica fume on mechanical properties of no-cement mortars

C B Cheah[1] and J Nurshafarina[1*]

[1]School of Housing, Building and Planning, Universiti Sains Malaysia, 11800 Penang, Malaysia

[*]Corresponding author: shafarinajasme@gmail.com

Abstract. Ordinary Portland Cement (OPC) is widely used as a binder material for various construction application and has been increasing in demand especially in developing countries. However, growing concern about the effect of OPC towards the environment either during production or transportation process captures the researcher's interest to come out with more sustainable binder material that has similar characteristic as OPC. This paper prescribed the study on the strength performance of no-cement mortar consist of waste materials such as ground granulated blast furnace slag, fly ash and rice husk ash partially replaced with silica fume up until 20% of total binder weight. Eleven different mix designs were fabricated with constant 0.5 water/binder ratio cured under ambient temperature. The mortar specimens of 40 x 40 x 160 mm prisms were prepared and test for their flexural and compressive strength at aged 7, 14 and 28 days after curing. Laboratory tested on strength performance studies indicate that different curing ages exhibit different silica fume replacement level to achieve their optimum strength. Maximum compressive strength achieved at 28 days is 27 MPa at 12% silica fume replacement level.

1. Introduction

Cement has been long used as the binder material in construction but the growing concern on the effect of its production that release carbon dioxide to the environment. Suitable alternative materials that can give sustainable benefit having cementitious properties such as ground granulated blast furnace slag (GGBS) or pozzolan such as metakaolin, fly ash and silica fume have widely studied as cement replacement materials [1,2]. According to Massazza [3], pozzolan can be distinguished into two different meaning, pyroclastic rock that happens in the place called Pozzuoli of Roman times ancient Puteoli around Rome. Secondly, it can be defined by all inorganic materials which release calcium hydroxide (Portland cement clinker) when mixed with calcium hydroxide source (lime) and water. Using industrial or agricultural by-product pozzolans as binder materials can give sustainable benefit toward environment such as reducing waste to be landfill and subsequently give space for other development. The study shows that all these by-product materials and pozzolans give better performance concrete but have relatively slow early hydration [3] and fully utilisation of these materials usually can be improved by adding a particular portion of alkali activator or geopolymer technology to accelerate the early hydration. GGBS is a slag by-product form production of pig iron in blast furnace that undergo rapid quenched in water before grinding to fine powder and has been used as binder material since Emil Langen found their latent hydraulic properties in 1862 [4]. The slag production was estimated to be 25 – 30% of crude (pig) iron production and based on that reference, the estimated iron slag output in 2017 was about 300 – 360 million tons [5]. Incorporating high volume GGBS as binder materials in concrete favour the heat of hydration compared to OPC concrete and beneficial for mass concrete application [6] and according to Ibrahim et al. [7], 70% replacement of OPC also improved mechanical properties with 28 days strength comparable with 100% OPC. Pulverised fuel fly ash (PFA) is a residue that flies out from combustion in electric power generating plant and collected by an electrostatic precipitator. Based on Malaysian Energy Commission [8], coal contributed to 42.5% electricity generation mix in 2016 behind gas with 43.5% and it was expected that coal share would increase to 63% exceeding gas in 2020 [9] which subsequently increase the PFA available. The pozzolanic properties of PFA improve the later strength and its spherical microstructure reduce water demand in concrete [10]. Other than that, paddy is the widely planted crop in Malaysia with estimated 674,332 hectares were planted in the year 2013 including those that planted twice a year [11]. Due to its low bulk density, the large dry volume of rice husk making the disposal process of this material quite challenging as it possesses rough and abrasive surfaces that are highly resistant to natural degradation [12]. In Malaysia, rice husk ash is majorly the residue of rice processing industry as it has been known primarily with its high silica content. Rice husk constitutes about 20% of the rice weight and contains approximately 50% cellulose, 25-30% lignin and 15-20% silica [13]. Since original rice husk ash is coarse, this material needs to be ground to work as a binder. Based on a study by Habeeb and Mahmud [14], increasing grinding time slightly increase the surface area and reducing the particle size. The high surface area of this material resulted from its microporous and multilayers surface. High surface area also may reduce in its workability due to their high water absorption property [15] thus contribute to the limited strength development [16]. RHA and silica fume can be an alternative source of silicon for alkali-activated materials. It was notable by Dembovska et al. [17] that silica fume enhanced the mechanical properties when used not more than 20% from the total binder and the study by Zareei et al, [18], the amount of micro silica added influence the performance of RHA as cement replacement in high strength concrete as it can create more dense concrete by improving the interface transition zone. Other than that, silica fume in ternary or quaternary blends enhance the compressive strength at an early age especially at higher water to binder (w/b) ratio [7]. Consequently, the present paper aims to a preliminary study of the mechanical strength of mortar by fully utilising by-products with cementitious and pozzolanic properties from various industry and replaced with silica fume to improve the mechanical performance of mortar. This subsequently eliminating the need of strong chemical activator and creating an environmental friendly binder.

2. Experimental program

2.1. Materials

The materials used in this study were ground granulated blast furnace slag, pulverised fuel ash, rice husk ash, silica fume and river sand. GGBS, the industrial by-product of iron industry was sourced from a local supplier, YTL Berhad with a specific gravity of 2.86 g/m^3 and specific surface area of 4.65 m^2/g. PFA was collected from local fuelled power plant precipitator unit. The PFA used in this study have a specific gravity of 2.8 g/m^3 and a specific surface area of 3.24 m^2/g. Other than that, the RHA also was sourced from local rice milling industry, Serba Wangi Sdn.Bhd. The rice husk ash was sieve passing 150 μm prior ground for 30 minutes in laboratory ball mill to achieve a comparable degree of fineness as cement and more than 90% of the ash may pass sieve size of 45 μm [19]. The specific gravity of RHA after been ground was 2.21 g/m^3. The densified silica fume (DSF) is filtered powder generated from silicon metals or high purity quartz - ferrosilicon metals and the one that used in this study has a specific gravity of 2.20 g/m^3 and specific surface area of 20,000 m^2/kg. The chemical composition of all binder materials tabulated in Table 1. Saturated surface dry quartzitic natural river sand was used as fine aggregate with the maximum aggregate size of 5 mm, graded as specified according to the grading limit as in BS 812-103: Part 2 [20]. The fine aggregates have a specific gravity of 2.65 g/m^3 and fineness modulus of 3.26. Water used sourced from locally supply tap water.

Table 1. The chemical composition of binder materials.

Chemical Compound	% by total mass			
	PFA	GGBS	RHA	DSF
MgO	5.94	5.01	0.71	4.60
Al_2O_3	17.61	12.59	0.26	0.27
SiO_2	43.22	32.62	90.18	84.00
P_2O_5	0.23	0.01	1.48	0.05
K_2O	1.31	0.32	3.97	2.70
SO_3	-	-	0.35	0.44
Cl	-	-	0.54	2.40
CaO	11.28	37.87	1.51	0.66
CuO	-	-	0.02	Trace*
ZnO	-	-	0.03	0.10
SeO_2	-	-	Trace[a]	-
Br	-	-	Trace[a]	-
Rb2O	-	-	0.01	0.02
SrO	-	-	Trace[a]	-
TiO_2	0.88	0.50	0.02	0.09
Cr_2O_3	-	-	0.02	-
MnO	0.14	0.25	0.12	-
Fe_2O_3	13.73	2.00	0.75	0.54
C	1.80	-	-	-
Na_2O	0.43	0.25	0.05	4.70

[a]Very small amount – not presented

2.2. Mix proportioning and mixing procedure

Mortar mixture proportion was design based on modified absolute volume method prescribed in ACI 211.1 [21] is given in Table 2 with each mix design nomenclature system of S*x* where *x* was the percentage of silica fume replaced from the total binder content. The binder to sand and water to binder ratio for all mix was remained constant at 1:3 and 0.5, respectively. Binder consists of GGBS, PFA, and RHA with a fixed ratio of 0.89: 0.09: 0.02 replaced with silica fume at an increment of 2% up to 20% of total replacement. The mortar was first prepared by dry mixing of binder materials and fine aggregate in the epicyclic mixer for 5 minutes and slow mixing speed. Quarter from total water was then introduced into the mix and mix for 1 minute. Half of the water was subsequently added and mix for a further minute before stopped to scrape inhomogeneous particles to the middle. Lastly, the rest of the water was added and further mixing for another 1 minutes before stopped, flow table test and placing the mixture into the mould in three layers. For better compaction, each layer was vibrated for 10 seconds on a vibrating table. Moulded samples were left cure in the mould for 24 hours under ambient temperature curing of 28 ± 5°C with relative humidity of 80 ± 5% prior to moist curing by wrapped in a plastic sheet in ambient temperature until testing days.

Table 2. Mix design proportioning.

Mix	GGBS	PFA	RHA	DSF	Sand	Water	w/b Ratio
	Mass (kg/m^3)						
S0	448	45	10	0	1511	252	
S2	439	44	10	10	1510	252	
S4	430	43	10	20	1508	251	
S6	420	42	9	30	1507	251	
S8	411	42	9	40	1505	251	
S10	402	41	9	50	1504	251	0.5
S12	392	40	9	60	1502	250	
S14	383	39	9	70	1501	250	
S16	374	38	8	80	1499	250	
S18	364	37	8	90	1489	250	
S20	355	36	8	100	1497	249	

2.3. *Testing methods*
The mechanical test conducted in this study consist of compressive and flexural strength test. From each batch of mortar mix design, a total of 3 prisms with 40 x 40 x 160 mm dimension were fabricated for both compressive and flexural strength test and tested based on procedures prescribed in ASTM C349-14 [22]. The result was taken based on the average of 3 prisms for flexural strength and 6 portions of broken prisms from the flexural test were then used for compressive strength test, noted in MPa.

3. Result and discussion

3.1. *Flexural strength*
The graph in Figure 1. shown a flexural strength result of no-cement mortar at 7, 14 and 28 days. At 7 days of curing, the strength steadily increased up to 4% (S4) replacement level which represented as only 2.85 MPa and notable decrease until the 10% replacement of silica fume. At similar curing age, strength was observed as the silica fume content increase by more than 10% (S12 – S20). Strength performance after 14 and 28 days of curing show similar fluctuated trend with optimum strength at 0% (4.66 MPa) and 10% (5.8 MPa) replacement level respectively. Accelerated strength gain also observed on all specimens except S4 which represent by 4% silica fume addition at14 days of curing. This confirmed on conclusion stated by [23] the pozzolanic effect of silica fume often occurs after 7 days.

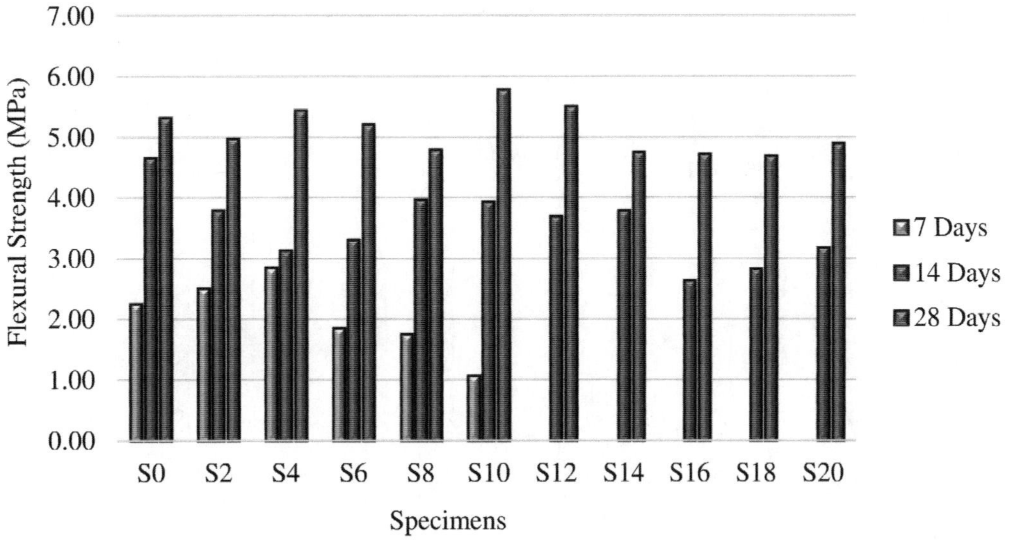

Figure 1. Flexural strength result.

3.2. Compressive strength

From Figure 2, it is shown that different curing age has different silica fume replacement percentage for as optimum compressive strength. At 7 days curing age, the compressive strength of mortar increased up to 4% replacement level of silica fume with 11.4 MPa and gradually decrease as the silica fume increase. The progression of strength can be considered as relatively slow as the pozzolanic reaction took longer time compared when used with OPC as these materials rely on calcium hydroxide from hydration of OPC to increase the rate of hydration [24, 25]. Moreover, slow hydration presented that does not contribute to the decrease in pore space as it still filled with water and resulted in low early strength. When the curing age reached 14 days, the graph seems to fluctuate with the highest compressive strength of 20.19 MPa at 10% replacement level. Accelerated strength gain observed on S0 to S2 at 14 days which represent 10% to 20% silica fume replacement. A similar trend also observed at 28 days of curing up to optimum compressive strength at 12% replacement level with 27.3 MPa but gradually decreasing as the replacement level increased. The increase in silica fume content more than the optimum value may cause the leftover to be act as filler materials rather than reactive mineral compound [26]. When dealing with pozzolanic powders, there might be a filler effect and the sum of filler and pozzolanic effect can be the net effect which contributed both mechanical and durability properties [27]. The rapid increase of strength can be observed from 7 – 14 days of curing. This is in agreement with the finding from Tan [28] which also show accelerated strength gain between the above mention curing aged.

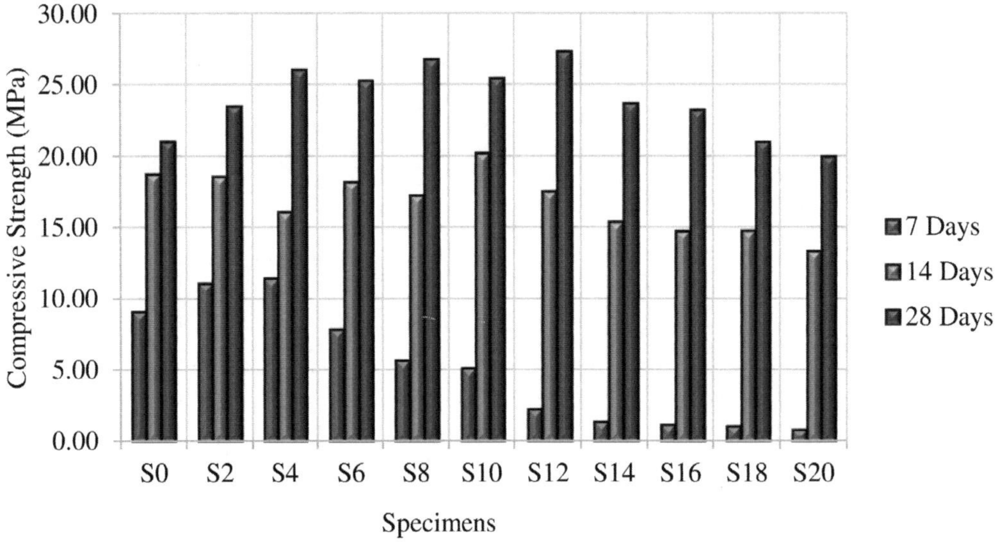

Figure 2. Compressive strength result.

4. Conclusion and recommendation

Based on the mechanical strength result, the no – cement mix containing 10 and 12% silica fume is suitable to be use and binder materials in mortar when early strength is negligible and strength up to 27.3 MPa at 28 days. Furthermore, it was noticeable that silica fume undergoes rapid strength gain between 7 to 14 days. However, further research needs to be conducted to increase the rate of hydration in the early ages and subsequently improve the mechanical strength of the mortar or concrete. Microstructure analysis of hybridisation of these binder materials also needs to be done in future research to study their hardening mechanism.

References

[1] Ramli M B and Alonge O R 2016 Characterization of metakaolin and study on early age mechanical strength of hybrid cementitious composites *Construction and Building Materials* **121** 599–611

[2] Habert G, Lacaillerie J B and Roussel N 2011 An environmental evaluation of geopolymer based concrete production: Reviewing current research trends *Journal of Cleaner Production* **19** 1229–38

[3] Massazza F 1998 Pozzolana and pozzolanic cements In *Lea's chemistry of cement and concrete* London

[4] Lang E 2002 Blastfurnace cements In *Structure Performance of Cement* London

[5] U.S. Geological Survey 2018 Mineral commodity summaries 2018: U.S. Geological Survey 200

[6] Han F 2017 Hydration heat of slag or fly ash in the composite binder at different temperatures. *Thermochimica Acta*, **655** 202–10

[7] Ibrahim A, El-Chabib H and Eisa A 2013 Ultrastrength flowable concrete made with high volume of supplementary cementitious materials *Journal of Materials in Civil Engineering* **25** 1579–86

[8] Malaysian Energy Commission 2017 2017 Malaysia Energy Statistic Handbook, Putrajaya

[9] Schernikau L 2016. Economics of the International Coal Trade, Cham, 197 – 198

[10] Thomas M 2007. Optimizing the use of fly ash in concrete. *Portland Cement Association* 24

[11] Department of Agriculture Malaysia, 2014. *Perangkaan Padi Malaysia (Paddy Statistics of Malaysia)*

[12] Hwang C L and Chandra S 1996 The use of rice husk ash in concrete In *Waste Materials Used in Concrete Manufacturing* 184–234

[13] Hisham B, Putrajaya R and Abdulaziz H 2010 Malaysian Rice Husk Ash – Improving the Durability and Corrosion Resistance of Concrete: Pre-review. *Concrete Research Letters*, **1** p 11

[14] Habeeb G A and Mahmud H B 2010. Study on properties of rice husk ash and its use as cement replacement material. *Materials Research* **13** 2 185–190

[15] Sathawane S H, Vairagade V S and Kene K S 2013 Combine effect of rice husk ash and fly ash on concrete by 30% cement replacement *Procedia Engineering* 51 35–44

[16] Chindaprasit P, Kanchande P, Sathonsawaphak A 2007 Sulfate resistance of blended cements containing fly ash and rice husk ash *Construction and Building Materials* **21** 1356–1361

[17] Dembovska L, Bajare D, Pundiene I and Vitola L 2016 Effect of pozzolanic additives on the strength development of high performance concrete *Procedia Engineering* 1–8

[18] Zareei S A, Ameri F, Dorostkar F and Ahmadi M 2017 Rice husk ash as a partial replacement of cement in high strength concrete containing micro silica: Evaluating durability and mechanical properties *Case Studies in Construction Materials* **7** 73–81

[19] Zain M F M, Islam M N, Mahmud F and Jamil M 2011 Production of rice husk ash for use in concrete as a supplementary cementitious material *Construction and Building Materials* **25** 798–805

[20] BSI, 1985 BS 812-103.1 Testing aggregates. Method for determination of particle size distribution. Sieve tests, London: *British Standard Institution*

[21] ACI Committee 211 2002 Standard practice for selecting proportions for normal, heavyweight, and mass concrete

[22] ASTM 2002, ASTM C349 Standard test method for compressive strength of hydraulic-cement mortars (Using portions of prisms broken in flexure), *ASTM International*

[23] Ömer Ö C and Sofyanlı Ö 2015 Influence of silica fume on mechanical and physical properties of recycle aggregate concrete *HBRC Journal* **11** 157-166

[24] Oner A and Akyuz S, 2007 An experimental study on optimum usage of GGBS for the compressive strength of concrete. *Cement and Concrete Composites* **29** 505–14

[25] Zhou X, Slater J R, Wavell S E and Oladiran O 2012 Effects of PFA and GGBS on early-ages engineering properties of Portland cement systems *Journal of Advanced Concrete Technology* **10** 74–85.

[26] Neville A 2006. Concrete: Neville' s Insights and Issues. *Concrete*, 314.

[27] Pedersen B M 2004. *Alkali-reactive and inert Fillers in Concrete. Rheology of fresh Mixtures and expansive Reactions*. Fakultet for ingeniørvitenskap og teknologi.

[28] Tan L.E 2017. *Advanced Self-Healing Geopolymer Composite Derived from Industrial By Product*. Universiti Sains Malaysia

Investigations on the chemical composition and tensile strength of steel bars in the Nigerian construction industry

Wasiu Ajagbe[1*], Abideen Ganiyu[2], Adefemi Adegbite[1] and Oluwasola Akodu[1]

[1]Department of Civil Engineering, University of Ibadan, Nigeria
[2]Department of Civil Engineering and Quantity Surveying, Military Technological College, Muscat, Oman

[*]Corresponding author: ajagbewas@gmail.com

Abstract. Steel reinforcement bar compliments concrete in reinforced concrete structures to provide the required strength, rigidity and durability. This study investigates the level of conformity of reinforced steel bars available in the Nigerian construction industry through investigations on the percentage of chemical composition and tensile properties in relation to relevant local and international standards. Samples from 12 different companies were considered in this research. The chemical analysis results discovered 16 elements, some of which are residual elements in minute quantities. 9 samples had carbon content of less than 0.27%, 17 samples had phosphorus content less than 0.055%, 16 samples had sulphur content less than 0.055%, 9 samples had copper content less than 0.25% and 17 samples had carbon equivalent of less than 0.54% which are all requirements of NIS 117. Similarly, most of the samples examined did not meet the requirements of NIS 117 and BS4449 in respect of steel bar diameter, characteristic strength, ductility, ultimate to yield strength ratio, or a combination of the parameters with the yield strength ranging from 462.02 N/mm^2 to 558.27 N/mm^2. An indication of the level of quality of steel bars available in Nigerian construction industry can be deduced from the results of this research.

1. Introduction

Steel reinforcement bar is a versatile constructional material widely used in the production of reinforced concrete. Reinforced concrete structure is designed on the principle that steel and concrete act together to withstand induced forces. Concrete has high compressive strength but low tensile strength. Steel bars, on the other, can resist high tensile stresses but will buckle when subjected to comparatively low compressive stresses. By providing steel bars within a concrete member predominantly in those zones which will be subjected to tensile stresses, an economical structural material can be produced which is both strong in compression and tension [1]. The properties of thermal expansion for both steel and concrete are approximately the same; this along with excellent bendability property makes steel the best material for reinforcement in concrete structures [2-4].

The incidences of building failures and collapses have become major issues of concern in the Nigerian construction industry. The frequencies of their occurrence and the magnitude of the losses in relation to lives and properties are now becoming very alarming [5-7]. It has been identified that the use of poor quality and substandard steel rods are among the causes of building failure in Nigeria [8-10]. The large-scale conversion of imported scrap in form of scrap rails, automobile scrap, defence

IOP Conf. Series: Materials Science and Engineering **513** (2019) 012026 doi:10.1088/1757-899X/513/1/012026

scrap, scrap from ship etc. of unknown or unsuitable quality into steel rebars, possibly under ineffectively controlled conditions, raises apprehensions about the quality and reliability of steel rebars used in the construction industry[11, 12].

In this research, the chemical composition of reinforcing steel bars used in the Nigerian construction industry was determined using optical emission spectrometry to compare the chemical composition with relevant standards and codes. Further, the tensile properties of the bars were determined and a check on the diameter of the tested reinforcement was done. In essence, the quality of the reinforcing steel bar used in the construction sites in Nigeria is being evaluated in this research.

2. Methodology

The reinforcing steel rod samples were collected from 3 different construction material markets at Ibadan namely; Iwo road, Agodi Gate and Iyana Church. These samples were obtained under 2 major sources namely; locally produced steel bars and imported products. Samples from 12 different steel mills were considered for the research. Table 1 shows the list of the steel mills and the diameter of the reinforcement sourced from each.

Table 1. Steel reinforcing bar samples.

S/n	Name of manufacturer	Label & Diameter (mm)
1	PJSC ArcelorMittal Kryvyi Rih, Ukraine	UKR(12,16)
2	African Foundries Limited Ogijo, Ogun	AFL(16,20)
3	Brazil Steel Mills, Brazil	BSM(12,16)
4	Eurotherm Steel Ife, Osun	ETM(12,16)
5	Landcraft steel Industries Limited, Lagos	LCI(16,20)
6	Mayor Engineering limited Ikorodu, Lagos	MYE(12)
7	Rashtriya Ispat Nigam Ltd India	RIL(12,16)
8	Hongzing Steel Company Limited Lagos	RSS(12,16)
9	Sankyo Steel Mills Company Limited, Lagos	SSM(16)
10	African Industries (Tiger TMT), Lagos	TGT(12,16)
11	Top Steel Nigeria Limited	TOP(20)
12	Turkey Reinforcing Steel Bar, Turkey	TKY(12,16)

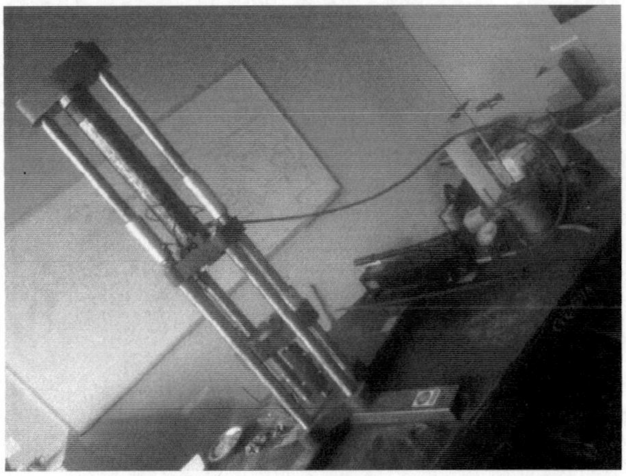

Figure 1. Tensile strength test using Universal Testing Machine.

IOP Conf. Series: Materials Science and Engineering **513** (2019) 012026 doi:10.1088/1757-899X/513/1/012026

The dimensional requirement for each sample for the chemical composition test in the form of a solid mass depends on the dimensions of the sample chamber. Each steel bar sample to be tested was cut into 50mm length which is the chamber size of LabSpark 750B Optical emission spectrometer, after which the surface to be tested was well grinded using a grinding machine so as to achieve smooth, flat and impurity free surface to be analyzed. LabSpark 750B Optical emission spectrometer was used in the determination of the chemical composition of the samples in this research. The test was carried out at the Quality Control Laboratory of Universal Steel Mill Ikeja Lagos. SM1000 TQ Universal Testing Machine (UTM), having a maximum load of 100 kN was used to carry out the tensile strength test [13]. The tensile strength test was carried out at material testing and research laboratory of Federal University of Technology, Akure. The UTM was connected to jack hammer which was used to applied load to the specimen at a rate of 0.02 mm/min (Figure 1).

3. Results and discussions

3.1. Chemical composition

Sixteen elements were discovered in this test.The main elements apart from Iron (Fe) were Carbon (C), Silicon (Si), Manganese (Mn), Sulphur (S), Phosphorus (P), and Copper (Cu). Other elements present in residual amounts were Chromium (Cr), Nickel (Ni), Niobium (Nb), Aluminium (Al), Boron (B), Tungsten (W), Molybdenum (Mo), Vanadium (V), and Titanium (Ti). The obtained results of the chemical composition with particular references to the main elements [C, Si, Mn, S, P & Cu] are presented in Table 2 along with their comparison with specifications of international standards; the BS4449 [14], ASTM A706 [15], ISO 6935-2 [16] and the Nigerian National Standard, NIS 117 [17]. Further analysis of the implications of the variations in the composition of each of the respective elements results are presented in sub-sections 3.11-3.1.6.

Table 2. Percentage chemical composition of elements and Code provisions.

	C				S				P				Cu		Si	Mn
Samples	Measured value	BS4449 [0.27%]	NIS 117 [0.24%]	ASTM A706 & ISO 6935-2 [0.33%]	Measured value	BS4449 & NIS 117 [0.055%]	ASTM A706 [0.053%]	ISO 6935-2 [0.048%]	Measured value	BS4449 & NIS 117 [0.055%]	ASTM A706 [0.053%]	ISO 6935-2 [0.048%]	Measured value	NIS 117 [0.25%]	Measured value	Measured value
$UKR_{(12)}$	0.191	✓	✓	✓	0.030	✓	✓	✓	0.014	✓	✓	✓	0.030	✗	0.08	0.554
$UKR_{(16)}$	0.292	✗	✗	✓	0.068	✗	✗	✗	0.041	✓	✓	✓	0.027	✗	0.298	0.907
$AFL_{(16)}$	0.145	✓	✓	✓	0.027	✓	✓	✓	0.016	✓	✓	✓	0.019	✓	0.077	0.515
$AFL_{(20)}$	0.199	✓	✓	✓	0.037	✓	✓	✓	0.038	✓	✓	✓	0.121	✓	0.185	0.584
$BSM_{(12)}$	0.178	✓	✓	✓	0.056	✗	✗	✗	0.069	✗	✗	✗	0.217	✓	0.259	0.666
$BSM_{(16)}$	0.233	✓	✓	✓	0.053	✓	✓	✓	0.049	✓	✓	✗	0.228	✓	0.212	0.681
$ETM_{(12)}$	0.316	✗	✗	✓	0.047	✓	✓	✓	0.044	✓	✓	✓	0.231	✓	0.25	0.572
$ETM_{(16)}$	0.437	✗	✗	✗	0.053	✓	✓	✗	0.056	✗	✗	✗	0.358	✗	0.202	0.528
$LCI_{(16)}$	0.212	✓	✓	✓	0.064	✗	✗	✗	0.068	✗	✗	✗	0.315	✗	0.273	0.794
$LCI_{(20)}$	0.324	✗	✗	✓	0.067	✗	✗	✗	0.057	✗	✗	✗	0.319	✗	0.288	0.662
$MYE_{(12)}$	0.341	✗	✗	✗	0.047	✓	✓	✓	0.042	✓	✓	✓	0.350	✗	0.18	0.892
$RIL_{(12)}$	0.247	✓	✗	✓	0.055	✓	✗	✗	0.054	✓	✗	✗	0.309	✗	0.23	0.559
$RIL_{(16)}$	0.376	✗	✗	✗	0.058	✗	✗	✗	0.046	✓	✓	✓	0.273	✗	0.183	0.679
$RSS_{(12)}$	0.415	✗	✗	✗	0.041	✓	✓	✓	0.035	✓	✓	✓	0.172	✓	0.142	0.384
$RSS_{(16)}$	0.345	✗	✗	✗	0.045	✓	✓	✗	0.029	✓	✓	✓	0.147	✓	0.138	0.546
$SSM_{(16)}$	0.338	✗	✗	✗	0.050	✓	✓	✗	0.049	✓	✓	✗	0.371	✗	0.246	0.938
$TGT_{(12)}$	0.277	✗	✗	✓	0.037	✓	✓	✓	0.044	✓	✓	✓	0.166	✓	0.224	0.561
$TGT_{(16)}$	0.220	✓	✓	✓	0.036	✓	✓	✓	0.039	✓	✓	✓	0.151	✓	0.242	0.612
$TOP_{(16)}$	0.373	✗	✗	✗	0.044	✓	✓	✗	0.048	✓	✓	✓	0.293	✗	0.252	0.693
$TKY_{(12)}$	0.210	✓	✓	✓	0.03	✓	✓	✓	0.034	✓	✓	✓	0.465	✗	0.233	0.669
$TKY_{(16)}$	0.205	✓	✓	✓	0.043	✓	✓	✓	0.032	✓	✓	✓	0.612	✗	0.153	0.638

3.1.1. Carbon. The principal determinant of the steel's strength and hardness is Carbon. From Table 2, 9 samples met the requirement in NIS 117 [17] of maximum carbon content of 0.27% while 10 samples met the requirement in BS4449 [14] of maximum carbon content of 0.24% and 14 samples met the requirement in ASTM A706 [15] and ISO 6935-2 [16] of maximum carbon content of 0.33%. Other researchers have also reported higher carbon content than the maximum requirement by the NIS standard [18-21]. The higher the carbon content the greater the hardness, the strength, hardness and wear resistance of the steel [22, 23]. However, ductility, weldability and toughness are reduced with increasing carbon content [20].

3.1.2. Sulphur. This is an undesirable impurity in steel, special effort is usually made to eliminate or minimize sulphur during steelmaking. Sulphur decreases notch impact toughness, reduces weldability, surface quality and decreases ductility. It generally appears as sulphide inclusions in the steel which decreases its strength. From Table 2, 16 samples were found to meet the requirement in NIS 117 [17] and BS4449 [14] of maximum sulphur content of

0.055% while 15 samples were found to meet the requirement in ASTM A706 [15] of maximum sulphur content of 0.053% and 11 samples were found to meet the requirement in ISO 6935-2 [16] of maximum sulphur content of 0.048%.

3.1.3. Phosphorus. Phosphorus increases steel embrittlement which reduces the toughness and ductility of the reinforcing steel. From Table 2, 17 samples were found to meet the requirement in NIS 117 [17] and BS4449 [14] of maximum phosphorus content of 0.055% while 16 samples were found to meet the requirement in ASTM A706 [15] of maximum phosphorus content of 0.053% and 14 samples were found to meet the requirement in ISO 6935-2 [16] of maximum carbon content of 0.048%. An Increase in the percentage of phosphorus leads to increased hardness and strength, and decreased ductility, thus making the bars brittle [24].

3.1.4. Copper. Copper is often found as a residual agent in steels. It is also added to produce precipitation hardening properties and increase corrosion resistance. From Table 2, 9 samples were found to meet the requirement of maximum copper content of 0.25% of NIS 117 [17]. All samples were found to meet the requirement in BS4449 [14] of 0.85%.

3.1.5. Silicon. Silicon is one of the principal deoxidizers for structural steel. It improves strength, elasticity, acid resistance and results in larger grain sizes; thereby, leading to greater magnetic permeability. From Table 2, all samples met the requirement in ASTM A706 [15] and ISO 6935-2 [16] of maximum Silicon content of 0.55% and 0.60% respectively.

3.1.6. Manganese. Manganese has a large impact on strength, ductility and hardenability. It helps to reduce oxides and also counteract the presence of Iron Sulphide. Manganese content also helps in improving the abrasive resistance of reinforcing steel. It is a very important determinant of the steel's carbon equivalent value. From Table 2, all samples met the requirement in ASTM A706 [15]and ISO 6935-2[16] of maximum Manganese content of 1.56%.

3.1.7. Carbon Equivalent. Carbon Equivalent (C_{eq}) is an empirical value in weight percent, relating the combined effects of different alloying elements used in the making of carbon steels to an equivalent amount of carbon (Equation 1). It was developed to give a numerical value for a steel composition which would give an indication of hardenability, hydrogen cracking susceptibility, and other properties that may be linked to hardness, such as toughness and strength of steel. Table 3 shows the carbon equivalent of the tested steel bars. The percentage composition of sample steel bars considered showed no convergence at any point between its constituent elements from samples from the same steel mill/company. For example, a company that produced two tested samples of different diameter showed different percentage composition for each sample, implying a negative signal with the production process, scrap treatment, quality control, personnel, equipment or their combination [11, 25].

$$C_{eq} = C + \frac{Mn}{6} + \frac{Cr + Mo + V}{5} + \frac{Ni + Cu}{15} \qquad (1)$$

Table 3. Carbon equivalent and Code provisions.

Samples	Carbon Equivalent	BS4449 [14] 0.52%	NIS 117 [17] 0.54%	ISO 6935-2 [16] 0.56%
$UKR_{(12)}$	0.29	✓	✓	✓
$UKR_{(16)}$	0.45	✓	✓	✓
$AFL_{(16)}$	0.24	✓	✓	✓
$AFL_{(20)}$	0.32	✓	✓	✓
$BSM_{(12)}$	0.35	✓	✓	✓
$BSM_{(16)}$	0.41	✓	✓	✓
$ETM_{(12)}$	0.47	✓	✓	✓
$ETM_{(16)}$	0.59	✗	✗	✗
$LCI_{(16)}$	0.42	✓	✓	✓
$LCI_{(20)}$	0.52	✓	✓	✓
$MYE_{(12)}$	0.54	✗	✓	✓
$RIL_{(12)}$	0.40	✓	✓	✓
$RIL_{(16)}$	0.56	✗	✗	✓
$RSS_{(12)}$	0.54	✗	✓	✓
$RSS_{(16)}$	0.48	✓	✓	✓
$SSM_{(16)}$	0.55	✗	✗	✓
$TGT_{(12)}$	0.41	✓	✓	✓
$TGT_{(16)}$	0.36	✓	✓	✓
$TOP_{(16)}$	0.55	✗	✗	✓
$TKY_{(12)}$	0.38	✓	✓	✓
$TKY_{(16)}$	0.38	✓	✓	✓

3.2. Mechanical properties

3.2.1. Cross-sectional area. Table 4 shows that the marketed diameters are greater than the measured diameters for all the reinforcing bars considered except for $UKR_{(12)}$, $AFL_{(20)}$, $RIL_{(16)}$ and $TOP_{(20)}$ whose average measured diameter is slightly higher than the marketed diameter. The average measured diameter of $MYE_{(12)}$ is the same with the marketed diameter. A very large discrepancy was observed on sample $RSS_{(16)}$ which is designated and sold in the market as 16mm bar against its average measured diameter of 14.47mm.

3.2.2. Effective cross-sectional areas and tolerances. In line with BS4449 [14] requirements which specify ± 6.0 % for ≤ 8mm bars and ± 4.5% for diameter greater than 8mm bars. It can be seen from Table 4 that the tolerances for some of the reinforcement bars irrespective of origin fall out of the specified range. 12 bars are out of range while 9 fall within the acceptable range. This indicates that the 11 bars that were out of range have varying diameters along the length which is not best for reinforcing bars. This should be carefully checked to ensure an average close diameter throughout the length.

3.2.3. Characteristic strength and percentage elongation. The characteristic strengths computed from the yield strengths and compared with the code requirements are shown in Table 5. From the table, it is clear that only 8 samples out of 35 bars tested met the minimum code provisions by BS4449 [14] with Ultimate to Yield Strength Ratio less than 1.15. It can be observed that some of the bar samples both local and foreign are below the minimum requirement. This could be as a result of cooling process which is a manufacturing fault within the line of production. However, when the ratio is high, it is not good either. It implies high carbon content which means the bars lack ductility.

Table 4. Comparison of measured and market diameters, cross-sectional areas and strength values

	Diameters (mm)			Area (mm^2)				Strength			
Samples	Market Assumed	Average Measured	Percentage Difference (%)	Measured Cross-Sectional	Nominal Cross-Sectional	NIS 117 Tolerance	Remarks	Yield Strength (N/mm^2)	Ultimate Tensile Strength (N/mm^2)	Ultimate to Yield Strength Ratio	Elongation (%)
UKR$_{(12)}$	12	12.05	-0.42	114.04	113.00	±5.7	✓	468.89	894.13	1.91	24.37
UKR$_{(16)}$	16	15.75	1.56	194.83	201.00	±7.6	✓	548.65	646.75	1.18	15.86
AFL$_{(16)}$	16	15.91	0.56	198.81	201.00	±7.6	✓	481.27	743.83	1.55	17.22
AFL$_{(20)}$	20	20.57	-2.85	332.32	314.00	±9.6	✗	490.89	588.21	1.20	23.46
BSM$_{(12)}$	12	11.07	7.75	96.25	113.00	±5.7	✗	468.61	735.39	1.57	17.61
BSM$_{(16)}$	16	15.13	5.44	179.79	201.00	±7.6	✗	481.02	504.73	1.05	21.57
ETM$_{(12)}$	12	10.89	9.25	93.14	113.00	±5.7	✗	462.02	593.83	1.29	24.29
ETM$_{(16)}$	16	15.13	5.44	179.79	201.00	±7.6	✗	486.66	645.53	1.33	10.64
LCI$_{(16)}$	16	15.68	2.00	193.10	201.00	±7.6	✗	558.27	662.54	1.19	5.15
LCI$_{(20)}$	20	19.53	2.35	299.57	314.00	±9.6	✗	481.33	574.00	1.19	14.65
MYE$_{(12)}$	12	12.00	0.00	108.80	113.00	±5.7	✓	462.02	732.73	1.59	25.05
RIL$_{(12)}$	12	11.81	1.58	109.54	113.00	±5.7	✓	471.64	539.39	1.15	13.04
RIL$_{(16)}$	16	16.45	-2.81	212.53	201.00	±7.6	✗	462.02	496.27	1.10	7.82
RSS$_{(12)}$	12	12.00	0.00	113.29	113.00	±5.7	✓	463.95	484.06	1.04	18.89
RSS$_{(16)}$	16	14.47	9.56	164.45	201.00	±7.6	✗	488.58	543.54	1.11	6.45
SSM$_{(16)}$	16	15.24	4.75	182.41	201.00	±7.6	✗	481.27	508.52	1.06	20.89
TGT$_{(12)}$	12	11.45	4.58	102.97	113.00	±5.7	✗	481.27	508.52	1.05	15.71
TGT$_{(16)}$	16	15.71	1.81	193.84	201.00	±7.6	✓	500.52	521.00	1.04	14.20
TOP$_{(16)}$	20	20.46	-2.30	328.78	314.00	±9.6	✓	471.70	529.78	1.12	6.37
TKY$_{(12)}$	12	11.36	5.33	101.36	113.00	±5.7	✗	468.89	894.13	1.64	14.96
TKY$_{(16)}$	16	15.71	1.81	193.84	201.00	±7.6	✓	500.52	806.18	1.61	30.59

4. Conclusion

The study employed micro-structure test to obtain the percentage elemental constituents of reinforcing steel bars in the Nigerian construction industry. Also, the tensile properties in terms of tensile strength, yield stress and elongation of the reinforcing bars were determined. In addition, the diameters of the bars available in the market were measured and compared to their nominal values.

Based on the results of the tests conducted, it was observed that most of the reinforcing steel bars in the Nigerian construction industry do not comply with the provisions of the percentage chemical composition stipulated in the NIS 117 [17], BS4449 [14], ASTM A706 [15] and ISO 6935-2 [16] standards. Further, there is inconsistency in percentage composition of all important elements for different sample sizes produced by the same steel mill in almost all the samples tested; this portends a laxity in the production process and a major cause of variabilities in the strength of reinforced concrete elements made from reinforcements of the same grade from the same source.

However, reinforcing steel bar samples tested complied with the carbon equivalent value prescribed in NIS 117 [17]. AFL reinforcing steel bars is the only bar that satisfied all the chemical composition requirements in the four codes used. A discrepancy in the nominal and measured diameter and cross-sectional area of most of the steel bars was also observed, the tolerance on effective cross-sectional area for some of the bars fall out of range due to non-compliance with the actual bars diameter in the BS4449 [14].

Furthermore, the characteristic strength values for most of the tested bar samples are low compared to the BS4449 [14] standards for high tensile steel. Most of the reinforcement bar samples did not complied with the minimum ultimate to yield strength ratio as specified by BS 4449 [14] provisions. The percentage elongation values for the tested samples did not comply with the requirements prescribed by the codes, such bars have low ductility and should not be used in reinforced concrete structures because they will not give warning prior to failure.

The research provided an insight on the variations of chemical composition and tensile properties of reinforcing bars in the Nigerian construction industry with respect to the provisions of available standards. The results affirm the view on the culpability of reinforcements in the recurring incidents of building failures and collapses in the Nigerian construction industry. The available reinforcements in the markets mostly do not meet the quality required in terms of chemical composition, dimensions, characteristic strengths and percentage elongation. The outcome of this research will assist with necessary data required for quality assurance and implementation of compliance measures by regulatory bodies on steel bars being used in the Nigerian construction industry.

References

[1] Neville A M 2011 *Properties of Concrete* (England: Pearson)

[2] Basu P C, Shyiamoni P and Roshan A 2004 Characterisation of steel reinforcement for RC structures: An overview and related issues *Indian concrete journal* **78** 19-30

[3] Baddoo N R 2008 Stainless steel in construction: A review of research, applications, challenges and opportunities *Journal of Constructional Steel Research* **64** 1199-206

[4] Rafi M M, Lodi S H and Nizam A 2014 Chemical and Mechanical Properties of Steel Rebars Manufactured in Pakistan and Their Design Implications *Journal of Materials in Civil Engineering* **26** 338-48

[5] Dimuna K O 2010 Incessant incidents of building collapse in Nigeria: A challenge to stakeholders *Global Journal of Researches in Engineering* **10** 75-84

[6] Fagbenle O I and Oluwunmi A O 2010 Building failure and collapse in Nigeria: The influence of the informal sector *Journal of sustainable development* **3** 268-76

[7] Joshua O, Olusola K O, Ayegba C and Yusuf A I 2013 *AEI 2013: Building Solutions for Architectural Engineering,* pp 296-305

[8] Ayininuola G and Olalusi O 2004 Assessment of Building Failures in Nigeria: Lagos and Ibadan Case Study' *African Journal of science and technology* **5**

[9] Alo F, Atanda P, Daniyan A, Abodunrin O and Oluwasegun K 2017 An Assessment of Imported and Local Constructional Steel in Nigeria: Analysis by One-Way ANOVA *International Journal of Materials Engineering* **7** 45-51

[10] Bamigboye G O, Olaniyi O A, Olukanni D O, Ede A N and Akinwumi I I 2017 Diameter Inconsistency, Strength and Corrosion Characteristics of Locally-Produced and Imported Steel Reinforcing Bars in Ilorin, Nigeria *International Journal of Engineering Research in Africa* **29** 90-7

[11] Balogun S, Esezobor D, Adeosun S and Sekunowo O 2009 Challenges of Producing Quality Construction Steel Bars in West Africa: Case Study of Nigeria Steel Industry *Journal of Minerals and Materials Characterization and Engineering* **8** 283-92

[12] Arum C 2008 Verification of Properties of Concrete Reinforcement Bars: Nigeria as a Case Study *Indoor and Built Environment* **17** 370-6

[13] Alabi A G F, Ayoadeb A O, Odusote J K and Akannid A A 2016 Assessment of Suitability of Nigerian Made Steel Bars for Structural Applications *The Journal of the Association of Professional Engineers of Trinidad and Tobago* **44** 17-23

[14] British Standards Institution 2005 BS4449-Carbon steel bars for the reinforcement of concrete—Weldable reinforcing steel—Bar, coil and decoiled product—Specification.

[15] American Society of Testing and Materials 2009 ASTM A706 -Standard Specification for Low-Alloy Steel Deformed and Plain Bars for Concrete Reinforcement,.

[16] International Standards Organisation 2007 ISO 6935-2-Steel for the reinforcement of concrete -- Part 2: Ribbed bars.

[17] Standards Organisation of Nigeria 2004 NIS 117-Standards on Steel Bars for Reinforcement of Concrete.

[18] Awofadeju A, Akanni A and Adeyemo B 2014 Evaluation of Locally Produced and Imported Reinforced Steel Rods for Structural Purposes in Nigerian Market *International Journal of Recent Development in Engineering and Technology* **3** 81-4

[19] Oyelade A, Fapohunda C and Dixon O 2013 Effects of Source of Reinforcement on Microstructure and Strength Characteristics of Reinforced Concrete Beams *Civil and Environmental Research* **3** 26-36

[20] Alabi A and Onyeji L 2010 Analysis and comparative assessment of locally produced reinforcing steel bars for structural purposes *USEP: Journal of Research Information in Civil Engineering* **7** 49-60

[21] Ocheri C and Ibe L O 2017 Comparative Assessment of Locally Produced Reinforcing Steel Bars for Structural Purposes: 12 mm Steel Bars from Delta Steel Company (DSC), Warri-Nigeria as a Case Study *Journal of Powder Metallurgy & Mining* **06**

[22] Lino R E, Marins Â M F, Marchi L A, Mendes J A, Penna L V, Neto J G C, Caldeira J H P and Costa e Silva A L V d 2017 Influence of the chemical composition on steel casting performance *Journal of Materials Research and Technology* **6** 50-6

[23] Salman A and Djavanroodi F 2018 Variability of chemical analysis of reinforcing bar produced in Saudi Arabia. In: *IOP Conference Series: Materials Science and Engineering*: IOP Publishing) p 012015

[24] Odusote J K and Adeleke A A 2012 Analysis of Properties of Reinforcing Steel Bars: Case Study of Collapsed Building in Lagos, Nigeria. In: *Applied Mechanics and Materials*: Trans Tech Publ) pp 3052-6

[25] Ezekiel A A and Silas O A 2017 Assessment of Suitability of Selected Reinforcing Bars Used for Construction in Nigeria *Journal of Multidisciplinary Engineering Science and Technology* **4**

IOP Conf. Series: Materials Science and Engineering **513** (2019) 012027 doi:10.1088/1757-899X/513/1/012027

Influence of porosity on the tensile strength of fine grained mortar containing fly ash, palm oil fuel ash and rice husk ash

Zalipah Jamellodin[1*], Hamidah Mohd Saman[2], Azmi Ibrahim[2] and Suraya Hani Adnan[3] and Nur Hafizah Abd Khalid[4]

[1]Faculty of Civil and Environmental Engineering, Universiti Tun Hussein Onn Malaysia, 86400 Batu Pahat, Johor, Malaysia.
[2]Faculty of Civil Engineering, UiTM Shah Alam, 40450 Shah Alam, Selangor.
[3]Faculty of Engineering Technology, Universiti Tun Hussein Onn Malaysia, 86400 Batu Pahat, Johor, Malaysia.
[4]School of Civil Engineering, Faculty of Engineering, Universiti Teknologi Malaysia, 81310 Johor Bahru, Malaysia

[*]Correspondence author: zalipah@uthm.edu.my

Abstract. The study is to identify the influence of porosity on the tensile strength of fine grained mortar containing fly ash (FA), palm oil fuel ash (POFA) and rice husk ash (RHA). The aim of this study is to reduce the consumption of cement due to its negative impact on the environment. The cement-to-sand ratio used in the mix design of fine grained mortar is 1: 2. The maximum size of fine grained sand used in this study is 0.6 mm. There are four (4) mix designs of fine grained mortar which include control mortar (100% OPC), 80% OPC + 20% FA, 90% OPC + 10% POFA and 80% OPC + 20% RHA. The selection of the mix design for each type of fine grained mortar was based on the mix designs of previous studies which produced the highest compressive strength. The tensile strength test and the porosity test were carried out in this study. The tensile strength test was performed on cylinder specimens measuring 100 mm diameter and 200 mm height according to ASTM C496. On the other hand, the porosity test was conducted on prism specimens measuring 40 x 40 x 160 mm. The results showed that fine grained mortar containing FA achieved the highest tensile strength of 5.9 N/mm^2 and a low porosity of 12.7%.

1. Introduction

The porosity structure in mortar greatly influences its performance. For example, porosity determines the rate at which aggressive ions can enter the mass and cause interference in the mortar. Porosity is related to cement, mineral mixtures and aggregate particles against water cement ratio as well as curing conditions. A reduction in the porosity of solid materials increases its strength in general, and the strength of cement-based materials in particular [1]–[3].

Fine grained mortar (FGM) is one of the cement-based materials which act as a binder for the repair and reinforcement of structural elements. FGM is unique as it contains fine grained sand with a maximum size of 0.6 mm. Due to its small grain size, it is possible to produce very thin concrete elements using FGM. On the other hand, the substitution of pozzolan material with cement can improve the strength and durability of mortar.

Nowadays, pozzolans from industrial and agricultural by-products such as fly ash, palm oil fuel ash and rice husk ash are commonly used as cement replacement. Calcium silicate hydrate is formed during the hydration between pozzolanic material and calcium hydroxide. This formation can reduce the pore size of crystalline hydration products, making the mortar microstructure more uniform besides improving its impermeability and durability. In addition, it can also increase the service life of a structure.

The substitution of pozzolanic material can reduce construction costs and reduce the negative impact towards the environment. Fly ash (FA) is a by-product of coal combustion in electric power generation plants whereas palm oil fuel ash (POFA) and rice husk ash (RHA) are agricultural by-products. The substitution of FA in FGM has been studied by previous researchers but research studies on the use of POFA and RHA in FGM are still lacking [4]–[6]. Thus, this study incorporated FA, POFA and RHA with 10% to 20% replacement of cement to examine its effects on the tensile strength and porosity of fine grained mortar. The objective of this paper is to determine the relationship between tensile strength and the porosity of FGM incorporated with FA, POFA and RHA.

2. Fine grained mortar

FGM can be presented as one of the new innovative technology binder systems in which fine sand with a grain size of 0.6 mm or less is used to produce high strength mortar [7]–[10]. The FGM mix shows that FGM has a high flow consistency that offers technical penetration to textiles. It is possible to obtain very thin concrete elements using FGM due to its fine grain size. It also has a high binder content when it is added with different pozzolanic additives and plasticizers [11]. The workability of cementitious material has been obtained by some compatible water reducing admixtures such as superplasticizers.

The use of a mixture of pozzolans has been shown to be advantageous owing to its synergic effects [12]. In this work, ordinary Portland cement, fly ash, palm oil fuel ash and rice husk ash are used as cement replacement materials in blended cement. Tensile strength and porosity tests are conducted to understand the mechanisms and uses of these types of waste in concrete.

The addition of fine pozzolanic particles cause large pore segmentation and increases the number of nucleation sites for hydration product precipitation in cement paste. This results in pore production and reduced calcium hydroxide content in the cement paste. The FGM containing FA results in lower porosity than FGM containing RHA [2].

At present, pozzolans have been widely used to replace cement in the construction industry in order to reduce costs and protect the environment [13], [14]. In addition, the use of pozzolans can help reduce waste disposal problems, improve mortar and concrete properties and lower the cost of cement production. The use of pozzolans is also associated with the current situation that demand for sustainable concrete and mortar performance [14].

Fly ash is the most common and widely used cement substitute for cement and mortar in the construction industry. FA is one of the by-products of coal combustion in electric power generating plants which is then collected from exhaust gases using electrostatic precipitators or bag filters [10]. Fly ash can be tan to dark grey, depending on its chemical and mineral constituents. The tan and light colours are colour usually associated with high lime content. Brownish colours are usually associated with iron content. Dark grey to black colour is usually caused by unburned carbon content. Fly ash colour is usually very consistent for each power plant and coal source [15], [16].

Palm oil fuel ash is a solid waste product by the palm oil industry. It is obtained in the form of ash from the combustion of oil palm bunches. POFA is produced from the combustion of palm oil biomass at 800 °C to 1000 °C after the palm oil extraction process. Palm oil biomass is used as fuel to facilitate power generation at palm oil mills. POFA contains a high amount of silica and silica oxide that meet the criteria of pozzolanic materials [7], [17], [18]. It has a high potential to be used as a substitute for cement or as a filler to produce strong mortar and concrete [19].

RHA is one of the agricultural by-products which is also widely used by the construction industry [20]. Well-burned RHAs can contribute significantly to mortar and concrete properties through pozzolanic reactions. The slow burning of rice husks under temperatures between 500°C to 700°C produces RHA with porous and cellular microstructure. RHA contains a high content of non-crystalline or amorphous silica with a high specific surface area and high pozzolanic reactivity [2], [21], [22]. Therefore, it is one of the pozzolanic materials that can be used as a substitute for cement because RHA particles have complex shapes which can break porous structures [9].

3. Experiment details

3.1. Materials

Fly ash was obtained from the thermal power plant called Kapar Energy Ventures in Kapar, Selangor. FA was oven dried at 105°C to 110°C for 24 hours and then ground by the grinder with 1-3% retained on sieve number 325. This process complies with the ASTM C618 requirements. POFA was obtained from a palm oil mill in Kluang, Johor. POFA was ground using a grinder for 15 minutes at a speed of 350 rpm to ensure 1-3% was retained on sieve number 325 (45μm) which complied with ASTM C618 requirements. Rice husks were collected from Muar, Johor. The rice husks were burnt at a temperature of 600°C for six hours. Before that, the rice husks were washed 5 times to clean them of all impurities. Then, the rice husks were oven dried at 105-110°C for 24 hours. Later, RHA was ground using a grinder with 1-3% retained on sieve number 325 which complied with ASTM C618 requirements.

3.2. Mix proportions

The mixture of the sample was designed to determine the suitability of mixed FGM ratios for obtaining the FGM with the highest strength. The mix design involved 4 samples with each pozzolan in the ratio of 1:2. All the samples were mixed uniformly according to the water-cement ratio obtained from the standard consistency test of cement and were cured for 28 days.

Four (4) series of mixed design proportions were adopted from previous studies by Z Jamellodin [10], and Yuslinda, [7]. The FGM specimens were prepared for the tensile strength test and the porosity test. OPCs were replaced with three pozzolans namely, FA, POFA and RHA at percentages of 20%, 10% and 20%, respectively by the weight of cement. The sand-to-binder ratio of 2 by weight and the water-to-binder ratio of 0.45 were used. The FGM mixed proportions with FA, POFA and RHA are shown in Table 1.

Table 1. The mix proportion by weight of FGM used in tensile strength and porosity test.

Mix No	Mix design proportion	Cement (%)	FA (%)	POFA (%)	RHA (%)	Fine grained sand	Water/binder
M1	OPC	100	-	-	-	2	0.45
M2	FA20	80	20	-	-	2	0.45
M3	POFA10	90	-	10	-	2	0.45
M4	RHA20	80	-	-	20	2	0.45

3.3. Determination of tensile strength

The tensile strength test was carried out according to ASTM C496 [23]. Cylinder specimens measuring 100 mm diameter and 200 mm height were prepared. After 1 day of casting, the specimens were demoulded and cured in water until the day of the test. The tensile strength of mortar specimens was determined at the age of 28 days. Three specimens were duplicated for each FGM mixed design proportion. The reported results were the average tensile strength for the three specimens.

IOP Conf. Series: Materials Science and Engineering **513** (2019) 012027 doi:10.1088/1757-899X/513/1/012027

3.4. Determination of porosity

Prisms specimens with dimensions of 40 x 40 x 160 mm were prepared for the porosity test accordance to ASTM D7063/D7063M-11. The porosity of mortar specimens were determined at the age of 28 days. After that, the prisms were dried at 100 ± 5°C until constant weight was achieved. The porosity can be calculated using equation 1 as follows:

$$P = [(W_a - W_d)/(W_a - W_w)] \times 100 \qquad (1)$$

P refers to vacuum saturated porosity (%),
W_a refers to the specimen weight in air of saturated sample (gm),
W_d refers to the dry weight of the specimen after 24 h in the oven at 100 ± 5 °C (gm) and
W_w refers to the specimen weight in water (gm).

4. Results and discussions

4.1. Tensile strength test

The tensile strength and porosity results of FGM containing FA, POFA and RHA according to their percentage by weight which are 20%, 10% and 20% respectively are shown in Table 2. According to the results, the tensile strength of FGM containing FA was the highest at 5.9 N/mm^2. Meanwhile, FGM containing POFA achieved the lowest tensile strength among all pozzolan materials which is of 5.5 N/mm^2. However, the tensile strength of FGM containing POFA was still higher compared to the tensile strength of control mortar which was only 5.4 N/mm^2. All three pozzolans were found to have increased the tensile strength of FGM.

This also indicates that with the replacement of cement with raw materials contribute to the strength of FGM. This is due to the fineness of the pozzolans which were combined with OPC and cured for 28 days to strengthen the bond between the particles. The high tensile strength could also be due to the increase in mortar microstructure that only happens through a pozzolanic reaction [7], [24]. The pozzolanic reaction occurred between silica or lime and calcium hydroxide during the hydration of OPC. This produced calcium silica hydrate (C-S-H) and calcium alumina hydrate which can cover cracks and pores [7].

4.2. Porosity test

Also can be seen the results of porosity in Table 2, FGM containing 20% FA has a lower porosity value of 12.7% compared with to the control FGM which has a porosity of 13.5%. Among all FGMs containing pozzolans, FGM containing POFA had the highest porosity value of 13.1%, followed by RHA with a porosity value of 12.9%. From the results, it can be concluded that FGM containing 20% of FA, 10% of POFA and 20% of RHA achieved a better performance than the control mortar in terms of porosity. This is because pozzolans such as FA, POFA and RHA are more effective in modifying pores and reducing the porosity of FGM [2].

Table 2. Tensile strength and porosity of FGM at 28 days.

Specimen no.	Symbol	Tensile strength (N/mm^2)	Porosity (%)
1	Control	5.4	13.5
2	FA20	5.9	12.7
3	POFA10	5.5	13.1
4	RHA20	5.6	12.9

IOP Conf. Series: Materials Science and Engineering **513** (2019) 012027 doi:10.1088/1757-899X/513/1/012027

4.3. Relationship between tensile strength and porosity

The relationship between tensile strength and porosity at the age 28 days was shown in Figure 1. It can be observed that the higher the tensile strength, the lower the porosity. It can be seen that cement replaced with 20% of FA, 10% of POFA and 20% of RHA achieved the highest tensile strength and the lowest porosity. This is because FA, POFA and RHA tend to modify pores more effectively than control mortar [2]. Meanwhile, the value of tensile strength for the control mortar was the lowest while its porosity was the highest among all the other specimens. The incorporation of pozzolans reduces the average pore size and modifies pores in FGM.

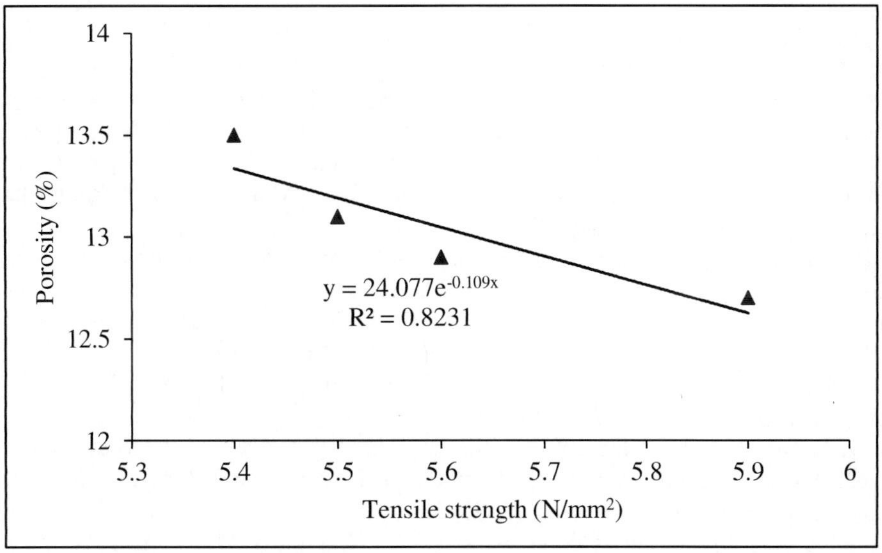

Figure 1. Relationship between porosity and tensile strength.

Based on the results, the pozzolans used in this research were found to be suitable for use in FGM. This may help create a green and sustainable construction industry. Besides that, incorporating FA, POFA and RHA into FGM slightly increases its strength. The relationship between tensile strength and porosity can also be shown using the R^2 value which was 0.8231 after a curing period of 28 days. Hence, this relationship was used in determining the tensile strength and porosity of FGM incorporated with FA, POFA and RHA as cement replacement.

5. Conclusions

This research was carried out to determine the tensile strength and porosity of FGM containing FA, POFA and RHA as cement replacement. From the results obtained, it can be concluded that the objectives of this study have been achieved.

The tensile strength of FGM containing FA, POFA and RHA as cement replacement was shown to be better than the tensile strength of the control mortar. The addition of 20% of FA, 10% of POFA and 20% of RHA is able to improve the tensile strength of FGM. FGM containing 20% of FA achieved the highest tensile strength of 5.9 N/mm^2, followed by FGM containing POFA and RHA with tensile strength values of 5.5 N/mm^2 and 5.6 N/mm^2, respectively. Meanwhile, the tensile strength of the control mortar was found to be the lowest.

The percentage of porosity for FGM containing FA, POFA and RHA as cement replacement was lower than that of the control mortar. FGM containing FA obtained the lowest percentage of porosity which is 12.7%, followed by RHA and POFA with percentages of porosity of 12.9% and 13.1%, respectively. This indicates that FGM containing FA, POFA and RHA were slightly effective in reducing the porosity of FGM.

When the value of R is close to 1, this means that there is a good relationship between tensile strength and porosity in FGM. Hence, it is verified that the addition of pozzolans is capable of improving tensile strength and reducing porosity in FGM.

Acknowledgments
The authors wish to acknowledge the Ministry of Education (MOE) and Universiti Tun Hussein Onn Malaysia for the educational support provided. They would also like to thank the laboratory staff at the Faculty of Civil and Environmental Engineering, for their guidance and assistance.

References
[1] X. Chen, S. Wu, and J. Zhou, "Influence of porosity on compressive and tensile strength of cement mortar," *Construction and Building Material*, vol. 40, pp. 869–874, 2013.
[2] P. Chindaprasirt and S. Rukzon, "Strength, porosity and corrosion resistance of ternary blend Portland cement, rice husk ash and fly ash mortar," Construction and Building Material, vol. 22, no. 8, pp. 1601–1606, Aug. 2008.
[3] I. Papayianni and M. Stefanidou, "Strength-porosity relationships in lime-pozzolan mortars," *Construction and Building Material*, vol. 20, no. 9, pp. 700–705, 2006.
[4] W. Y. Wan Yusof, S. H. A. Adnan, Z. Jamellodin, N. S. Mohammad, W. Y. W. Yusof, S. H. A. Adnan, Z. Jamellodin, and N. S. Mohammad, "Strength Development of Fine Grained Mortar Containing Palm Oil Fuel Ash as a Partial Cement Replacement," *Applied Mechanics and Materials*, vol. 774, pp. 964–968, 2015.
[5] N. S. Mohammad, S. H. Adnan, Z. Jamellodin, and W. Y. Wan Yusof, "Performance of Rice Husk Ash as a Partial Cement Replacement in Fine Grained Mortar," *Applied Mechanics and Materials*, vol. 774, pp. 969–973, 2015.
[6] Z. Jamellodin, H. Mohd Saman, S. H. S. H. Adnan, N. S. N. S. Mohammad, W. Y. W. Y. Wan Yusof, H. M. Saman, S. H. S. H. Adnan, N. S. N. S. Mohammad, and W. Y. W. Y. Wan Yusof, "Compressive and Flexural Strength of Fine Grained Mortar Containing Rice Husk Ash: A Review," *Adv. Mater. Res.*, vol. 1051, pp. 757–762, 2014.
[7] Z. Jamellodin, H. Mohd Saman, S. H. Adnan, N. S. Mohammad, W. Y. Wan Yusof, M. S. Hamidah, S. H. Adnan, N. S. Mohammad, and W. Y. Wan Yusof, "Strength Development of Fine Grained Mortar Containing Fly Ash and Rice Husk Ash," *Applied Mechanics and Materials*, vol. 753, pp. 182–188, 2015.
[8] T. Brockmann, "Mechanical and Fracture Mechanical Properties of Fine Grained Concrete for Textile Reinforced Composites," RWTH Aachen, 2005.
[9] G. C. Isaia, "Physical and pozzolanic action of mineral additions on the mechanical strength of high-performance concrete," vol. 25, pp. 69–76, 2003.
[10] M. R. Karim, M. F. M. Zain, M. Jamil, and F. C. Lai, "Significance of Waste Materials in Sustainable Concrete and Sustainable Development," in *International Journal of Biotechnology and Environmental Management*, 2011, vol. 18, pp. 43–47.
[11] R. Karim, M. Hossain, M. Nabi, N. Khan, M. Fauzi, M. Zain, M. Jamil, and F. C. Lai, "On the Utilization of Pozzolanic Wastes as an Alternative Resource of Cement," pp. 7809–7827, 2014.
[12] R. Cheerarot and C. Jaturapitakkul, "A study of disposed fly ash from landfill to replace Portland cement," vol. 24, pp. 701–709, 2004.
[13] G. Sua-Iam and N. Makul, "Utilization of high volumes of unprocessed lignite-coal fly ash and rice husk ash in self-consolidating concrete," *J. Clean. Prod.*, vol. 78, pp. 184–194, 2014.
[14] Sooraj VM, "Effect of Palm Oil Fuel Ash (POFA) on Strength Properties of Concrete," *International Journal of Scientific and Research Publications*, vol. 3, no. 6, pp. 2250–3153, 2013.

[15] D. Galau and M. Ismail, "Characterization of Palm Oil Fuel Ash (POFA) from Different Mill as Cement Replacement Material," 2010.
[16] W. Tangchirapat, C. Jaturapitakkul, and P. Chindaprasirt, "Use of palm oil fuel ash as a supplementary cementitious material for producing high-strength concrete," *Construction and Building Material.*, vol. 23, no. 7, pp. 2641–2646, Jul. 2009.
[17] S. K. Antiohos, V. G. Papadakis, and S. Tsimas, "Rice husk ash (RHA) effectiveness in cement and concrete as a function of reactive silica and fineness," *Cement and Concrete Research*, vol. 61–62, pp. 20–27, 2014.
[18] Z. Jamellodin, N. Nor Azhar, M. A. Jamaly, M. A. Majid, N. Salleh, N. A. Abdul Hamid, N. Ismail, S. H. Adnan, and N. N. Abdul Rashid, "Strength of Blended Cement Mortar Containing Palm Oil Fuel and Eggshell Ashes," in *Advanced Research On Computer, Civil, Electrical, And Petroleum Engineering International Conference*, 2017.
[19] C. Jaturapitakkul, J. Tangpagasit, S. Songmue, and K. Kiattikomol, "Filler effect and pozzolanic reaction of ground palm oil fuel ash," *Construction and Building Material.* vol. 25, no. 11, pp. 4287–4293, 2011.
[20] ASTM, "ASTM C496-96 - Standard Test Method for Splitting Tensile Strength of Cylindrical Concrete Specimens," 1996.
[21] P. Chindaprasirt, S. Homwuttiwong, and C. Jaturapitakkul, "Strength and water permeability of concrete containing palm oil fuel ash and rice husk–bark ash," *Construction and Building Material*, vol. 21, no. 7, pp. 1492–1499, Jul. 2007.

IOP Conf. Series: Materials Science and Engineering **513** (2019) 012028 doi:10.1088/1757-899X/513/1/012028

Deferred maintenance of buildings: A review paper

M N Yasin[1,2*], R Mohamad Zin[2], M Y Hamid[3], M A Zakaria[1] and R Deraman[1]

[1] Faculty of Civil & Environmental Engineering, Universiti Tun Hussein Onn Malaysia, 86400 Parit Raja, Johor, Malaysia
[2] Faculty of Civil Engineering, Universiti Teknologi Malaysia, 81310 Skudai, Johor, Malaysia
[3] Universiti Teknologi Mara, Bandar Seri Iskandar, 32610 Seri Iskandar, Perak, Malaysia

[*]Corresponding author: irazam79@gmail.com

Abstract. The building must be maintained to keep it in optimal operation in giving service and it is vital and also it can be influenced towards the building performance. Therefore, immediately after construction of the building project is completed, maintenance start to set in. The conducive environment that provided for the buildings will give a good performance and significant impact. Therefore, building maintenance is crucial in making sure the buildings can deliver a good service or function, but in certain reason building maintenance and repair activities that were not performed when they should have been will be deferred. Deferred maintenance occurs because there is no efficiency in managing building maintenance. Deferred maintenance has become a pressing issue in many countries and have adverse effects to the organization (stakeholders and users). Hence, this study is intended to provide definitions, issues and challenges of deferred maintenance of the buildings.

1. Background

Nowadays, it's become a great challenge to ensure that all the building's condition and quality are at the top so that their serviceability could be maintained and can fulfil their function requirements [1]. Generally, once buildings have been completed they expected to perform on function in certain period of time. Proper maintenance of the buildings is important and need to be given a priority by the organization [2]. Thus, a building must be maintained to keep it in optimal operation in giving service and it is vital and also it can be influence towards the building performance [3]. [3] also mentioned, building maintenance and building safety are closely related to each other. Therefore, there are a need to create a conducive environment that support all activities in the buildings. Thus maintenance is crucial in making sure the buildings can deliver a good service or functional. In Malaysia maintenance of buildings were not different from other facilities which are preventive, corrective and condition based but currently most of the building rely heavily on corrective maintenance [3]. Hence, can be said that organizations in Malaysia are still managing their buildings in the traditional ways [4].

2. Introduction

Building construction is a human activity that existed a long time ago [5] and is one of the biggest industries in the world today [6] and building maintenance is one part of it that need to be carried out after the completion of the building. Buildings require maintenance in order to create a conducive environment and meet the physical needs that supports and stimulates the activities. Therefore to meet the purpose and keep the buildings usable [7], there are a need to make sure all the buildings are working in a good performance condition [8]. With this conducive environment that provided to the buildings will give a good performance and significant impact. Hence, in order to optimise the users it should never be exposed to the risks [9]. Nowadays, [10] points out that one of the major problems for the countries is inadequate maintenance. According to [11], more than 90% of the life cycle of building required maintenance after completion of construction and 75% approx. of the total expenditure on life cycle cost of building is due to maintenance. Hence, this proves that repair and maintenance is a critical element in organization and expected work will become more important and will be increased in the future and it carries dire consequences if ignored. Top management and maintenance organisation are aware of the physical appearance of their respective buildings, installed systems and services within the buildings appear to be deteriorating.

3. Definition of building maintenance

Building maintenance is defined as the preservation of a building so that it can serve its intended purpose [12]. In addition, [13] state that British Standard 3811 (1984), has defined building maintenance as a combination of any actions carried out to retain an item in, or restore it to, an acceptable condition. In placing more emphasis, [14] mentioned that The Chartered Institute of Building, CIOB (1990) defined maintenance as work undertaken in order to keep, restore, or improve every facility i.e. every part of a building, its services and surrounds, to an agreed standard determined by the balance between the need and the available resources. For some people, building maintenance is a complex process and involves large financial resources but if implemented efficiently, it can save money other than the potential for building life. It is practically impossible to produce buildings which are maintenance free [15]. Therefore, immediately when construction of buildings project are completed, maintenance start to set in. Building maintenance assist a lot in retaining the economic value of building and also to extend the building's lifespan The maintenance of the building should be taken into account and carried out during the phase which follows the construction phase until the operational phase [16].

4. Definition of deferred maintenance

Basically the term of deferred maintenance is repair work that have been postponed to a future date beyond the recommended service [17]. Deferred maintenance refers to maintenance and repair activities that were not performed when they should have been. The repair and deficiencies that are unfunded, unavailability of sufficient funds and human resources [18] and any kind of reason or issues at the fiscal year on a planned or unplanned basis and are deferred or postponed to a future fiscal year until the issues are solved. Deferring maintenance work can reduce the overall life of facilities and may lead to higher costs in the long term [19]. Basically deferred maintenance occurs because there is no efficiency in managing building maintenance. Building maintenance management practice which is not in the right practice will give the bad impact towards the building and the environment [20]. According to [21] deferred maintenance is physical depreciation, or loss in value of a building, that results from postponed maintenance to the building. This type of depreciation normally is curable by making the necessary repairs and improvements. It is sometimes called curable physical depreciation or curable obsolescence. According to [22] the deficiency in maintenance management system also makes organizations hard to totally record facilities defect periodically can contributed to deferred maintenance. In placing more emphasis, [23] mentioned that, deferred maintenance often referred to as unfunded or unaccomplished maintenance or backlog, has not been a popular or frequently discussed topic in the literature. Furthermore, in an observation based on study by [23] also opined that, there are four basic analytic steps were identified for the full reporting of deferred maintenance:

- Assessment of the condition of the building and facilities.
- Determination of a minimum acceptable condition level for each type of buildings and facilities.
- Estimation of the cost to bring those buildings and facilities back to acceptable condition; and
- Estimation of the consequences of deferring maintenance.

5. Issues of deferred maintenance

Deferred maintenance have become a pressing issue in many developing countries and have adverse effects on quality service and performance of buildings, stakeholders and users. The related issue of deferred maintenance whereas organisation needs to identify factors causes and implementing programs to reduce deferred maintenance in future [14]. Reducing facilities deterioration and sustain functional facilities in support of institutional objective and mission is also a requirement. There is a challenge for organisations to keep the aesthetic value to give attraction to the buildings. With deferred maintenance the stress on organisation to hold it all together, there is a need for find a way to prioritize and achieve sustainable building agenda. In addition, [24] also opined that the issues of deferred maintenance can be summarized as:

- No linkages between strategic and operational planning in piecemeal approach for capital planning.
- Chronic in resource shortage.
- Inadequate management policies and practices as well as internal politics.
- Unrealistic financial planning.
- Misunderstood and misapplied needs assessment methodologies and tools.
- Lack of performance measures.

6. Challenges of deferred maintenance

Deferred maintenance is a mistake that comes from presenting part of the appropriate category of needs to improving existing physical conditions by combining the annual life cycle reforms for its buildings and systems that have reached the end of life as well as enhancement and improvement needs. According to [24], there are two (2) major challenges in addressing deferred maintenance:

- Consistent and commonly applied a definition of deferred maintenance.
- A capital planning through process identifying and integrating of all campus capital needs.
- Addressing deferred maintenance is a fundamental responsibility of the maintenance management professional. It is required to understanding of the definitions and methodology in developing reliable and convincing capital planning process.

7. Overview of causes of deferred maintenance

Existing studies such as [25] investigating the research of poor facility management in the Public Schools of Ghana. Empirical discoveries was used in gathering and analysing the data obtain from the research. The researcher found that several causative factors contributed to the poor facility management. However, only fifteen (15) causes direct relate to deferred maintenance. The causes are lack of funds, lack of management; lack of knowledge, lack of training; lack of planning; neglect towards importance of maintenance; lack of expertise; building age; do not have rule, standard practice, guideline and implementation mechanism of building maintenance; lack of qualified and professional maintenance/ facility managers; poor maintenance culture; attitude of deferred maintenance; low priority on maintenance financial planning and capital budgeting; difficult to maintain because of poor designs of buildings and budgetary restriction on maintenance expenditure. The researcher [26] conducted a study of maintenance management sourcing strategies and the condition of tertiary institution buildings in Lagos and Ogun States, Nigeria. A structured questionnaire was designed and used as the principal instrument for collecting data for this study. This study found that the right chosen a strategy of maintenance sourcing procurement either insourcing (in- house) or outsourcing is capable of assisting to prevent the occurrence of deferred maintenance and consequently improved maintenance services towards well maintained buildings. Thus, in the other word the accuracy of the procurement selection will be a cause to the deferred maintenance. According to [27] attempted to assess the maintenance and rehabilitation prioritization of school buildings using Knapsack Problem. Data were gathered through a case study to apply knapsack algorithm within dynamic programming framework. The researcher addressed the main factor inhibits the quality of the buildings in their research is limited budget. This factor is related to the causes of deferred maintenance. Further, [28] was carried out a research to identify the factors militating against effective maintenance of tertiary institutions' buildings. A questionnaire and interview were carried out, 4 potential causes related to deferrred maintenance were identified. The causes are lack of funds, lack of training, unavailability of skilled appointed maintenance personnel and lack of understanding of the importance of maintenance. The researcher [7] who carried out assessment of building maintenance management practice for Abubakar Tafawa Balewa University Teaching Hospital, Bauchi found that in his study by using questionnaire to gather information. There were several factors contributing greatly to the poor management practice. However, 23 causes that related to deferred maintenance are budget constraints; poor maintenance management; lack of awareness among the maintenance staff; little priority on maintenance; lack of quality and qualified maintenance manager; unqualified, ill- trained and unprofessional personnel; ineffective planning; lack of knowledge; lack of resources; administration system faults; lack of maintenance culture; lack of emphasis on training; lack of training and continuing education in efficient; absence of planned maintenance programme; complexity of design; Unstable organizational structure; poor strategies for maintaining; absence of regulations/legislations; lack of sustainable policies; political interference; lackadaisical attitude of senior management staff; lack of understanding of the importance of maintenance; and poor attitude of maintenance team.

Another researcher [11] has carried out a study on Factors for Maintenance Priority in Malaysian University. A review of literature was incorporated as a means of generating data for the research. The study conducted through literature review to provide a general guideline for the factors that are considered in setting maintenance priorities. These several factors are important and need to give attention where they can affect and become a causes of deferred maintenance. The findings discovered that the causes are lack of funds; stakeholders (policies, goal & objectives); and political interference. The researcher [3] conducted regarding interval of routine maintenance and maintenance performance: a literature review. The research by an extensive review of literature published in 1987 to 2014 is performed for the purpose of this research implementation of maintenance strategy. The results from study suggested the most important causes of deferred maintenance are ineffective planning; lack of knowledge; poor management; and neglect towards importance of maintenance. This claim can also be supported by [29] conducted a research on establishing the status of building maintenance operations of Nigerian Universities based on the European Construction Institute Total Quality Management Measurement Matrix. The questionnaire was employed in generating the data required for the research. Some of their findings revealed and indicated that in order to achieve acceptable quality management in their building maintenance practices these factors need to be improved the prosperity of organization; making work more enjoyable; need commitment to overcome resistance; training awareness, education and skills; clear policy, objective and strategies; effective planning; the existence of rules, standard and procedures; adequate resources; communications and team work for improvement; and good ethics and attitude of maintenance team. Hence, in the research of deferred maintenance, the causes are contrary to the factors of this study. However, it still contributes to the causes of deferred maintenance. Furthermore, [20] carried out a study on the development of building maintenance management best practice in Malaysia: a review. The study discovered that, there are a lot of elements that must be considered to develop the building maintenance management best practices. Five major aspects that have become the focal points for previous scholars are strategic planning, resource management, information management, user involvement, and performance control. However, from this highlighted aspect come out a few factors that can be related to the causes of deferred maintenance. The causes are lack of funds; unrealistic financial planning and management; poor of management; ineffective planning; lack of emphasis on training; unavailability skilled of maintenance personnel; poor attitude of maintenance team; information and Communication technology; absence of regulations/ legislations/ procedure/ standard practice/ guideline/ manual; lack of commitment; lack of human resources; lack of financial control; unstable organizational structure; and poor strategies for maintaining. In observation based on study of [30] directed their research efforts towards maintenance practice for school buildings in Malaysia. A survey questionnaire identified problems was conducted in performing maintenance work of this research. Their major findings revealed that there are several causes can be related to deferred maintenance and the most causes are insufficient budget allocation for maintenance; lack of efficient management; and do not have sufficient knowledge. Another study related to deferred maintenance was carried out by [14]. Data was gathered through review of literature. The reseachers concluded that maintenance which is deferred because of insufficient funding; poor of management; ineffective planning; lack of understanding of the importance of maintenance; neglect towards importance of maintenance; lack of expertise; unforeseen expenses; building age; misunderstood needs assessment methodologies and tools; lack of performance measure; lack of quality and qualified professional maintenance/ maintenance managers; lack of maintenance culture; low priority on maintenance financial planning and capital budgeting; and lack of resources. Research conducted by [31] on towards effective management of university education in Nigeria. Their study were identified inadequate funding for maintenance activities and mismanagement of the funds available as the underlying factors responsible for the decay and deferred maintenance. However, these claims also contended by [32] directed their research efforts on identifying maintenance issues in Malaysia. A literature review, with the support of interviews conducted to generate the much needed data for the study. The findings of their research uncovered the causes that can be related to the deferred maintenance are limited budget; lack of management; ineffective planning; lack of quality and qualified professional maintenance/ maintenance managers; chronic in resource shortage; poor of ethics; and ambiguous contract. The study that can be summarized from the academic literature of research on contributing causes of deferred maintenance of buildings shows in Table 1.

Table 1. Causes of deferred maintenance of buildings.

Causes	Wuni et al 2018	Faremi et al 2017	Thohir et al 2017	Nwacuku et al 2016	Aliyu et al 2016	Chong et al 2016	Au-Yong et al 2016	Abdulazeez et al 2015	Nah et al 2015	Ropi & Tabassi 2014	Hamid et al 2010	Ekundayo & Ajayi 2009	R. Mazalan & Mohammad 2009
Lack of funds	X		X	X	X	X			X	X	X	X	X
Poor management	X				X		X		X	X	X		X
Lack of knowledge & skills	X				X		X	X	X	X			
Lack of emphasis training	X			X	X			X	X				
Ineffective planning	X				X		X	X	X				X

IOP Conf. Series: Materials Science and Engineering **513** (2019) 012028 doi:10.1088/1757-899X/513/1/012028

Neglect towards importance of maintenance	X					X			X	
Lack of expertise	X								X	
Building age	X								X	
Do not have rule, standard & regulation	X			X			X	X		
Lack of quality & qualified	X		X	X					X	X
Poor culture	X			X					X	
Lack of ethics, Attitude & commitment				X			X	X		X
Low priority	X			X					X	
Poor design & budget restriction	X			X						
Procurement accuracy		X								X
Lack of understanding			X	X						
Lack of resources				X			X	X	X	X
Administration system faults				X						
Absence planned program				X						
Unstable organizational structure				X			X	X		
Lack of sustainable policies, objective & strategies				X	X			X		
Political interference				X	X					
Lack of communication & teamwork							X	X		

8. Conclusion

Addressing deferred maintenance is a fundamental responsibility of the organization professional. Required is an understanding of the definitions, issues and challenges of deferred maintenance to develop a credible and persuasive the planning process. Deferred maintenance liability requires documentation of the causal factors, and shall include the impacts need to be developed and follow by methodology and contribute to meeting the changes and challenges of an organisation. This study has made and provided a platform for addressing future research into a process protocol for deferred maintenance. This literature review set out to identify all relevant published journal articles from the past ten years (2018-2008), focusing on issue, practice, management, effective management, total quality management matrix, performance and priority of maintenance.

IOP Conf. Series: Materials Science and Engineering **513** (2019) 012028 doi:10.1088/1757-899X/513/1/012028

References

[1] Yasin MN, Zin R, Abdullah AH, Mahmad MS, Hasmori MF 2017 The most common problem facing by the maintenance department: A case Study between Universiti Tun Hussein Onn Malaysia (UTHM) and Universiti Teknologi Malaysia (UTM) In: IOP Conference Series: Materials Science and Engineering

[2] Yasin MN, Abdullah AH, Singh B, Abd Hamid AR 2016 Maintenance Organization at Universiti Tun Hussein Onn Malaysia Int Conf Ind Eng. 2014

[3] Au-Yong CP, Ali AS, Chua SJL 2016 Interval of Routine Maintenance and Maintenance Performance: A Literature Review MATEC Web Conf. 66

[4] Zakiyudin Z, Fathi S, Rambat S, Tobi U, Rejab N 2014 Building Maintenance Management in Malaysia: An Overview

[5] Yasin MN, Khalid H, Wahab NSN 2018 A Study of Potential Retroffiting Existing Sultan Ibrahim Heritage Building to Green Building Adv Sci Lett. 24(5):3213–3216 (4)

[6] Yasin MN, Abdullah AH, Hamid HA, Shahirah N, Wahab N 2017 Potential Retrofit of Existing Sultan Ismail Parliament Building, Johor to Green Building MATEC Web Conf 103

[7] Aliyu A, Bello A, Muhammad S, Singhry M, Bukar G 2016 An Assessment of Building Maintenance Management Practice for Abubakar Tafawa Balewa University Teaching Hospital, Bauchi

[8] Bidi NK, Ayob MF 2015 Investigation of Quality of Cost Data for Life Cycle Cost Analysis in. In: 14th Management in Construction Researchers' Association (MiCRA 2015) p 1–15.

[9] Khalil N, Kamaruzzaman SN, Baharum MR 2016 Ranking the indicators of buildng performance and the users' risk via Analytical Hierarchy Process (AHP): Case of Malaysia Ecol Indic. Elsevier Ltd; 71:567–76

[10] Ebi K, Paul C, Veronic A 2014 Maintenance Culture and Sustainable Economic Development in Nigeria: Issues, Problems and Prospects Int J Econ Commer Manag United Kingdom II(11):1–11

[11] Chong A, Wing K, Mohammed AH, Abdullah MN 2016 Factors for Maintenance Priority in Malaysian University Sains Humanika p 1–5

[12] Arditi D, Nawakorawit M 1999 Issues in Building Maintenance: Property Managers' Perspective J Archit Eng. p 117–32

[13] Chan E 2014 Building Maintenance Strategy: A Sustainable Refurbishment Perspective. Univers J Manag 2(1):19–25

[14] Hamid Y, Alexander K, Baldry D 2010 The Cause and Effects of Deferred Maintenance on Higher Education Buildings In: 7th International Postgraduate Research Conference, as part of 4th International Research Week 2007 (IRW) p 78–88

[15] Ajetomobi OO, Olanrewaju SB 2015 Evaluation of the factors affecting housing maintenance and its probable solutions Int J Latest Res Eng Technol 1(4):59–64

[16] Gajzler M 2013 The support of building management in the aspect of technical maintenance Procedia Eng. Elsevier B.V 54:615–24

[17] Council of The Great City Schools 2014 Reversing the Cycle of Deterioration in The Nation's Public School Buildings

[18] Wuni IY, Boafo HK, Dinye RD 2017 Critical Drivers and Consequences of Poor Facility Management in the Kumasi Metropolis of Ghana J Resour Dev Manag. 37(2017):39–49.

[19] Oig-a- RN 2016 Report Number OIG-A-16-06

[20] Nah N, Abdullah S, Razak A 2015 The Development of Building Maintenance Management Best Practice in Malaysia: A Review Adv Environ Biol. p 97–9

[21] Fakhrudin H, Suleiman Z, Talib R 2011 The need to implement Malaysia's Building and Common Property Act 2007 (Act 663) in building maintenance management J Facil Manag. 9(3):170–80

[22] Ismail Z-A, Kasim N 2014 Maintenance Management Practices For Building Facility: A Case Study Int J Eng Res Appl. 3(4):2248–9622

[23] Hatry HP, Liner BE 1994 Issues in Deferred Maintenance
[24] Kaiser HH 2004 Reviewing the State of Deferred Maintenance APPA Cent Facil Res.
[25] Wuni I, Yeboah A, Boafo H 2018 Poor Facility Management in the Public Schools of Ghana; Recent Empirical Discoveries J Sustain Dev Stud. 11(1):1–30
[26] Faremi J, Adenuga O, Ameh J 2017 Maintenance management sourcing strategies and the condition of tertiary institution buildings in Lagos and Ogun States Nigeria Ethiop J Environ Stud Manag. 10(1):64–74
[27] Thohir MA, Sangadji S, As S 2017 Maintenance and Rehabilitation Prioritization of School Buildings Using Knapsack Problem Int Conf Sci Appl Sci. 2(1):236–48
[28] Nwachukwu L, Agu E, Orji S, Okolie K 2016 Factors Militating against Effective Maintenance of Tertiary Institutions ' Buildings in Enugu Urban Nigeria Int J Sci Technoledge. 4(10):175–80
[29] Abdulazeez A., Abbas A, Ibrahim A 2015 Establishing the Status of Building Maintenance Operations of Nigerian Universities based on the European Construction Institute Total Quality Management Measurement Matrix Innovare J Eng Technol. 3(1):3–8
[30] Ropi RM, Tabassi AA 2014 Study on Maintenance Practices for School Buildings in Terengganu and Kedah Malaysia MATEC Web Conf. 3:[2261-236X, vol:10
[31] Ekundayo HT, Ajayi IA 2009 Towards effective management of university education in Nigeria Int NGO J. 4(8):342–7
[32] Raja Mazlan RM, Mohammed M 2008 Identifying Maintenance Issues in Malaysia In: ASOCSA2008

IOP Conf. Series: Materials Science and Engineering **513** (2019) 012029 doi:10.1088/1757-899X/513/1/012029

A comparative of building condition assessment method used in Asia countries: A review

Muhd Zubair Tajol Anuar[1], Noor Nabilah Sarbini[1*], Izni Syahrizal Ibrahim[1], Mohd Hanim Osman[1], Mohammad Ismail[1] and Ma Chau Khun[1]

[1] Department of Structures & Materials, School of Civil Engineering, Faculty of Engineering, Universiti Teknologi Malaysia, 81310 UTM Johor Bahru, Johor, Malaysia.

[*]Corresponding author: noornabilah@utm.my

Abstract. This paper presents a comparative study of methods building condition assessment used in Asian countries. The following methods were compared: a Malaysia method to assess buildings condition, an Indian, Hong Kong and Singapore method about buildings condition assessment and the assessment methods. The comparative analysis included three tasks. First, every approach was described separately. Then, the main features of the methods were compared. Finally, some guidelines to improve or strengthen the Malaysia method were drawn. Main differences of the methods are the objectives and scope of the assessment, the assessment criteria, the type final results obtained and type of surveyor. The main similarities are that the assessment is carried out mainly by visual inspection method, the scope of inspection had to assess structural element and the tools developed for its implementation. Malaysia method should maintain the present assessment model, suggests a website to manage the entire assessment procedure such as document form related to assessment, checklist and guideline. Thus, the summary of this paper shall improvise Malaysia method from perspective defect process or factor, Non Destructive Test (NDT) test procedure and selection of the concrete repair material.

1. Introduction

Building Condition Assessment (BCA) is an effective tool for determining the physical state or condition of buildings. According to Yacob *et al* it is necessary to assess the condition of the building as it reflects the physical state and performance of the building from defects and failures. Defects and failure such as crack and corrosion is a problem faced in any building especially aging building [1]. Defects and failures of a building are often driven by inaccuracy in designing phase manufacturing imperfections damaged materials, unsuitability of materials used in constructions, contractors did not follow the final construction design, or any of these combinations. [2]. Not only that, climates construction site, materials used in constructions, purpose of a building altering its functions, building maintenance, faulty design, dishonesty, and unprofessional supervision also contribute to defects and failures [3]. However, in Malaysia, the main current maintenance management issue is low service quality [4]. The problem is due to failures in the planning of maintenance strategies or methods caused by a few factors, such as deficiency of awareness of maintenance strategies, insufficient performance standards, deficiency of building performance monitoring data, failure to provide adequate design advice and performance - based planning and other [5]. As a result, the buildings cannot sustain any longer and are unsuitable for the use and function of the building.

Good management of building and infrastructure like planning, implementation and maintenance are critical element in sustainability in construction [6]. There are two types of maintenance activities; scheduled maintenance and condition-based maintenance. However according to N. Tirpude *et al*, both activities requires inspection to examine quality of a building before proceeding to the next step. Output and details from building inspection will later be used to make decisions about repair work [2].

Evaluating the condition of a building could be tough. It requires skills, proper tools and time. The evaluation/assessment of a building's condition should cover different structural components such as beam, column, slab and other. The method to conduct building condition assessment is visual inspection. The data on defects obtained at the on- site assessment is important for the maintenance decision [7].

The aim of this review paper is to differentiate methods assessing buildings' condition between Malaysia and other few Asian countries such as India, Hong Kong and Singapore. These countries are selected because nearest to Malaysia which have good in practising maintenance work. The comparison method made between Malaysia and few Asian counties because to differentiate it's and making some improvements on Malaysia method. The improvement that is content verifies the conditions of the building or supports decisions on rehabilitation, repair or maintenance.

Thus, this review paper present two major parts – the differences and similarities between methods; and suggestions that might reinforce Malaysian assessment method.

2. Building condition assessment guideline

2.1. Guideline of building condition inspection for existing building in Malaysia (PWD)

To implement the building inspection work, the building inspector needs general guide from the guideline. The guideline also creates consistency in the field of inspection and finally preparation of reports [1]. This guideline describes the methods and responsibilities of the Ministry / Department / Agency involved in conducting inspections for the purpose of rating existing buildings [8]. In 2013, the Public Works Department developed a standard assessment of the conditions of building components and services.[1]. The purpose state in Malaysia (PWD) method is to ensure the building inspectors perform inspection fits with the inspection prescribed guidelines.

2.2. Handbook on repair and rehabilitation of India, Nirman Bhawan (CPWD)

Deterioration of reinforced concrete (RC) by chloride, carbon dioxide, freeze-thaw, sulphate, acid attacks, and etc. is a natural phenomenon which could not control. This phenomenon was overlooked by practising engineers. Admittedly, since civil engineers in India still in the beginner phase especially in the safety, repairs, rehabilitation, strengthening and retrofitting of the existing RC structures, they need establishments and professional circulation as guidance [9].

Therefore, having the Central Public Works Department (CPWD) of India and other organisations Repairs & Rehabilitation on board, their experiences and expertise are shared in the handbook [9].

The book consists of deterioration factors, investigative strategies, design standards, determination criteria to repair materials and techniques, guidelines in selecting specialist, short-listing of contractors and preparation for contract agreements for repair/restoration contracts. The purpose of this method is to lead the organization practicing in the construction and maintenance of buildings that have been faced with the problem of deterioration in RC buildings [9].

2.3. Professional guide to building inspections in Hong Kong (HKIS)

This guide is issued not solely to implement the Mandatory Building Inspection Scheme (MBIS) but to surveying professionals to assist them to conduct building inspections. Then, also act as a general guide for inspecting pre-1980 residential and composite buildings in Hong Kong [10].

This guideline was published in 2013 by Building Surveying Division (BSD) of the Hong Kong Institute of Surveyors (HKIS). This guideline consists of forecasting and analysis – usually preparations and planning for inspection of a buildings and defect analysis method can be found in residential and composite buildings.

Distinctive construction methods and features are also available – regulations, practice codes, construction techniques, layout and materials commonly found in age buildings. [10]. The main aim of this procedure is to provide professional guidance on the survey of residential and composite buildings in Hong Kong prior to 1980, it can also be used as a reference for those preparing inspection reports under Building (Inspection and Repair).

2.4. Periodical structural inspection of existing buildings guidelines in Singapore (BCA)

The Building and Construction Authority (BCA) is a National Development Ministry agency that promotes the development of an excellent built environment for Singapore. "Community activities are provided with the setting of "Buildings environment" referring to buildings, structures, and infrastructure in our environment [11]. This guideline was published in 2012 by The Building and Construction Authority (BCA).

Periodic structural inspections of existing buildings were carried out in 1989 when the Building Control Act was promulgated. Part V of the Building Control Act and the Building Control Act stipulates the requirements for periodic structural inspections of existing buildings. There are two stages of inspection which are visual inspection and full structural investigation [6 , 7]. The purpose of this method is to guide structural engineer carry out scope of inspection that need to be assessed.

3. Comparative study of methods

There are four methods of building assessment which be analysed the main characteristics of the methods studied such as Malaysia (PWD), India (CPWD), Hong Kong (HKIS) and Singapore (BCA) standard. The primary features of the methods are summarised in Table 1. The following sections analyse the consequences of each comparison of features.

3.1. Objective of method

For countries as stated above have their objectives in defining the condition of a building to achieve the similar global inspection standards for existing buildings, respectively. Thus, providing rating and information relating to the defects assessed with verifying the condition of the building [1]. Other objectives are to ensure the practicality, safety, and comfort of the building [8].

There are four main objectives in Indian Method. The objective of a building structure's condition survey is to identify the causes and sources of distress. Then also determine the extent of distress caused by corrosion, fire, earthquake or any other reason. The other objectives are residual strength and rehabilitation of the structure, the priority of the distressed elements according to the significance of the repairs and finally selecting and planning an effective remedy [9].

The objectives of the Hong Kong method is to verify the status of the building and its work; conditions and performance of building components and installations. Formulate a close investigation or maintenance and management plan and examine and diagnose the building's symptoms and defects, if any, so that a close investigation or repair, rectification or demolition scheme can be established [10]. The Singapore method also aims to confirm that structural defects due to lack of maintenance are detected and corrected early in order to keep the buildings structurally work and function.[12].

3.2. Inspection method and scope

In order to assess the building, each country has their preferences and guidelines. The preferences and guidelines refer to the methods of assessment. Malaysia, Indian and Hong Kong methods state the assessment/inspection methods are as visual inspection with simple tools and instruments such as camera with flash, magnifying glass, binoculars and gauge for crack width measurement but Singapore method state only visual inspection.

The scope inspection of Malaysia method are split into a few elements such as architecture & civil, mechanical, electrical and external work. Architectural & Civil elements consisting of basic systems, structures, exterior parts, roofs, interior structures, staircases, ceilings, fixtures and even building structures. Mechanical elements consisting piping systems, air conditioning systems, distribution systems, fire prevention systems and elevators. Electrical elements consisting lighting, communication and security systems. External work components consisting roads, outdoor water reticulation, sewerage systems, retaining walls, sewage treatment plants, and landscape [8].

The scope inspection of Hong Kong method split into four element such as external, structural, fire safety and drainage system [10]. The scope inspection of Indian and Singapore only focus on structural element which are reinforced concrete, prestressed concrete, and steel[11,9]

3.3. Distress to be assess

The Malaysian method contains a list of defects or distresses, simply describing the possible cause of the defect. The defect in Malaysia's method is leakage, rust, skin, fungus, wet, bent, settlement, etc [8].

Singapore method has only a list of defects, including are cracks, excessive deflection, connection failure, instability, floor settlement, foundation settlement, tilt, spalling concrete, corrosion of steel, termite infestation, dry & wet rot timber, etc. [11].

The Hong Kong method contains a list of defects that are categorized according to elements such as external elements; (1) tiles and rendering loose or missing; (2) cracks; (3) bulging, bowing, separation and delamination; (4) corrosion of metal parts embedded in external finishes; and (5) spalling. Comment defect on structural element; (1) spalling; and (2) corrosion of reinforcement bars. Fire safety defects such as defective fire doors (broken doors or widened gaps), defective building components resistant to fire, missing exit signs, inadequate lighting and Blocked or obstructed means of escape. Corrosion, leakage, deformation, displacement, damage and blockage of drainage pipes and traps are the defects in the drainage system [10].

The Indian method has a chapter explaining the cause of deformation or distress and the initial concrete damage, which is divided into four categories, such as Construction Deficiency Physical. (High w/c Ratio, Inadequate curing, Poorly Graded Aggregates); Chemical (Chloride, Sulphate, and Reactive aggregates); Design deficiency (Wrong assessment of design Loads, shrinkage, thermal and structural behaviour); and Environmental physical (Heating/ Cooling and Abrasion of Surface). This method also has a crack type and its illustration pattern [9].

3.4. Assessment criteria

The Malaysia and India methods studied use different assessment criteria, except Hong Kong and Singapore method. This method assessment criteria used by Hong Kong and Singapore method is measured by the defects recorded but there is no classification of level defect or condition.

The method evaluation criteria used by the Malaysian method are measured by the registered defects and have a five - point colour - coded building rating. The aggregate score is then used to generate a numerical rating between 1 and 5, where 1 indicates optimal condition and 5 indicates critical conditions [1,8]. Thus, Indian method is measured by the defects recorded and have Classes of Damage and Repair Classification Repair The aggregate class is then used to produce a numerical rating from 0 to 4, where 0 no structure distress and 4 major structure loss [9].

3.5. Findings

The difference finding of these studies methods:
The Malaysian method provides a defect parameter for the building condition and maintenance priority action level in the form of matrix analysis for each component of the evaluated building. [1].

The rules of interpretation of the Indian method and divides the repair classification into five classes for each part of the building. This method also provide Non- Destructive Test (NDT) such as Insitu Concrete Strength (Rebound Hammer Test & Ultrasonic Pulse Velocity), Chemical Attack (Carbonation Test & Chloride Test), Corrosion Potential Assessment (Cover-Meter & Half Cell

IOP Conf. Series: Materials Science and Engineering **513** (2019) 012029 doi:10.1088/1757-899X/513/1/012029

Method), Fire Damage Assessment (Thermo-Gravimetric Analysis & Differential Thermal Analysis) and Structural Integrity/Soundness Assessment (Radiography & Impact Echo Test)[9].

Concrete Strength Assessment is the evaluation of concrete quality and it is necessary to estimate the quantitative value of concrete strength to assess a structural member's load carrying capacity. Chemical tests are concrete chemical analysis that can provide very useful information on the cause of concrete failure. Corrosion Potential Assessment is to evaluate the status of corrosion actively in the embedded steel. Fire damage assessment is an assessment of the qualitative and quantitative composition of a sample of concrete and determines the extent of the deterioration in fire - prone concrete. Structural Integrity/Soundness Assessment is an assessment to determine deterioration such as discontinuities, cracks, depth of cracks and detect hidden damage [9].

Beside NDT, India method also explain about selection of repair material for concrete such as use polymer modified cement to recover the damaged cover concrete because their better connection with substrate and reinforcement. Fusion bonded epoxy powder coatings (FBEC) and IPN coatings are used to protect bars against corrosion in concrete reinforcement structures. The system consists of a fibre reinforcement layer that is wrapped on the exterior surface of the retrofitted structural element [9].

The method Hong Kong provides reasons for defects in structural elements and external finishes. The preparation of any plan for rectification or repair work should be established on the results of thorough research and robust diagnoses. The defects in the structural elements are classified. into a four × three matrix [10].

Singapore method result based on visual inspection and judgment & recommendation of professional engineer appropriate actions to be taken based signs of significant structural deterioration. There are no calculation or present specific result of defect [11].

3.6. Implementation tools

The main implementation tools are:-

- Checklist : All four methods have a checklist to guide and record surveys;
- Instructions guideline: The instructions of Malaysia, Hong Kong and Singapore standard present a list of the most common defects for structure components. The guidelines for India include design or pattern of common defects for several structure element.
- Website: The Singapore method has a website (https://www.bca.gov.sg) which is to manage the whole assessment process.

IOP Conf. Series: Materials Science and Engineering **513** (2019) 012029 doi:10.1088/1757-899X/513/1/012029

Table 1. Summary of the main features of the assessment methods.

	Malaysia(*(PWD)*	India *(CPWD)*	*Hong Kong (HKIS)*	*Singapore (BCA)*
Objective of method	Providing rating and information relating to the defects.	Identify causes and sources of distress	check the status of the building and conditions & performance of the building's	Ensure structural defects are detected due to lack of maintenance
Inspection and method Scope	Assessment methods are as visual inspection with scope; Architectural & Civil Mechanical, Electrical and External work	Assessment methods are as visual inspection with scope on structural element	Assessment methods are as visual inspection with scope external, structural, fire safety and drainage system	Assessment methods are as visual inspection with scope on structural element
Distress assessed/section	List of defects or distresses, simply describing the possible cause of the defect	Cause of deformation or distress and the initial concrete damage. List of crack type and its illustration pattern	List of defects that are categorized according to elements such as external, structural, Fire safety defects and defects in the drainage system	Only a list of defects
Assessment criteria	Defects recorded and have the five-point colour- coded building rating	Defects recorded and have Classes of Damage and Repair Classification Repair	Defects recorded but there is no classification of level defect or condition	Defects recorded but there is no classification of level defect or condition
Implementation tools	Checklist and Instructions guideline	Checklist and Instructions guideline	Checklist and Instructions guideline	Checklist ,Instructions guideline and Website
Finding of method	Building condition defect parameter and priority maintenance action level in the form of a matrix analysis	Subdivide the repair classification in five classes in general. This method also offers non - destructive testing (NDT) and the selection of the concrete repair material.	The causes of structural elements and external flaws and preparation of any plan for rectification or repair work should be established on the results of thorough research.	Visual inspection and judgment & recommendation of professional engineer appropriate actions

4. Conclusion

The main differences and similarities in the methods analysed indicates that the methods studied had different objectives and scopes. They also differ in respect of the section / defect assessed, the assessment criteria and the type of final results obtained. However, some similarities were found between the methods studied. From these four methods, the assessment is mostly carried out by visual inspection, the scope of inspection had to assess structural element and the tools established for its implementation. In addition, the recommendations that can be drawn to improve the Malaysia method (PWD) as follow;

- The inspection method are consistent, attributes assessed, attributes criteria and surveyor. The increase in the number of element would consume more time for the survey without significant purposes to the accuracy of the result.
- Suggested a website or blog to easily access and manage the entire assessment procedure such as document form related to assessment, checklist and guideline.
- Improvement on content of guideline such as NDT test procedure and illustrations of common defects for some structure component.
- Suggested a chapter in Malaysia standard would have a selection of the concrete repair material and explained deeply about process or factor of defect.

Acknowledgments

The authors are thankful to Universiti Teknologi Malaysia under grant No. (Q.J130000.7751.4J343) for the completion of this work.

References

[1] S. Yacob, A. S. Ali, and A. C. Peng 2016 "Building Condition Assessment : Lesson Learnt from Pilot Projects,". **00072**, p.7.
[2] C. A. Adi Irfan, M. T. Norngainy, J. Suhana, A. R. Mohd Zulhanif, and Y. Hafsah 2014 "Jurnal Teknologi Full paper Building Condition Assessment for New Houses : A Case Study in Terrace," *J. Teknol.* **70**, (1), pp. 43–50
[3] N. Ahzahar, N. A. Karim, S. H. Hassan, and J. Eman 2011 "A study of contribution factors to building failures and defects in construction industry," *Procedia Eng.* **20** 249–255
[4] S. Nizam Kamaruzzaman and E. Marinie Ahmad Zawawi 2010 "Development of facilities management in Malaysia," *J. Facil. Manag.* **8** (1), 75–81
[5] C. P. Au-yong, A. S. Ali, and F. Ahmad 2014 "Preventive Maintenance Characteristics towards Optimal Maintenance Performance : A Case Study of Office Buildings," no. September, pp. 1–6
[6] C. P. Au-Yong, A. S. Ali, and F. Ahmad 2014 "Significant Characteristics of Scheduled and Condition-Based Maintenance in Office Buildings," *J. Perform. Constr. Facil.* **28** (2) 257–263
[7] A. Straub, "Dutch standard for condition assessment of buildings 2009 " *Struct. Surv.* **27** (1) 23–35
[8] Public Work Department, "Guideline of Building Condition Inspection for Existing Building," 2013.
[9] CPWD, "Handbook on repair and rehabilitation of RCC buildings (CPWD)," *Cent. Public Work. Dep. (CPWD), Gov. India, New Delhi.*, p. 498, 2002.
[10] C. Buildings and H. Kong, "Professional Guide to Building Inspections Professional Guide to Building Inspections," 1, 1980.
[11] BCA, "Periodic Structural Insection of Existing Buildings," 1–14, 2012.
[12] BCA, "of Existing Buildings Building Owner ' S Guide," 2012.

IOP Conf. Series: Materials Science and Engineering **513** (2019) 012030 doi:10.1088/1757-899X/513/1/012030

Bambusa blumeana fiber as erosion control geotextile on steep slopes

S B Valle[1*], R D Albay[1] and A M Montilla[1]

[1]Civil Engineering Department, University of Batangas, Batangas City, 4200, Philippines

[*]Corresponding author: Siddhartha.valle@ub.edu.ph

Abstract. Accelerated erosion has been a threat on the agronomic productivity of arable lands, especially in the tropics with prevalent high-velocity winds and seasonal rainfalls. To prevent further soil degradation, geotextiles are installed as soil cover to provide interim protection until the soils are stabilized with vegetation as a long-term slope protection. This study examined the mechanical properties of a locally available fiber source to assess its viability as geotextile fabric for steep slopes. A woven natural geotextile was fabricated from the fibers of thorny bamboo ("*kawayang tinik*"). Ropes of 6-mm and 8-mm diameters were produced from the bamboo fibers, and subjected to tensile strength tests in accordance to ASTM D4268. Runoff simulations were carried out to compare the sediment yield between 60° slopes with and without geotextile netting. Sediment yield analysis showed significant reduction in topsoil erosion from the geotextile-covered slope. The bamboo geotextile performed effectively in mitigating slope failure in barren slopes.

1. Introduction

Soil degradation by accelerated erosion is of global interest due to its severe adverse impacts on environmental sustainability and agronomic productivity [1]. More than 10 million hectares per year of the world's arable land has been lost to erosion, reducing the viable cropland for food production [2]. Erosion-caused food production losses have been seen as most critical in developing countries in the tropics and subtropics, including Southeast Asia, where erosion potential is high due to heavy rainfalls [3].

To promote vegetation growth and improve the shear resistance of soils against surface runoff and wind exposure, a self-propagating vegetal cover is found to be more suitable in steep slopes highly susceptible to erosion rather than more conventional methods such as bench terracing [4]. Slopes are covered with erosion control netting of geosynthetic material, providing a barrier over the soil to protect the seeds and the soil until the soils are stabilized with vegetal cover [5]

Geosynthetics is the umbrella term in civil, environmental and agricultural engineering pertaining to flexible, planar, polymeric fabrics applied for slope reinforcement and improved stability of structures such as earth fills. Forming the largest group are geotextiles, permeable geosynthetics comprised solely of textiles, which may either be natural or synthetic, woven or non-woven. Upon soil application, a geotextile netting functions as a filter, drainage and slope protection, allowing steeper cut slopes and less grading in hillside terrains.

Most synthetic geotextiles consist of a polymer from the polyolefin, polyester or polyamide family. While generally stable to microbial degradation and to acidic/alkaline conditions, polymers are subject to thermal oxidation and to photooxidation under ultraviolet radiation. provide greater stability is provided through the incorporation of antioxidants and UV stabilizers [6, 7]. Over time, however, the polymers and their additives may degrade into microplastic particles, which under various conditions enhance leaching into soils, as additives are often not covalently bound to the polymer backbone [8]. Plant uptake of these particles may cause interference in the photosynthetic system, as in the case of algae.

The presence of non-biodegradable residues has drawn concerns over the environmental sustainability of polymer textile covers. To address these concerns, natural fibers have been utilized as substitutes to synthetic fibers in applications where low density, low cost, high strength and high modulus are required [9, 10]. Among the known natural fibers are jute, coir and palm [5, 11, 12, 13]. These fibers provide immediate erosion control, as they are readily available, making the geotextiles simple and cost-effective to manufacture. Natural geotextiles also significantly increase soil fertility thereby enhancing aboveground biomass production [14].

Due to their location within the Pacific Ring of Fire, the Philippine islands are of volcanic origin, creating a series of mountain slopes, interrupted by cultivated plains and terraces. Highland slopes are made vulnerable to erosion by monsoon rainfalls and tropical cyclones from adjacent Pacific Ocean. Geosynthetic technologies are being sought to reduce the construction of concrete slope protections, which tend to undermine the agricultural viability of land.

In line with the goals of sustainable development to utilize locally available resources, this study investigates the performance of geotextile netting, made from woven bamboo fiber, in mitigating slope erosion against rainfall runoff. Bamboo is one of the strongest and fastest-growing woody plants on earth, growing at a rate of up 4 cm in an hour. Bamboo belongs to a subfamily of flowering perennial evergreen plants in the grass family *Poaceae*. There are 62 identified bamboo species in the Philippines [15]. Annual bamboo production in the country is estimated to be about 39,200 to 52,700 hectares.

Fibers used in the production of the geotextile were obtained from one of the most abundant species of bamboo in the country, *Bambusa blumeana*, locally known as *Kawayang Tinik* ("thorny bamboo") [15]. The UNDP-FAO Bamboo Research and Development Project (1987) listed *B. blumeana* as an economically important bamboo species, owing to its reliable mechanical performance. It is a natural composite material that has higher tensile strength than many alloys of steel, and also has higher compressive strength than many mixture of concrete and higher strength-to-weight ratio, making bamboo one of the most durable materials used in the construction industry.

Bamboo is a lingo-cellulosic composite, composed of fibers in vascular bundles, and matrices, thin walled cells around vascular bundles, vessels and sieve tubes in vascular bundles [16]. Due to their high flexibility, bamboo fibers are extensively used in the production of ropes and woven fabrics.

While most studies on geotextiles have demonstrated the effectiveness of natural fibers in reducing erosion and subsequent slope degradation on small or intermediate slope gradients (normally $\leq 30°$), steeper slopes need immediate practical engineering solutions [17]. In this regard, this study performed surface runoff simulations to evaluate the efficiency of woven bamboo geotextile in reducing surface erosion on steep barren slopes.

2. Methodology

2.1. Preparation of bamboo cellulose

Fresh culms of *kawayang tinik* were collected from Bagalangit Mabini, Batangas. The stalks were split into strips then pounded to turn into pulps. Cellulose fibers were extracted from the pulps by hydrolysis alkalization. The crushed bamboo pulps were soaked for three hours in an aqueous solution of 15-25% sodium hydroxide. Bamboo fibers were combed out from the culms. The fibers were then water-washed to remove excess NaOH, then air-dried for 24 hours. Upon drying, the bamboo fibers were spun mechanically using an electric wheel rotating at 300 rpm to produce ropes in diameters of 6 mm and 8 mm.

2.2. Tensile strength test

The tensile strengths of the 6-mm and 8-mm ropes were assessed according to ASTM D4268 Rope Tensile Strength Test using a universal testing machine (UTM). The average normal strength of the 6-mm and 8-mm diameter ropes were compared to the natural breaking strength of Manila hemp set by the US Federal Specification TR-605B, Type M Class 1, Amendment 3.

2.3. Surface runoff analysis

A surface run-off test was performed to simulate a run-off scenario on a steep slope with water as the principal agent for erosion and weathering [18]. Soil control mats were weaved from the 6-mm diameter ropes. Each mat is produced as 1000-mm by 600-mm open-meshes with 5-cm square spacing.

Soil samples were prepared in two soil core test boxes with plane dimensions of 1000 mm in length and 600 mm in width, and a surface slope of 65°. Soil samples were assumed to be homogenous, with uniform distribution of fine-grained soils, silt and clay. Soil compaction was performed every 10-in depth.

The test boxes were placed side by side, with one of the test slopes fully covered with geotextile mat, while the other test slope was left barren. Water was introduced into the system through a perforated pipe hung horizontally 2 feet above the peak, running perpendicular to the slope.

Surface run-off from the test box flowed through a gutter placed at the bottom of the slope, which was directed to a bucket serving as the watershed catchment. The volume of the run-off collected in each bucket was measured every 10 minutes over the duration of an hour.

To determine the sediment concentration of the run-off, deposits were filtered by gravity using a canvas cloth. The dry weight, m_{dry}, of the sediments was taken after oven-drying the filter residue until its moisture content is at most 5%. The sediment concentration was computed as the ratio of the dry weight of the sediments in the runoff from the covered slope, $m_{dry,covered}$, to the volume of the covered slope runoff, ***V***.

The sediment yield was assessed as the rate of sediment removal from the watershed per unit area during a specified period of time, and calculated as:

$$Sediment\ Yield = \frac{m_{dry}}{Surface\ Area \times Time} \tag{1}$$

The performance efficiency of the bamboo geotextile netting in mitigating erosion was evaluated as:

$$Net\ efficiency = \frac{m_{dry,uncovered} - m_{dry,covered}}{m_{dry,uncovered}}\% \tag{2}$$

where $m_{dry,uncovered}$ is the dry weight of the sediments eroded from the uncovered slope.

3. Results and discussions

3.1. Tensile strength of bamboo fiber ropes

To evaluate their mechanical strength, bamboo rope samples were subjected to tensile tests. Figure 1 presents the result of tensile strength test on three samples of a 6-mm diameter bamboo fiber rope. The ropes were able to carry a mean load of 428.68 N at an average ultimate normal stress of 10.65 MPa. Rope elongation averaged 37.20 mm before the ultimate normal stress was reached. The factors of safety for all samples are above 1.0, with a mean of 2.14. This indicates that the average strength of the 6-mm bamboo rope is 2.14 times greater than the standard breaking strength of a commercial manila hemp. The US Federal Specification set a safe load for the 6-mm diameter manila hemp at 200 N.

IOP Conf. Series: Materials Science and Engineering **513** (2019) 012030 doi:10.1088/1757-899X/513/1/012030

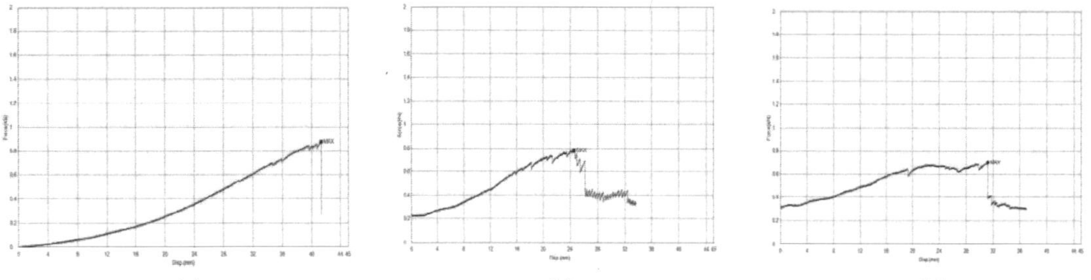

(a) (b) (c)

Figure 1. Applied normal load vs. elongation on 3 samples of 6-mm diameter bamboo ropes.

Table 1. Tensile strength of 6-mm diameter bamboo rope samples.

Sample	Load (N)	Elongation (mm)	Average Strength (MPa)	Ultimate Strength (MPa)	Rupture Strength (MPa)	Factor of Safety (MPa)
1	512.76	41.50	10.20	30.06	30.06	2.56
2	410.87	33.86	14.53	28.12	13.27	2.05
3	362.40	36.25	7.21	27.06	12.56	1.81
Mean	428.68	37.20	10.65	28.41	18.63	2.14

For the 8-mm diameter bamboo ropes, the Federal Specification requires a safe load of 330 N in manila hemp. Results of the tensile strength test show that the samples resisted loads up to 786.84 N.

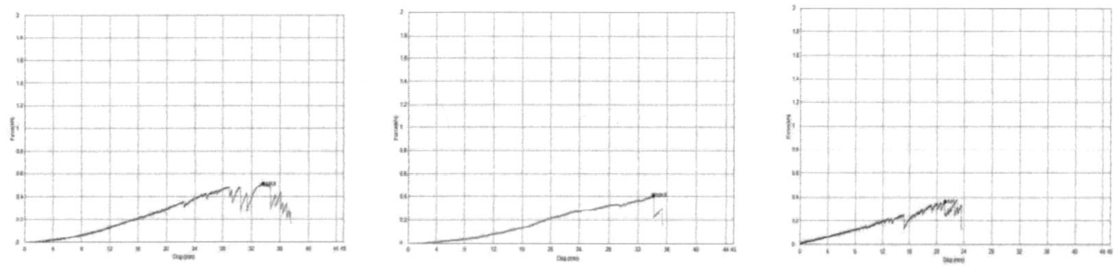

(a) (b) (c)

Figure 2. Applied normal load vs. elongation on 3 samples of 8-mm diameter bamboo ropes.

Table 2. Tensile strength of 8-mm diameter bamboo rope samples.

Sample	Load (N)	Elongation (mm)	Average Strength (MPa)	Ultimate Strength (MPa)	Rupture Strength (MPa)	Factor of Safety (MPa)
1	878.97	37.86	17.49	10.25	3.58	2.66
2	778.52	35.34	27.54	8.16	3.68	2.36
3	703.02	23.90	24.86	7.66	3.38	2.13
Mean	786.84	32.37	23.30	8.67	3.55	2.38

Ultimate strength reached around 8.67 MPa before declining to rupture limits of 3.55 MPa. The average rope elongation is at 32.37 mm. As with the 6-mm diameter ropes, all samples of 8-mm ropes yielded a factor of safety above 1.0, averaging at 2.38. This signifies that that the bamboo ropes were able to carry loads up to 2.38 times greater than the standard tensile strength requirement for manila hemp.

3.2. Sediment yield analysis

Rainfall simulations were performed to compare erosion rates between bare and geotextile-covered slopes. Sediment concentrations in runoffs from the two test slopes are recorded in Table 3.

Table 3. Runoff sediment concentration, in g/L.

Test Slope	Time Elapsed, in minutes					
	10	20	30	40	50	60
Uncovered (UTS)	0.476	17.362	37.805	70.127	*Slope failure*	*Slope failure*
Covered (CTS)	0.205	5.226	4.781	21.617	17.250	14.561
%diff	*56.93%*	*69.90%*	*87.45%*	*69.17%*	-	-

Mass concentration in runoff from the uncovered test slope (UTS) increased dramatically as the soil gets more saturated over time. UTS mass flow was around 188.66 g/hr compared to 63.64 g/hr CTS mass flow. Faster rate of soil saturation in UTS resulted to weaker shear stresses between soil layers. Before the 40-minute mark was reached, UTS was fully saturated, and its bottom soil strata were loose. Surface runoff subsequently washed away the remaining sediments.

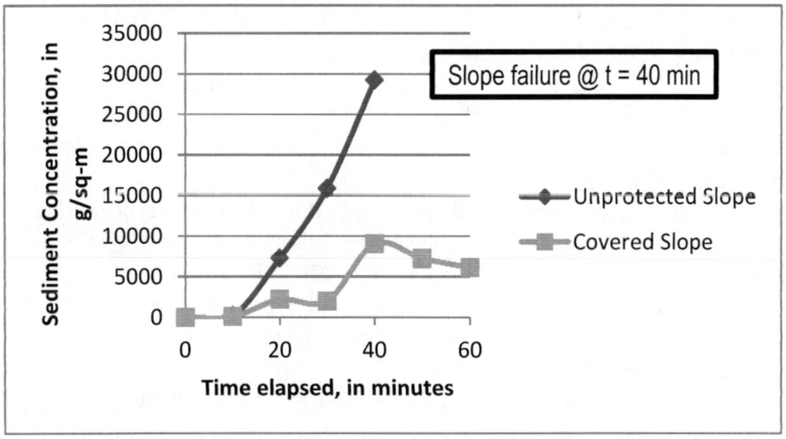

Figure 3. Runoff sediment concentration projected in grams per square meter.

At t=40 min, UTS sediment yield was projected to be around 29.21 kg/sq-m, while CTS sediment yield was 68.92% lower at 9.08 kg/sq-m. Forty minutes under constant rainfall, the cumulative UTS mass runoff was 125.77 g/L. With the cumulative CTS mass concentration was 31.83 g/L, the bamboo geotextile was evaluated to have a net cover efficiency of 74.70%.

4. Conclusions

Tensile strength tests and surface runoff simulations were performed to assess the viability of bamboo fiber geotextile for erosion control netting on steep barren slopes. The tensile strengths of the bamboo fiber ropes passed the safe load standards for woven natural ropes based on the acceptable breaking strength of manila hemp.

IOP Conf. Series: Materials Science and Engineering **513** (2019) 012030 doi:10.1088/1757-899X/513/1/012030

Results of the sediment yield analysis showed that the bamboo geotextile prevented significant mass runoff from the topsoil during the rainfall simulations. The bamboo geotextile was effective in mitigating surface erosion and subsequent slope failure.

To validate the results of this study, sediment runoff analysis on actual slopes is recommended. A wide-width strip test should also be performed to cover the measurement of tensile strength and elongation of bamboo geotextile, including the assessment of the initial modulus, offset modules, secant modulus and breaking toughness.

Acknowledgement
This research was made possible through a grant from the University of Batangas Center for Publications, Research, Linkages and Liaison (UB-CPReLL).

References

[1] Lal R 2001 Soil Degradation by Erosion *Land Degradation and Development* **12** 519–39
[2] Pimentel D 2006 Soil Erosion: A Food and Environmental Threat *Environment, Development and Sustainability* **8** 119–37
[3] Labrière N, Locatelli B, Laumonier Y, Freycon V, and Bernoux M 2015 Soil erosion in the humid tropics: A systematic quantitative review *Agriculture, Ecosystems & Environment* **203** 127–39
[4] Fattet M, Fu Y, Ghestem M, Foulonneau M, Nespoulous J, Le Bissonnais Y, et al 2011 Effects of vegetation type on soil resistance to erosion: Relationship between aggregate stability and shear strength *CATENA* **87** 60–9
[5] Lekha K 2004 Field instrumentation and monitoring of soil erosion in coir geotextile stabilised slopes—A case study *Geotextiles and Geomembranes* **22** 399–413
[6] Cooke T and Rebenfeld L 1988 Effect of chemical composition and physical structure of geotextiles on their durability *Geotextiles and Geomembranes* **7** 7–22
[7] Rollin A, and Lombard G 1988 Mechanisms affecting long-term filtration behavior of geotextiles *Geotextiles and Geomembranes* **7** 119–45
[8] Wiewel B and Lamoree M 2016 Geotextile composition, application and ecotoxicology—A review *Journal of Hazardous Materials* **317** 640 - 55
[9] Tran L, Fuentes C, Truong Chi T, Nguyen Minh T, Van Vuure A and Verpoest I 2015 Investigation of microstructure and tensile properties of porous natural coir fibre for use in composite materials *Industrial Crops and Products* **65** 437 - 45
[10] Defoirdt N, Biswas S, De Vriese L, Ngoc Tran L, Van Acker J, Ahsan Q, et al. 2010 Assessment of the tensile properties of coir, bamboo and jute fibre *Composites Part A: Applied Science and Manufacturing* **41** 588 - 95
[11] Choudhary N 2013 Jute Geotextiles as Substitute to Synthetic Geotextiles *Advanced Materials Research* **821** 85-9
[12] Davies K, Fullen M and Booth C 2006 A pilot project on the potential contribution of palm-mat geotextiles to soil conservation *Earth Surfaces Processes and Landforms* **31** 561-9
[13] Bhattacharyya R, Fullen M, Davies K, and Booth C. 2009 Utilizing palm-leaf geotextile mats to conserve loamy sand soil in the United Kingdom *Agriculture, Ecosystems & Environment,* **130** 50-8
[14] Bhattacharyya R, Yi Z, Yongmei L, Li T, Panomtaranichagul M, Peukrai S, et al 2012 Effects of biological geotextiles on aboveground biomass production in selected agro-ecosystems *Field Crops Research* **126** 23-36
[15] Roxas C 1998 *Bamboo research in the Philippines.* International Plant Genetic Resources Institute. Kunming and Xishuanbanna, Yunnan, China
[16] Amada S, Ichikawa Y, Munekata T, Nagase Y and Kirigari N 1993 *Proceedings of the 5th Bioengineering Div. Conference* p. 62 Japan: Society of Mechanical Engineering

[17] Álvarez-Mozos J, Abad E, Giménez R, Campo M, Goñi M, Arive M, et al 2014 Evaluation of erosion control geotextiles on steep slopes. Part 1: Effects on runoff and soil loss *CATENA* **118** 168-78
[18] Luo H, Zhao T, Dong M, Gao J, Peng X, Guo Y, et al 2013. Field studies on the effects of three geotextiles on runoff and erosion of road slope in Beijing, China *CATENA* **109** 150-6

IOP Conf. Series: Materials Science and Engineering **513** (2019) 012031 doi:10.1088/1757-899X/513/1/012031

Numerical modelling of concrete bearing strength for different heights of concrete blocks

N A Yahya[1,3*], M Dhanasekar[3], M R Md Zain[1], C L Oh[1] and S W Lee[2]

[1] Faculty of Civil Engineering, Universiti Teknologi MARA, Shah Alam, 40450, Selangor, Malaysia.
[2] Faculty of Civil Engineering, Universiti Teknologi MARA, Johor, Pasir Gudang Campus, 81750, Masai, Johor, Malaysia.
[3] Science and Engineering Faculty, Civil Engineering and Built Environment, Gardens Campus Point, Queensland University of Technology, QLD, 4001, Australia.

[*]Corresponding author: azmi_216@yahoo.com

Abstract. Bearing strength of concrete is one of important design characteristics in transmit the bearing force in structural supports such as column corbel, bridge bearing system, concrete connection, anchorage in post-tension members and other types of structure supports. Concrete bearing strength is depending on ratio of unloaded-to-loaded area and compressive strength of concrete. However, the effect of concrete block height on bearing strength of concrete is not included. Hence, the effect of different heights on concrete bearing strength were investigated using numerical modelling. The three-dimensional finite element model (FEM) for nonlinear analysis of concrete bearing were developed and analyzed in ABAQUS/Explicit. The FE models were validated based on pre-existing experimental results. The confinement effect and structural ductility of concrete blocks with different heights were evaluated. The FE results indicate that the level of internal confinement and the structural ductility of concrete blocks were affected significantly as increasing of concrete block heights. The bearing force for the concrete block with 150mm high was significantly dropped to 55% as compare to control specimen of 50mm high.

1. Introduction

The evaluation of bearing strength characteristic is crucial especially when designing the concrete transfer members such as concrete footing. The concrete bearing behaviour is relying on the contact mechanism between the steel bearing plate and the concrete surface. The steel bearing plate widely used as load dispersing plate for load transfer and intermediate supports for beam corbel, bridge bearing system and concrete anchorage system. Failure of concrete bearing is not an option as stability of structure is critical at their support base. The concrete bearing strength for unreinforced concrete block is depending on the unloaded to loaded area ratio and concrete compressive strength as stated in the Australian standards AS5100.5 [1]. For unreinforced concrete support, the bearing stress at concrete surface shall be taken as minimum value of

IOP Conf. Series: Materials Science and Engineering **513** (2019) 012031 doi:10.1088/1757-899X/513/1/012031

$$\phi 0.85 f_c^{'} \left(\frac{A_4}{A_3} \right)^{1/2} \text{ or } \phi 2 f_c^{'} \tag{1}$$

where A_4 is unloaded area, A_3 is loaded area, $f_c^{'}$ is concrete compressive strength, ϕ is 0.6 for bearing.

Based on the formula [1], the effect of concrete block height is clearly not considered in the concrete bearing strength prediction. Other standards [2] also has the similar scenario. Earlier studies on concrete bearing was focused on the effect of different strength of concrete grade and the different unloaded-to-loaded area ratios [3 - 8]. In industry practise, the height of concrete support was determined depending on their design purpose [9]. In spite of important parameter for the concrete height in concrete bearing strength, only limited fundamental studies was available in literature. In previous studies, the ratio of height to width for concrete specimen is close to 1. Hence, the finding of these research may not reflecting the structural performance for low or high ratio of shallow and slender concrete blocks respectively. This present research was undertaken to extend fundamental explanation on bearing strength of concrete due to change of concrete block heights.

2. Methodology

The advanced software widely used in civil engineering especially due to limited data obtained from experiment. It leads some of complex behaviour of structural member is not easily explained. In the present study, FEM were examined the effect of concrete block height on concrete bearing strength in ABAQUS/explicit. The FE analysis is based on the structural response of the concrete block subjected to bearing force when loaded by steel bearing plate on concrete surface.

The quarter scale of three-dimensional (3D) FE model was developed to optimise the computational time. The size of FE model was determined based on the Australian Standard AS5100.5 for the expected bearing force of 3500kN. The size of concrete block and steel bearing plate was designed for 400mm by 400mm and 300mm by 300mm respectively. Five different heights of concrete blocks have been investigated through series FE models. The range of 50mm to 150mm for selected heights of concrete blocks were selected as considered for typical heights of concrete bearing footing in industry practise.

For the material model, behaviour for concrete (50MPa) was based on the Concrete Damage Plasticity Model (CDPM) (refer Table 1). For the contact behaviour, "Surface-to-Surface" contact was selected to stimulate the contact interaction due to frictional force and hard surface between concrete block and steel bearing plate. The contact surface is based on master to slave surface as shown in Figure 1.

Table 1. CDPM for concrete with compressive strength of 50MPa.

Compression behaviour		Tension behaviour	
Stress (MPa)	Crushing strain ($\times 10^6$)	Stress (MPa)	Cracking strain ($\times 10^6$)
15.00	0	1.99	0
20.20	74.73	2.83	33.33
30.00	98.84	1.87	160.42
40.30	154.12	0.86	279.76
50.01	761.53	0.22	684.59
40.24	2557.55	0.06	1086.73
20.24	5675.43		
5.26	11733.11		

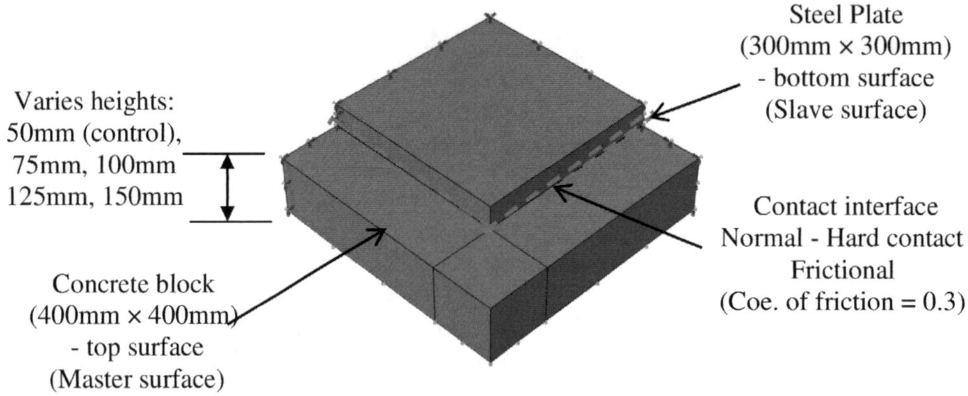

Figure 1. One-quarter model of concrete pedestal (without meshing).

In mesh sensitivity studies, the FE model was discretized for different mesh sizes to determine optimize mesh size that minimize the computational time. The optimise mesh size was obtained at 5mm. For the boundary condition, the surface of concrete block at the base surface was restrained in vertical direction. Due to symmetric loading and geometrical, quarter size model was developed based on plane of symmetry as shown in Figure 2. For loading condition, the incremental displacement was applied up to bearing failure.

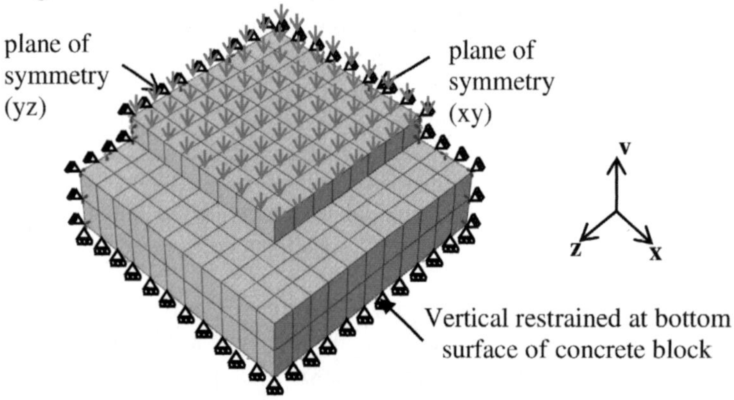

Figure 2. A quarter size of concrete bearing model.

For model validation, concrete blocks for 101.6mm and 203.2mm in three different ratios of concrete-to-steel area (A_c/A_s) of 2, 4 and 6 are considered. The concrete block has constant surface dimensions of 203.2 × 203.2mm for all specimens. Table 2 shows a series pre-existing database based on the experimental work [10]

Table 2. Experimental results [10].

A_c/A_s	height, h (mm)	Confinement effect, f_b/f_{cu}
2	203.2	1.27
2	101.6	1.37
4	203.2	1.63
4	101.6	1.75
6	203.2	1.92
6	101.6	2.00

Concrete Compressive strength, f_{cu} = 55.5 MPa

Table 3. Finite element.

Specimen No.	(A_c/A_s)	Block height, h (mm)	Ultimate load, F_{ult} (kN)	Bearing strength, $f_b = F_{ult} / A_s$ (MPa)	Confinement f_b/f_{cu}
Fe-1	2	203.2	1260.00	60.75	1.22
Fe-2	2	101.6	1436.60	70.56	1.41
Fe-3	4	203.2	849.20	82.27	1.65
Fe-4	4	101.6	902.20	87.42	1.75
Fe-5	6	203.2	610.80	88.54	1.77
Fe-6	6	101.6	746.80	108.25	2.16

Concrete Compressive strength, $f_{cu} = 50$ MPa

The accuracy of FE results in predicting of confinement effect of concrete have been evaluated by comparing with the experimental results [10]. In the present study, the confinement effect is referring to the ratio of bearing stress at the concrete surface (f_b) to compressive strength of concrete (f_{cu}). Table 3 shows the results of FE models for different ratio of 2, 4 and 6 for steel bearing plate size 144.02mm × 144.02mm, 101.6mm × 101.6mm and 83.06mm × 83.06mm respectively. Note the constant concrete surface of 203.2mm × 203.2mm was used in all FE models.

There is a significant different of compressive strength between experiment (55MPa) and modelling (50MPa). Hence, the results have been normalized by applying two important dimensionless parameters:

a) The confinement effect is the ratio of bearing pressure divided by compressive strength of concrete (f_b/f_{cu})
b) The ratio of concrete surface divided by surface area of steel plate (A_c/A_s)

It can be seen Table 2 and 3 that the confinement effect (f_b/f_{cu}) for both experimental and FE results are consistence.

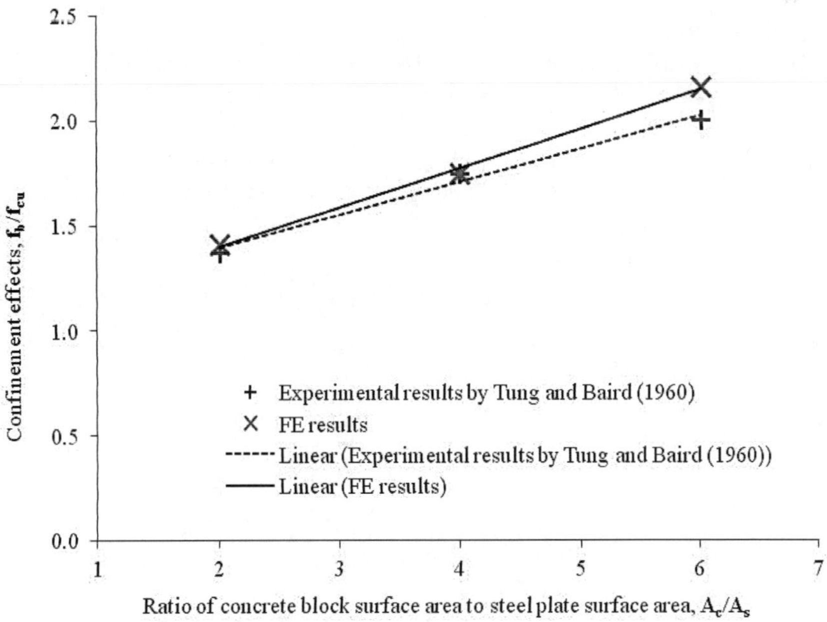

Figure 3. FE and experimental [10] for height concrete block of 101.6mm.

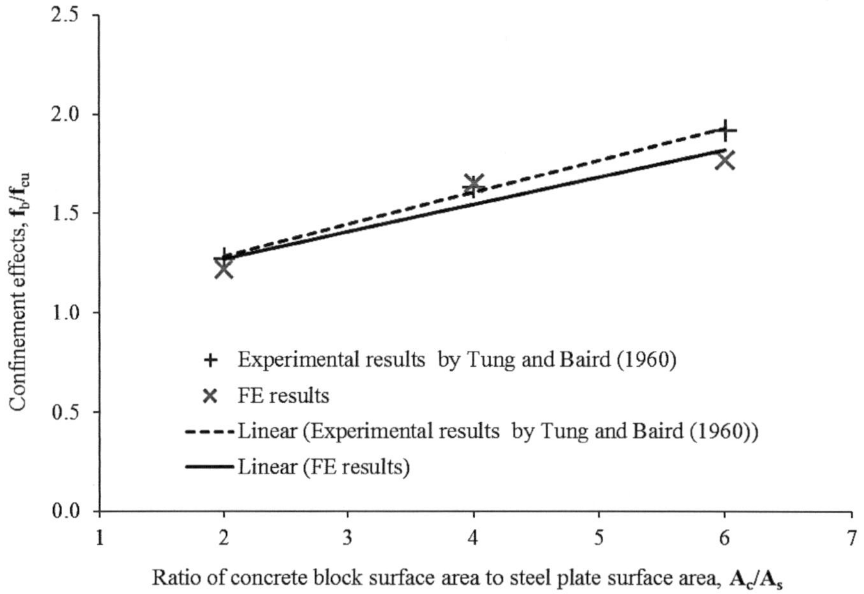

Figure 4. FE and experimental [10] for height concrete block of 203.2mm.

Figure 3 and 4 show the significant results on the trends of confinement effects for different bearing ratios (A_c/A_s) in different heights of concrete block under bearing load. It proved that FE models able to capture the reasonable trend of confinement effect due to contact interaction between steel bearing plates for different height of concrete blocks. It was found that the trend remained the consistent irrespective of height of concrete blocks. The prediction of confinement effect of concrete (f_b/f_{cu}) under bearing force was in good agreement with previous experimental results [10] for all different ratios of (A_c/A_s).

3. Results and discussion

Series FE models were evaluated based on control model of 50mm high and further examine for different height of concrete blocks, i.e. 75, 100, 125 and 150mm. In the present study, the structural performance of concrete blocks under bearing load was evaluated based on their ultimate load and structural ductility. The ductility was calculated based on maximum displacement divided by displacement at yielding. The results of FE models for load-displacement relationship are shown in Figure 5.

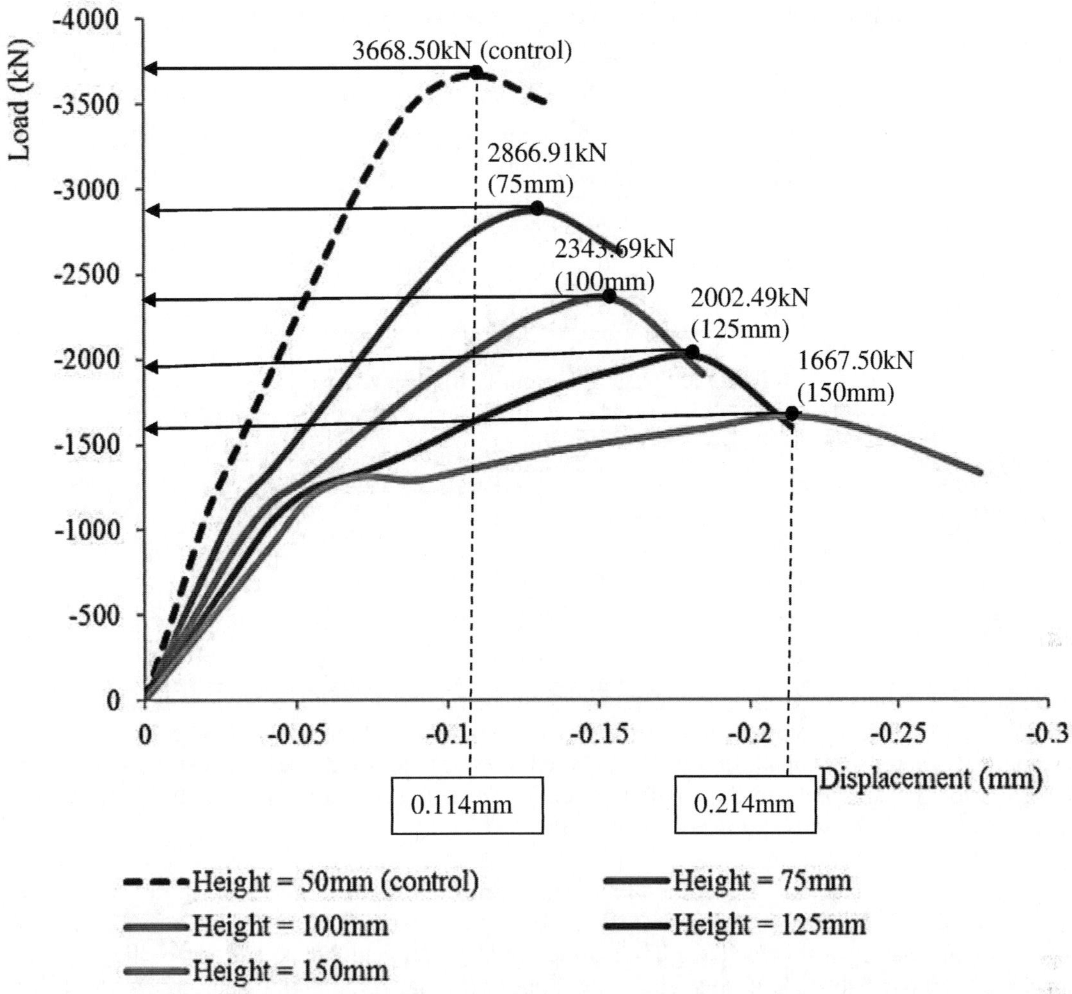

Figure 5. Load-displacement relationships for all FE models.

Figure 5 shows that the ultimate load of control specimen significantly dropped from 3668.50kN to 1667.50kN when the concrete block has increased their height from 50 mm to 150 mm. It was notified that the concrete block with 150mm height has reached the ultimate load was delayed at displacement of 0.214mm as compared to control specimen at displacement of 0.114mm. It was proved that changing of the height of concrete block have significant effect on the bearing strength and ductility of concrete block under bearing force. These results were further analysed in Table 4 and 5.

IOP Conf. Series: Materials Science and Engineering **513** (2019) 012031 doi:10.1088/1757-899X/513/1/012031

Table 4. Structural performance of concrete blocks under bearing force.

Height	Displacement at max. load (mm)	Ultimate load (kN)	Percentage reduced (%)	Vertical stress at the sharp edge (MPa)
50mm (Control)	0.114	3668.50	-	37.66
75mm	0.132	2866.91	22	33.51
100mm	0.157	2343.69	36	28.86
125mm	0.184	2002.49	45	27.46
150mm	0.214	1667.50	55	25.59

Table 5. Structural ductility for all models.

Height	Displacement at yield point (mm)	Max. displacement (mm)	Ductility	Stiffness at yield point ($\times10^3$Nmm^{-1})
50mm (Control)	0.064	0.131	2.04	43750
75mm	0.078	0.157	2.01	27215
100mm	0.080	0.184	2.30	21975
125mm	0.090	0.214	2.37	16683
150mm	0.095	0.277	2.92	13888

Table 4 shows that the ultimate load for concrete block with 150mm height was reduced up to 55% as compared control specimen (50mm). It was noted that the highest vertical stress at the sharp edge is 37.66MPa for control specimen due to highest contact area. For height of concrete block 150mm, the vertical stress is significantly dropped to 25.59MPa.

Table 5 indicates that increase of the block height has highest ductility due to high penetration of small steel bearing plate on the concrete surface. However, in term of stiffness, the height of concrete block 150mm was lowest which explaining the slender concrete block has lower bearing force capacity as compared to shallow concrete block. It indicates that at the high displacement, the shallow concrete block can resist higher bearing force as compared slender concrete block. In design stage, perhaps more precaution is needed especially when dealing with slender concrete blocks.

4. Conclusions

The FE models have examined the structural responses of concrete bearing for different heights of concrete blocks. The following observations are made based on data generated from the predicted results;

1. The height of concrete block is an important design parameter in determining the bearing capacity of concrete blocks. However, none of existing standard or design equation in previous national standards has taken account this parameter. It was observed that the increasing of concrete block height has significantly reduced the load bearing capacity of concrete.
2. It was noted that the ultimate load for the concrete block with 150mm high was dropped up to 55% of control specimen for 50mm high. The increase of height of concrete block has improved the ductility of the structure but the stiffness is significantly reduced which explaining the slender concrete block has low bearing force capacity as compared to shallow concrete block.

Acknowledgements
The authors gratefully acknowledge the financial supports from Universiti Teknologi MARA (UiTM) Shah Alam, Selangor, Malaysia under Research Grant 600-IRMI/DANA KCM 5/3/LESTARI (126/2017). Author would like to dedicate this paper to Puan Sri Datin Prof Dr Hanizah Abd Hamid for her huge contribution toward this study.

References

[1] AS5100.5 2004. Australia Standard for Concrete Structures, Standards Australia Limited.
[2] ACI Committee 318, Building Code Requirements For Structural Concrete (ACI 318-99) and Commentary (318R-99), America Concrete Institute, Farmington Hills, Mich., 1999, pp. 391.
[3] Roberts-Wolimann, C. L., Banta, T., Bonetti, R. & Charney, F. 2006. Bearing strength of lightweight concrete. *ACI Materials Journal,* **103**(6), pp. 459.
[4] Zhou, W., Hu, H. & Zheng, W. 2013. Bearing capacity of reactive powder concrete reinforced by steel fibers. *Construction and Building Materials*, **48**, pp.1179-1186.
[5] Bonetti, R., Roberts-Wollmann C. L., & Santos, J. T. 2014. Bearing Strength of Confined Concrete. *ACI Structural Journal*, **111**(6), pp. 1317.
[6] Yahya, N. A., & Dhanasekar, M. 2014, Explicit finite element modelling of bridge girder bearing pedestals in *The 23rd Australasian Conference on the Mechanics of Structures and Materials (ACMSM23)*, Dec 9-12, Southern Cross University, Byron Bay, Australia.
[7] Md Zain, M. R., & Yahya, N. A. 2017, , *Pertanika Journal of Science and Technology*, **25**(S), 67-76.
[8] Scheffers, C. A., Sri Ravindrarajah, R., & Reinaldy, R. 2011. Bearing Strength of CFRP Confined Concrete. Paper presented at the Advances in FRP Composites in Civil Engineering: *Proceedings of the 5th International Conference on FRP Composites in Civil Engineering (CICE 2010)*, Sep 27–29, 2010, Beijing, China.
[9] Yahya, N. A., & Dhanasekar, M. 2017, Strategies for mitigation of the failure of concrete pedestals supporting bridge girder bearings, PhD Thesis, University Technology of Queensland.
[10] Au, T., & Baird, D. L. 1960. Bearing capacity of concrete blocks. *Journal of the America Concrete Institute,* **56**(3), pp. 869-880.

Experimental study on the bearing capacity of concrete blocks loaded through square and circular steel plates

Mohd Raizamzamani Md Zain[1*], Norrul Azmi Yahya[1], Lee Siong Wee[1], Oh Chai Lian[1] and Balqis Md Yunus[1]

[1]Faculty of Civil Engineering, Universiti Teknologi MARA, Shah Alam, Selangor, Malaysia

[*]Corresponding author:raizam@salam.uitm.edu.my

Abstract. The bearing capacity of concrete loaded through steel plate has been the subject of numerous studies, play as an important element that need to be considered in designing many types of concrete members such as foundation structure, the end bearing zone of prestressed post-tensioned concrete beams and even for concrete pedestal in bridges. This paper presents an experimental investigation into the bearing capacity of concrete blocks axially loaded through square and circular steel plates. The relative area of the surface of concrete block specimen to the area of the bearing plate is one of the most significant factors influencing the bearing capacity of concrete blocks. Thus, the main purpose of this study is to determine the maximum bearing capacity and observe the failure modes of concrete blocks concentrically loaded through square and circular steel plates. A total of twenty-one (21) numbers of specimens with three different types of concrete blocks were experimentally investigated. There are three (3) specimens of plain concrete blocks, nine (9) specimens of concrete blocks laterally reinforced with ties (loaded and tested under three different sizes of square steel plate) and nine (9) specimens of concrete blocks laterally reinforced with ties (loaded and tested under three different sizes of circular steel plate). The use of steel plate helps in determining the bearing capacity when concentrically loaded on top of the specimen. The results of the experimental testing herein were compared with the bearing capacity values calculated based on previous researches and international standards. Experimental results demonstrate that concrete blocks laterally reinforced with ties loaded and tested under larger relative area of the steel plate gave lower value of bearing capacity compared with other reinforced specimens. Meanwhile, the failure pattern of concrete blocks under the bearing load was found to vary with the increase in sizes of steel plates.

1. Introduction

At present, the demand of using concrete is increasing especially in construction fields due to its versatility and easy availability of its constituent materials. Nowadays, concrete is being applied for many functions and there is uncountable of innovation of concrete around the world. Concrete is a mixture of cement, sand, coarse aggregate made of crushed stone and water. With the use of reinforced concrete, cracking strength can be enhanced. As demonstrated and pointed out by previous studies ([1], [2], [3], [4] and [5]), several circumstances like vertical and radial cracks were observed on the unreinforced concrete blocks due to high compressive loading. Among the approach to obtain minimal numbers of cracking failure was by putting steel plate on the top surface of concrete block as its application can be observed on many civil engineering works like concrete bridges, concrete footing, anchorage for pre-stress member and others. The term used for the strength of that concrete block towards the load applied on it was load bearing capacity.

The load-bearing capacity was first examined by [6] followed by [7], [8], [9] and [10]. The researchers found that there was a formation of an inverted pyramid under the loading bearing plate and expressed a formulation for concrete bearing capacity based on that observation. The different sizes and shapes of steel plates shows a different kind of crack failure for the concrete blocks. The enveloping concrete gives an effect to the higher compressive strength of concrete at bearing area as stated by [2], [11] and [12]. The reason was due to the local pressure that exerted on the concrete structural members.

Even though previous investigations do not specify all the pertinent aspects of the bearing problem if unreinforced concrete been chosen, there was a need to investigate the effect of having difference sizes and cross-sectional shapes of steel plate on top of plain and reinforced concrete blocks. The understanding on the behaviour and generation mechanism of the bearing capacity of concrete blocks was still rather crude and lack of comprehensive research for the design purposes of concrete members. Therefore, in view of this and keeping in mind the usefulness of bearing capacity to the design of structural member, experimental investigations were performed to evaluate the ultimate bearing capacity of concrete blocks concentrically loaded via different cross-sectional shapes and sizes of steel plates and observe the possible failure modes of the concrete.

2. Theoretical developments

The development of mathematical formulation has been the subject of numerous numbers of researchers. Shelson (1957) developed an empirical formula to obtain the bearing strength of concrete [8]. The researcher stated that the maximum bearing strength of concrete, f'_{cc}, be conservatively estimated as the product of the concrete strength, f_c', and a factor equal to $\sqrt[3]{(A_2/A_1)}$, where A_2 is denoted as gross area of the concrete foundation and A_1 as concrete bearing area or bearing plate area as in equation (1).

$$f'_{cc} = f_c'\left(\sqrt[3]{\frac{A_2}{A_1}}\right) \tag{1}$$

Then, [13] derived an expression for ratios of A_2/A_1 ranging from 1 to 40 as in equation (2).

$$\frac{f'_{cc}}{f_c'} = 1 + \frac{4.15}{\sqrt{f_c'}}\left(\sqrt{\frac{A_2}{A_1}} - 1\right) \tag{2}$$

where, f'_{cc} represents the concrete bearing strength, f_c' denotes as the compressive strength of concrete (MPa), A_1 represents the bearing plate area and A_2 represents the area of the concrete. Later, [14] expressed an equation by taken into account the analysis of upper bound limit based on the proposed formula by [13] as in equation (3).

$$\frac{f'_{cc}}{f'_c} = 0.8\left[1 + 12.5\frac{f_t}{f'_c}\left(\sqrt{\frac{A_2}{A_1}} - 1\right)\right] \tag{3}$$

In which, f_t represents the concrete strength in tension. ACI-318 proposed an expression to estimate the bearing strength of concrete [15] as in equation (4).

$$f'_{cc} = 0.85\phi_c f'_c\left(\sqrt{\frac{A_2}{A_1}}\right) \tag{4}$$

where, ϕ_c represents the capacity reduction factor for concrete in bearing and $0.85f'_c$ as the concrete compressive strength under sustained loads.

In addition, to estimate the bearing capacity of concrete blocks, [12] proposed an expression as in equation (5).

$$P_P = 0.8A_1\left(f'_c + 12.5f_t\left(\sqrt{\frac{A_2}{A_1}} - 1\right)\right) \tag{5}$$

In which, P_p represents the maximum bearing strength of concrete, f_t denotes the strength of concrete in uniaxial tension and is equal to $0.33\sqrt{f'c}$ (MPa), f'_c as the compressive strength of concrete (MPa) and 0.8 as the proposed reduction factor for sustained load.

3. Methods

A series of experimental investigations have been conducted to examine the bearing capacity of plain and reinforced concrete blocks and observe its possible mode of failure. The test was carried out subjected to concentric compression load. Detailed description on the specimen preparation, experimental testing and test set-up were elaborated as on the following sub-topics.

3.1. Material properties

In order to provide optimum strength and durability, water-cement ratio of 0.46 was used. There are two types of bearing steel plates were used, square and circular cross-sectional shapes as depicted in figure 1. The square steel plate having dimensions of 50 mm x 50 mm, 100 mm x 100 mm and 150 mm x 150 mm in cross section and 10 mm thickness. Meanwhile, circular steel plates having diameters of 50 mm, 100 mm and 150 mm with the same thickness as square shape.

(a) Square

(b) Circular

Figure 1. Square and circular bearing steel plate.

3.2. Specimen preparation

Reinforced concrete (concrete consist of reinforcement bar embedded in the concrete) was chosen as the concrete block specimens while plain concrete acting as a control specimen. All concrete block specimens with the characteristic strength of 30 N/mm^2 were cast from a concrete mix as depicted in table 1. Then, the concrete was poured into the prepared mould.

Table 1. Proportion of concrete per m^3.

Water, kg/m^3	Cement, kg/m^3	Fine aggregate, kg/m^3	Coarse aggregate, kg/m^3	
			10 mm	20 mm
37.3	80.8	108.8	88.3	176.5

Twenty-one (21) numbers of concrete blocks including plain concrete and reinforced concrete blocks were prepared. From a batch of concrete, six 150 mm × 150 mm concrete cubes were cast as control specimens. The concrete cubes were tested up to the failure for their compressive strength after being stored in a curing tank for 7 and 28 days to examine the mechanical properties of the mix. A total of twenty-one (21) numbers of concrete block specimens were cast and tested up to failure subjected to concentric compression load after 28 days achieving its designated age. The specimens were concentrically loaded through square and circular steel plates. Detailed configurations of the concrete block specimens were shown as in table 2 and figure 2.

Table 2. Detail of test specimens.

Specimen designation	Block Cross section	Block height	Plate cross section (10 mm thickness)	Remarks
PC-01, PC-02, PC-03	200 mm x 200 mm	200 mm	-	Plain concrete blocks
PC-SQ50-01, PC-SQ50-02, PC-SQ50-03			50 mm x 50 mm	Reinforced concrete blocks
PC-SQ100-01, PC-SQ100-02, PC-SQ100-03			100 mm x 100 mm	
PC-SQ150-01, PC-SQ150-02, PC-SQ150-03			150 mm x 150 mm	
PC-CS50-01, PC-CS50-02, PC-CS50-03			50 mm diameter	
PC-CS100-01, PC-CS100-02, PC-CS100-03			100 mm diameter	
PC-CS150-01, PC-CS150-02, PC-CS150-03			150 mm diameter	

Note that:
PC-01 denotes the un-reinforced (plain) concrete block with the characteristic strength of 30 MPa of first specimen.
PC-SQ50-01 denotes the reinforced concrete block with the size of square steel plate (50 mm x 50 mm in cross-section) of first specimen.
PC-CS50-01 denotes the reinforced concrete block with the size of circular steel plate (50 mm diameter) of first specimen.

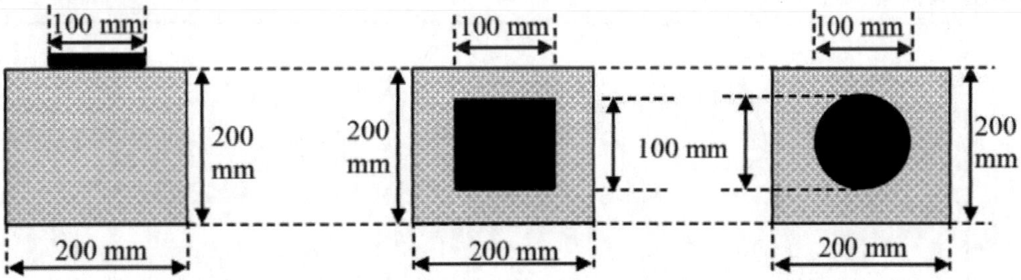

(a) Front view **(b)** Top view (square) **(c)** Top view (circular)

Figure 2. Detail cross-sectional dimensions of the concrete specimens.

3.3. Experimental details and test set-up

The bearing surface of the machine and specimen was cleaned and any loose grit or other extraneous materials from the surfaces of the specimen that will be in contact with the platens were removed. The main purpose was to ensure smooth contact between the bearing plate and concrete surface. The excess moisture from the surface of the concrete block specimen also was cleaned before placing in testing machine.

The top platen of the testing machine bore directly on the entire area of the steel bearing plate. The piston was lowered gently to the upper surface of the concrete block specimen using a lever. The specimens were loaded continuously until failure. The test set-up for the specimens is shown in figure 3. A linear variable differential transformer (LVDT) was used to determine cross-head displacement under the applied bearing load. The maximum load and any crack occur on the concrete block specimens were recorded and observed.

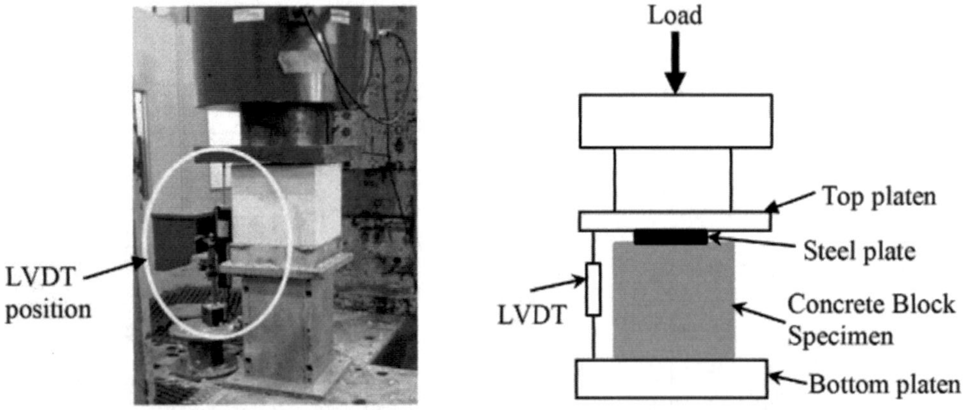

Figure 3. Test set-up and position of LVDT.

4. Results and discussion

A relationship between the bearing load and displacement, results of bearing strength test and its comparison with previous literatures were discussed on the following sub-topics.

4.1. Load bearing capacity-deformation relationship

The maximum load capacity and displacement values were attained using data acquisition system, recorded by the computer. The maximum load bearing capacity and deformation relationship were plotted with regards to the experimental measurements for twenty-one (21) numbers of concrete block specimens as shown in figure 4.

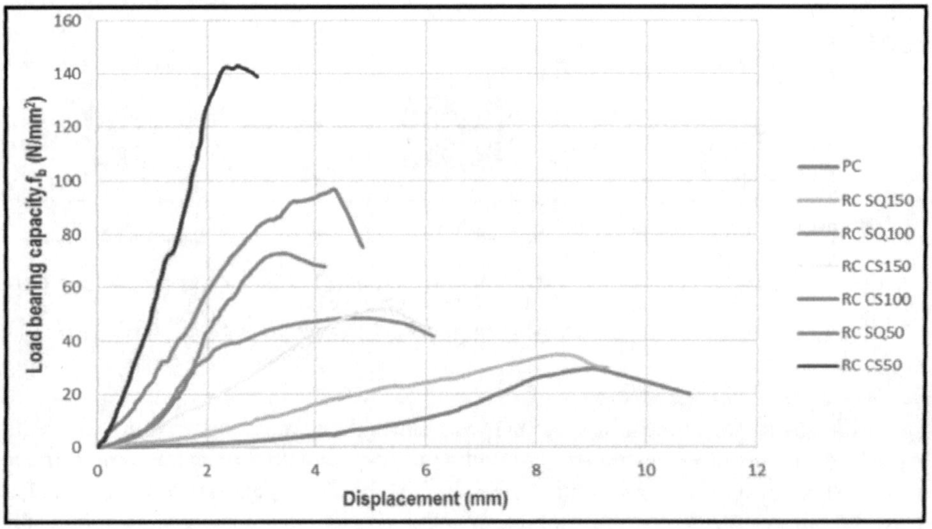

Figure 4. Load bearing capacity-deformation relationship.

With regards to the results shown in figure 4, for the unreinforced concrete blocks denoted as PC-01, PC-02 and PC-03, its average maximum value of load bearing capacity was recorded as 0.033 kN/mm^2 with 9.34 mm deformation. Reinforced concrete block specimens loaded through square steel plate denotes as RC-SQ50-01, RC-SQ50-02 and RC-SQ50-03 gave the average maximum value of bearing capacity as 0.116 kN/mm^2 with 3.42 mm deformation. In addition, for reinforced concrete block specimens denotes as RC-SQ100-01, RC-SQ100-02 and RC-SQ100-03, its average maximum value of load bearing capacity was recorded as 0.051 kN/mm^2 with 4.12 mm deformation. For reinforced concrete block specimens denotes as RC-SQ150-01, RC-SQ150-02 and RC-SQ150-03, 0.039 kN/mm^2 with 9.30 mm deformation was recorded.

Meanwhile, for reinforced concrete block specimens loaded through circular steel plate denotes as RC-CS50-01, RC-CS50-02 and RC-CS50-03, its average maximum value of load bearing capacity was recorded as 0.155 kN/mm^2 with 2.60 mm deformation. Reinforced concrete block specimens denotes as RC-CS100-01, RC-CS100-02 and RC-CS100-03 gave the average maximum value of bearing capacity as 0.081 kN/mm^2 with 4.04 mm deformation. For specimens denotes as RC-CS150-01, RC-CS150-02 and RC-CS150-03, 0.052 kN/mm^2 with 5.40 mm deformation was recorded. The result of each type of specimen and its comparison with previous studies were then presented in table 3 and figure 5.

Table 3. Experimental results of bearing capacity.

Specimen designation	Average Ultimate Load (kN)	Average Bearing Capacity (kN/mm^2)	Average deformation (mm)
PC	1339.95	0.033	9.34
RC-SQ50	290.42	0.116	3.42
RC-SQ100	513.83	0.051	4.12
RC-SQ150	897.82	0.039	9.30
RC-CS50	303.32	0.155	2.60
RC-CS100	636.36	0.081	4.04
RC-CS150	927.11	0.052	5.40

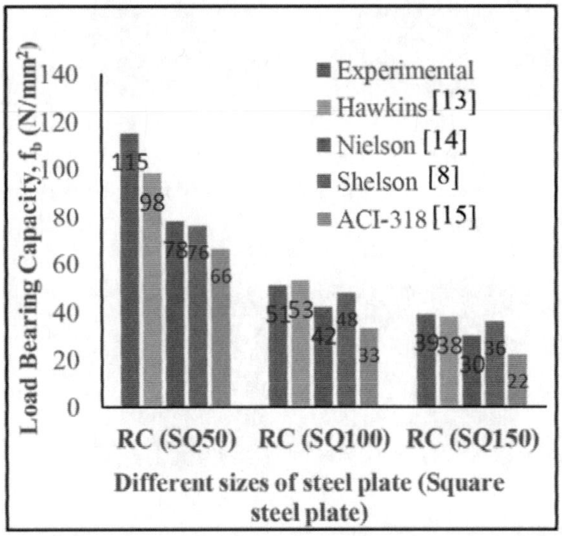

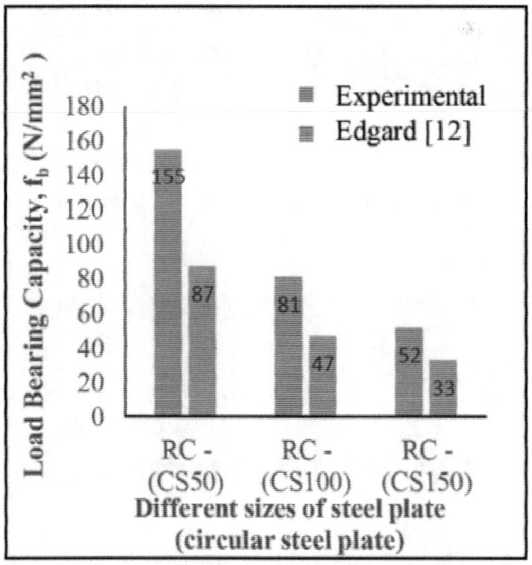

Figure 5. Comparison for concrete block specimens loaded through square and circular steel plates.

Experimental investigations indicate that concrete block specimen having smaller size of circular steel plate (50 mm in diameter) gave higher maximum value of bearing capacity compared to other concrete block specimens loaded through circular steel plate. The same behaviour can be observed for

IOP Conf. Series: Materials Science and Engineering **513** (2019) 012032 doi:10.1088/1757-899X/513/1/012032

the concrete block specimens loaded through square steel plate. Thus, in other words, as the length or diameter of the bearing plate increased, decreasing the relative area of the surface of concrete block to the area of the steel bearing plate and the ultimate bearing capacity decreases. Based on the value of specimen deformation, experimental results demonstrate that specimen loaded through smaller size of steel plate (either square or circular steel plates) gave lower values of deformations and higher values of bearing capacities.

4.2. Mode of failures

Among the type of failure modes that have been observed during the experimental testing is in the forms of localized damage, which can be detected at the outer edge of the contact area. It has revealed that these observations especially for unreinforced concrete blocks are comparable with previous studies ([3], [4], [16] and [17]) where the vertical cracks, non-explosive crack, edge crack and inverse pyramid failure have been detected at the outer edge of contact area, as shown in figure 6 until figure 12.

(a) (b) (c) (d)

Figure 6. Mode of failures for unreinforced concrete blocks subjected to concentric loading; (a) Vertical crack, (b) Non-explosive crack, (c) Inverted pyramid shape and (d) Conical shape.

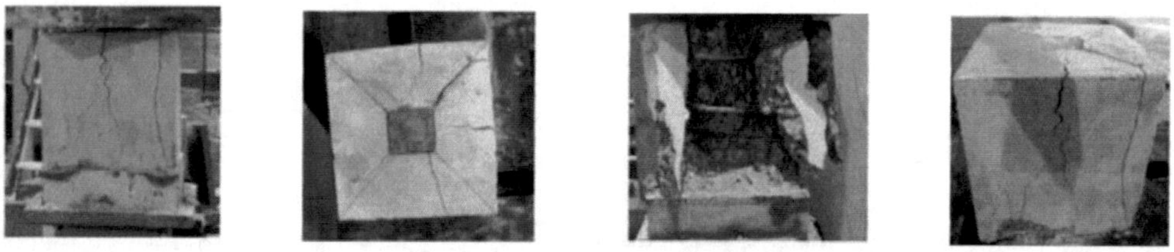

(a) (b) (c) (d)

Figure 7. Mode of failures for reinforced concrete blocks concentrically loaded through 50mm x 50mm square steel plates; (a) Vertical crack, (b) Radial crack, (c) Brittle crack and (d) Splitting along the vertical planes.

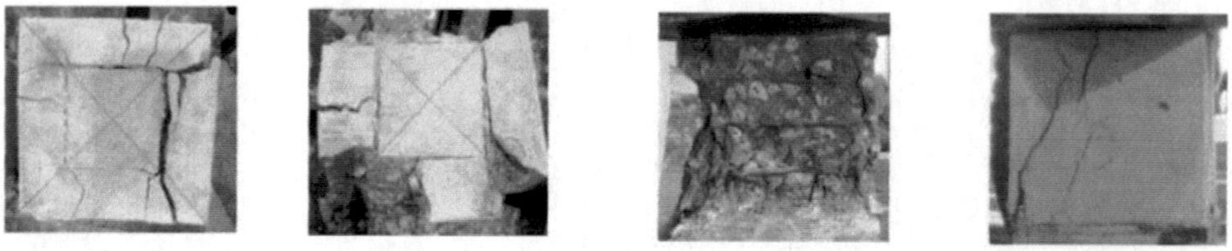

(a) (b) (c) (d)

Figure 8. Mode of failures for reinforced concrete blocks concentrically loaded through 100mm x 100mm square steel plates; (a) Splitting wedge, (b) Edge crack, (c) Edge crack outside lateral bar and (d) Vertical crack.

(a) (b) (c) (d)

Figure 9. Mode of failures for reinforced concrete blocks concentrically loaded through 150mm x 150mm square steel plates; (a) Vertical crack, (b) Splitting wedge, (c) Brittle crack and (d) Edge crack.

(a) (b) (c)

Figure 10. Mode of failures for reinforced concrete blocks concentrically loaded through 50mm diameter circular steel plates; (a) Radial crack, (b) Brittle crack and (c) Edge crack outside plate area.

(a) (b) (c)

Figure 11. Mode of failures for reinforced concrete blocks concentrically loaded through 100mm diameter circular steel plates; (a) Medium splitting wedge, (b) Edge crack and (c) Crack outside lateral bar.

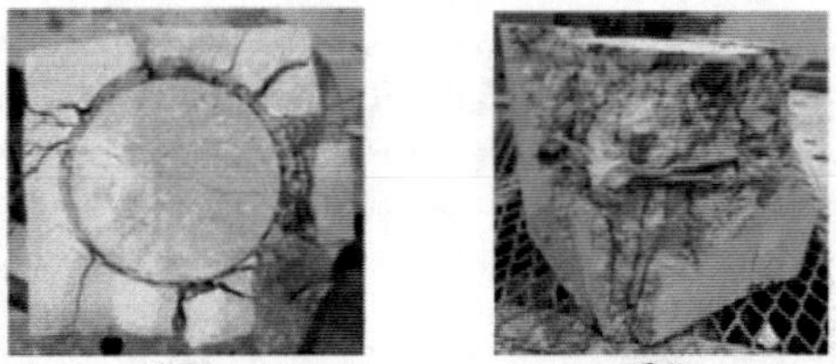

(a) (b)

Figure 12. Mode of failures for reinforced concrete blocks concentrically loaded through 150mm diameter circular steel plates; (a) Large splitting wedge, (b) Formation of small inverted cone under plate area.

Based on figure 6, the failure modes for plain concrete blocks obtained are vertical crack, non-explosive crack, inverted pyramid shape and conical shape. The initial failure in the form vertical crack can be observed as in figure 6(a) where the concrete starts to fail. Then, when the tensile stress at the top of specimen achieved its ultimate stresses and exceed the tensile strength of the concrete specimen, the formation of inverted pyramid and conical shape were detected. On the other hand, the type of failure modes for reinforced concrete block specimens loaded through various cross-sectional dimension of square steel plate are presented as in figure 7 up to figure 9. The formation of crack start with the similar phases such as vertical crack and then, splitting wedge failure, followed by edge crack until it brittle. These occurs because of the ultimate load of the concrete was attained and the tension of the concrete at the maximum value. But, there is no formation of inverted pyramid or in the form of conical shape for reinforced concrete blocks due to the presence of steel reinforcement bar.

Meanwhile, figure 10 until figure 12 shows the failure modes of reinforced concrete blocks loaded through different sizes of circular steel plate. All reinforced concrete block specimens indicate the behaviour of radial and edge cracks. As they were progressively loaded under concentric loading, radial crack and edge crack developed at the corners of the concrete blocks. The type of failure for reinforced concrete block specimens loaded via circular steel plate (150mm in diameter) shows the formation of small inverted cone under plate area and large splitting wedge. When the load reaches its limiting value, a small inverted cone can be observed on the block specimens. Large splitting wedge punched out from beneath the steel bearing plate as can be observed in figure 12(a).

5. Conclusions

The conclusions of the present study were drawn as follows:

1) For all reinforced concrete block specimens, the ultimate bearing capacity decreases related with decreasing value in ratio of unloaded area to loaded area.
2) Concrete block specimens loaded through smaller size of steel plate (either square or circular steel plates) gave lower values of deformations and higher values of bearing capacities.
3) Concrete blocks specimens loaded through circular steel plate gave higher values of bearing capacities compared to specimens loaded via square steel plate due to the reason that stress concentration for circular steel plate quite uniform throughout the contact surface.
4) High stress concentration at the shape edge of the square steel plate contribute to the lower values of bearing capacities for concrete block specimens loaded through square steel plate compared to specimens loaded via circular steel plates.

Acknowledgements

The authors would like to acknowledge the Ministry of Higher Education under Research Acculturation Grant Scheme (RAGS) (RAGS/1/2015/TK01/UITM/03/1) for providing financial support to perform this study.

References

[1] Escobar-Sandoval, D., Andren, S.W., & Gary, F.D. 2006. Concentrically loaded circular steel plates bearing on plain concrete. *Journal of Structure Engineering*, 132(11), pp. 1784-1792.

[2] Ince, R., & Arici, E. (2004). Size effect in bearing strength of concrete cubes. *Construction and Building Materials*, 18, pp. 603-609.

[3] Kameswara Rao, C.V.S., & Swamy, R.N. 1974. Bearing strength of steel fibre reinforced concrete. *Build. Sci.*, 9, pp. 263-268.

[4] Mohd Raizamzamani M.Z. & Norrul Azmi, Y. 2017. Experimental Study on Bearing Strength of
Concrete Blocks under Concentric Compression Load. *Pertanika J. Sci. & Technol.* Vol. 25 (S). pp. 67 – 76.

[5] Rodolfo B., Carin L. R. & Juan T. Santos. 2014. Bearing strength of confined concrete. *ACI Structural Journal*, pp. 1317 – 1327.

[6] Baushinger, J. 1876. Tests with Blocks of Natural Stone. *Mechanisch und Technischen Laboratorium der Kgl. Technischen Hochschule*, Munich, 6, 13.

[7] Meyerhof, G.G. 1953. The bearing capacity of concrete and rock. *Magazine of Concrete Research*, pp. 107-116.

[8] Shelson, W. 1957. Bearing capacity of concrete. *Journal of the American Concrete Institute*, 54(5), pp. 405-414.

[9] Au, T., & Baird, D.L. 1960. Bearing capacity of concrete blocks. *Journal of the American Concrete Institute*. 56, pp. 869 – 880.

[10] DeWolf J.T. 1978. Axially Loaded Column Base Plates. *Journal of the Structural Division ASCE*,
Vol. 104. No. ST4, pp. 781-794.

[11] Ahmed, T., Burley, E., & Rigden, S. 1998. Bearing capacity of plain and reinforced concrete loaded over a limited area. *ACI Structural Journal*, pp. 330-340.

[12] Edgard D. Escobar-Sandoval, Andrew S. Whittaker, M. ASCE, & Gary F. Dargush (2006), Concentrically Loaded Circular Steel Plates Bearing on Plain Concrete. *Journal of Structural Engineering, ASCE*, pp. 1784-1792.

[13] Hawkins, N.M. 1968. The bearing strength of concrete loaded through flexible plates. *Magazine of Concrete Research*, 20(63), pp. 95-102.

[14] Nielsen, M. 1999. Limit analysis and concrete plasticity. *CRC Press*, LLC, Boca Raton, Fla.

[15] ACI Committee, Structural Building Code, & ACI Committee 318 2005. *Building code requirements for structural concrete (ACI 318-05) and commentary (ACI 318R-05),* 2nd Edition, Farmington Hills, MI:American Concrete Institute.

[16] Al-Sahawneh, E.I., Amjad, A.Y., Hassan, R.H., & Khair Al-Deen, B. 2013. A proposed model at
failure stage to assess the bearing stress of normal weight concrete. *International Journal of Engineering Research and Application*, 3, 793-802.

[17] Zhou, W., Hu, H., & Zheng, W. 2013. Bearing capacity of reactive powder concrete reinforced by steel fibers. *Construction and Building Materials*, 1179-1186.

IOP Conf. Series: Materials Science and Engineering **513** (2019) 012033 doi:10.1088/1757-899X/513/1/012033

Investigating the safety cognition of construction personnel based on safety education

Erita Mazwin Mazlan[1*], Mohd Hanim Osman[2] and Muhammad Sukri Saud[3]

[1]Centre of Engineering Education, Universiti Teknologi Malaysia, Skudai, Johor Bahru, 81300, Malaysia
[2]Forensic Engineering Centre, Universiti Teknologi Malaysia, Skudai, Johor Bahru, 81300, Malaysia
[3]Faculty of Education, Universiti Teknologi Malaysia, Skudai, Johor Bahru, 81300, Malaysia

[*]Corresponding author: eritamazlan@gmail.com

Abstract. Competency in safety is important for construction personnel and it is compulsory for all construction personnel in Malaysia to attend safety training/courses. A literature review of the recommended safety module revealed gaps in evaluating the effectiveness of safety cognition among construction personnel. Therefore, this paper investigates the safety cognition based on safety education. A structured, self-administered questionnaire was designed and used to assess the level of safety cognition in safety education. The results show the safety cognition of construction personnel is still at a moderate level and there is a difference level of safety cognition among construction personnel on Occupational Safety module.

1. Introduction

Safety in construction sites is of utmost importance which makes it fundamental to distinguish the competency level, training and knowledge among Malaysian's construction personnel. This is necessary because incompetency and inadequacy in these areas among construction personnel may increase the risk factors in the occurrences of accidents, incidents, injuries, fatalities and loss of properties on construction sites [1-4]. It is worthy to mention that there are a total of 763 cases of construction accidents filed in Malaysia between 2007 to 2012 and 422 or a concerning 55% of the number involved fatality accidents as recorded by the Department of Occupational Safety and Health Malaysia, Ministry of Human Resources [5]. Therefore, this whopping figure further fortifies the requirement for construction personnel to have adequate safety cognition in construction safety. Safety cognition and the level position of construction personnel are interrelated behavioural displays which affect any form of incidents in construction. Realizing the importance of safety in the construction industry, the National Institute of Occupational Safety and Health (NIOSH) conducts safety education to deliver the necessary knowledge of construction personnel. Upon getting the necessary safety education, construction personnel are expected to have more in-depth knowledge on safety which should be demonstrated in their daily practices at work. This in turn should improve their abilities to identify any possible risks and hazards connected to the workplace.

The sample of the study consisted of 150 construction personnel at various positions such as the Site Safety Supervisor, Safety Health Officer and Safety Health Manager who had participated in safety education. The study uses Snowball-sampling to obtain representative samples. In order to obtain a comprehensive set of data, this study uses quantitative approaches. This method includes surveys with two sets of questionnaires that are self-designed by researchers in the form of an objective questionnaire closed-ended questions with 3 multiple choice options.

1.1. Safety Education for Construction Personnel

Implementing safety education among construction personnel is an important element in promoting a safe working environment. Safety training and education are two important topics in general construction safety. Safety education is a learning process through which construction personnel acquire knowledge, information, cognitive capacities and the transfer of norms, values and modes of behavior. While safety training is a technique for developing human resources through the enhancement of the skills of workers and their ability to cope with workplace hazards. The significant index scores of construction safety risk factors in Australia and China found that "low levels of safety education/training" was the biggest differences scores at 0.22 with the China (0.46) and Australia (0.24) in general terms is a high risks for construction site workers [6]. Construction workers usually rely on the safety cognition of construction personnel to deliver the safety expertise he has learned in safety education. [7] in his study found that education was an essential element for those who hold positions in safety as they were able to create a healthy and safe environment in keeping with the safety knowledge and skills it received. It is without doubt that construction personnel need to have all the necessary knowledge for effective safety practices while getting the required and continuous safety education.

The execution of a safety modules in safety education was developed for construction personnel to foster a safety culture at construction site. The safety modules contain 4 major modules which comprises of OSH Management, OSH Legislation, Occupational Safety and Occupational Health. OSH Management module in which participants are required to acquire an understanding of the application of OSH management system while OSH Legislation module provides specific legal requirements on major considerations of construction safety and health. However, Occupational Health and Industrial Hygiene module, specifically aims for participants to control an environmental factors in the workplace that may cause sickness, impaired health or discomfort will affect to the workers. Meanwhile the Occupational Safety module is to create a safer workplace by providing workers with the knowledge and tools to enable them to realise the fortunes and to exercise safely. By attending this safety education, construction personnel will become a competent person. Therefore, this study aims to investigate the safety cognition of construction personnel based on safety education.

2. Methodology

2.1. Selection of construction personnel

The study uses snowball-sampling to obtain representative samples which involved 150 construction personnel. The participants are Safety Health Officer, Safety Health Managers and Site Safety Supervisors who are of technical background with extensive working experience in construction and are equipped with safety knowledge.

2.2. Questionnaire

A questionnaire-based survey was carried out in this study comprising of 3 major parts. These parts include the following; Part 1 requires the participants to provide their demographic information; Part 2 evaluate the safety cognition of participants towards construction accident; and Part 3 attempts to collect information on the safety cognition of participants in minimising construction accident.

The questions for Part 2 and Part 3 were developed based on a multiple questions because it is the most common and widely used assessment tool for the measurement of knowledge, ability and complex learning outcomes [8,9] and extensively used [10]. Safety modules provided by NIOSH were used as a reference to develop the multiple choice questions. The questionnaire contained 50 close ended questions which include 4 major modules namely OSH Management, OSH Legislation, Occupational Health and Occupational Safety.

Part 2 involves some theory questions to seek the safety cognition of construction personnel towards construction accidents through 14 close-ended questions with two possible answers 'true' and 'false'. For each right answer, one (1) point was awarded. Part 3 attempts to assess the level of safety cognition among the construction personnel through 36 close-ended questions regarding practical work in minimising construction accident. The construction personnel will also be required to organize their answers in the order of their preferences and their scores will be dictated by the preference that they have assigned in their answers. The setting of the individual's acceptance level is referred to response criterion. The correct acceptance is referred to as a hit and will be analysed.

3. Analysis of data

The knowledge score for each participant in Part 2 and Part 3 were calculated by dividing the sum of correct answers by the total number of correct responses. Mean score was used to know the level of safety cognition among construction personnel on the content of safety modules. The Kruskal-Wallis test is a non-parametric test that compares three or more unpaired group and do not require data to be distributed normally. Because of that, to identify safety cognition gaps, Kruskal-Wallis was used to compare the percentage of correct replies to each question across the three types of construction personnel. Values with $p < 0.05$ were considered statistically significant.

4. Results

4.1. Sample characteristics

150 construction personnel participated in this study which comprising of 21% were Site Safety Supervisor, 69% Safety Health Officer, and 10% were Safety Health Manager. This figure is made up of 85% male and 15% female participants respectively. A majority of 50% participants were 26-35 years old while those between the ages of 36 to 45 years old were 23%, 16% of the participants were above 45 years old and 11% of the participants were below 25 years old. It is worthy to note that 60% of the participants had less than 5 years of work experience, with 31% of them had between six to fifteen years of work experience, and 9% had between sixteen to twenty five years of work experience. It is also worthy to note that majority of participants had attending the safety education from National Institute of Occupational Safety and Health (NIOSH) (68%), 5% of the participants from RANACO Education & Training Institute(RANACO), with 11% of them from Industrial Skills Enhancement Programme(INSEP), 4% from KLIA Training & Research Centre and 12% participant were from Melaka Industrial Skills Development and Entrepreneurship Centre(MISDEC).

4.2. Knowledge score within different construction personnel

Table 1 shows the descriptive statistics of the safety modules show the average competence level of safety cognition among construction personnel were at a moderate level of not exceeding mean score 3.66. The data analysis also showed that the highest average score was Occupational Safety module (mean = 3.49) and lowest score was Occupational Health module (mean = 2.72). While for the sub topic, the general area fencing (mean = 4.91) has the highest mean score followed by the indoor air quality (mean = 1.34) with the lowest mean score. The mean score between 1.00 and 2.33 shows the score for the variable sub-component or the variable's dimension is low, while for the mean score between 2.34 and 3.66, the mean score of the variable sub-component or the variable dimension is researched at the medium and the mean score of 3.67 to 5.00 shows a high mean score for the variable sub-component or the variable dimension studied.

Table 1. Descriptive statistics on the main topics and sub topics of the safety modules for construction personnel.

Modules	Mean Score	Competency Level
OSH Management	**3.07**	
Incident prevention	2.43	Moderate
Risk Assessment	4.22	High
Emergency response preparedness & recovering planning	1.43	Low
Incident investigation and corrective action	3.43	Moderate
Auditing	4.63	High
Management Review	2.28	Low
OSH Legislation	**2.80**	
Introduction to OSH Legislation	2.63	Moderate
OSH Act and its Regulations (DOSH)	3.67	High
Regulations on Health in the FMA 1967	1.44	Low
Regulations on Safety in the FMA 1967	4.12	High
Employment Act 1955 and SOCSO Act 1969	3.27	Moderate
Introduction & overview of other related, legislation & codes	1.65	Low
Occupational Health	**2.72**	
Overview of occupational health	2.78	Moderate
Basic principles of industrial hygiene	2.57	Moderate
Classification of chemicals, packaging & labelling	2.09	Low
Chemical hazad management	4.10	High
Ventilation	2.53	Moderate
Indoor air quality	1.34	Low
Mental workload & shift work	3.66	Moderate
Occupational Safety	**3.49**	
Housekeeping and physical arrangement	1.88	Low
Machinery hazard	4.44	High
Safe guarding of machinery	4.07	High
Safe machine operation	4.88	High
Safe of hand tools and portable power tools	4.12	High
Maintenance of machine guard & tools	4.61	High
General area fencing	4.91	High
Transport safety	2.55	Moderate
Manual handling	4.87	High
Mechanical handling	4.57	High
Pressurised equipment	1.38	Low
Construction safety	2.82	Moderate
Fire safety	2.15	Low
Office safety	2.31	Low
Confined space	2.82	Moderate

4.3. Identifying knowledge gaps

In order to test the difference in safety cognition among construction personnel on the content of the safety modules, Kruskal-Wallis test analysis has also been used. The value of $p < 0.05$ is set as the confidence level in determining the likelihood of differences in safety cognition among construction personnel.

IOP Conf. Series: Materials Science and Engineering **513** (2019) 012033 doi:10.1088/1757-899X/513/1/012033

Kruskal-Wallis test analysis in Table 2 shows the total difference level in safety cognition among construction personnel based on safety modules. The findings showed that there was no significant difference level in safety cognition among construction personnel against safety modules, $p > 0.05$ with $p = .175$. Analysis data discussions are as follows: -

4.3.1. OSH management. Kruskal Wallis test analysis for OSH Management Module shows that there is no significant difference level in safety cognition among the construction personnel for OSH Management module. It can be seen through the value of Kruskal Wallis statistical test (Chi-Square) is .28 with a significant level at .868 ($p > 0.05$). Higher mean score was obtained from Site Safety Supervisor's knowledge of 77.44 compared with two other positions namely Safety Health Officer and Safety Health Manager which recorded their mean score was 75.68 and 70.27.

4.3.2. OSH legislation. Results obtained in identifying safety cognition gaps among construction personnel have generally revealed that there is no significant difference level for the OSH Legislation module. It can be seen through the value of Kruskal Wallis statistical test (Chi-Square) is 5.40 with a significant level at .067 ($p > 0.05$). Meanwhile, safety cognition of Site Supervisor recorded a higher mean score of 89.35 compared with two other positions namely Safety Health Officer and Safety Health Manager which recorded their mean score of 70.14 and 84.03.

4.3.3. Occupational health. Safety cognition of Site Safety Supervisor stated a higher mean score of 78.35 while Safety Health Officer and Safety Health Manager which recorded their mean score was 75.75 and 67.83. The Kruskal-Wallis analysis test against the Occupational Health module shows the value of Kruskal-Wallis statistical test (Chi-Square) is .610 with a significant level at .737 ($p > 0.05$). This result means that there is no significant difference level in safety cognition among construction personnel.

4.3.4. Occupational safety. Safety Health Officer recorded a higher mean score of 80.16 compared with Site Safety Supervisor of 71.77 and Safety Health Manager of 50.90. However, the data found in the analysis of Occupational Safety module shows that there is a significant difference level in safety cognition among construction personnel. It was identified that the value of Kruskal-Wallis statistical test (Chi-Square) is 6.26 with a significant level at .044 ($p < 0.05$).

Table 2. The difference in the safety cognition among construction personnel based on safety modules.

Modules	Position	N	Mean Rank	Value of Kruskal-Wallis	*df*	*p*	Total Mean
OSH Management	Site Safety Supervisor	31	77.44	.28	2	.868	
	Safety Health Officer	104	75.68				
	Safety Health Manager	15	70.27				
OSH Legislation	Site Safety Supervisor	31	89.35	5.40	2	.067	
	Safety Health Officer	104	70.14				
	Safety Health Manager	15	84.03				0.175
Occupational Safety	Site Safety Supervisor	31	78.35	.610	2	.737	
	Safety Health Officer	104	75.75				
	Safety Health Manager	15	67.83				
Occupational Health	Site Safety Supervisor	31	71.77	6.26	2	.044	
	Safety Health Officer	104	80.16				
	Safety Health Manager	15	50.90				

Notes : Significant at $p < 0.05$ is set as a difference level in safety cognition

IOP Conf. Series: Materials Science and Engineering **513** (2019) 012033 doi:10.1088/1757-899X/513/1/012033

5. Discussion

This survey provides significantly important information regarding the safety cognition of construction personnel in the construction industry holding different portfolios such as Site Safety Supervisor, Safety Health Officer and Safety Health Manager based on safety education. It is generally assumed that by acquiring safety knowledge, construction personnel can enhance their skills and gain more positive attitude while increasing their safety cognition level in performing their jobs. Construction personnel with high cognitive efficiency can support employees in yielding high quantity of work by appropriately addressing safety related issues.

The descriptive statistics used in this study are aimed to know the level of safety cognition among construction personnel through mean score. Results gathered on the responses among construction personnel showed that many of the construction personnel who participated in the study have moderate knowledge of employees' safety. They need to go through the refreshing safety courses to ensure that their knowledge in dealing with safety at the construction site can be effectively addressed. The results of this study are in line with the study conducted by [11,12] which found that the safety cognition of construction personnel in terms of awareness and behavior is still moderate.

The inference statistics used in this study are aimed to investigate the safety cognition based on safety education among construction personnel through Kruskal-Wallis test. If significant value is less than 0.05, it means construction personnel have a difference level in safety cognition with the content of safety modules. Occupational Safety modules showed a difference in safety cognition among construction personnel on Occupational Safety modules $p < 0.05$. The results of this study are not consistent with the findings of the study by [12] found that the level of safety cognition among construction personnel in Occupational Safety such as reporting unsafe situations in the workplace without doing any action is at a moderate high (mean = 3.96), ensuring competent workers are eligible to perform work (mean = 4.15), perform work according to a safe procedure (mean = 4.02) and able to identify hazard at work (mean = 3.92).

Meanwhile results gathered on the responses from identifying the level safety cognition gaps in OSH Management, OSH Legislation and Occupational Health found that mean score of Site Safety Supervisor knowledge level was high. This demonstrates that Site Safety Supervisor plays the most important role in safety at the site. Hence they need to have a good level of safety cognition. This coincides with a study by [13] found that Site Safety Supervisors are more likely to have a significant impact upon safety, compared to top Managers and Safety Managers.

6. Conclusion

This study concludes that the safety cognition of construction personnel based on safety education are at moderate level. However there is a difference in safety cognition among construction personnel on Occupational Safety module. The results are contradicting to the literature, that the safety cognition of construction personnel are at moderate high in practice safety measures.

References

[1] Argote L 1999 *Organizational learning: creating, retaining and transferring knowledge* (Boston, USA : Kluwer Academic Publishers)

[2] Ben-Simon et al 1997 *A comparative study of measures of partial knowledge in multiple choice tests* Applied Psychological Measurement **21** pp 65–88

[3] Department of Occupational Safety and Health Retrieved 5 March 2014 from http://www.dosh.gov.my/index.php?option=com_content&view=article & id = 795 : occupational-accidents-statistics- 2012&catid=458&Itemid=695&lang=en.

[4] Dingsdag D D P et al 2008 *Understanding and defining OHS competency for construction site positions : worker perceptions* Safety Science **46(4)** pp 619–633

[5] E Gronlund 1993 *How to make achievement tests and assessments* (5th ed. Allyn and Bacon)

[6] Goh S G 2002 *Managing effective knowledge transfer: an integrative framework and some practice implications.* Journal of Knowledge Management **6(1)** pp 22-30

[7] Langford D Rowlinson, S and Sawacha E 2000 *Safety behavior and safety management: its influence on the attitudes of workers in the UK construction industry*. Eng. Constr. Arch. Manage **7(2)** pp133-140

[8] Lingard H Cooke T, Blismas N 2012 *Do perceptions of supervisors' safety responses mediate the relationship between perceptions of the organizational safety climate and incident rates in the construction supply chain* Constr. Manage. Econom **138(2)** pp 234–241

[9] Merwe H Van Der 2015 *Quality assuring multiple-choice question assessment in higher education.* South African Journal of Higher Education **29(2)** pp 279–297

[10] Rosliza Osman Noorhasimah Awang et al 2015 *Level of awareness on behaviour-based safety (BBS) in manufacturing industry towards reducing workplace incidents*, International Journal of Education and Research **3(1)**. Available at: www.ijern.com 1.0

[11] Schindler M and Eppler M J 2003 *Harvesting project knowledge: a review of project learning methods and success factors* International Journal of Project Management **21(3)** pp 19-228

[12] Sulastre Mat Zin and Faridah Ismail 2011 *Employers' behavioural safety compliance factors toward occupational, safety and health improvement in the construction industry* in ASEAN Conference on Environment-Behaviour Studies, Savoy Homann Bidakara Bandung Hotel, Bandung, Indonesia, 15-17 June 2011 pp 742–751. doi: 10.1016/j.sbspro.2012.03.081

[13] X W Zou P and Zhang G 2009 *Comparative study on the perception of construction safety risks in China and Australia* Construction Engineering and Management pp 620–627

IOP Conf. Series: Materials Science and Engineering **513** (2019) 012034 doi:10.1088/1757-899X/513/1/012034

Flood risk reduction initiative: A service learning project by engineering students

Lam Foong Sin[1*], Lee Jer Vui[1] and Ngeow Yeok Meng[1]

[1]LKC Faculty of Engineering and Science, Department of Civil Engineering, Universiti Tunku Abdul Rahman (UTAR), Jalan Sungai Long, Bandar Sungai Long, Cheras, 43000 kajang, Selangor, Malaysia

[*]Corresponding author: lamfs@utar.edu.my

Abstract. This study examined the influence of a small-scale service learning experience on engineering undergraduates. The aim of this study is to test the potential of service learning to develop social responsibilities, soft-skill competencies, engineering design self-efficacy and risk attitude of engineering students. Our results suggested that conducting a service learning flood risk reduction initiative increase learning and help to develop the necessary hard and soft skills to address community needs.

1. Introduction

Disaster education is usually not taught in higher education except in disaster management and healthcare course [1]. The effectiveness of disaster-related subjects in undergraduate programmes other than programmes in environmental studies and healthcare has not been properly evaluated. Post-disaster response and recovery provide an excellent learning opportunities for students [2]. Aftermath of disasters create windows of opportunity for change, promote sustainable development and disaster risk reduction [3]. Sixty four percent of total world population exposed to floods each year live in South Asia [4]. With increased vulnerability due to rapid urbanization, environmental degradation and poor governance in developing nations in this region, it is important to educate our citizens who may be deployed or volunteer to help in the post relief and recovery efforts. Even though natural disasters like flood cannot be prevented, we can reduce the risk of impact of disaster by taking preventive measures. Disaster education and early warning system can reduce the disruptive impact of disaster on communities.

The advancement and convenience of mobile apps for travel purposes, budget-travel tools and attractive mobile deal promotions increase the number of budget youths travelling abroad [5,6]. Asia is an immensely popular travel destination for young, budget-minded travellers but the Asia-Pacific region is also one of the most disaster-prone areas in the world. Therefore this increases the level of exposure of youth visiting disaster-prone Asian countries with frequently occurring natural disasters such as earthquake, tsunami, typhoons and flood. It is through educating youth and students of higher education in disaster risk reduction that students travelling abroad to high risk locations understand the risks they may face and what they need to do to mitigate disaster risks. Students need to be made aware of the need to keep safe, disaster early warning services and evacuation plans as part of disaster risk reduction in the event of occurrence of future disasters. Disaster education also will benefit the disaster-stricken people they may happen to help at the disaster sites as tourists.

The inclusion of disaster risk reduction in educational institutions and school curricula emphasized the implications of risks associated with disaster and create awareness in people to be mindful of policy or development that may lead to the creation of an environment with higher vulnerability to disaster risks. Therefore, schools students need to be taught risk identification and evaluation. Project Learning Tree developed an education model called Focus on Risk which emphasize risk education for students in Grade 9th to 12th using a variety of learning activities and tools [7].

Accreditation Board for Engineering and Technology (ABET) outcomes set forth by Engineering Accreditation Council Malaysia (EAC) include 3 types of skill development and learning outcomes: psychomotor, cognitive and affective domains [8]. One of the outcomes include developing knowledge, skills and attitude to assess environmental impact of engineering projects and risk identifications and apply risk treatments to ensure safety, reduce uncertainties and negative impacts. It is not sufficient to develop intellectual skills in risk analysis and management but to place an equally important emphasis on "risk attitude" [9]. Hilson et. al [10] define "risk attitude" as the "the chosen response to perception of significant uncertainty". Students must be able to sense, be aware of and identify possible risks and evaluate trade off where possible. Higher education must continue to review effective educational tools to provide opportunities for risk management learning and risk attitude cultivation.

Engineering students need to apply critical thinking skills to real world problems and recognize the importance of risk management in engineering design, development of policy and organizational responses and decisions [11]. Therefore, it is becoming increasingly apparent that capacity building and education training in the field of risk management is needed to avoid unforeseen risks and contribute to sustainable development and robust society. For this reason, risk management and education need to be included in higher education courses through which students can be trained as risk managers. To reach a large number of student university-wide, courses that expose students to safety engineering and disaster risks are needed. Learning approaches such as problem-based learning, case study, internship, project based field exercise, service learning and specialized modules by integrating concepts of disaster risks into their existing course and syllabi have been adopted [1,9,12]. In Japan, Osaka University designed a new environmental risk management training to train environmental risk managers by cooperating with other local universities through faculty exchange, credit exchange system, joint seminars and research collaborations [9].

In New Zealand, the University of Canterbury in Christchurch responded to 2010-2011 earthquake by introducing a service learning programme that involved students in community recovery projects [13]. Students from Stanford University and Syiah Kuala University jointly collaborate with the Indonesian Ministry of Marine Affairs and Fisheries to support community rebuilding after the devastating Indian Ocean tsunami 2004 as part of Stanford University international service learning project [14]. The presented examples have involved students provided service to the community allowing students to apply knowledge to solve real life community problems and deepen their understanding of natural disaster risks.

Service learning has been defined as "a form of experiential education in which students engaged in activities that address human and community needs together with structured opportunities intentionally designed to promote student learning and develop reciprocity and reflection are key concepts of service learning" [15]. Exposure and participation in service have the potential to stimulate helpful behavior and increase in social responsibility attitude of students. Service experience in risk and disaster education is able to empower youth to serve and support disaster recovery efforts. Participation in service allow students to have hands-on experience of the impact they can have on others when dealing with local natural disasters and with an increased continued commitment to volunteer in the onset of future disaster and not feel helpless [1].

In this research, we explore whether a small scale (16 students) service learning experience had positive impacts on student development of social responsibilities, soft skill development, risk attitude development and engineering design self-efficacy. It was a learning experience for the engineering students because students are able to achieve one of the course objectives by applying critical and problem-solving skills in designing flood mitigating prototypes for school students. Engineering students had the opportunity to use engineering design process through which hands-on demonstration models were designed adopting the various design stages from identification of problems, conduct background research, develop possible solutions, selecting the best options, construct a prototype, test, redesign and evaluate design solutions and communicating thè solutions to school students [16]. Engineering students also had the opportunity to develop non-technical skills such as communication and teamwork, project management skills and increased sense of responsibilities to serve the society. It was a service experience because students worked with schools to build resilient communities and raise STEM awareness in school students.

Flood hazard in Malaysia contributes to 98.7% of annual average losses compared to other hazard. The northeast monsoon season from November to March brings heavy rain to the east coast of Peninsula Malaysia. In December 2014, prolonged rainfall resulted in disastrous flooding particularly in the East Coast of Malaysia. Thousands of houses and businesses were inundated – characterised as the worst in Malaysia natural disaster – hundreds of thousands were forced into relief centres [17]. He recommended disaster awareness programme be conducted for those affected by monsoon flooding. The United Nations has also called for disaster risk reduction to be included in school curricula around the world to raise disaster preparedness and awareness among students and their immediate family members [18]. For this reason, we conducted flood risk reduction science activities for primary school children because children has the ability to influence disaster preparedness knowledge and behavior of their parents [19]. Researchers have reported children positively influenced their parents in learning basic computer skills, transfer of environmental knowledge, attitudes and behavior [20, 21].

2. Methodology

In this service learning project, engineering students worked with 60 students from Year 5 and Year 6 in a school during school hours at a school Sekolah Kebangsaan Batu Kapur in Temerloh. UTAR students made a) a hydraulic pump to transfer water from low lying location to a higher location b) a watershed and floodplain model c) a floating home. Primary school students collect data to explore flood proofing homes and how to better prepare for flooding. The hands-on science demonstration project as part of the engineering students' community project was designed to develop primary school students' interest in science and other STEM (Science, Technology, Engineering & Math) content area and to prepare students for STEM-related career and to help students better prepare for flooding.

It was a learning experience for the engineering students because students are able to achieve one of the course objectives by applying critical and problem-solving skills in designing flood mitigating prototypes for school students. It was a service experience because students worked with schools to build resilient communities and raise STEM awareness in school students.

2.1. Course and participants

The context for the disaster risk reduction service learning was a soft-skill undergraduate course entitled Community Project. The course objectives are 1) to enhance students' practical experience in organizing and preparing community project 2) to develop social responsibilities skills by expanding network 3) to serve the community with knowledge and skills acquired by students in their curriculum. Sixteen engineering undergraduates out of 120 students from Universiti Tunku Abdul Rahman (UTAR) students registered for the module offered to take up the disaster risk reduction service learning experience.

2.2. Procedures

Students participating in the service learning project attended few sessions of meetings: 1) two 3 hours sessions of lecture facilitated by Community Project lecturer to learn about principles of community development, community asset mapping and project management 2) meetings to discuss the logistics of the project 3) the 3 hours hands-on flood science demonstration and activities in school 4)

IOP Conf. Series: Materials Science and Engineering 513 (2019) 012034 doi:10.1088/1757-899X/513/1/012034

debriefing session facilitated by lecturer in which students were asked to share their service experiences.

2.3. Measures
Five point likert scale survey was used to quantify level of agreement on the achievement of course learning objectives and to assess the impact of the service learning project on development of risk attituded, engineering design skills, social responsibility and soft-skills in engineering students. All 16 students responded. The questions for this study were adapted and based on evaluations of service learning projects at other engineering departments of institute of higher education. Students were asked free response questions and ordinal questions with following responses: "Not at all", "To a small extent", "To a moderate extent", "To a large extent", "To a very large extent" which were coded from 1-5 respectively.

We developed a five-item likert scale measure of based on applying soft skills based on Robles [22], and development of social responsibilities based on Canney & Bielefeldt [10]. We assessed students' belief how well they can use engineering design process using engineering design self-efficacy through five-item likert scale based on Carberry et.al [24]. Self-efficacy refers to an individuals' judgment of their capability to organize and execute courses of action for a given task [25]. The risk attitude survey was adapted based on Carvalho, Dong, Tumer, & Van Bossuyt [26].

3. Results and discussion
The learning objectives were divided into 3 subcategories: applying soft skills, developing social responsibilities, and using engineering design skills in building flood-mitigating prototypes for school students. As summarized in Figure 1-3, the service learning community project has measurable impacts on student attitudes in all three aspects of learning objectives. In addition to achievement of the three course learning objectives, one of the objectives of this project is to cultivate 'risk attitude'. A summary of the data can be found in Figure 4. Out of the four course objective categories, student development in the area social responsibilities and soft-skills appears to be performing better than other areas. See Figure 1-4.

When asked about the service learning experience in the free response question section, students reported the following: 1) "the enthusiasm shown by the school children had made the service learning more enjoyable." 2) "Doing background research of what types of prototypes to be designed and built for hands-on demonstration was a very good experience. I enjoyed learning from others" 3) "meeting the needs of community have brought great satisfaction" 4) "learning new skills to serve the community is fun"

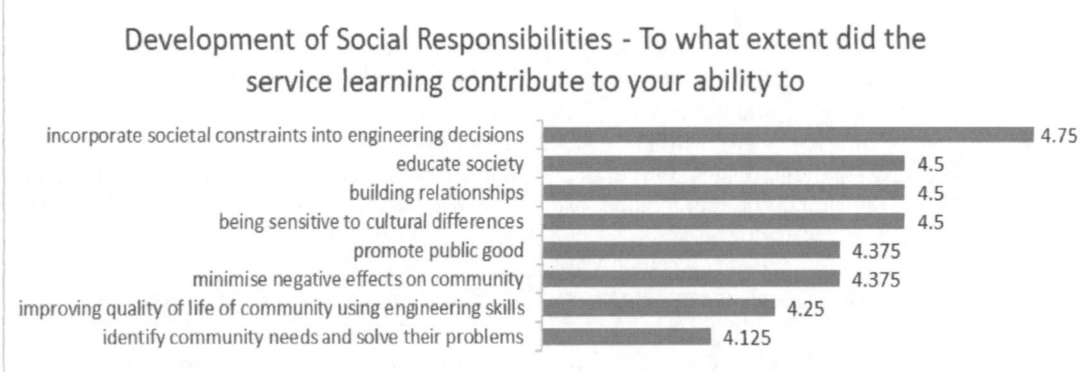

Figure 1. Service learning contribution to development of student social responsibilities.

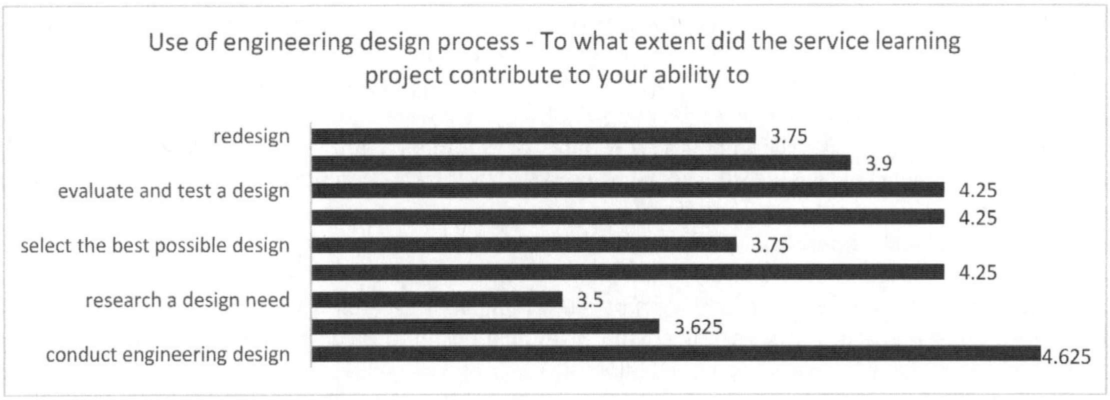

Figure 2. Service learning contribution to engineering design self-efficacy.

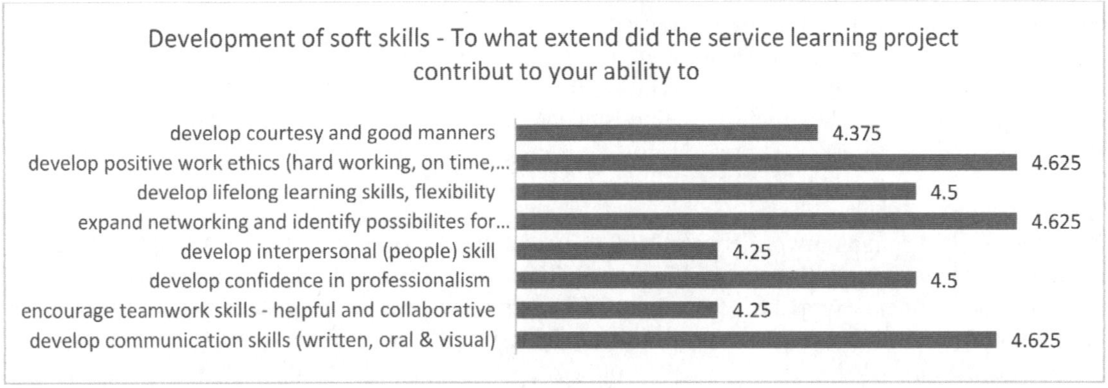

Figure 3. Service learning contribution to development of student soft-skills competence.

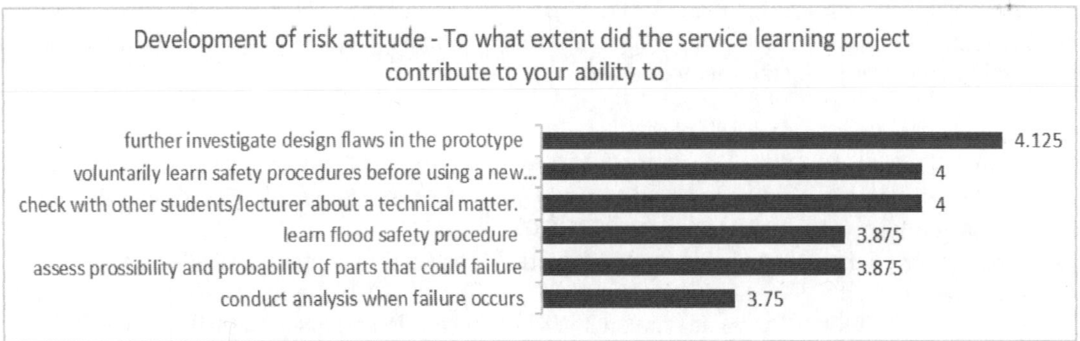

Figure 4. Service learning contribution to development of student risk attitude.

Students participated reported an increase in their social responsibilities' skills and soft-skill competences. This finding is consistent with previous work. Simon & Cleary [27] studied students who had service learning experience with community partners and found that service learning contributed to students' personal and social development [27].

Student cultivation of risk attitude and engineering self-efficacy had the lowest rating. Guntzburger [12] cited that there was a surprisingly low publications on risk management in engineering education. He further commented that it is essential to develop safer engineering practices and improve systems' safety in engineering education. Since students are first year engineering students with little design experience, it is expected that engineering design self-efficacy was low because engineering design self-efficacy is dependent on engineering experiences. Professors and engineers have significantly higher engineering design self-efficacy compared to undergraduates and individuals can build up their self-efficacy through opportunities that provide engineering design mastery skills [25].

We suggest some improvements to the course delivery. Students enrolled in this community project soft-skill based course had undergone a few lectures of community project management but they did not receive training in engineering design process to tackle complex societal problems and sustainability design methods. It is desirable to include the service learning approach into academic curriculum and student research projects involving students and faculty members from different multidisciplinary background. Students can hone their design skills in developing prototypes by interactions with non-academic stakeholders by identifying and researching design needs (lowest score in engineering self-efficacy as shown in Figure 2). Actual design projects in the real world involves a variety of stakeholders from diverse background [28].

Students need opportunities to learn out-of-classroom skills, creativity and problem-solving skills not only during normal times but also after disasters. This project leverage service learning and community project for disaster risk education complementing other service learning for disaster risk reduction initiatives and post disaster recovery.

We find that the service component of the project provided meaningful service to the school students and enhance their interest in STEM subjects and increased flood science knowledge and disaster preparedness. The primary school students were enthusiastic of the university students' efforts.

4. Conclusion

The objectives of community project soft skill course offered at Universiti Tunku Abdul Rahman was to develop student soft skill and social responsibilities. As compared to the usual community social work as practical field experience, we suggested two groups of engineering students to do something different by implementing a service project that teach primary school students disaster risk education. Students gained knowledge and information on hazards, risk and vulnerability while at the same time applying science and engineering to build resilient communities. Reducing environmental risks requires both policy and local individual action. Disaster preparedness is not just the responsibility of the government and disaster managers but also of every member of the community. With a knowledge of risks and vulnerability, young engineers can better prepare themselves to meet societal needs and to volunteer in the aftermath of disaster.

References

[1] Rivera JD, & Miller DS. Disaster Vulnerability Education: A New Focus on Disaster Education Across the Curriculum. *Journal of Applied Security Research*, 2008. 4(1–2), 60–67. https://doi.org/10.1080/19361610802210202

[2] Brundiers K, Eakin, H, Awotona A, & Sarewitz D. Disasters as Opportunities for Change Towards Sustainability. 2016.Retrieved from https://repository.asu.edu/attachments/178527/content/Brundiers_asu_0010E_16622.pdf

[3] Birkmann, J, Buckle P, Jaeger J, Pelling M, Setiadi N, Garschagen M, & Kropp *J. Extreme events and disasters*: a window of opportunity for change? Analysis of organizational, institutional and political changes, formal and informal responses after mega-disasters. Natural hazards, 2010. 55(3), 637-655.

[4] Kafle. Disaster risk management systems in South Asia: natural hazards, vulnerability, disaster risk and legislative and institutional frameworks. *Journal of geography and natural disasters*, 2017. 7(3), 1-10.

[5] Blomgren E, & Ljungström, S. Youth Tourism:–Impacts on places from a consumer perspective. 2018.

[6] Choi K, Wang Y, & Sparks B. Travel app users' continued use intentions: it's a matter of value and trust. *Journal of Travel & Tourism Marketing*, 2018;1-13.

[7] Covitt BA, Gomez-Schmidt C, & Zint MT. An evaluation of the risk education module. *The Journal of Environmental Education*, 2005; 36(2), 3-13.
[8] Aziz AA, Noor MMM, Ali AA, & Jaafar MS.A Malaysian outcome-based engineering education model. *International Journal of Engineering and Technology*, 2005; 2(1), 14-21.
[9] Morioka T, Saito O, Yamamoto Y, Okano M, Yabar H, & Matsui T. New risk management training programs in higher education in Japan - A comparative study and a challenge by Osaka university. In *Journal of Risk Research*. 2007; https://doi.org/10.1080/13669870701342926
[10] Hillson D, Murray-Webster R. Understanding and managing risk attitude. In Proceedings of 7th Annual Risk Conference, held in London, UK 2004 Nov (Vol. 26).
[11] Prokofieva EN, Erdyneyeva KG, Galushkin AA, Prokopyev AI, Prasolov VI, Ashmarina, SI, Kubiatko M. Risk based ecological economics to engineering students. *Eurasia Journal of Mathematics, Science and Technology Education*. 2018; https://doi.org/10.12973/ejmste/80903
[12] Guntzburger, Y. Ethics Empowerment For Better Managing Industrial Risks; Analysis And Experimental Investigations In Engineering Education And Practice. 2017.
[13] Hayward BM. Rethinking resilience: reflections on the earthquakes in Christchurch, New Zealand, 2010 and 2011. *Ecology and Society*, 2013; 18(4).
[14] Suckale J, Saiyed Z, Hilley G, Alvisyahrin T, Muhari A, Zoback M Lou, & Truebe S.Adding a community partner to service learning may elevate learning but not necessarily service. *International Journal of Disaster Risk Reduction*, 2018; 28, 80–87.
[15] Jacoby Barbara. Service-learning in higher education: Concepts and practices. San Francisco: Jossey-Bass. 1996.
[16] Hutchison MA, Follman DK, Sumpter M, & Bodner GM. Factors Influencing the Self-Efficacy Beliefs of First-Year Engineering Students. *Journal of Engineering Education*, 2006; 95(1), 39–47. https://doi.org/10.1002/j.2168-9830.2006.tb00876.x
[17] Omran A, Schwarz-Herion O & Bakar, AA. Factors Contributing to the Catastrophic Flood in Malaysia. In The Impact of Climate Change on Our Life. 2018; pp. 33–55. Singapore: Springer Singapore.
[18] Selby D, & Kagawa, F. Disaster Risk Reduction in School Curricula: Case Studies from Thirty Countries. 2012. Retrieved from https://www.unicef.org/education/files/DRRinCurricula-Mapping30countriesFINAL.pdf
[19] Pfefferbaum B, Pfefferbaum, RL, & Van Horn, RL. Involving children in disaster risk reduction: the importance of participation. *European Journal of Psychotraumatology*, 2018;
[20] Duvall J, & Zint M. A Review of Research on the Effectiveness of Environmental Education in Promoting Intergenerational Learning. *The Journal of Environmental Education*, 2007; 38(4), 14–24. https://doi.org/10.3200/JOEE.38.4.14-24
[21] Williams S, McEwen LJ, & Quinn N. As the climate changes: Intergenerational action-based learning in relation to flood education. *Journal of Environmental Education*. 2017.
[22] Robles, MM. Executive perceptions of the top 10 soft skills needed in today's workplace. 2012.
[23] Canney NE, & Bielefeldt, AR. Validity and reliability evidence of the engineering professional responsibility assessment tool. *Journal of Engineering Education*, 2016; 105(3), 452-477.Business Communication Quarterly, 75(4), 453-465.
[24] Carberry AR, Lee HS. & Ohland MW. Measuring engineering design self-efficacy. *Journal of Engineering Education*, 2010; 99(1), 71-79.
[25] Royalty A. Design-based Pedagogy: Investigating an emerging approach to teaching design to non-designers. Mechanism and Machine Theory. 2018 Jul 1;125:137-45.
[26] Carvalho L, Dong A, Tumer, IY, & Van Bossuyt, D. On Measuring Engineering Risk Attitudes 2011. https://doi.org/10.1115/DETC2011-47106
[27] Simons L, & Cleary B. The Influence of Service Learning on Students' Personal and Social Development. College Teaching, 2006. 54(4), 307–319.

[28] Deininger M, Daly SR, Sienko KH, Lee JC. Novice designers' use of prototypes in engineering design. Design studies. 2017 Jul 1;51:25-65.

IOP Conf. Series: Materials Science and Engineering **513** (2019) 012035 doi:10.1088/1757-899X/513/1/012035

Contractual issues for Building Information Modelling (BIM)-based construction projects: An exploratory case study

Ahmad Huzaimi Abd Jamil[1] and Mohamad Syazli Fathi[2*]

[1,2]Razak Faculty of Technology and Informatics, Universiti Teknologi Malaysia, Level 7, Razak Tower Jalan Sultan Yahya Petra, Kuala Lumpur, 54100, Malaysia
[1]Faculty of Industrial Management, Universiti Malaysia Pahang, Lebuhraya Tun Razak, Gambang, Kuantan, 26300, Pahang, Malaysia
[2]Occupational Safety and Health (OSHE) Unit, Universiti Teknologi Malaysia, Jalan Sultan Yahya Petra, Kuala Lumpur, 54100, Malaysia

[*]Corresponding author: syazli@utm.my

Abstract. Building Information Modelling (BIM) enables close collaboration among project stakeholders to visualize what is to be built in a simulated environment by identifying any potential design, construction or operational issues. Despite the fact of its well-documented benefits, the extensive integration of BIM throughout the project lifecycle remains sporadic. The conventional contracts used in the Malaysian construction industry were established before the development of BIM. As BIM becoming ingrain in the delivery process, the inadequacies of these existing contracts have become visible. This paper presents the results of an exploratory case study using Qualitative Content Analysis dedicated to a BIM project and discusses a procurement approach of BIM to establish the contractual context by addressing the legal and contractual impediments namely; intellectual property, liability and process related risks-allocation in a Malaysian public construction. This paper also outlines the reasons why the implementation of BIM will foster the evolution of integrated contract delivery methods.

1. Introduction

Both positive and negative outcomes have over the years been observed in the area of building construction. Such outcomes have been associated with the Building Information Modelling (BIM). Building information modelling (BIM) promises outstanding results of the construction process. The results can be determined through the measurement of the yardstick and an international benchmark for efficiency in Architectural, Engineering, and Construction (AEC), and host of other building services [1,2]. The process of BIM can enhance knowledge sharing for information of a building or facility. It forms a strong fundamental for decision making throughout the life cycle phase from the conceptual design to demolition [3,4]. BIM is a useful multidisciplinary integrated source of information technology that posts benefits and disputes in the construction industry. This model is well recognized as fragmented, adversarial and possibly significant for or during the process of cultural revolution [2,5]. The legal issues of BIM are considered challenges that need to be addressed accordingly. One of the issues is the need to establish contractual arrangements to accomplish BIM positive outcomes [5]. The current laws, regulations and the contractual model lack relevant provisions for the integrated design and the information versions of all stakeholders collaborated in the construction project could not cooperate extensively [3,4]. It has significant impacts on how project stakeholders collaborate and the legal implications that can emerge from any disputes.

BIM has become the potential risk to construction projects contracts because there is no legal status of electronic information and electronic versions could not be filed, which results in duplication of information gathering work [4]. It is essential to form the viable approach of contract document by outlining the goals of implementing BIM and the way it benefits a project [5,6]. The outline of this paper is as the followings; section 2 aims to provide a theoretical background that are minimized administratively or contractually relating to BIM procurement. Section 3 establishes the research methodology that details out the significant aspects of the proposed research. Subsequently, section 4 analyses and synthesizes the case study supported by the literatures towards developing a new framework and strategy to contract delivery may necessarily need to expand.

2. Theoretical background of BIM

BIM is usable for multiple purposes, which requires the source of interrelating information by providing the structure, format and relevant content [7] where the model and the data can be extracted and reused to prevent rework or the need to reproduce work. Kuiper et al.[5] indicates that there appears probable risk scenario for owners' responsibility in building/infrastructure assets with regards to traditional contracting delivery methods and to be applied at each stage of the built environment cycle. Alternatively, an industry may differentiate the terms between 'Design Intent BIM' (known as Design BIM) and 'Construction BIM'. In particular, 'Design Intent BIM' encompasses the information commonly originated by designers or consultants, while 'Construction BIM' incorporates the mechanical, architectural and structural details for assembly and installation as prescribed by trade contractors [2,3].

2.1. Comparisons between AIA E202, consensus docs 301, NSW Government GC21 construction contract and contract for complex projects (CPC2013)

Based on the literatures reviewed, there are similarities between the two BIM contracting forms (AIA E202 and ConsensusDOCS BIM 301 addendum). The documents are among the widely used standard form contracts in the global construction industry. The forms have provisions for model ownership, copyrights, information and model management, and model standards are also addressed. However, the contents and the level of details are different. McAdam [4] justifies that the documents published by the American Institute of Architects (AIA E202) contains less detail than ConsensusDOCS 301 in many elements. For instance, both forms require an information manager (or model manager), but "its obligations are delineated in more detail" in the ConsensusDOCS 301 [4]. Some importance differences identified between the two standard forms are as follows:

- ConsensusDOCS 301 considers a liability waiver for using the model information, and it provides licenses to use the intellectual property. AIA E202 separates the process of using BIM from contents of models. In the process portion of the document, the process and responsibilities for model ownership, the model standard, model management, conflict coordination and model archiving are clearly explained in the protocol.
- Levels of Detail (LOD) are defined for the modelled elements in handling the model substance [5,8].
- In ConsensusDOCS 301, project participants have to develop a BIM Execution Plan (BEP), but AIA E202 does not have such a requirement [1,4].

Meanwhile, the New South Wales GC 21 2nd edition contract document represents a typical construction contract and is widely used in New South Wales, Australia for educational, corrective services and infrastructure projects. It is a standardized contract that has been developed in consultation with a wide range of industry stakeholders including contractors, subcontractors, suppliers and client groups [9]. However, from the findings of the literature, a number of recommendations are tabled that address the perceived legal and contractual issues associated with BIM implementation [4,5]. In a similar vein, the Chartered Institute of Building's Complex Projects Contract (CPC 2013) have been suffered from a release date prior to the United Kingdom's publication of key documents to support BIM and consequently, key documents are not present in the contract[10]. Far too little attention has been paid that this novelty in BIM implementation within global and Malaysian context raises distinct legal and contractual issues, the notion of multidisciplinary collaboration related to intellectual property (IP) from the stance of BIM are depicted, for example, design and digital IPs, especially those models build with consolidation [5]. Meanwhile, comprehensive considerations of BIM's legal issues in contract documents namely; security of data confidentiality, safeguarding data and liability related to IP issue are strongly urged, particularly for an equal and seamless information dissemination, due to the complexity of integrating BIM-use regulations and terms within projects [11,12]. These issues could lead to uncertainty that hinders effective collaboration among project participants from early design and continues through facility management (demolition and reconstruction).

2.2. Implications of BIM on procurement
Procurement can be described as an organizational setting, that is vital in managing the designing process of a constructing a project. In most organizational forms, it is widely employed to facilitate building demands, dispersion of roles, jobs, and risks [13]. Further investigation, in terms of the benefits and drawbacks of BIM were observed by Holzer [14] by means of employing procurement approach in Australia. It was argued that the implementation of BIM was best fit particularly for integrated project delivery (IPD) procurement model. The characteristics of BIM in several contract procurement practices adopted in the industry are delineated in the following subsections [5,8,14].

2.2.1. Construct only
The 'Construct Only' approach is the commonly utilized contract papers in regulating projects within the construction industry, which appears to distinguish the roles from risks. Even though it does not represent the best mode of communication between BIM contractors and facility management (FM) consultants, the 'design-bid-build' method employed in this type of procurement helps add to the aspect of competition in bidding.

2.2.2. Design and construct (DnC)
Clients have the tendency to request for lower risks by means of appointing a Head Contractor who will supervise the designing and building phases. Consequently, The DnC is the most well known approach in medium and magnate projects. Moreover, DnC is beneficial due to the transparency of data flow as the specialists can trail their BIM models that are employed by others. Nevertheless, it has to be noted that there is no guarantee of flawless BIM transition from consultants to contractor or to others.

2.2.3. Managing contractor / ECI
Among the allied project players, the opportunities for BIM collaboration are relatively higher with regards to the method of 'Managing contractor / Early Contractor Involvement (ECI). It is also worthy of note that at the early conceptual design, such an approach integrates dual phases and contractors will supervise the costing matters.

2.2.4. Public private partnership (PPP)
The PPPs refer to project finance and not as a method of delivery. The PPP is widely adopted for massive projects involving infrastructure that is similar to the DnC approach through the integration of designing and constructing services in the contracts, together with the FM services. Therefore, BIM is employed to collaborate in reducing the lifecycle costs towards maximizing the project sustainability [5].

2.2.5. Alliancing / integrated project delivery (example: Project Alliance)
Based on the contractual correlation concept, numerous studies were identified that Alliancing or IPD enables displaying smart digital data sharing amongst all project players owing primarily to the elimination of legality. These relational contracts are comprised of procurement, which is communal, including impartial benefits and drawbacks of project stakeholders for which data sharing is considered as transparent throughout all parties.

IOP Conf. Series: Materials Science and Engineering **513** (2019) 012035 doi:10.1088/1757-899X/513/1/012035

3. Research methodology

The focus of this paper is on BIM contract documents conceptualization and implementation. Databases were searched for construction projects articles using the key words, 'building information modelling (BIM)', contracts, legal risks, issues, and procurement with further subcategories. The inclusion criteria of studies were searched in the following electronic databases namely: Elsevier, Emerald insight and ACM Digital Library. Every study was independently peer reviewed and its suitability was determined with the agreement of acceptance standards. The literatures were analysed to identify the implications that are practical for BIM execution, especially upon developing a contract form for BIM-based construction projects.

This case study relied on a Malaysian public construction project procured under a design and build (D&B) contract with the consultants being hired by the main contractor and the contractor holding the single contract with the client. The collaboration in BIM-based construction networks from contractual aspects was evaluated to enhance the project lie cycle throughout the organizational. The first step of the empirical research consists of a review of the project documents the project team made available to the researchers. The reviewed documents include the project's BIM execution plan, the project organization chart, and the appendices dedicated to the BIM processes. Some other documents related to the BIM modelling activities have also been reviewed by the researchers. The exploratory case study is performed with aims at extending the understanding of BIM-based contracting that are considered as complex. It is used as "a sound and sensible first step" when extensive empirical research has not been yet dedicated to the topic of interest [15]. The aim of the study reported in this paper is not to generate final or definitive evidences, but to suggest some hypotheses to be investigated later, in future works. The investigation adopted an exploratory case study strategy using Qualitative Content Analysis as a means of generating recommendations for amending contracts for BIM implementation rather than testing some form of theory for BIM implementation in the contractual and legal context. The framework for coding in a content analysis uses the thematic topics derived from the literature review, which is comprised of a hierarchical order of domains (thematic distinctions) and subcategories, as delineated in section 4. Besides, the framework of coding is driven by data, where domains will be added while performing the coding process based by using Nvivo 12 on the extent of the definitions attributed to the classifications.

Following the project documents review, semi-structured interviews were then carried out to deepen various aspects of the implementation of BIM in this project. The aim was to confront the project actors with certain inconsistencies between the planned practices (from the project documentation). A total of six project participants were interviewed. The topics discussed mainly revolved around the organization of work in the project, the management and the sharing of information in the BIM processes, related contractual challenges, and the potential individual recommendations each interviewee could provide. Moreover, the key aims and objectives of this paper are to provide a better understanding of BIM to conceivably evolve the construction industry's avenue of approach to the current challenges of contract delivery methodologies in both theoretical and practical aspects. The challenges involve people, policy, process, and technology that determine the stakeholder roles, their information rights and liabilities, their BIM access (read, write) or their obligation to provide a special functionality or data output [16].

4. Findings of the case study

Besides considerable benefits of BIM for project stakeholders, there are several risks and barriers to implementing BIM. However, for the purpose of this study, these hindrances are categorized according to specific aspects namely; BIM legal and contractual impediments, intellectual property (IP), liability and process-related risks. The results include a conceptualization of typical barriers to collaboration in BIM-based construction networks and how to deal with the identified barriers. Among the findings of the exploratory case study, the trade contractors were asked how much they think BIM has contributed to their works. The project participants acknowledge the added value of BIM in reducing the average delay of a construction period, and the additional construction cost. This exploratory case study revealed the effective digital collaboration in construction through the case study of a BIM-based hospital construction project. A set of significant key factors namely; legal and contractual impediments with BIM implementation, intellectual property (IP), liability, process related risk allocation were identified in promoting digital collaboration, including change agents, BIM contracts, a cloud computing infrastructure, and new roles and responsibilities different from the traditional ones.

4.1. Legal and contractual impediments with BIM deployment

From a theoretical point of view, the BIM execution plan places a particular emphasis on the responsibilities of the BIM managers. These responsibilities are mainly related to the management of the models' content, quality control and 3D coordination. In practice, the architecture firm and the MEP (mechanical, electrical and plumbing) firm have designated dedicated actors (other than the project manager) to hold this role in their organization. The BIM manager designated by the structural engineering firm seems to have a more general role that can be likened to that of a project manager. He is assisted by a BIM coordinator who is responsible for internal BIM model management. In practice, the BIM managers' roles are not really similar from one discipline to another. The BIM manager designated by the architecture firm has both technical and managerial roles. He is responsible for the weekly upload of the architectural models and their integration with the other models. He defines in advance the elements to be checked for intra- and interdisciplinary clash detection. However, the unclear contract clause regarding such the design professional could recognize when information needed to be preserved was perceived as having a significant impact by the interviewed project participants. Significant efforts have been devoted on the implementation of BIM in the AEC sector to boost the development of Information and Communication Technology (ICT) in Malaysia and other countries [17]. However, Steward et al.[16] stresses that many industry players are unable to gain the benefits of BIM besides technological and cost hurdles from the context of BIM legal issue [1,18]. Research on BIM legal challenges has focused on two primary aspects. The first aspect is to identify the legal issues [1,18,19] and potential challenges in the construction industry. The second aspect is to develop a procurement framework for delivering a BIM in the light of collaborative work-practices [1,5]. Nevertheless, the time-lapse of previous research is quite long where the current contract documents fall off behind technological advancements that do not extensively consider the interrelation between legal issues and the procurement approach faced by the project stakeholders [4,5,16]. Therefore, there is the necessity and feasibility of establishing contracting document management system that allow the deployment of an integrated BIM.

4.2. Intellectual property (IP)

The protection of intellectual property is crucial for project stakeholders to identify and define in any event that requires consideration to regulate the ownership of IP (and any licensing arrangements, either actual or implied) in a BIM contextual setting [5]. Particularly, for this case project, the client regularly has access as it is being developed. However, the client of the project confirms that there is no explicit contractual provision on how extensive care must be taken into consideration to ensure that the intellectual property rights are not obscured due to the transparent and collaborative setting of model development. To distinguish IP in the model, areas that need considerable attention include element parts, trade secrets such as construction techniques and sequencing embedded data or databases, file formats in an integrated system, and private information [4,18]. Given that the demand for BIM is increasing, Azhar et al. [3] asserted that the development of more sophisticated platforms is imperative to enhance the level of achievement by delivering IP in a collaborative environment, then it increases the complexity of IP allocation. As far as the authors are concerned, significant efforts have proposed the need to define IP in integrated BIM models and the virtual collective working environment. Due to a lack of determination of ownership of the BIM data, the protection of ownership is needed regarding copyright laws and other legal channels [3,5]. When BIM is adopted by the project stakeholders or contracting parties, it is essential for them to develop specific agreements or contractual provisions to address any contractual interfaces, BIM-related requirements, and the limitations of contractual documentation in the current Malaysian market. The best solution for the

IOP Conf. Series: Materials Science and Engineering **513** (2019) 012035 doi:10.1088/1757-899X/513/1/012035

disagreement over copyright issues is to set forth the documents ownership rights, responsibilities, authorized users, and sensitive information in integrated project delivery (IPD) [2,5,11].

4.3. Liability

Professional Indemnity (PI) is a primary concern for design authorities who are involved in a BIM collaborative atmosphere [19]. These authorities are obliged for the design contributions to non-professionals, including automatic changes by software. The promising multidisciplinary platform where information is provided by an author or BIM contributor (as a recognized professional) and adopted (or relied upon) by other BIM contributors or another party were acknowledged as part of a process related risk [3]. The issue could arise in contract negligence (either as a duty to a client, contractor/consultant or to a third party) which was not foreseeable [5,18]. Moreover, issues related to the quality control processes were notified by multiple practitioners. For the structural engineering firm, too many quality control processes described in the execution plan that were not adapted to their situation. They are required for an effective quality control as described would take about four weeks, which is totally unrealistic which such a fast track project where the models were continually evolving. The participants adopted the collective idea of advancing the models and performing interference detection at planned moments when the appropriate elements had been modelled properly. Another aspect of quality control is related to the contractor who, according to the execution plan process, had the responsibility to audit the models and their compliance to the Uniformat standard. This process, planned to be managed by the contractor, raised some contractual liability questions. While it was not possible to validate these assertions during the case study, some stakeholders suggested that BIM managers should be responsible for interdisciplinary quality control. Therefore, it is necessary to develop collaborative and integrated contracting methodologies to avoid disputes among project teams involved. Hence, future research should highly consider the design delegation to non-professionals and subcontractors, as well as the effect of software on design updates.

4.4. Process-related risks allocation

Legal, contractual and organizational risks are different forms of risks associated with creating, using and managing BIM-related information. "It is a great effort to have outlined the processes to follow but they are natural and too obvious. [...] Ideally, it would have been useful to show the information requirements at some given moments. Here, nothing is detailed enough", said a BIM manager. Beyond the formalization issues, one of the main criticisms practitioners made about the execution plan is that it does not take into account the particular need in a fast track project for different disciplines to coordinate with each other. "We tried to apply the processes proposed in the BIM execution plan. But the fast track project is not linear with only 60% similarity" noted a BIM coordinator. During the semi-directed interviews, all the interviewed participants acknowledged that the processes recommended by the execution plan were too theoretical, too general and were not adapted to the project. "The processes in the BIM execution plan remain very theoretical, and there are many project participants are not knowledgeable with these documents to work on a daily basis, despite the fact that the recommended processes show the main principles of the project", said a BIM manager. There was a sense of understanding the BIM execution plan specification amongst interviewees whereby the proposed processes are seen as difficult to apply and not close enough to the project's reality. It is encouraging to compare this issue with a study conducted by Lee et al. [20] who suggested that the importance of coordination for safeguarding provisions which could be pre-determined within the timeline of BIM deliverables. Such provisions provide clear procedures, processes, and timing of submittals, and determine the roles of the design coordination manager and the data security manager in establishing the model security and BIM data-base.

5. A way forward for BIM-based contracting strategy

Upon exploring and reviewing the significant challenges that are confronting all of BIM-based project stakeholders, the question arise, "What needs to be done?" In this study, various aspects, which are of interest in terms of contractual issues with regards to specific BIM functionalities and contract delivery procurement methods and recommendations for guiding BIM management process were explained. According to the integrated project delivery (IPD) system, the existence of project procurement systems from the practical, technical and industrial advancements alter the project's delivery to implement effective BIM [5,6,16]. American Institute of Architects (AIA), the Associated General Contractors of America (AGC) and Consensus Docs work on respective contractual guidelines and publish contract samples [6,20,21] in enabling BIM project stakeholders (including contracting parties) to plan, design, review, program, cost or manage projects. However, the standard practice of International Federation of Consulting Engineers (FIDIC) contract documents in Malaysia, the contractor constructs the works; designed, civil, mechanical, and electrical and construction works following the design provided by the employer [22]. Nonetheless, issues relating to legal uncertainties in BIM implementation and fee specifications in the local and international construction industry remain unresolved. Since many processes in FM and deconstruction are not aligned with BIM, there are no specific contracts developed and standardized for integrated practices to date [11,16,17].

Although the updated BIM content is crucial for maintenance, retrofit, and deconstruction planning, the onus of the model and content management during the maintenance has not been addressed by previous studies and legal frameworks [3,5]. The rapid development of new technology combined with the lean principles have also guided the project teams to eliminate waste during the construction process [4,5]. Therefore, it is essential to develop new approaches to and strategy for BIM contractual document frameworks. When these tools are used properly, they can change the way an owner makes a contract with a project team.

6. Conclusion

Based on an exploratory case study, this paper explored some obstacles when deploying the integrated BIM contract documents. Several elements such as legal and contractual impediments, intellectual property, liability and process-related risk should be considered in the current projects that are delivered for contractual and administrative purposes, which are expected to align with the optimal use of BIM. It is worthy of note that it depends on the dimension being considered for investigation. For instance, the primary focus can be either on the possible influence of an approach on the BIM technology, which features transformative power. Such a power may not only impact on the design process but on the construction process as a whole. It may also equally impact on the pull of a transformative approach to management to employ this design technology for transcending current constraints for performance. The applicable approach is vital as solution of the perceived fragmented and isolated approach in the current construction practices. This may also represent a new frontier for innovation in the construction industry of Malaysia. In fact, the integrated BIM contracting provides significant opportunities that are yet to be discovered in this particular industry.

Acknowledgment

This work was financially supported by Universiti Teknologi Malaysia Fundamental Research Grant Scheme (UTM-FRGS Grant No. 4F951) under the Ministry of Education, Malaysia.

References

[1] Ashcraft, Howard W. 2008. "Building Information Modeling: A Framework for Collaboration." *In Society of Construction Law International Conference*, **28** (5).

[2] Chen, Ke, Weisheng Lu, Yi Peng, Steve Rowlinson, and George Q. Huang. 2014. "Bridging BIM and Building: From a Literature Review to an Integrated Conceptual Framework." *International Journal of Project Management* **33**(6). Elsevier Ltd: 1405–16.

[3] Azhar, Salman, Malik Khalfan, and Tayyab Maqsood. 2015. "Building Information Modelling (BIM): Now and beyond." *Construction Economics and Building* **12** (4) 15–28.

[4] McAdam, B., 2010. Building information modelling: the UK legal context. *International Journal of Law in the Built Environment* **2**(3) 246-59.

[5] Kuiper, Ilsa, and Dominik Holzer. 2013. "Rethinking the Contractual Context for Building Information Modelling (BIM) in the Australian Built Environment Industry." *Australasian Journal of Construction Economics and Building* **13** (4): 1–17.

[6] Steward, R.A. and Mohamed, S., 2003, "*Integrated Information Resources: Impediments and Coping Strategies in Construction*"; The Australian Centre for Construction Innovation, University of New South Wales, Sydney.

[7] Wang, Xiangyu. 2012. "*BIM Handbook: A Guide to Building Information Modeling for Owners, Managers, Designers, Engineers and Contractors*." Australasian Journal of Construction Economics and Building **12** (3): 101.

[8] Aibinu, A. and Papadonikolaki, E. (2016), "BIM implementation and project coordination in design-build procurement", *in Proceedings of the 32nd Annual ARCOM Conference*, 5-7 September 2016, Manchester, UK, pp. 15-24.

[9] NSW Department of Commerce, 2012 Overview and summary of the specific GC21 provisions.Sydney: NSW Government.

[10] Gibbs D.J., Emmitt, S.B. and Ruikar BArch, K., 2015. BIM and construction contracts – CPC 2013's approach. Institution of Civil Engineers Publishing, 168(Level 0), pp.285–293.

[11] Abd Jamil, A.H. and Fathi, M.S., 2018. Contractual challenges for BIM-based construction projects: a systematic review. *Built Environment Project and Asset Management*, pp. 1-15.

[12] Hafsi, I., 2017. Building Information Modelling Impact on Contracts 1. PM World Journal, VI (Xii), pp.1–8.

[13] Patil, N.A. and Laishram, B.S. (2016), "Sustainability of Indian PPP procurement process", *Built Environment Project and Asset Management*, **6**(5) 491-507.

[14] Holzer, D. (2015), "BIM for procurement-Procuring for BIM", paper presented at 49th International Conference of the Architectural Science Association: *Living and Learning: Research for a Better Built Environment (ANZAScA 2015),* 2-4 December 2015, Melbourne, Australia, pp. 237-246.

[15] R.T. Ogawa, B. Malen, Towards Rigor in Reviews of Multivocal Literatures:Applying the Exploratory Case Study Method, Review of Educational Research. 61 (1991) 265–286.

[16] Smith, Peter. 2014. "BIM Implementation - Global Strategies." *Procedia Engineering* **85** 482–92.

[17] Becerik-Gerber, Burcin, Farrokh Jazizadeh, Nan Li, and Gulben Calis. 2012. "Application Areas and Data Requirements for BIM-Enabled Facilities Management." *Journal of Construction Engineering and Management* **138**(3) 431–42.

[18] Larson, D. and Golden, K., 2007. Entering the brave new world: an introduction to contracting for building information modeling. Mitchell Law Review, **75**, pp.75-108.

[19] RIBA. (2013) RIBA Plan of Work Overview. London. Retrieved August 2013,

[20] Lee, C.Y., Chong, H.Y. and Wang X., 2018. Enhancing BIM Performance in EPC Projects through Integrative Trust-Based Functional Contracting Model. *Journal of Construction Engineering and Management*, **144**(7) 1–6.

[21] ConsensusDocs, ConsensusDocs Guidebook, 2013.

[22] Zakaria, Zarabizan, Syuhaida Ismail, and Aminah Md Yusof. 2013. "An Overview of Comparison between Construction Contracts in Malaysia : The Roles and Responsibilities of Contract Administrator in Achieving Final Account Closing Success." *Proceedings of the 2013 International Conference on Education and Educational Technologies (EET 2013)*, July16-19, 2013, Rhodes Island, Greece, 34–41.

IOP Conf. Series: Materials Science and Engineering **513** (2019) 012036 doi:10.1088/1757-899X/513/1/012036

A review on wooden formwork for concrete casting

Mohamad Shazwan Ahmad Shah[1*], Norhazilan Md. Noor[1], Ahmad Beng Hong Kueh[1] and Mohd. Nasir Tamin[2]

[1] School of Civil Engineering, Faculty of Engineering, Universiti Teknologi Malaysia, 81310 UTM Johor Bahru, Johor, Malaysia
[2] School of Mechanical Engineering, Faculty of Engineering, Universiti Teknologi Malaysia, 81310 UTM Johor Bahru, Johor, Malaysia

[*]Corresponding author: mohamadshazwan.ahmadshah@gmail.com

Abstract. This paper discusses about the advantages of using wooden formwork to cast concrete. The type of wood focused in this paper is BBCC plywood. The discussion on the wooden formwork arises when it is being compared to the steel formwork. The fact that steel-type formwork has bigger virtues over wooden formwork is undeniable. However, it should be understood that in making decision and judgement on formworks application, one must consider few important factors like cost and suitability. Wooden formwork has its own uniqueness and by some modifications, wooden formwork could be as good as steel formwork or even better from it. The advantages of using wooden formwork, specifically plywood-type will be explained from three perspectives; the cost, performance, and safety of wooden (plywood) formwork. Based these three perspectives, the paper presents positive arguments and facts to accentuate the wooden formwork if to compare to the steel formwork, with regard to the situation and environment of their usage. Based on the discussion in the paper, wooden (plywood) formwork is as dominant as steel formwork, especially if it is used in research and laboratory environment.

1. Introduction

Formwork functions as a mould by casting the concrete into the anticipated structural design in the project [1]. After a certain period of time depending on the duration of concrete gaining a minimal strength to withstand, the formwork will later be disassembled [2]. Similarly, industries use formwork to fulfil specific requirements of each project. Formworks are made up of different materials depending on several potent factors in construction such as cost, time, and quality of the concrete components. As long as the formwork criteria are fulfilled, then the formworks are considered operational. The basic criteria that need to be satisfied for formwork are quality or buildability, safety, and cost [3,4].

In some sense, material of formworks chosen should be suitable with the projects or jobs to be catered. For example, wooden formworks are suitable to mould smaller structure and light-weight. Wooden formwork costs much lower than steel formworks. Steel formworks on the other hand although they are costly but they could be re-usable for many times. However, in construction industry or in research, cost really plays an important measure in a successful project [3].

Thus, in this paper the discussion will generally discusses on the advantages from the perspective of cost and quality of concrete components using wooden formwork, in term of academic research.

1.1. Timeline in developments of formwork

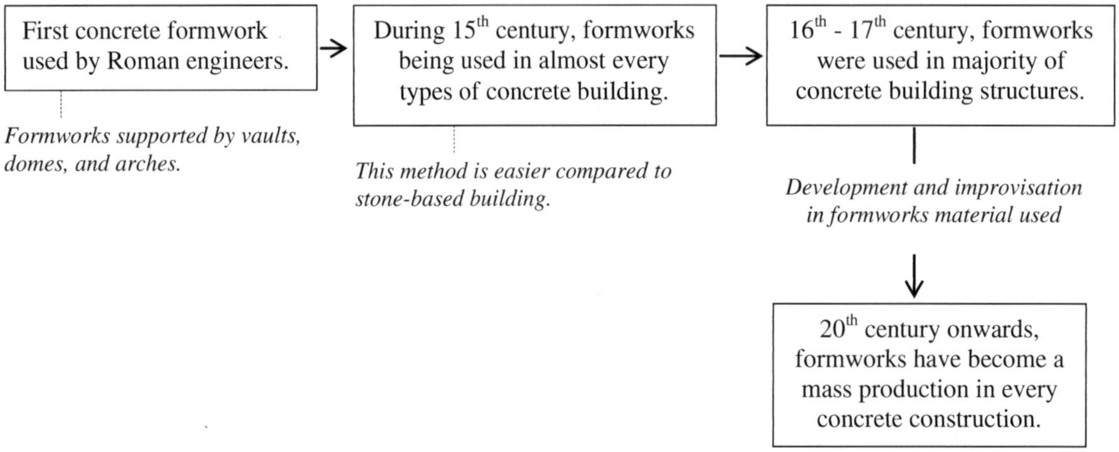

Figure 1. History of formworks [5].

2. Steel formwork and wooden formwork review

Materials used to make formworks are dependent on several factors as mentioned earlier. Thus, in this paper only two most used materials will be layout – steel and wooden formworks.

Table 1. Advantages and disadvantages of steel and wooden formwork respectively.

Steel Formwork	**Wooden Formwork**
When properly maintained, makes steel formwork an effective long-term solution for construction needs [6,7].	Cheaper in the short term [6].
Steel formwork does not warp or absorb moisture from poured concrete [6].	For larger or more time-consuming jobs, wood is a less ideal choice due to its tendency to warp, shrink, and swell [6,7].
High ability to carry heavy load [7].	Must be braced safely and effectively, which can be problematic on larger jobs sites [6].
Easy to be fixed [7].	Time spent cutting and drilling wood forms can impact labour costs, which can offset the lower cost of the material [6].
Uniform size and surface [7].	wooden formwork can be used only five to ten times before it needs to be discarded and replaced [6].
Limited size or shape.	Easy handling because it is light weight [7].
Excessive loss of heat.	Easy to disassemble [7].
Smooth surface would give problems for finishing process.	Damaged parts can be replaced with new one[7].
High strength [7]	Very flexible [7].
	The cost of wooden formwork proven to be cheaper in certain cases [8].

IOP Conf. Series: Materials Science and Engineering **513** (2019) 012036 doi:10.1088/1757-899X/513/1/012036

Disadvantage of wooden formwork absorbing moisture might be actually an advantage to the moulded concrete. When the concrete is hardening, at the same time the moisture is being absorb by the wooden formwork minimally, it might prevent bubbles and void to form which actually the voids will weaken the concrete. However, as one knew about the weakness of using wooden formworks, the wooden formwork could be covered by cellotape making it as good as water-resistance material. Again, covering the formworks with cellotape is workable for smaller size of concrete especially in research works. It will be not practical if the covering method used in industry involving with bigger concrete structures.

For research purposes which normally uses smaller specimens, it is wiser to use wooden formwork for more cost-effective. Mentioning about cost-effective or reducing cost, not only the material itself has lower price – but say if the dimensions of the beam differs from one batch to another, thus the previous formworks can be used by simply add a wood-piece and screw or nail them together so that the formworks sized as needed. On the other case, if steel formworks are quite costly to fabricate few dimensions steel formworks. On top of that, the cost will be even higher if the order is just to cater small amount of concrete mix.

In research, normally the size of concrete specimen can be scaled-down based on proven facts and the testing specimens are not as many as the real building structure components. Thus, as mentioned steel formworks are more suitable and effective in term of cost and reusability for large projects and construction [9]. Meanwhile, as for academic research purposes, to cast concrete, wooden formworks are good enough.

3. Benefits in Using Wooden Formwork to Cast Concrete

3.1. Costing of wooden formwork

Formwork might consumes a huge percentage of total cost of a project – the cost of formwork itself sometimes might be more than the cost concrete and steel bar combined [3,10]. From a client perspective, having a lower total project cost with higher quality are always preferable. But there are always three parameters which influence the profile of a project; time, cost, and quality or performance [11] – it could be shown in a triangle below:-

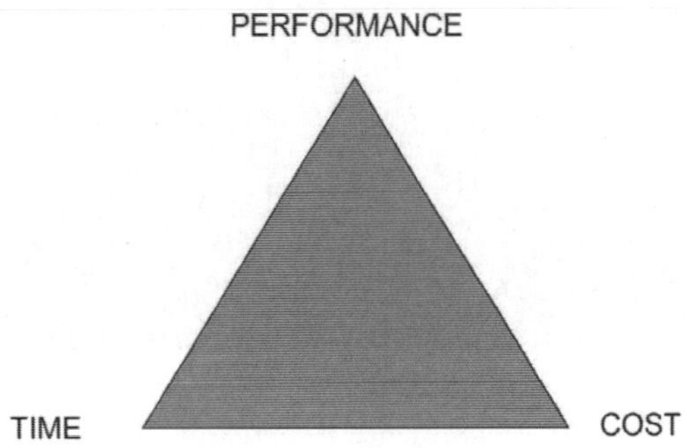

Figure 2. The project triangle [12].

In simple words explaining the "Project Triangle", if the project needs to be fast and cheap meaning that reducing time and cost respectively, the quality or performance might not be that promising. On the other hand, if the project wants to have a quick completion with good structures which also means decreasing time and higher performance, the cost will eventually be higher or vice versa [13].

There are few methods in reducing cost of formworks – one of it is considering the minimum size and least weight of permanent concrete structure. Although formworks were meant to support the temporary concrete structure after casting, this methods somehow jeopardize the permanent structure of concrete [3]. Scaling down the concrete structure (although the strength was proven high) does have critical arguments among experts. Generally, the size of concrete components should not be simply subsidies and diminished because there were numerous researches testified changing the concrete specimen size may results instability in strength [14,15,16]. Henceforth, the cost should not be decreased by risking the size or concrete used in structures [17,18].

Discussing on formwork costing, it could be divided into three which are (i) cost of the material, (ii) cost of labour – fabricating, erecting, and dismantling the forms, and (iii) cost of equipment involved to grip the formworks [19,20,21].

Thus, especially in Malaysia with limitation in term of financial support in engineering research, for huge quantity of specimens, the usage of wooden formwork seems to be more feasible and practical [22]. For smaller projects or research, wooden formworks are appropriate [7,8].

3.2. Performance of wooden formwork

Wood that was often used to build formworks are lumber, plywood, consist of several types of plyform [23]. To be specific, the discussion is meant to understand BBCC plyform type of formwork and equivalent. Plyform is a engineered plywood functions as forming concrete structures [3]. This type of plyform or generally known as plywood, is suitable to be used as formworks essentially because it can withstand high pressure [23].

Figure 3. BBCC plywood.

Plywood is consist of few odd layers of ply might be with the same or different grade of woods, glued internal and externally making it a good formwork – and the glue is definitely water-proof [3]. Plywood usually sized rectangular 4-ft width and 8-ft length right angle.

IOP Conf. Series: Materials Science and Engineering **513** (2019) 012036 doi:10.1088/1757-899X/513/1/012036

Figure 4. 5-layer typically about 18-mm thickness BBCC plywood.

In any event, formworks should able to hold the concrete poured in it. For the wooden formworks, it must be hard or rigid enough to avoid the wood from damaging while containing concrete [24]. The shape of plywood as mentioned in previous paragraph is efficient making it resisting bending, shear, and deflection [25].

As a temporary structure, formworks should also have certain capabilities to cater minimal live load and dead load, while awaiting the concrete structures becomes harden [26]. Thus, before ordering the timber/wooden formworks, there are several factors that have to take into account such as load-duration, wet-service, beam stability, temperature and buckling adjustment factors. Wooden formwork design and recommendations might be referred from American Forest & Paper Association in National Design Specification (NDS) for Wood Construction year 2005 [3]. Specifically for plywood, the strength and manufacture, it always obey the Voluntary Product Standard PS 1-95 approved by The Engineered Wood Association.

Temperature and moisture of concrete mix poured in also have significant effect to the wooden formworks and concrete mix itself [27]. Moisture if being in contact with the hygroscopic material like normal timber will easily swell which causes internal stress and damage [28]. In the other hand, temperature might not be a direct factor for plywood. However, temperature might enhance chemical process and biological activities on the plywood, which indirectly harms them [29].

Figure 5. Examples of swollen and warped wood due to being exposed to humid environment and moisture, respectively.

Henceforth, to ensure the stated incidents prevented, all BBCC plywood for formwork purposes must be treated and coated, internal and external respectively [30]. These issues will not occur if the project uses steel formworks. However, again, wooden formwork has bigger advantages in term of workability, cost-saving, and insulation. On top of that, tested concrete strength results between concrete moulded by steel and plywood has only small difference, which about only 5-7% [31].

3.3. Safety of wooden formwork

In year 1997, OSHA has reported that 5.83% of accidents happened due to the flaw and failures in either formworks or construction temporary structures itself. From the given percentage, 21.2% calamities were involving wooden framing and formworks [32].

As mentioned in earlier section, wooden formworks has limited cycles of usage – ergonomic studies revealed that majority cases involving wooden formworks were due to the repetitive works such as sawing, lifting and hammering which leads to an obvious vibration occurred onto the wooden formworks [33].

Thus, some bodies have made some researches in order to cater the safety issues in using wooden formworks for constructions. The researches include incurring few important factors in calculating the criteria needed by wooden formworks. The factors are such as load duration, bearing area, moisture, flat application, safety factor for formwork accessories, repetitive factor of member utilization, manufactured wood products, and adjustment factor for plywood stresses [34].

However, the detailed calculations and factor considerations are mainly meant to the construction industry which involves with real scale and big structures. Whereby, in research where the specimens are down-scaled and lighter, precautions are consumed but not as critical as on-site manner [35].

4. Conclusion

As the construction industry develops with many technologies and advancements, nowadays wooden formworks have been modified to cater problems as said above. Plywood has undergo numerous improvements such as having a higher strength and panel shear, more flexible, made to resist moisture, chemical, impact and fire, and better insulation.

References

[1] M. N. M. Nawi, M. A. O. Mydin, E. M. Elias, I. N. M. Shaharanee, and M. N. Yusoff 2014 "Suitability of IBS Formwork System in Malaysian Construction Industry," *Aust. J. Basic Appl. Sci.* **8** April, pp. 231–238

[2] H. Lingard, G. Gilbert, and P. Graham 2001 "Improving Solid Waste Reduction and Recycling Performance using Goal Setting and Feedback," *Constr. Manag. Econ.* **19**(8), pp. 809–817

[3] R. Peurifoy and G. Oberlender 1996 *Formwork for Concrete Structures*, Mcgraw-Hil. McGraw Hill Professional

[4] U. K. Essays, "Origin And Applications Of Formwork Construction Essay 2013. [Online]. Available:https://www.ukessays.com/essays/construction/origin-and-applications-of-formwork-construction-essay.php?vref=1. [Accessed: 05-Dec-2016].

[5] BOFU Construction Material Co. Ltd., "History of Formwork," *BOFU Plastic Formwork System*. [Online]. Available: https://www.bofuform.com/2017/04/05/history/. [Accessed: 14-Oct-2018].

[6] Forming America Ltd., "Concrete Forms — Steel Vs. Wood," *Forming America Ltd.*, 2017. [Online]. Available: https://www.formingamerica.com/concrete-forms-steel-vs-wood. [Accessed: 22-Feb-2017].

[7] G. Mishra, "Types of Formwork (Shuttering) for Concrete Construction and its Properties," *The Constructor*, 2015. [Online]. Available: https://theconstructor.org/building/formwork-shuttering/types-of-formwork-shuttering/3767.

[8] S. Deb, "Modern Concrete Formwork Systems: Types, Reuse, Reliability & Risk Assessment," *The Masterbuilder*, India, pp. 92–97, Mar-2015.

[9] R. Das, I. Bhattacharya, and R. Saha 2016 "Comparative Study between Different Types of Formwork," *Int. Res. J. Adv. Eng. Sci.* **1**(4), pp. 173–175
[10] J. P. Wacker 2010, "Use of Wood in Buildings and Bridges," in *Wood Handbook - Wood as an Engineering Material*, Centennial., U.S. Dept. of Agriculture, Forest Service, Forest Products Laboratory, 2010, pp. 17.1–17.13.
[11] S. Burtonshaw-Gunn, "Financial Assessment and Performance of Projects," in *Risk and Financial Management in Construction*, First Edit., Gower, Ed. London: Routledge (Taylor & Francis Group), 2009.
[12] S. McFadyen, "Cost, Performance and Time," *myElectrical Engineering*, 2009. [Online]. Available: https://myelectrical.com/notes/entryid/17/cost-performance-and-time. [Accessed: 06-May-2009].
[13] R. Keefe, "Project Priorities Perspective – Time and Performance versus Cost," *Successful Endeavours Pty Ltd*, 2009. .
[14] J.-H. Kim, S.-T. Yi, and J.-K. Kim 2012 "Size Effect of Concrete Members applied with Flexural Compressive Stresses," *Int. J. Fract.* **126** (1) 79–102
[15] Y. Che, S. Long Ban, J. Yu Cui, G. Chen, and Y. Pu Song 2011 "Effect of Specimen Shape and Size on Compressive Strength of Concrete," *Adv. Mater. Res.* **163** 1375–1379
[16] Z. P. Bazant and K. Xu 1991 "Size Effect in Fatigue Fracture of Concrete," *ACI Mater. J.*, **88**, no. (4) 390–399.
[17] A. Poursaee 2014 *Corrosion Sensing for Assessing and Monitoring Civil Infrastructures*. **1**. Woodhead Publishing Limited
[18] M. A. Murray 2009 "Protection Systems for Concrete Structures," in *Failure, Distress and Repair of Concrete Structures*, Woodhead Publishing Limited, pp. 270–281.
[19] M. K. Hurd 2005, *Formwork for Concrete*, 7th Editio. Farmington Hill, Michigan: American Concrete Institute, 2005.
[20] K. K. Humphreys and P. Wellman, *Basic Cost Engineering*, Third Edit. United States of America: CRC Press, 1995.
[21] G. R. Smith and A. S. Hanna 1993 "Factors influencing Formwork Productivity," *Can. J. Civ. Eng.*, **20** (1), pp. 144–153
[22] "Formwork in Construction-Applications of Shuttering," *CivilDigital*, 2018. [Online]. Available: https://civildigital.com/formwork-construction-types-applications-shuttering.
[23] APA - The Engineered Wood Association, "Concrete Forming: Design/Construction Guide," p. 1, 2012.
[24] K. M. Nemati 2007, "Temporary Structures - Introduction to Concrete Formwork and Vertical Formwork Design," in *Formwork for Concrete*, Winter Qua., no. 6, University of Washington, pp. 1–35.
[25] M. Setareh and R. Darvas 2016 "Rectangular Beams and One-Way Slabs," in *Concrete Structures*, Springer, Cham, pp. 37–149.
[26] Joint Technical Committee TM/8, "AS/NZS 2269:1994 - Plywood (Structural)," 1995.
[27] P. Klieger 1958 "Effect of Mixing and Curing Temperature on Concrete Strength," *Am. Concr. Inst.* **54** (6), pp. 1063–108.
[28] J. Ashley-Smith 1999, *Risk Assessment for Object Conservation*, 1st ed. London: Taylor & Francis Group
[29] S. Slade, "Temperature And Humidity," Australia, 2003.
[30] L. Hu, Z. L. Chen, and M. J. Zhan 2011 "Preliminary Study on Fire-retardant-treated Plywood," *Eucalypt Sci. Technol.* **2**, p. 001
[31] M. Arslan, O. Şimşek, and S. Subaşı 2005 "Effects of Formwork Surface Materials on Concrete Lateral Pressure," *Constr. Build. Mater.* **19** (4), pp. 319–325
[32] X. Huang and J. Hinze 2003 "Analysis of Construction Worker Fall Accidents," *J. Constr. Eng. Manag.* **129** (3), pp. 262–271
[33] W. Y. Har, "Ergonomic Studies on Formwork Carpentry," *Appl. Ergon.*, no. 3, pp. 683–686,

IOP Conf. Series: Materials Science and Engineering **513** (2019) 012036 doi:10.1088/1757-899X/513/1/012036

202AD.

[34] M. I. Hamakareem, "Wooden Concrete Formwork Design Criteria with Calculation Formulas," *theconstructor.org*, 2018. [Online]. Available: https://theconstructor.org/building/wooden-formwork-design-calculation-concrete/16697/.

[35] A. Singh and S. Singh 2005 "The Connection between Academia and Industry," *Mens sana monographs*.

IOP Conf. Series: Materials Science and Engineering **513** (2019) 012037 doi:10.1088/1757-899X/513/1/012037

Implementation of eco-costs per value ratio (EVR) on construction waste management in Shah Alam, Malaysia

S K Arumugam[1*], R Muhamad[2] and K Yahya[3]

[1, 2] Razak School of Engineering and Advanced Technology, Universiti Teknologi Malaysia, Kuala Lumpur, Malaysia
[3] Faculty of Civil Engineering, Universiti Teknologi Malaysia, Skudai, Malaysia

[*]Corresponding author: sharan.kumar@spsetia.com

Abstract. The construction industry in Malaysia which is rapidly growing is indirectly contributing towards the economic influx to the country. However, the waste generation in construction is increasing proportionally with new construction developments. The current construction methods focus on minimal recycling process where most construction wastes are dumped at landfills because construction companies are required to set aside a higher preliminary cost for proper waste disposal. This study focuses on creating a more potential outcome in decreasing the construction waste by implementing eco-costing in development for the construction waste generated from the beginning stages of a project. Therefore, this study discusses the construction waste management using the eco-costing per value ratio (EVR) assessment. The EVR assessment for eco-costing were targeted to the construction waste products from Shah Alam. The analysis included 5 selected sample sites in Shah Alam conducted within the year 2013 to 2017. The findings indicated that EVR index ranged between 0.054-0.1225 for conventional and semi industrialised building system (IBS) method based on five-selected sample of sites. This range signify on current EVR index which will be compared with Malaysian EVR benchmarking in result and discussion. Furthermore this result is useful to evaluate EVR index between residential for construction waste management towards sustainable development.

1. Introduction

Construction industries consume a large number of raw materials [1]. The types of raw materials used in the industry include sand, aggregates, water, clay, rock, wood and manufacturing goods such as cement, timber, bricks, plaster cement, tiles, drywall, concrete, rebar, ceiling, and metal. As a result, the residual raw material from such a large amount of these resources results in a significant amount of construction waste, which impacts the environment. The total environmental impact of a construction product during its entire life cycle can be determined using the Life Cycle Analysis (LCA) method. EVR is an LCA based technique to analyse the utilisation designs, business procedures, and design outline choices regarding eco-effective esteem creations. According to Firman et al. [2], EVR is used to benchmark products and service systems. The EVR is a marker to sustainable and unsustainable utilization designs, where the eco-cost is an indicator for the natural contamination of products, while the esteem is the cost paid for the product in the market. The EVR ratios can be used to indicate the level of environmental pollution. As such, poor waste management leads to hazardous environmental impacts and financial losses as the companies eco-costs become greater [3] [4]. The EVR method was adopted in this study because the construction waste management at the selected project sites still has room for waste minimisation improvement. Hence, EVR can be used as a tool to reduce and monitor the construction waste throughout the contract period.

Based on the analyses, the selected five projects were not within the range of benchmarks. The EVR benchmark for the Malaysian construction industry is within 0.0024 – 0.0028 for typical multi-storey projects applying conventional or partial semi-IBS, and up to 0.0014 for projects utilizing full IBS system or projects with exceptionally good waste management awareness and practice [2].

2. Materials and method

Five project sites (Project A – Project E) in Shah Alam constructed from 2013 until 2017 were selected for the study, which mainly includes residential low rise. The construction involved a wide range of contractors employing conventional and sustainable building materials in their construction systems. The construction method in Malaysia has been based on the conventional timber formwork (including plywood) for many years. However, although the IBS system has started gaining acceptance and encouragement from the government, only a limited number of larger contractors have opted to use it [2]. Majority of IBS systems employed include metal formwork system (steel or aluminium), drywall panels and precast concrete system (for columns, beams, slabs, and wall) which relatively produce less waste. It is important to note that this study is focused on evaluating the eco-costs for waste generated during the construction phase only, and not the full life cycle.

The major construction materials which were taken into account include concrete, timber, reinforcement bars, bricks and blocks, tiles, and plaster. Whereby, waste generated during construction activities at superstructure phase or known the above groundwork was considered for this assessment because waste generated from substructure or below ground and during foundation works were considered minimum consisting mostly of soil. Moreover, the construction projects covered in the data collection process were from the low rise residential, conventional and the sustainable building material adopted systems within the Shah Alam area. The eco-costs considered in this study included; (i) the purchased cost and delivery cost, (ii) labour cost, (iii) wastage disposal cost, and (iv) landfill cost. Besides, cost waste generation rate (WGR), waste index (WI) and waste level (WL) were also identified in this analysis. Major data extracted included (i) Gross Floor Area (GFA), (ii) material order quantities, (iii) material work done quantities from Bills of Quantity (BQ), (iv) construction debris disposal trip record, (v) purchase and delivery costs, and (vi) costs associated with waste generation and total project cost (contract sum). This study assessed the eco-costs of produced wastes.

The first step in quantifying the carbon footprint is to determine the amount of wastage generated and wastage level for each specified material as described below:

$$\text{Wastage level},(W_l) = \frac{\text{Total wastage},(T_w)}{\text{Total work order},(T_{wo})} \times 100\% \ [2] \tag{1}$$

where,

Wastage level (W_l) = The percentage of wastage between projects
Total wastage (T_w) = Total wastage amount (RM) during the contract period
Total work order (T_{wo}) = Quantity (RM) material work order

Meanwhile, the calculations for each cost are as in the following section.

2.1. *The purchased cost and delivery cost*

$$\text{Purchased cost and delivery cost } (\%) = \frac{\text{Total wastage cost},(T_{wc})}{\text{Purchased cost},(P_c)} \times 100\% \tag{2}$$

where,
Purchased cost and delivery cost (%) is calculated to justify the percentage between total wastage cost and purchased cost.

$$Purchased\ cost\ and\ delivery\ cost = Total\ wastage\ cost,\ (T_{wo})\ x\ purchased\ cost,\ (P_c)\ [2] \quad (3)$$

where,

Total wastage cost, (T_{wc}) = Total wastage amount (RM) during the contract period
Purchased cost, (P_c) = Total material purchased and delivery cost

The price of materials used was obtained from the contract department of Setia Alam. However, the price for some materials that were not listed was obtained from various resources. The unit cost in the price list for each material was inclusive of the delivery unit cost.

2.2. *The labour cost for construction waste disposal*

$$Labour\ cost,\ (L_c) = Total\ labour\ needed,\ (T_{ln})\ x\ Total\ duration,\ (T_d)\ (hour)\ x\ Cost\ per\ hour,\ (C_{ph})\ x\ Total\ days\ required,\ (Td_r)\ x\ Total\ contract\ duration,\ (Tc_d)[2] \quad (4)$$

where,

Total labour needed, (T_{ln}) = Quantity manpower required
Total duration, (T_d) (hour) = Total working hour for site clearance
Cost per hour, (C_{ph}) = Salary per hour
Total days required, (Td_r) = Total duration required
Total contract duration, (Tc_d) = Total contract period

Based on information obtained from Environmental Waste Management (EMS) reference documents, it was clear that the number of labours required per week for housekeeping and waste handling varied between projects and the cost per labour also varied for general construction building-worker in the Klang Valley.

2.3. *The wastage disposal cost*

Total waste generated by project (month's contract) $\mathbf{m^3}$

$$(v\ x\ n) = w \quad (5)$$

where,

Truck volume ($\mathbf{m^3}$), (**v**) = the volume of waste bin
Total number of loads for waste disposal, (**n**) = Total trip of disposal
Total waste generated by project per month, (**w**) = the volume of waste bin x Total trip of disposal

$$(t\ x\ c) = Total\ wastage\ disposal\ cost \quad (6)$$

where,

Cost per trip, (**c**) = Cost per single trip
Total Trip, (**t**) = Total trip of disposal

2.4. *The waste index*

Waste index is calculated to justify the index rate by total waste generated between projects divided with gross floor area of a project as shown below:

$$Waste\ index,\ (W_i) = w\ /gross\ floor\ area\ (GFA)\ [5] \quad (7)$$

where,

Gross floor area (GFA) = Total built-up per unit x total units [2]

IOP Conf. Series: Materials Science and Engineering **513** (2019) 012037 doi:10.1088/1757-899X/513/1/012037

The cost per trip data provided for project A was manipulated to fit the other 4 project costs as it was assumed to be the same as project A following the range of length. This is because the cost per trip is relatively the same since the location of the dumping ground was within a specific range. However, the cost still depends on the size of the waste bin, falling within the range of RM 150-RM 250. Besides, data for the number of trips was extracted from the Environmental Waste Management (EMS 14001) index calculation.

Total waste volume can be calculated as pyramidal and rectangular shape as below if the construction waste material not placed at a designated area.

Pyramidal shape waste

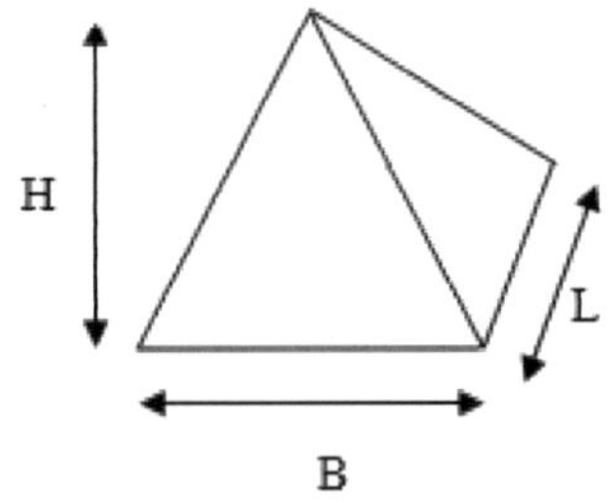

Rectangular shape waste

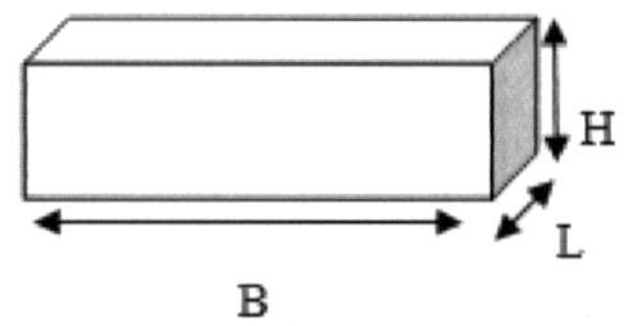

$$\textit{Total volume of waste, } (vw) = (L x B x H)\text{[6]} \quad (8)$$

where,

L = length
B = width
H = height

$$\textit{Landfilling cost} = \textit{Total waste volume x Cost per volume } \text{[2]} \quad (9)$$

This above-mentioned equation (9) was also used by [6] [7] in their studies to evaluate the volume of waste produced at site. Based on the data obtained from Perbadanan Pengurusan Sisa Pepejal and Pembersihan Awam (PPSPPA), the landfilling cost was assumed to be approximately RM 200 per m3 of waste [2]. The data for total waste volume was also extracted from the Environmental Waste Management (EMS 14001, 2015) index calculation. This total waste volume required to justify the overall construction waste produced by selected five project sites. Thus, EVR can be described as the total eco-costs divided by the total value of the project as seen below.

$$\textit{Waste generation rate, (WGR)} = \frac{\textit{total wastage amount per month, } (T_{wa})}{\textit{Gross floor area (GFA)}} \times 100\% \text{ [8]} \quad (10)$$

where,

Total wastage amount per month, (m^3) = Total volume of waste monthly basis
Gross floor area (GFA) = Total built-up per unit x total units

2.5. EVR ratio

$$EVR = \frac{\textit{Eco-costs (RM)}}{\textit{Total project cost (RM)}} \text{ [2]} \quad (11)$$

where,

Eco costs = Purchased cost and delivery cost (RM) + Total Wastage Disposal Cost (wc) (RM) + Total Labour Cost (RM) + Landfilling cost (RM)

Total project cost (RM) = Contract Sum (RM)

Although this study did not fully include all attributes proposed by [4], it was expected that the costs considered in this study should represent the eco-costs for every project objectively. The attributes that were not included in the scope of the study are the cost of impacts, cost of emission from equipment, and cost of depreciation of equipment.

Figure 1 illustrates the process of construction waste disposal at the selected project site. Waste quantity or volume obtained by truck trip.Based on the data collected, the waste classification and justification were only made for superstructures since substructure waste products were of low to moderate level. A total of nine types of waste materials were identified such as rebar, concrete, timber formwork, bricks, plaster cement, tiles, drywall, metal deck roofing, and ceiling. With the large consumption of materials at constructions currently, the implementation of EVR can help developers to keep the construction waste produced during the contract period under control. This would lead to sustainable construction which creates a sustainable building environment and human health through efficient use of resources. Table 1 summarises the spreadsheet on EVR calculation for the selected construction projects in this study.

Figure 1. Construction waste disposal process (adapted from EMS 14001 records).

Table 1. EVR calculation based on five-project site.

Project	A	B	C	D	E
Type of building	Low-rise residential 72 units (Double Storey)	Low-rise residential 127 units (Double Storey)	Low-rise residential 108 units (Double Storey)	Low-rise residential 48 units (Triple Storey)	Low-rise residential 80 units (Double Storey)
Construction Method	Semi IBS (drywall for partition),shear wall for party wall and conventional method for rest	Semi IBS (drywall for partition),shear wall for party wall and conventional method for rest	Semi IBS (drywall for partition),shear wall for party wall and conventional method for rest	Semi IBS(,shear wall for party wall) and conventional method for rest	Semi IBS (drywall for partition),shear wall for party wall and conventional method for rest
Total Construction	December 2016-March	October 2016-January 2018 (	October 2016-January 2018 (	December 2016-April	November 2016-February

IOP Conf. Series: Materials Science and Engineering **513** (2019) 012037 doi:10.1088/1757-899X/513/1/012037

Period	2018 (15 months)	15 months)	15 months)	2018 (17 months) (on-going)	2018 (15 months)
GFA (m2)	136,800	208,280	189,000	163,200	133,600
Contract Sum (RM)	14,217,543.46	24,152,992.65	19,103,112.53	18,864,702.25	12,601,026.77
Wastage %	Rebar & BRC 10%,Concrete grade 25 (5%),Timber formwork 10%, Bricks 10%, Plaster cement 25%, Tiles 10%, Drywall 5%, Metal deck rofing 6% and Ceiling 5%	Rebar & BRC 12%,Concrete grade 25 (6%),Timber formwork 100%, Bricks 10%, Plaster cement 13%, Tiles 9%, Drywall 5%, Metal deck rofing 4% and Ceiling 5%	Rebar & BRC 13%,Concrete grade 25 (5%),Timber formwork 100%,Bricks 8%, Plaster cement 13%, Tiles 9%, Drywall 5%, Metal deck rofing 4% and Ceiling 5%	Rebar & BRC 8%,Concrete grade 25 (7%),Timber formwork 100%Bricks 8%, Plaster cement 9%, Tiles 10%, Drywall 0%, Metal deck rofing 4% and Ceiling 3%	Rebar & BRC 8%,Concrete grade 25 (7%),Timber formwork 100%, Bricks 8%, Plaster cement 9%, Tiles 10%, Drywall 2%, Metal deck rofing 4% and Ceiling 3%
Wastage Level (%)	10.39%	13.15%	14%	11.17%	22%
Total Unit and Delivery Lost (RM)	764,850.42	1,756,357.03	1,170,626.14	793,390.90	1,225,036.43
Total Unit and Delivery Lost (%)	0.09%	0.13%	0.14%	0.11%	0.22%
(vw)	25,920.00	42,120.00	51,840.00	48,384.00	69,120.00
(wc) (RM)	30,000.00	48,750.00	60,000.00	56,000.00	80,000.00
(WGR) (%)	0.0126	0.0135	0.0183	0.0185	0.0323
Waste index (wc/GFA) (%)	0.0146	0.0156	0.0212	0.0214	0.0374
Total Labour Cost (RM)	33,750	39,375	61,200	36,000	46,080
Landfilling cost (RM)	72,000.00	117,000.00	144,000.00	134,400.00	192,000.00
Total eco-costs (RM)	900,600.42	1,961,482.03	1,435,826.14	1,019,790.90	1,543,116.43
Material Lifespan	Timber formwork (4 - 5 times), shearwall panel 2 sets	Timber formwork (4 - 5 times), shearwall panel 4 sets	Timber formwork (4 - 5 times), shearwall panel 4 sets	Timber formwork (4 - 5 times), shearwall panel 2sets	Timber formwork (4 -5 times), shearwall panel 4 sets

EVR = Eco Costs (RM) /Total project cost (RM)EVR (x10-2)	**6.3344**	**8.1211**	**7.5162**	**5.4058**	**12.25**

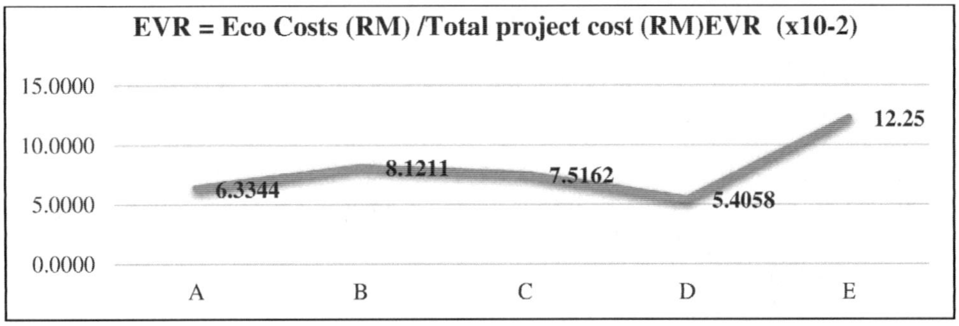

Figure 2. EVR ratio between five projects.

Based on Figure 2, the EVR index for Project D (0.054) was identified as the lowest among the other projects. The calculated EVR index values can serve as a contributing factor to the betterment of the waste management at the project areas by its developers. Furthermore, the waste index tabulated in Table 1 for Project A to Project E ranges from 0.014% to 0.040%. These index values will aid in monitoring the total waste disposal cost during any stages of the construction period. Besides, the project developers will be able to establish a better resources management plan towards sustainable development.

3. Conclusion

Construction waste management is an important element in construction for a sustainable development. The intention of this study was to develop a sustainable waste eco-cost model based on case study approach during the construction stage. The detailed assessment of the waste disposal cost saving between the conventional and sustainable building materials was performed using EVR method. The project D that adopted semi industrialised building system (IBS) with combination of shear wall and conventional method obtained better EVR index of 0.054 compare to other projects. However, for overall assessment of EVR index there are few steps of calculation required as above. The assessment provides efficient solutions to handling construction wastes which leads to sustainable construction waste management. Poor waste management can lead to unsafe environmental impact along with financial losses, the increase of the eco-costs for any project [13]. The increase in total waste produced from each material will increase the eco-costs for a project with lower gross floor area (GFA), hence, elevates the EVR index. Therefore, to lower the EVR index, total waste produced should be reduced indefinitely. The lower the EVR index, the better it is for society.

Acknowledgements
The first author is an Engineering Doctorate Student from UTM Razak School of Engineering and Advanced Technology who is partially supported by KPT-MyPhD Ministry of Higher Education.

IOP Conf. Series: Materials Science and Engineering **513** (2019) 012037 doi:10.1088/1757-899X/513/1/012037

References

[1] Arshad H, Qasim M, Thaheem M J and Gabriel H F 2017 Quantification of material wastage in construction industry of Pakistan: An analytical relationship between building types and waste generation *J. Constr. Dev. Ctries.* **22** 19-34
[2] Firman M A, Hassan C R C, Meriam S N, Zalina M N, Nazziera M S and Khairulzan Y 2012 Eco-costing per value ratio assessment of construction waste: A study in Klang Valley, Malaysia *J. Environ. Res. Dev.* **7** 99-106
[3] Hultman J and Corvellec H 2012 The European waste hierarchy: From the sociomateriality of waste to a politics of consumption *Environ. Plan.* **44** 2413-27.
[4] Yahya K and Boussabaine A H 2006 Eco-costing of construction waste *Manag. Environ. Qual. Int. J.* **17** 6-19
[5] Masudi A F, Hassan C R C, Mahmood N Z, Mokhtar S N and Sulaiman N M 2011 Construction waste quantification and benchmarking: A study in Klang Valley, Malaysia *J. Chem. Chem. Eng.* **5** 909-16
[6] Zainun N Y, Rahman I A and Rothman R A 2016 Mapping of construction waste illegal dumping using geographical information system *IOP Conf. Ser. Mater. Sci. Eng.* vol 160 (Bristol, UK: IOP Publishing) p 012049
[7] Lau H H, Whyte A, Law P L 2008 Composition and characteristic of construction waste generated by residential housing project *Int J. Environ. Res.* **2** 261-8
[8] Mei M H and Fujiwara T 2016 A survey of construction and demolition waste in Malaysia, mixed-use development *J. Faculty Environ. Sci. Technol. Okayama University* **21** 1-2
[9] Masudi A F, Hassan C R C, Mahmood N Z, Mokhtar S N and Sulaiman N M 2010. Quantification methods for construction waste generation at construction sites: A review *Advanced Mater. Res.* **163** 4564-9
[10] Masudi A F, Hassan C R C, Mahmood N Z, Mokhtar S N and Sulaiman N M 2012 Waste quantification models for estimation of construction and demolition waste generation: A review *Int. J. Glob. Environ. Issues* **12** 269-81.
[11] Begum R A, Siwar C, Pereira J J and Jaafar A H 2009 Attitude and behavioral factors in waste management in the construction industry of Malaysia. *Resour. Conserv. Recycl.* **53** 321-8.
[12] Mahayuddin S A, Pereira J J, Badaruzzaman W H W and Mokhtar M B 2008 Construction waste management in a developing country: Case study of Ipoh, Malaysia *WIT Trans. Ecology Environ.* **109** 481-9
[13] Sangita G R and Verinder K 2011 A novel approach to improve road quality by utilizing plastic waste in road construction *J. Environ. Res. Dev.* **5** 1036-42.
[14] Akber M Z, Thaheem M J and Arshad H 2017 Life cycle sustainability assessment of electricity generation in Pakistan: Policy regime for a sustainable energy mix *Energy Policy* **111** 111-26.
[15] Hultman J and Corvellec H 2012 The European waste hierarchy. From the sociomateriality of waste to a politics of consumption. *Environ. Plan.* **44** 2413-27.
[16] Poon C S, Yu A T and Jaillon L 2004 Reducing building waste at construction sites in Hong Kong *Constr. Manag. Econ.* **22** 461-70.

IOP Conf. Series: Materials Science and Engineering **513** (2019) 012038 doi:10.1088/1757-899X/513/1/012038

Experimental study on flexural behaviour of partially encased cold-formed steel composite beams using rebar as shear connector

Musab N A Salih[1*], Mahmood Md Tahir[2], S Mohammad[3], Y Ahmad[3], A Sulaiman[3], P N Shek[2], Achmad Abraham[4], Muhammad Firdaus[4], K M Aminuddin[4] and Khadavi[2]

[1]Ph.D Student, Civil Engineering Department, Faculty of Engineering, Universiti Teknologi Malaysia, 81310, Johor Bahru, Malaysia
[2]UTM Construction Research Centre (UTM CRC), Institute for Smart Infrastructure and Innovative Construction (ISIIC), School of Civil Engineering, Faculty of Engineering, Universiti Teknologi Malaysia, 81310 Johor Bahru, Malaysia
[3]Faculty of Engineering, Universiti Teknologi Malaysia, 81310, Johor Bahru, Malaysia
[4]Ph.D Student, Civil Engineering Department, Faculty of Engineering, Sriwijaya, Inderalaya, 30862, Indonesia

[*]Corresponding author: nasmusab@yahoo.com

Abstract. This paper focuses on the composite action of two samples each consisting of two composite beams and a composite slab with bent up rebar as a shear connector. The specimens were knowingly adapted to have different shear transfer capacities along the interfaces with different geometries of the composite beams. By using the four-point bending tests of the two full-scale specimens, the authors experimentally evaluated the flexural behaviour of the partially encased composite beams. Therefore, this paper presents experimental works on the structural performance of composite beams consisting of cold-formed steel (CFS) section with self-compacting concrete (SCC) by means of the shear connection mechanism of the proposed using deformed reinforcement bar. The results have shown that the theoretical value of flexural capacities designed agrees well enough with the experimental results. Based on the present beam tests results it was realised that the new composite system showed slightly higher results than predicted with ultimate moment ratios between 0.89-0.91% and 0.83-1.08% for the 250 mm and 150 mm beam respectively. According to the experimental bending test results, the composite section can reach the ultimate strength without compression buckling failure or local shear when the proposed shear connector was presented in the composite action. Analysis and design of composite beams are studied to validate the present test results.

1. Introduction

CFS is used widely in ships, mining head frames, steel and concrete bridges, gantry cranes and aircraft. But there are numerous reasons why CFS structures must be given more attention in their analysis and design. In CFS beams the shear stresses and strains are moderately higher than that those in a solid beam. This arises questions into the usual hypothesis of bending theory, that is, that plane sections remain plane over the entire cross-section of a box girder. CFS is also susceptible to local buckling if the in-plane stress reaches their critical values. This means that CFS structures must be designed against both local buckling and overall buckling. Therefore, infilling a box section with concrete will redistribute stresses which leads to enhancing the resistance of the CFS section to local and global buckling."

Composite beams are one of the most widely used members in the construction industry and have been extensively used for its reliability, rapid erection, cost-effectiveness, and strength. However, one property they all have in common is that they are very light compared with alternative structures and, therefore, they are used extensively in long-span bridges and other structures where weight and cost are prime considerations. Recently, researchers have shown an increasing interest in the application of cold-formed steel in the construction industry by utilizing it in composite sections to provide lighter weight of the structural system. Therefore, to bring back symmetricity and decreases lateral-torsional buckling effect a boxed section was created by orienting two CFS lipped C-channel sections toe-to-toe as in Figure 1. These two advantages retain by the system, encourage the use of CFS sections in a wider range of structural applications [1,2,3].

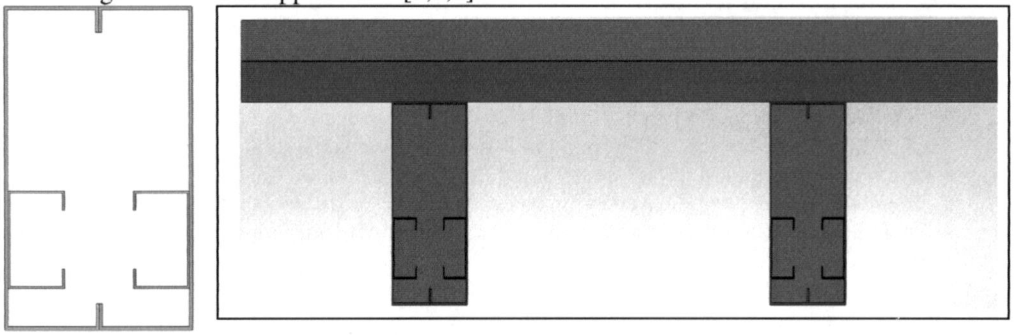

Figure 1. CFS box section.

It was at first customary to design the steelwork to carry the whole weight of the concrete slab and its loading, but by about 1950 the development of shear connectors had made it practicable to connect the slab to the beam, and so to obtain the T-beam action that had long been used in concrete construction. Composite construction takes advantage of the qualities of concrete and steel. The steel beams and profiled steel sheeting act as permanent falsework and formwork for the wet concrete. Once cured, the concrete can provide all or most of the compression force needed for bending resistance of the floor slab or beams. The steel beam provides all the tension force for bending resistance and, due to the stabilizing effect of its connection to the concrete slab, as much compression force as is called for. The functionality of the composite slab to endure loads is governed by the interaction level of the two main structural parts which called the degree of shear connection. consequently, whenever the connection between these two substances is improved, the stiffness and capability of the composite slab enhanced [4,5].

The development of new types of composite beams and shear connectors has become increasingly popular in the past decades and the effectiveness of applying CFS sections to beams to either help reduce the costs or eliminate the construction difficulty associated with the shear connectors or prevent CFS failure modes.

In the present research, a new composite beam with CFS infilled with self-compacting concrete has been experimentally and analytically studied to achieve higher strength and ductility, as well as to yield a more economical design purpose. A full-scale four-point load testing has been done to evaluate the performance of the innovative shear connector for the proposed composite structure illustrated in Figure 2.

In the new proposed composite system, each concrete slab was connected to two parallel CFS infilled beams using deformed bent-up rebar. system comprises of a corrugated steel deck to support a cast-in-place self-compacting concrete slab with transverse reinforcement on top of two CFS beams parallel to each other each consisted of two toe-to-toe C-sections and a bent-up rebar as a shear connector to is link the CFS beam and the composite slab as a way to offer vertical interlocking and horizontal shear resistance. Self-drilling fasteners are used for connecting all the steel parts. Thus, the key part of an effective composite structure comes from a novel shear connector within the concrete slab, cold-formed steel joists, fasteners, and metal deck so that all structural components in this composite floor system are acting together as one unit to transfer the applied loads without failing [5].

a) Full Scale

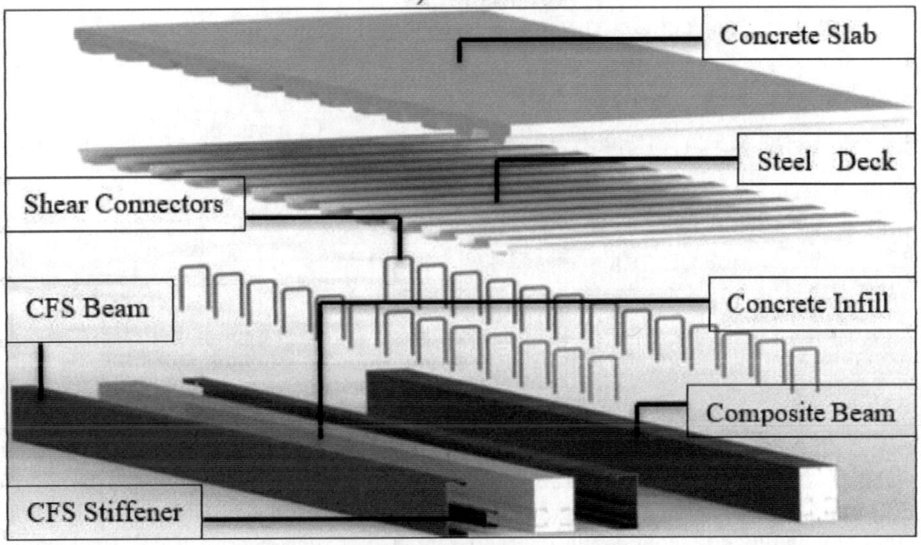

b) Exploded view

Figure 2. The proposed composite system.

2. Material and methods

2.1. Material

Material used in this study are CFS of lipped C-channel section with the width of 150 mm, the depth of 250 mm, and lipped depth of 20 mm and a thickness of 2.4 mm with a yield strength of 450 N/mm^2 is used for the proposed encased composite beam with a modulus of elasticity taken as 210 kN/mm^2. The CFS beam material will be formed into Box-profile (Toe-to-Toe) linked with top and bottom arc welding. The steel properties were obtained according to BS EN 10002-1 using INSTRON 600 DX as shown in Table 1 [6]. Three samples were collected from the web of the CFS profile and three more samples were collected from steel reinforcement. The proposed shear connector made from deformed bent-up rebar of 12 mm diameter with a steel grade S460, the shear connectors was embedded in the SCC through a CFS flange that has been cut to install the shear connectors. BRC welded wire mesh

IOP Conf. Series: Materials Science and Engineering **513** (2019) 012038 doi:10.1088/1757-899X/513/1/012038

A142 with a diameter of 6 mm with a spacing of 200 mm in both directions was used as reinforcement mesh and SCC of grade 40 N/mm^2 at 28 days respectively are the material used in this study.

SCC was a ready-mix from LAFARGE company and was cast in the laboratory, without the need of extra formwork. The profile steel deck yield strength of 350 MPa, the nominal thickness is 1.0 mm and profiled sheeting rib height of 50 mm, the composite sample dimensions are showed in Table 2 [7,8,9].

Table 1. CFS coupon tensile test results.

Properties N/mm^2	Thickness mm	Web		Flange		Average	$\frac{f_u}{f_y}$
		Wb1	Wb2	Fln1	Fln2		
Yield Stress (f_y)		570	571	546	555	560.50	
Ultimate Stress (f_u)	2.4	639	649	623	636	636.75	1.14
Elastic Modulus (E_s)		188000	187000	176000	177000	182000	

Table 2. Composite Specimen Dimensions.

		Slab		CFS Beam			Shear Connectors		
Specimen ID	Length	Width (mm)	Depth (mm)	Thickness (mm)	Depth (mm)	Transverse Spacing (mm)	Number	Size (mm)	Height (mm)
TT-250-12-12	4000	1500	100	2.4	250	750	13	12	75
TT-150-12-12	4000	1500	100	2.4	150	750	13	12	75

2.2. *Test setup loading protocol and instrumentation*

All full-scale composite beam specimens were tested using hydraulic jack machine with a maximum capacity of 1000 kN. The specimen was subjected to four-point bending load test where the load was applied 1000 mm from the supports. A load cell is used between the jack and spreader beam in order to measure the applied load by connecting the load cell to the data logger. The load was applied to the specimen through a spreader beam that receives the load from the jack and transfers it to the concrete slab. A strip of gasket rubber under the point loads is used to absorb any imperfections in the concrete slab surface and to provide uniform contact surface between the concrete slab and the point load.

The sample was fixed in a simply supported position. To measure the vertical deflection at mid-span of the beam specimen, deflections of the specimen were measured using six linear variable displacement transducers (LVDT's) positioned at the critical zones at loading points and mid-span of the test specimen as in Figure.3. Slips were also observed using two LVDT's attached to the centreline of the concrete slab faces at the support's sides i.e. roller and pin. All LVDTs were connected to data logger where the displacements were measured in "mm". Strain gauges were installed in the critical zones at the bottom flange of the steel beam as well as the top surface of the concrete slab at the maximum shear and bending areas.

After the test specimen was setup and instrumented, the specimen was loaded up to 15-20% of its expected ultimate capacity and waited for 5 minutes, then unloaded to make sure that the specimen and instrumentation were settled. All the readings of load, LVDTs and strain gauges were zeroed to insure equilibrium. Subsequently, the load was applied to the specimen by loading increment with a pause of 5 minutes between each increment. The reason of using this mode is to provide enough time for observing and visualizing the specimen until failure. The load was increased until the ultimate capacity of the specimen. The specimen considered failed where there was a considerable deformation of the test specimen was spotted or a substantial drop in the applied load [10,11,12].

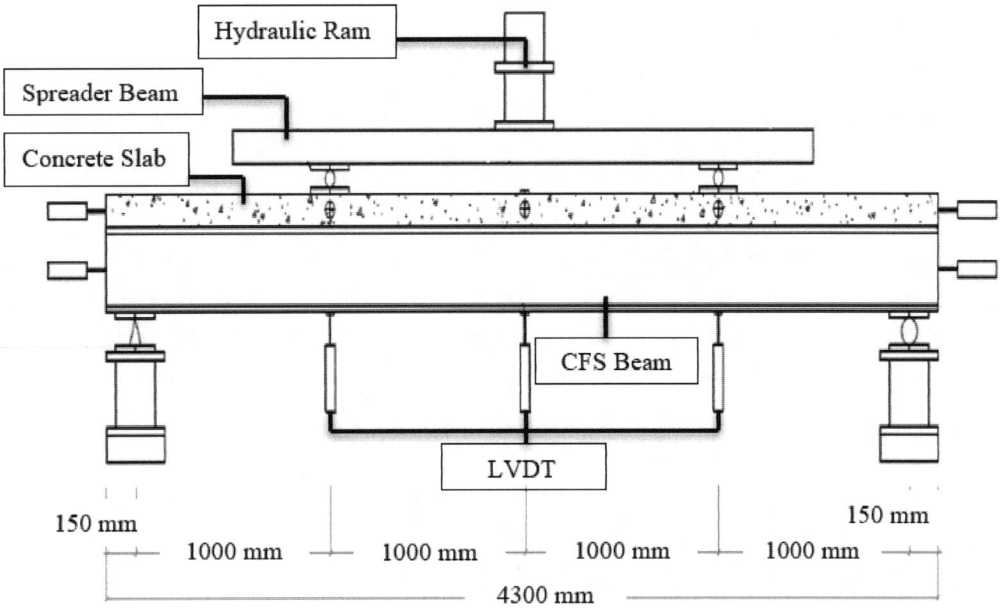

Figure 3. Schematic diagram of a test arrangement.

3. Experimental results and validation

Both samples failed in the same manner due to a combination of high shear force and bending moment at critical sections (loading points and support zones). The failure mode originated by the creation of longitudinal cracks along the line of shear connectors on the surface of the concrete slab. As loads increased, new cracks developed in the concrete slab towards the loading points. The concrete start cracking at the top surface of the slab near the support zone with and extending transverse crack, followed by crushing of concrete, local buckling of CFS at loading zone and a clear expansion of the support transverse crack due to the combination of maximum moment and maximum shear. The load was monotonically increased until failure occurred.

Table 2 shows a comparison between the predicted ultimate moment capacity values following the EC 4 and EC 3 and the ultimate moment capacity obtained in the experiments [13,14,15,16]. The predicted values using both stress block method and interpolation method. The predicted values by the interpolation method gave more accurate results compared to the stress block method.

The load carrying capacity of the tested specimens or Load against mid-span deflections curves for specimens is shown in Figure 4. The ultimate loads (Pu, exp.) of the specimens were 713.6 kN and 314.1 kN with a 45.5 mm and 71 mm recorded mid-span deflections at an ultimate loads level for the 250 mm and 150 mm beam respectively.

From table 2 and figure 4, it's noticeable that the difference in section capacity between the two specimens is high, the 250 mm beam carry the higher load with lower mid-span deflection while the 150 mm beam has significantly higher deflection than the ultimate for the ultimate limit state. The failure modes of the specimens were local buckling of CFS and cracks of the concrete slab as shown in Figure 5.

IOP Conf. Series: Materials Science and Engineering **513** (2019) 012038 doi:10.1088/1757-899X/513/1/012038

Table 2. Experimental and theoretical capacities of composite specimen.

Specimen ID	*TT-250-12-12*	*TT-150-12-12*
f_{ck} at test day (N/mm^2)	60.55	60.55
Ultimate load P_u, exp (kN)	713.6	314.1
Mid-Span Deflection at P_u, exp (mm)	45.5	73.62
Ultimate Moment M_u, exp (kNm)	356.8	157
Interpolation Method M_u, theory (kNm)	390.04	145.28
M_u, exp /M_u, theory	0.91	1.08
Stress Block Method M_u, theory (kNm)	401.18	188.23
M_u, exp /M_u, theory	0.89	0.83

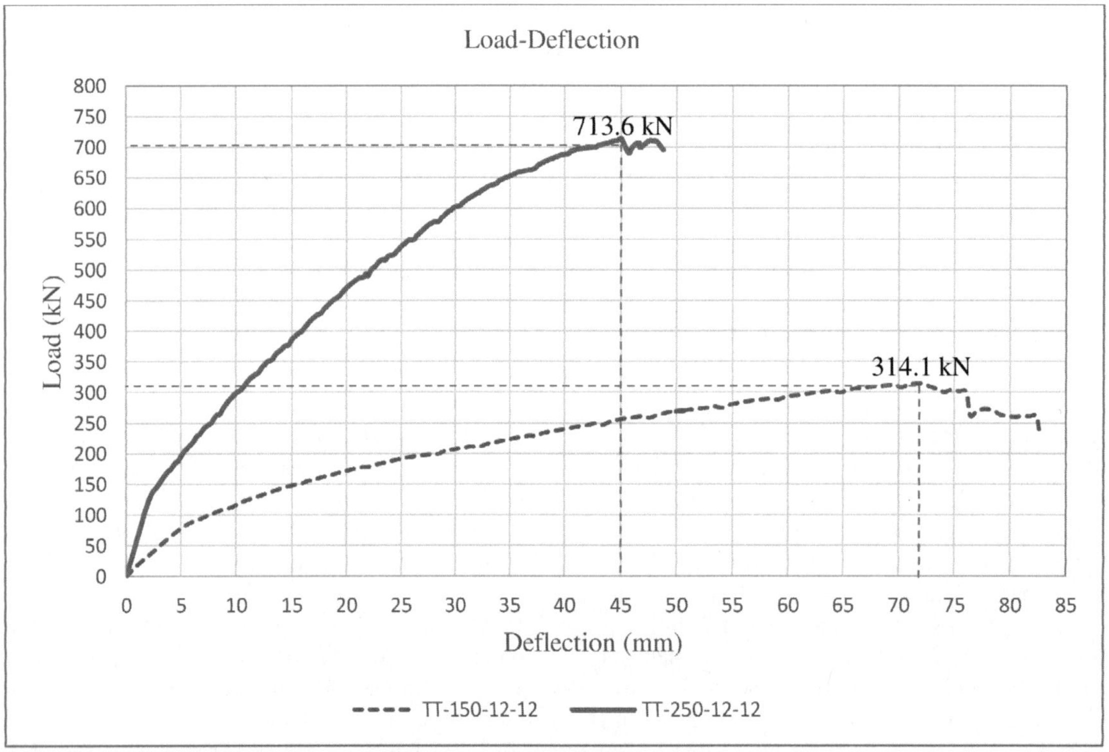

Figure 4. Load-deflection curve.

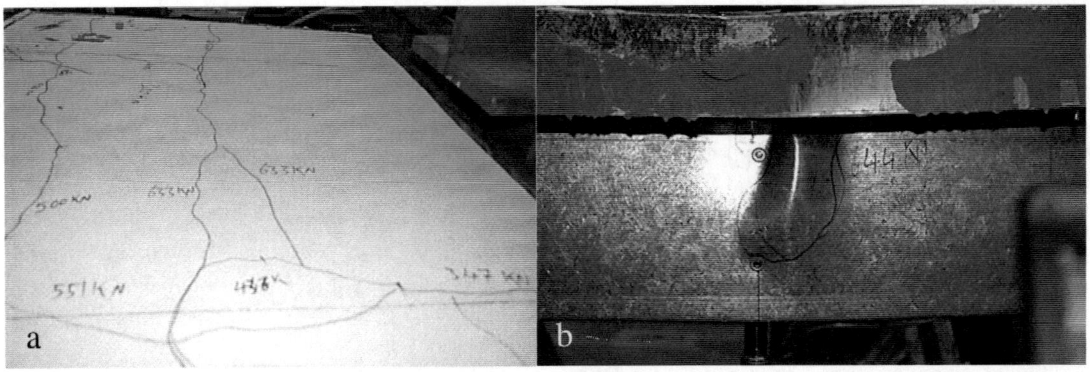

Figure 5. Failure modes of composite specimens, (a) transverse and longitudinal cracks in concrete slab surface, (b) local buckling on CFS web.

4. Conclusions

Based on the results from the experimental and analytical work completed in this research, the following significant conclusions may be drawn:

1. By looking at the mid-span deflection, the 250mm beam appears to have stiffer behavior compared to the 150 mm leading to higher section capacity.
2. The infilled CFS beam showed higher capacity compared to normal slender sections by reaching to the yield stress.
3. The proposed analysis and design methods herein have been found to be able to predict the ultimate strength capacity of the proposed composite system by using the procedure adopted from EC4.
4. Using bent-up rebar as shear connector gives a high degree of composite action ratio which means attaining higher yield strength of the composite section. It's because the continuous shear connector has helped move upward the neutral axis of the composite section, thus prevented the top flange of CFS to buckle under compression.
5. The experimental specimens of the partially encased composite beams consistently showed ductile flexural behavior. The high flexural strength and ductility observed in each specimen suggest that the partially encased composite beam can be adopted in future practice.
6. The partially encased steel beam is an efficient technique for enhancing its flexural capacity.

Acknowledgments

The help, support, and collaboration to all members of the team are greatly acknowledged. The author also acknowledges the support of Universiti Teknologi Malaysia – Construction Research Centre (UTM-CRC) and Institute for Smart Infrastructure and Innovative Construction (ISIIC) for the funding grant number 4B235 for research and development under the supervision of Prof. Mahmood Md. Tahir and the help of Dr. Shek Poi.

IOP Conf. Series: Materials Science and Engineering **513** (2019) 012038 doi:10.1088/1757-899X/513/1/012038

References

[1] Dubina D, Landolfo R, Ungureanu V 2011. *Design of cold-formed steel structures.: Eurocode 3:*
design of steel structures. Part 1-3, Design of cold-formed steel structures. Eccs.

[2] Hsu CT, Punurai S, Punurai W, Majdi Y 2014 Jul 15. New composite beams having cold-formed
steel joists and concrete slab. *Engineering structures.*; 71:187-200.

[3] Salih MN, Tahir MM, Mohammad S, Ahmad Y 2018 Nov 24. Behavior of Boxed Cold-Formed Steel as Composite Beam with Rebar as Shear Connector. In International Congress and Exhibition" *Sustainable Civil Infrastructures: Innovative Infrastructure Geotechnology*" pp. 324-332. Springer, Cham.

[4] Yu WW 2000 Jun 26. *Cold-Formed Steel Design*. John Wiley & Sons.

[5] Lee YH, Tan CS, Mohammad S, Md Tahir M, Shek PN 2014. Review on cold-formed steel connections. *The Scientific World Journal.*

[6] Ahmad S, Masri A, Saleh ZA 2018 Sep. Analytical and experimental investigation on the flexural
behavior of partially encased composite beams. *Alexandria engineering journal.* 1;57(3):1693-712.

[7] EN BS.206-9 2010, Concrete—Part 9: *Additional Rules for Self-Compacting Concrete (SCC).* British Standard Institution, London.

[8] EN BS.12390-3 2009. *Testing hardened concrete—part 3: Compressive Strength of Test Specimens.* British Standard Institution, London.

[9] EN BS.10002-1 2001. *Tensile testing of metallic materials. Method of test at ambient temperature.* British Standard Institution, London.

[10] Tahir MM, Lawan MM, Saggaff A, Mirza J 2016. Influence of Shear Connector Size on Ultimate
Strength in Composite Construction with Cold-Formed Steel Channel Lipped Section. *Rev. of Industrial Engineering Letters*;3(1):1-0.

[11] Kvočák V, Kožlejová V, Dubecký D 2012 Jan 1. Analysis of encased steel beams with hollow cross-sections. *Procedia Engineering*: 40:223-8.

[12] Kvočáka V, Drab L 2012 Jan 1. Partially-encased composite thin-walled steel beams. *Procedia Engineering*; 40:91-5.

[13] EN BS.1994-1-1 2004. *Design of Composite Steel and Concrete Structures-Part 1-1: General Rules and Rules for Buildings.* British Standard Institution, London.

[14] EN BS.1993-1-5 2006. *Design of Steel Structures. Plated Structural Elements.* British Standards Institution, London.

[15] EN BS 1993-1-1 2005. *Design of steel structures. General rules and rules for buildings.* British Standard Institution, London.

[16] EN BS 1993-1.3 2004. *Design of steel structures. General rules. Supplementary rules for cold-formed members and sheeting.* British Standard Institution, London.

IOP Conf. Series: Materials Science and Engineering **513** (2019) 012039 doi:10.1088/1757-899X/513/1/012039

Comparison of behavior between hollow and composite K-joints under sustained loading and corrosion

S Saleh[1*], M A Saqif[1] and F Ramiz[1]

[1] Department of Civil Engineering, University of Asia Pacific, Dhaka, Bangladesh

[*]Corresponding author: shameer.saleh@uap-bd.edu

Abstract. Concrete filled steel tubular (CFST) truss structures are widely being used in bridges and transmission towers all over the world. Design of connections is one of the most critical issues in truss structures; and it becomes even more complicated when corrosion causes loss of outer steel tube section. However, behavior of composite joints is among the least understood topics in structural engineering today. Previous attempts (numerical and experimental) to study the behavior of composite joints have focused mainly on the effect of sustained loading. But the widespread use of CFST structures in harsh, marine environment necessitates observing the performance of composite joints under corrosion alongside long-term loading. The aim of this paper therefore is to study the combined effect of chloride corrosion and sustained loading on circular composite K-joints through finite element analysis (FEA). The results thus obtained have been presented and verified against experimental observations of previous researchers. For a side by side comparison with composite K-joints, FEA has also been established for circular hollow section (CHS) K-joints. In addition to these, failure modes and load-deformation characteristics of both the composite and hollow K-joints have been thoroughly investigated under various practical loading cases and different corrosion scenario. From numerical analyses, it has been observed that corrosion in infilled chord leads to only 4% loss of joint strength whereas, corrosion of equal intensity in hollow braces decreases the joint capacity by 35%. Finally, joint strength and ductility have been predicted based on the full-range numerical analyses. Concrete infills have been found to enhance the joint strength by 2.5-3.0 times.

1. Introduction

For very long span bridges and tall transmission towers, lightweight hollow steel tubular sections are preferred over much heavier concrete sections. However, the inherent weakness of hollow steel sections to buckling and corrosion limits their use. It is apparent that both concrete and steel have inherent weaknesses that limit their applicability as structural members. Concrete is weak in tension and its behavior is affected by creep and shrinkage, whereas steel is prone to buckling and extremely susceptible to moisture. Steel-concrete composite structures utilize the strength of both the materials while significantly limiting their weaknesses. In concrete filled steel tubular (CFST) structures the infilled concrete prevents the outer steel tube from buckling while confinement of concrete by the outer tube results in higher ductility, greater section capacity and better fire and seismic resistance [1]. Consequently, improved performance and greater durability has led to the prevalence of CFST structures in marine environments (Figure 1). A fully rational understanding of the behavior of composite joints is necessary to design CFST structures. Still the amount of research dedicated to studying the behavior of composite joints is fairly low as of today.

Several researchers in the past have studied the behavior of single composite members subjected to corrosive environment alongside sustained loading [2-4]. In those studies CFST members exhibited significantly better performance compared to CHS members. CFST members could resist a much higher load without inward buckling due to the presence of infill concrete. Also CFST members exhibited greater cross-sectional ductility and the members underwent far less load redistribution due to corrosion compared to CHS members. The behavior of the composite joints is greatly influenced by several cross sectional parameters such as steel to concrete ratio at composite chord (α), brace to chord diameter ratio (β), chord radius to thickness ratio (γ) and brace to chord thickness ratio (τ); which are affected during the corrosion process [5].

Figure 1. A bridge with circular CFST K-joints.

In this paper, the behavior of CFST K-joints has been investigated and their performance has been evaluated against CHS K-joints under sustained loading and various cases of chloride corrosion. A FEA model is established and verified against previously published experimental observations. The model is then utilized to perform mechanism analysis on both types of joints. Findings from full-range numerical analysis including load-deformation pattern and stress distribution have been presented and finally the joint strength is predicted.

2. Finite element analysis (FEA) model

Numerical models of CHS and CFST K-joints were developed using the finite element package ABAQUS to account for the combined effects of sustained loading and chloride corrosion. Dimensions and properties of K-joint samples in this research are adopted from experimental studies previously conducted [6]. The model is a further development of numerical behavior analysis of composite joints carried out in [7]. The combined effect of sustained loading and corrosion has been replicated by modifying the material properties in finite element model and by adopting 'Model change – Remove elements' interaction.

2.1. Development and configuration of K – joint

In the FEA model, two different types of material properties were selected for steel in joint tube and in end-plates. An elastic-plastic model with five distinct stages based on a previous study [8] was implemented to represent the steel behavior. For CFST K-joints, a damage-plasticity model was implemented to account for the core concrete plasticity behavior under both short-term and long-term loading with corrosion. The long-term stress-strain relationship was modified according to previous studies presented in [8] to reflect the time dependent factors of infilled concrete; such as creep and shrinkage. For the purpose of this paper, CHS-6 and CFST-6 with a brace and chord thickness of 6 and 10 mm respectively were chosen [6]. For CFST joint, the chord was infilled with concrete, whereas the braces were hollow.

The steel tube component of the joint has been modeled with 8-node linear 3-D solid element with reduced integration (C3D8R) to consider the effective section loss. Other components of the joint such as the infill concrete and end-plates are also simulated using the same element type. A proper meshing technique is crucial for the convergence and accuracy of FE simulation as composite K-joints are relatively complex in shape with particular local profiles. Structured meshing technique has been adopted where possible in order to achieve a proper element shape. However, the connection part of chord-brace tube was too complex to assign structured meshing due to irregularity of the plane. For the sake of simplicity, sweeping meshing technique was applied in the complicated connection area (Figure 2). A mesh convergence study was carried out to identify a suitable mesh size which provides tolerable simulation accuracy in reasonable computation time.

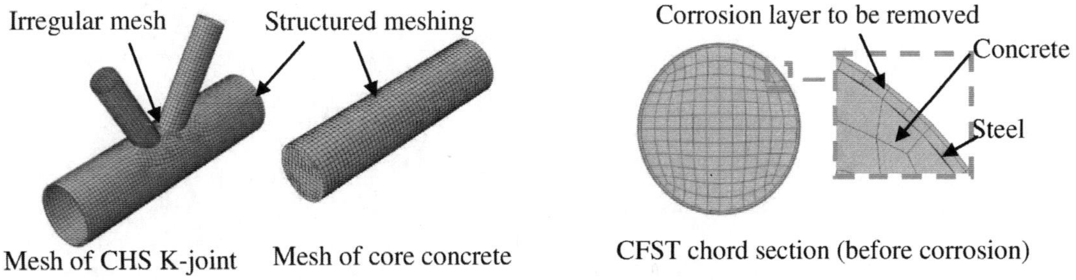

Figure 2. Meshing technique of CHS and CFST K-joint modeling.

2.2. Interface, boundary conditions and simulation of long-term loading with corrosion

The boundary conditions considered in this research were developed based on previously conducted studies [7, 9 and 10]. Axial static loading was applied in K-joints and the practical scenario was reflected by assessing three different boundary conditions, i.e., (1) compression load at chord, (2) compression load at brace and (3) tensile load at brace. Figure 3 illustrates the three distinct boundary conditions. Tie constraints were produced to simulate interaction between steel joint and end-plates. The contact interface between core concrete and steel tube in CFST joint was formed between two matching surfaces and by defining behavior along two orthogonal directions. These two surfaces are allowed to be separated from each other, whereas penetration into one another is not permissible. A 'Hard contact' model in normal direction and 'Coulomb friction model' in the tangential direction are adopted.

Corrosion is a fairly complex and uncertain natural phenomenon; and no numerical model has been able to replicate corrosion effects precisely. Therefore, current researches suggest considering corrosion at a constant rate which has a long-term uniform effect on outer steel of the joint tube [3, 4]. In this FEA model, the mass reduction and diameter loss of steel tube associated with corrosion has been simulated uniformly by a 'Model change – Remove elements' interaction. Corrosion is applied to

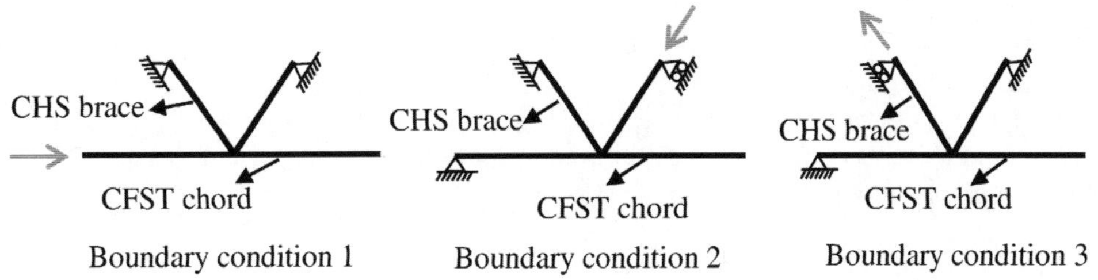

Figure 3. Typical boundary conditions considered for the study.

the model in four steps. First, partitioning the member affected by corrosion and defining corrosion set element; second, meshing of tube with several elements according to the corrosion thickness; third, creating a separate 'corrosion step' and selection of the corrosion region to deactivate in this step; and finally, adopting the new stress-strain relationship to represent steel-concrete confinement in corroded tubes (for CFST joints). Modified stress-strain relationship was applied by using the updated reduced dimensions of steel tubes under corrosion.

IOP Conf. Series: Materials Science and Engineering **513** (2019) 012039 doi:10.1088/1757-899X/513/1/012039

3. Verification of the FEA model

For verification of the results, two distinct approaches were used. One is comparing the load-displacement relation from numerical analysis with that of the experimental investigation carried out in [6]; the other is comparing the failure modes.

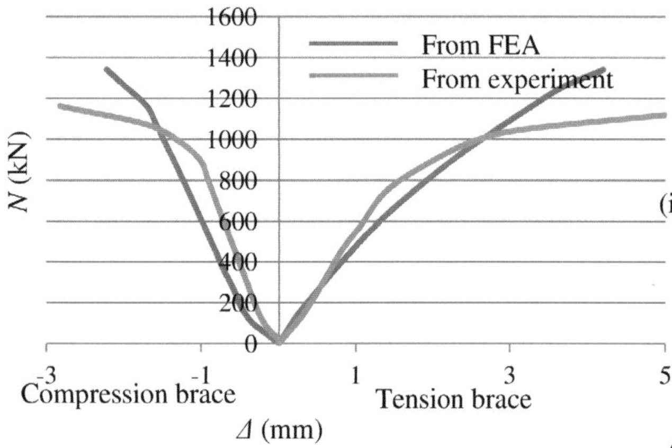

Figure 4. Comparison of the $N - \Delta$ relationship between experimental values and FEA.

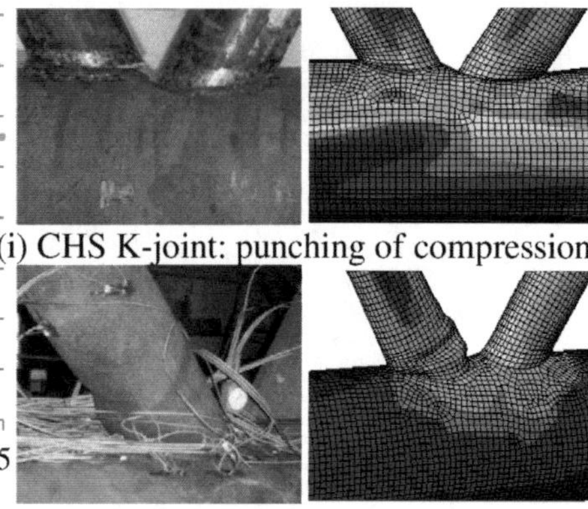

(i) CHS K-joint: punching of compression brace

(ii) CFST K-joint: outward local buckling in compression brace

Figure 5. Comparison between the experimental and numerically predicted failure modes.

It is apparent from Figure 4 that the results from FEA agree to the experimental values to a large extent. Chord stiffness has remained almost similar in FEA, but the ultimate load carrying capacity has exceeded the experimental value by almost ten percent. This can be attributed to ignoring the effect of welding at the brace-chord intersection in FEA. FEA has also depicted the failure patterns of CHS and CFST K-joints fairly accurately as can be seen from Figure 5. Failure of CHS K-joints occurs due to punching of the compression brace into the chord. However in CFST K-joints, infill concrete at the chord prevents punching failure. Outward local buckling of the compression brace along with bulging at the connection area is observed similar to the experimental observation. From the comparisons, it can be said with certainty that FE models are capable of reflecting the actual behavior of CHS and CFST K-joints. Hence further analysis run on the models considering sustained loading and chloride corrosion will be a near representation of the actual scenario.

4. Analytical behavior

4.1. Typical failure modes for CHS K-joints

Two failure modes are common with CHS K-joints (Figure 6). Since there is no infill concrete at the chord, the compression brace is likely to punch into the chord. At higher loads local buckling of chord takes place alongside punching of the compression brace.

4.1.1. Mode A: Punching in the chord

Punching of the compression brace occurs when moderate load is applied at the chord. Punching of brace is noticed at 20 mm chord displacement. Mises stress at the chord-brace connection is found to be around 384 MPa. Moderate chord plastification is noticed with corresponding Mises stress in the range of 260-320 MPa. No considerable local buckling is observed at the chord away from the connection.

4.1.2. Mode B: Punching in the chord combined with local buckling

This combined mode of failure was obtained when applied load at chord tube was significant. Noticeable plastic deformation of compression brace is observed along with severe punching, with maximum Mises stress of 497 MPa at critical brace region. Plastification of compression brace is observed at stress level 280-350 MPa, while Mises stress at tension brace remains at moderate level (250-280 MPa). At time of compression brace failure, hollow chord tube fails due to inward buckling away from the connection. Buckling of the chord occurs at around 50 mm chord displacement; and stress at chord region remains at moderate level (270-320 MPa).

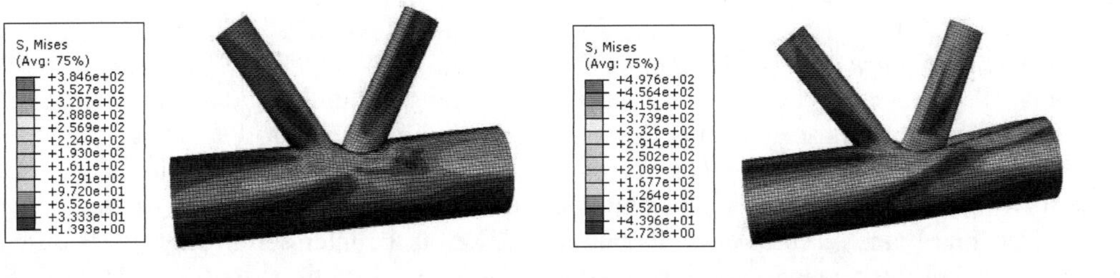

(a) Punching in the chord (b) Punching in the chord combined with local buckling

Figure 6. Typical failure modes for CHS K-joints.

4.2. Typical failure modes for CFST K-joints

Presence of infill concrete in the chord makes CFST K-joints bear higher load than CHS K-joints as compression brace can no longer punch into the chord. Depending on different boundary condition, three types of failures have been observed.

4.2.1. Mode A: Plastification of chord

When moderate amount of load is applied at the chord, this failure mode is noticed (Figure 7). A significant portion of the chord undergoes plastification at the chord-brace intersection area as the Mises stress stands to be 431 MPa. Inside the chord, the outer region of the core concrete receives the maximum stress (around 58 MPa) on the areas where chord plastification is prominent. At the braces the Mises stress stays in the 280-340 MPa range and with no observable local buckling.

(a) Overview of the joint (b) Inner concrete (cut view)

Figure 7. Chord plastification of CFST K-joints.

The infill concrete in the chord increases the resistance of the CFST K-joint since the core concrete prevents the inward buckling of the compression braces, a phenomenon very common to CHS joints.

4.2.2. Mode B: Chord plastification combined with local buckling of the brace

This failure mode can be observed when a large load is applied at the chord end (Figure 8). At the onset of failure, the chord tube at the connection region reaches plastic state, and then noticeable local buckling of the compression brace takes place.

(a) Overview of the joint (b) Inner concrete (cut view)

Figure 8. Chord plastification and local buckling of brace of CFST K-joints.

IOP Conf. Series: Materials Science and Engineering **513** (2019) 012039 doi:10.1088/1757-899X/513/1/012039

High stress concentration is found at the region of local buckling as the hollow brace section fails at Mises stress of 497 MPa. Tension brace and chord remain moderately stressed (310-350 MPa) when the joint fails. Only the outer region of core concrete receives stress at the joint (nearly 40 MPa), inner regions remain almost unstressed.

4.2.3. Mode C: Premature local buckling of the brace
Initially a small region in the tension brace enters the plastic stage in this mode. However, as loading continues, localized stresses accumulating in the compression brace lead to yielding in the connection region. Unlike the first two failure modes, here the entire compression brace fails in local buckling as large plastic deformation sets in while the chord remains elastic (Figure 9). Mises stress at the critical sections of the brace tube reaches 493 MPa but the stresses at the intersection stays within the elastic limit. At the critical intersection region, core concrete is stressed up to around 55 MPa. Stress at the concrete surface in this mode is similar to Mode A and approximately 38% higher than Mode B.

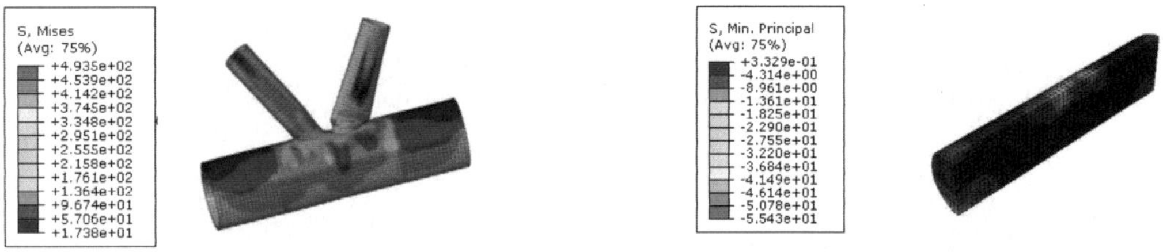

(a) Overview of the joint (b) Inner concrete (cut view)
Figure 9. Premature local buckling of brace of CFST K-joints.

Failure modes B and C are quite similar except for chord plastification in the former one. Compared to CHS joints, CFST joints are stronger since punching of compression brace into the chord is prevented by the infilled concrete. For both CHS and CFST joints, compression brace is found to be the critical element.

5. Full range analysis

5.1. Typical full-range load-displacement relationship
Full-range analyses of CHS and CFST K-joints have been performed using FEA models. Figure 10 demonstrates the typical load *(N)* versus displacement *(Δ)* relationships of CFST K-joints under short-term loading and long-term loading with corrosion.

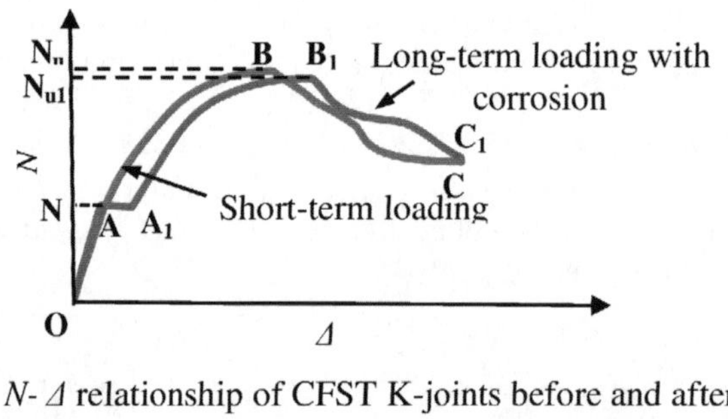

Figure 10. Typical *N- Δ* relationship of CFST K-joints before and after corrosion.

Curve O-A-B-C shows the short-term loading condition before corrosion has taken place. It is apparent from the curve that CFST K-joints exhibit excellent ductile behavior as the joint fails long after attaining its ultimate capacity (N_u). Curve O-A-A_1-B_1-C_1 displays the *N-Δ* relationship of CFST K-joints under long-term loading and corrosion. This curve has four distinct parts. At elastic stage (O-A), the response is elastic up to application of the load N_l. At corrosion stage (A-A_1), the load N_l is kept constant and uniform corrosion is applied, thereby creating the combined effect of sustained loading and corrosion. Corrosion is applied in the form of loss of thickness in outer steel tube which results in deformation growing up to A_1. Time dependent behaviors of concrete, e.g. creep and shrinkage also manifest at this stage. Corrosion not only gives rise to stress redistribution between steel and concrete but it also adversely affects the confinement of core concrete. The following elastic-plastic stage (A_1-B_1) is marked by continued loading up to ultimate joint capacity (N_{u1}). Significant loss of stiffness and ultimate strength is prominent when corrosion takes place alongside long-term loading. At descend stage (B_1-C_1), deformation keeps increasing while the axial load decreases. The declining limb of the curve reflects the degree of corrosion and the extent of composite action in the CFST chord. A noteworthy finding is that, effect of corrosion on the declining stage is much less severe than the effect of composite action. Overall performance of CFST K-joints is significantly better under long-term loading combined with corrosion compared to CHS K-joints.

5.2. Effects of different corrosion patterns

Full-range analysis of CHS K-joints in Figure 11(a) shows significant reduction of joint capacities as corrosion occurs. A 13% reduction in strength is noticed for a 30% thickness loss due to corrosion either in chord or brace. When corrosion occurs throughout the joint, the joint capacity decreases even further. For a thickness loss of 30% in steel tubes, joint capacity reduction is noticed as 26%. From load-displacement relationship in Figure 11(b) of CFST K-joints, it is evident that corrosion in chord has an insignificant effect on the performance of the joints in terms of ductility and ultimate strength. When the chord thickness decreases by as much as 30%, the corresponding loss of joint capacity is a mere 4%. The infilled concrete improves the resistance of the chord; and thereby prevents premature local buckling, leading to a stronger and more ductile joint behavior. As opposed to corrosion in chord, the joint capacity is severely impaired when brace member is corroded. For a brace thickness loss of 30% due to chloride corrosion, joint capacity reduction of nearly 35% has been observed. Besides reduction in strength, brace corrosion also decreases the ductility of the joints. The chord displacement at ultimate load observed for chord corrosion is 35mm which falls drastically to just 18mm when brace is exposed to corrosion.

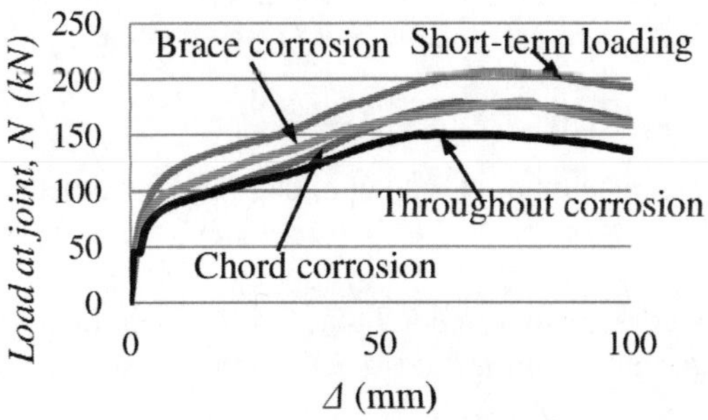

(a) CHS K-joints

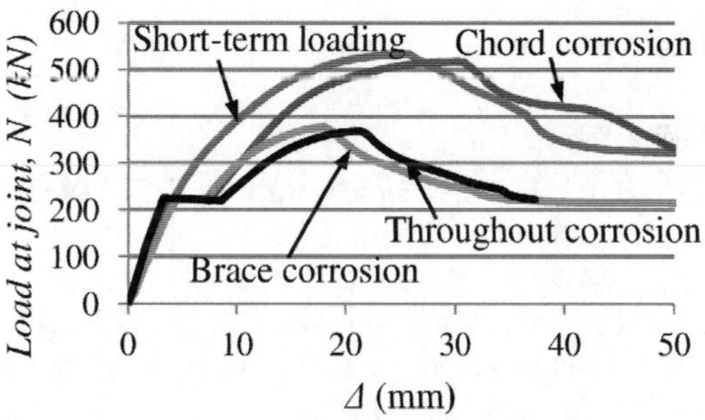

(b) CFST K-joints

Figure 11. *N-Δ* relationship of K-joints for different corrosion conditions.

Ultimate load at failure for different boundary conditions and corrosion patterns has been presented in Table 1. It can be seen that infilled concrete improves joint capacity significantly. CFST joint strength is found to be around 250 – 300% compared to CHS joints.

Table 1. Comparison of CHS and CFST joint capacities under various corrosion conditions.

Corrosion condition	Load at chord (kN)		Load at compression brace (kN)		Load at tension brace (kN)	
	CHS 6	CFST 6	CHS 6	CFST 6	CHS 6	CFST 6
Before corrosion	205	533	223	594	259	522
Chord only corrosion	178	517	212	583	215	498
Brace only corrosion	178	376	210	397	207	369
Throughout corrosion	150	368	159	393	198	357

As corrosion sets in, joint capacities reduce regardless of the boundary condition. Corrosion throughout the joint has almost the same effect as corrosion only in the braces of CFST joints. From the full-range analysis, it can be concluded that regardless of boundary condition and corrosion case, the hollow compression brace is the critical member in K-joints.

6. Conclusion

Based on the research conducted in development of this paper, following conclusions can be drawn:
(1) FEA of composite joints can be used to verify experimental observations of previous researchers, as the results obtained from ABAQUS agree to the experimental investigations to a large degree.
(2) Two distinct failure modes for CHS and three for CFST K-joints have been observed for long-term loading with corrosion. Hollow compression brace has been identified as the critical member in the connection regardless of the boundary condition.
(3) From full-range analysis, it is apparent that corrosion of chord has only moderate influence on the behavior of the joints. However, corrosion on the braces affects the joint performance significantly.
(4) Punching shear is the predominant mode of failure in CHS joints. Punching starts at low displacement and although load carried by the chord keeps increasing during FEA, joint had already failed in punching, exhibiting poor ductility.
(5) Load carrying capacity of CFST K-joints starts to decrease when interaction between steel and concrete is affected, i.e. high degree of composite action leads to higher joint strength.
(6) Infill concrete does not significantly contribute in bearing the loads applied to CFST K-joints. Rather its function is to delay the yield propagation and avoid premature buckling. Resistance and ductility of the joints improve greatly when concrete infills are used inside hollow steel tubes.

References

[1] Han L H, Li W and Bjorhovde R 2014 Developments and advanced applications of concrete-filled steel tubular (CFST) structures: members. *Journal of Constructional Steel Research.* **100**(1): 211-28.

[2] Han L H, Hou C and Wang Q L 2012 Square concrete filled steel tubular (CFST) members under loading and chloride corrosion: Experiments. *Journal of Constructional Steel Research.* **71**: 11-25.

[3] Hou C, Han L H and Zhao X L 2013 Full-range analysis on square CFST stub columns and beams under loading and chloride corrosion. *Thin-Walled Structures.* **68**: 50-64.

[4] Han L H, Hua Y X, Hou C and Wang Q L 2017 Circular concrete-filled steel tubes subjected to coupled tension and chloride corrosion. *Journal of Structural Engineering.* **143**(10) :04017134-1.

[5] Wardenier J, Packer J A, Zhao X L and Vegte G J 2010 *Hollow sections in structural applications, Delft: Bouwen met Staal.*

[6] Huang W, Fenu L, Chen B and Briseghella B 2015 Experimental study on K-joints of concrete-filled steel tubular truss structures. *Journal of Constructional Steel Research.* **107**:182-93.

[7] Saleh S, Hou C, Han L H and Hua Y X 2018 Numerical behavior of composite K-joints subjected to combined loading and corrosive environment. *12th International Conference on Advances in Steel-Concrete Composite Structures ASCCS 2018* (Valencia-Spain).
[8] Han L H, Tao Z and Liu W 2004 Effects of sustained load on concrete-filled hollow structural steel columns. *Journal of Structural Engineering, ASCE*. **130**(9): 1392–404.
[9] Song Q Y and Han L H 2010 FE analysis of composite tubular K-joints subjected to static loading. *Tubular Structures XIII – Young (ed);* (Hong Kong: The University of Hong Kong) ISBN 978-0-415-58473-9.
[10] Hou C, Han L H and Mu T M 2017 Behaviour of CFDST chord to CHS brace composite K-joints: Experiments. *Journal of Constructional Steel Research*. **135**:97-109.

IOP Conf. Series: Materials Science and Engineering **513** (2019) 012040 doi:10.1088/1757-899X/513/1/012040

Identifying occurrences of accident at work place in terms of occupational safety on roads and bridges infrastructure in Indonesia

Mairizal[1,2] , Edrizal[1,3] , Mohammad Ismail[1*] and Rosli Mohamad Zin[1]

[1] School of Civil Engineering, Faculty of Engineering,Universiti Teknologi Malaysia, 81310, Johor, Malaysia

[2] Program of Industrial Engineering, Faculty of Industrial Technology, Universitas Pamulang, 15417, Banten, Indonesia

[3] Post Graduate of Civil Engineering, Universitas Bung Hatta, 25133, Padang, West Sumatera, Indonesia

[*]Corresponding author: mohammad@utm.my

Abstract. The economic growth of a country will be greatly influenced by road and bridge infrastructure that serves as a connecting flow of goods and human transportation. There are still some road and bridge construction projects in Indonesia that show the occurrence of accidents because lack of awareness by the management of the adverse effects of safety. Therefore, the author tries to find solutions to the improvement measures in minimizing the occurrence of occupational accidents by providing safety knowledge to the Project Managers and related staff, to be more concerned about the safety and the environment around the project. Based on data collected from the Ministry of Public Works and Indonesian People's Housing, there are several work accidents in toll road projects in Indonesia from August 2017 to February 2018. When viewed from several of these accidents, it can be concluded that the main cause is the lack of knowledge of the management staff about the need to enforce safety regulations, especially in the work of lifting object to the upper structure. Therefore, a commitment is needed to implement serious occupational safety regulations from all stakeholders who play a role in developing this infrastructure project.

1. Introduction

The Indonesian government has in recent years put in place a robust institutional framework to support its infrastructure plans. In the last year it has announced 13 economic policy packages ("deregulation packages") focusing on the deregulation of investment and tax incentives. The government expects these deregulation packages to improve Indonesia's competitiveness and help to attract investment by cutting bureaucracy and providing greater legal and business certainty. Overall, total government infrastructure spends in Indonesia increased by a substantial 51% from IDR 139 trillion (US$ 11.7 billion) in 2014 to IDR 209 trillion (US$ 15.5 billion) in 2015. While this represents significant progress in pursuing the government's ambitious infrastructure expansion plans, it was still below the planned increase of 63% [1].

Improving infrastructure is critical to stimulating inclusive growth in Indonesia. Inadequate infrastructure has hampered growth and poverty reduction, and poses significant obstacles to doing business in Indonesia. Poor road infrastructure contributes to logistics cost in Indonesia of 24% of GDP. Food security is affected by poor irrigation infrastructure: since 2005, irrigated land with good infrastructure has decreased from 78% to less than 50% today. Access to basic urban water and sanitation services lags far behind other middle-income countries. In 2014, just 80.7% of the urban population had access to improved water sources, while 61.06% had access to improved sanitation facilities. Such gaps in basic infrastructure provision are the result of both low government infrastructure investment and under-utilization of the capital budget [2].

Development of road and bridge infrastructure in Indonesia through the Ministry of Public Works and Public Housing (PWPH) in the Strategic Plan of the Ministry of PWPH 2015-2019, the activities will pursue time completion, cost effective and good quality [3]. However, by neglecting the rules of safety and health in the implementation of such work, it seems that previously, there were some events that resulted in work accidents in several major project sites, whereby the safety factor appears ignored in the project management. In the management of work safety, it is clearly seen that the competence factor of workers and employees are poorly managed.

"Safety First" should be a key element of project integration management because it affects all of the project management knowledge areas outlined in *A Guide to the Project Management Body of Knowledge* (*PMBOK® Guide)*. "Safety First" must be integrated into all aspects of the project to ensure success. This integration needs to begin with project scope management. The project manager needs to ensure that the project charter adequately identifies the importance of developing and maintaining a safe working environment for the project team. Each project manager should insist that a project-level Safety Plan is developed implemented and enforced. The project specific plan should be a subset of the corporate safety plan of either the client or the implementing organization. This top-tier project document will provide a framework for the implementation of safety procedures and safe work practices for all project members and sub-contractors. The project manager must believe that an injury-free work environment is possible [4].

Sistem Manajemen Kesehatan, Keselamatan Kerja dan Lingkungan (SMK3L) is an inseparable part of the labour protection system in Indonesia [5] and also for construction services work in order to minimize and avoid everyone from the risks of moral or material loss, lost of working hours, and the safety of people and the surrounding environment. This will be able to support an improvement which has effective and efficient project performance in the development process of the infrastructure projects. Road and bridge infrastructure projects are at high risk in terms of safety, especially flyovers and bridges that involved many workers and machine tools. Therefore, it is necessary to manage SMK3L in those projects by Project Manager which has integrated system together with their assistants who have experienced, expertise and care about work safety.

Recently, the case of work accidents that happened to construction workers on development projects in DKI Jakarta, was widely reported by the media. In the latest news, one of the main girder of Jakarta Becakayu toll road project was collapse, 7 workers are seriously injured. The previous tragedy occurred in the Double Double Track (DDT) project of Jalan Matraman Raya claiming the lives of 4 workers and other workers injured by the collapse of one of the concrete transport cranes. In January 2018, accident at the Light Rapid Transit (LRT) project on the Pulogadung line also seriously inflicted five workers [6].

Based on data from Badan Penyelenggara Jaminan Sosial (BPJS) Ketenagakerjaan the number of work accident cases continues to decline. In 2015 there were 110,285 work accidents, while in 2016 there were 105,182 cases, which decreased by 4.6%. Until August 2017 there were 80,392 cases [7].

Infrastructure development projects in Indonesia is still very weak in implementation of safety rules, even though there are several regulations such as Occupational Safety and Accidents (OSA/K3) in the regulation of Law no. 1 of 1970 on Occupational Safety [8]. Government Regulation no. 50 of 2012 on the Implementation of Health and Safety Management System [9], and Law No.13 of 2003 on Employment which also discuss about Health and Safety [10].

The spirit of the Indonesian government in boosting infrastructure development, hampered by number of work accident from the workers accordingly. In fact, the construction sector becomes a very important part in the development of a State. The Construction Services sector which is one of the main sector that support the Indonesian economy, and employ a significant number of labour. Based on the research note from the team of Economist of the Bank Mandiri, where the number of construction workers in the year 2017 amounted to about 7.2 million people [11]. On the other hand, the construction services sector is considered to be one of the sectors that are at high risk for workplace accidents.
For the type of occupational accidents [12], mentions that there are two types of jobs in the dangerous construction sector, namely work carried out at height and excavation work. For work at height the accident occurred in the form of workers falling from a height and for excavation work, inhaled toxic gas. Occupational accident risk research in the apartment building projects undertaken by [13] shows the greatest risk is the lifted materials falling from height, electrocuted, hit by equipment, and workers fall from height. With all the issues highlighted above, it is therefore timely for the authors to seriously look at this issue.

2. Methodology research

Every construction company should have involved the project managers and his/her team work who are experienced in carrying out good work, which does not only refer to time, cost and quality. However, mastery of occupational health and safety issues is a mandatory requirement to be implemented. In terms of enforcing regulations that are requirements that have been set out in a work contract and several regulations, the role of the project manager most important crucial.

2.1. Collection and analysis of data

In the context of this study, the collection of articles from the mass media related to workplace accidents was carried out on toll road and bridge construction projects that occurred in several regions in Indonesia within 1 year. The article taken is an online newspaper article obtained by Google's search. Key words used are: 'Infrastructure construction accidents', 'Toll road work accidents', 'Overpass work accident', 'Dead workers', and 'Workers fall'.

This study only collected a number of articles related to work accident cases in toll road and bridge infrastructure construction projects that occurred in Indonesia from August 2017 to February 2018. These data were classified and written in the form of Ms. Words and given name based on the date, month and year of the event. At the initial stage, by gathering news online using various google search engines with keywords, there were 15 articles on project work accidents which are toll road and bridge infrastructure construction projects were found.

3. Results and discussion

In general, occupational accidents or structural failures in the construction phase are mostly caused by a lack of discipline on the part of workers. The desire to speed up projects is one of the reasons why certain stages are passed without adequate care. According to Handbook of Occupational Health and Safety, construction safety is measured by the compliance of project implementers toward various stages, starting at planning and continuing through the entire life cycle of project. The design must be equipped with quality assurance in the method of implementation, proper work equipment and materials that comply with safety standards. The contractor selected in the bidding process is responsible for safety and preventing accidents — thus reviewing a company's history of safety should be part of selecting a contractor. The next stage is the implementation stage with zero accidents. In this case, efforts are necessary to supervise standard operating procedures, work equipment and materials used. In the last stage, inspections are critical to assessing the project's safe completion [14].

Based on data collected from the Ministry of Public Works and Indonesian People's Housing [15] and some news within various media [16, 17, 18, 19, 20, 21, 22], There are several work accidents occurring in toll road projects in Sumatera and Java Island from August 2017 to February 2018 as follows (Table.1)

Table 1. Construction work accident data of infrastructure in Indonesia.

No.	Dated	Project Title	Location	Kind of accident	Cause of accident
1.	Aug 04, 2017	Palembang Light Rail Transit	Palembang	Falling from the height	Negligence of officers
2.	Sept 22, 2017	Bogor-Sukabumi Toll Road	Bogor	Failed of lifting	Negligence of officers
3.	Oct 16, 2017	Jakarta Light Rail Transit	Jakarta	Failed of lifting	Negligence of officers
4.	Oct 26, 2017	Bogor Outer Ring Road	Bogor	Failed of lifting	Negligence of officers
5.	Oct 28, 2017	Toll Road Pasuruan Probolinggo	Pasuruan	Failed of lifting	Negligence of officers
6.	Nov 3, 2017	Jakarta Mass Rapid Transport	Jakarta	Failed of lifting	Negligence of officers
7.	Nov 16, 2017	Jakarta-Cikampek Toll Road	Bekasi	Failed of lifting	Negligence of officers
8.	Dec 09, 2017	Manggarai-Jatinegara DDT	Jakarta	Scaffolding collapse	Negligence of officers
9.	Dec 09, 2017	Ciputra Pinggan Bridge	Banjar	Failed of lifting	Negligence of officers
10.	Dec 30, 2017	Pemalang-batang Toll Road	Pemalang	Failed of lifting	Negligence of officers
11.	Jan 02, 2018	Depok-Antasari Toll Road	Jakarta	Failed of lifting	Negligence of officers
12.	Jan 22, 2018	Jakarta Light Rail Transit	Jakarta	Failed of lifting	Negligence of officers
13.	Feb 04, 2018	Manggarai-Jatinegara DDT	Jakarta	Failed of lifting	Negligence of officers
14.	Feb 05, 2018	Soekarno-Hatta Airport Main Road	Banten	Embankment sliding	Negligence of officers
15.	Feb 20, 2018	Becakayu Toll Road	Jakarta	Pier collapse	Negligence of officers

Referring to all the accident event data shown in Table 1 above, it can be classified against the causes of accidents and the types of accidents that occur as summarised in Table.2:

Table 2. Classification of causes accidents.

No.	Kind of accident	Total	Cause of accident
1	Falling from the height	1 case	Negligence of officers
2	Failed of lifting	11 cases	Negligence of officers
3	Scaffolding collapse	1 case	Negligence of officers
4	Embankment sliding	1 case	Negligence of officers
5	Pier collapse	1 case	Negligence of officers

Looking at some of the above incidents, it is clear that the safety aspects of work are somewhat marginalized in order to pursue the progress of work demanded by the time of completion known to the public. There is a need for commitment from all stakeholders who play a role in the development of these infrastructure projects, so that the same focus is reminded each other about the risk of danger that will occur in the work place. The construction company is one of the main executor in this work should be very careful in the management of work safety that refers to the safety regulations that are referred to in the work.

Table 2 shows that, out of 5 cases, there are 4 types of accidents it is generally caused by the negligence of the safety officers in monitoring and preventing work accidents. However, eleven cases type of accident was caused by the process of lifting objects in construction structures.

Besides that, almost all accidents were related to the problem of lifting work using cranes, which many must pay attention to (Crane, Lifting Accessories, Scaffolding, Rigging Plan, etc).

4. Conclusions

Based on the description in the section above, the parties involved in the construction work activities seemed do not understand the meaning of work safety starting from the project manager and the related staff involved in these projects. Where in every activity they must have observed the potential hazards that will arise and be reminded through tools-box meeting, safety briefings and safety induction. And also put up the safety signs, banners, posters and reminder which is easy to see and read.

Also, it is required that every party involved in regulating and overseeing the performance of national construction service companies in Indonesia, to conduct an assessment of the project manager and all of his employees in implementing occupational health and safety programs in his projects.

In addition, related institutions that play a role in providing competency test certification for occupational safety officers in construction companies, to be even more stringent in certifying their assessments.

Acknowledgement

The authors gratefully acknowledge RMC Universiti Teknologi Malaysia (grant No. 16H94) and the Ministry of Higher Education (grant No. 4F528), for providing financial support and assistance for the present research.

References

[1] PWC Indonesia 2016, *Indonesian infrastructure : Stable foundations for growth* (Jakarta, PWC)
[2] ADB Indonesia 2016, *Republic of Indonesia: Accelerating Infrastructure Delivery through Better Engineering Services Project* (Jakarta, ADB)
[3] Kementerian PUPR 2015, *Rencana Strategis Kementerian Pekerjaan Umum dan Perumahan Rakyat 2015-2019* (Jakarta, PUPR)
[4] Terrel, MS 200, *The project manager's role as safety champion* (Houston, PMI)
[5] Pemerintah Indonesia 2012, *Peraturan Pemerintah Nomor 50 tentang Penerapan Sistem Kesehatan dan Keselamatan Kerja* (Jakarta, PP)
[6] EVN 2018, *Potret Kecelakaan Kerja Konstruksi di Indonesia* (Jakarta, TURC)
[7] BPJS Ketenagakerjaan 2018, *The Minister of Manpower and Transmigration Encourages Local Governments to Make Occupational Safety and Health Commitments (K3) in their Regions* (Batam, Kemenaker)
[8] Departemen Tenaga Kerja 1970, *Undang-undang keselamatan kerja Nomor 1* (Jakarta, Depnaker)
[9] Pemerintah Indonesia 2012, *Peraturan Pemerintah Nomor 50 tentang Penerapan Sistem Kesehatan dan Keselamatan Kerja* (Jakarta, PP)
[10] Departemen Tenaga Kerja 2003, *Undang-undang ketenagakerjaan Nomor 13* (Jakarta, Depnaker)
[11] Danang Sugianto 2017, *Rampant Development, The Number of Construction Workers Is Down 500,000 workers* (Jakarta, Detik Finance)
[12] Reini D. Wirahadikusumah 2007, *Challenges of Occupational Safety and Health Problems in Construction Projects in Indonesia* (Jakarta, Researchgate.net)
[13] Wicaksono I.K, Singgih. L 2011, *Occupational Health and Safety Risk Management at the Puncak Permai Apartment Construction Project in Surabaya* (Surabaya, Resits)
[14] Muhammad, ZR 2018, *The problems with the Indonesia's infrastructure projects* (Jakarta, The Diplomat)
[15] Kementerian PUPR 2018, *Catatan Kecelakaan Kerja pada Proyek Infrastruktur Jalan dan Jembatan* (Jakarta, PUPR)
[16] Liputan-6 2018, 3 *Kecelakaan Proyek Infrastruktur Jakarta di awal 2018,* (Jakarta, SCTV)
[17] Agung Supriyadi 2018, *14 Kecelakaan Kerja Konstruksi tahun 2017-2018,* (Jakarta. Katigaku.top)
[18] Metro-sindonews 2017, *Jasa Marga beberkan penyebak kecelakaan Crane pengangkut VMS,* Jakarta-Sindonews)
[19] Detiknews 2017, *Insiden Tiang LRT Timpa Rumah, Polisi sudah periksa operator,* (Jakarta, Detik.com)
[20] Dani P 2018, *11 Kasus Kecelakaan Kerja terjadi dalam 6 bulan*, (Jakarta, Kompas)
[21] Hanang Y 2017, *Begini Kronologi Lengkap Kecelakaan Proyek Tol Pasuruan-Probolinggo yang Tewaskan 1 Orang,* (Jakarta-TribunSolo)
[22] Harwanto 2018, *5 Fakta dibalik maraknya kecelakaan proyek pembangunan era Jokowi-JK,* (Jakarta-Merdeka.com)

IOP Conf. Series: Materials Science and Engineering **513** (2019) 012041 doi:10.1088/1757-899X/513/1/012041

Development of hybrid life cycle inventories (HLCI) database for embodied energy and carbon intensities of Malaysian construction materials

Wan Mohd Sabki Wan Omar[1*]

[1]School of Environmental Engineering, Universiti Malaysia Perlis, 02600 Arau, Perlis, Malaysia.

[*]Corresponding author: wansabki@unimap.edu.my /sabki.wanomar@griffithuni.edu.au

Abstract. This paper investigates the completeness of process, input-output (I-O) and hybrid life cycle inventory (LCI) analysis and their applicability as applied to the Malaysian construction materials. The process, I-O, and hybrid analysis were evaluated at the basic materials and products level to identify the gaps between models and further incorporated into the material and product inputs of a selected building. It was found that the largest differences in embodied energy (EE) and embodied carbon (EC) intensities of the materials and products between I-O and hybrid analysis accounted for up to 65.90% and 62.08%, respectively. Further analysis of the EE and EC intensities of the building revealed that the hybrid analysis were significantly higher than I-O analysis estimated to be 44.39% and 43.57% respectively. On average, the EE and EC intensity values using I-O analysis were lower than hybrid analysis due to inherent limitations in this model. Although using I-O data was considered not representative due to inherent limitation, it was considered applicable for energy analysis due to lack of process data in Malaysian building industry.

1. Introduction

The United Nations Environmental Programme [1] currently reported that building construction in developed and developing countries consumed more than one third of total energy use and emitted greenhouse gas (GHG) emissions. Thus, this sector has significant opportunities for GHG emissions reduction that contribute to the global warming. Malaysia is among fast developing countries in the Asian region. Rapid development in human resources and infrastructures is the focal point for the Malaysian government to achieve developed country status in the year 2020. In fulfilling this vision, the requirements on energy sectors (e.g. natural gas, coal, petroleum products, and electricity supply) are compulsory not only for Malaysia, but also other developing countries.

In Malaysia, the energy consumption is increasing rapidly because of economic growth and development since year 2000 to 2010. It can be seen that the energy demand was dominated by the transport sector at 40.5%, the industrial sector at 38.6%, the residential and commercial sector at 13.1%, the non-energy sector at 7.3%, and the agricultural sector at 0.5% [2]. The energy consumed by commercial and residential buildings was accounted for about 13% of total energy consumption and 48% of electricity consumption. The energy-intensive industries such as cement, ceramic, iron and steel were predicted to be major consumers for the future [3].

The energy consumed by the building consists of embodied energy and operational energy. Generally, EE is defined as the energy consumed in all activities required for a process. It is categorised into two components, namely direct energy and indirect energy. Direct energy includes energy consumed on-site during building construction and other energy purchased. Indirect energy consists of the energy embodied in goods and services, as well as energy in upstream boundary of construction processes [4]. Meanwhile, EC is a measure of the carbon dioxide emissions associated with the building construction. This includes carbon emissions during the processing of raw materials, manufacturing of the building materials and components, and transporting to site and fabricating the building components [5].

For building energy and emissions in Malaysia, previous studies mainly focused on operational energy. Due to the comprehensive analysis of operational energy, EE analysis methods become a significant component of the energy rating systems for operational energy use [6]. However, there is lack of studies on the applicability of these methods to the Malaysian construction industry. Recently, there has been inconsistency and wide variation of EE and EC intensities in LCI data, due to the variability, uncertainty and incompleteness of the system being studied. Treloar [7] pointed out factors contributing to these problems, which include completeness of system boundaries, selection of analysis methods (i.e. process, I-O, or hybrid analysis), regional variability, and reliability of data from government and agencies.

Therefore, this research is part of an iterative process in LCI analysis to develop HLCI and evaluate the applicability of I-O and HLCI model in the Malaysian construction materials. While previous researches investigated the key factors and issues contributing to the variations in EE and EC analysis, this research aims to evaluate the applicability of these models in terms of its completeness and accuracy and its significant influence on estimating building's total embodied impact.

2. Life cycle inventory analysis

Life cycle inventory (LCI) analysis involves the collection of data and calculation procedures to quantify relevant inputs (i.e. materials and energy) and outputs (i.e. emissions and waste) of a product system [8]. LCI analysis is a second phase in life cycle assessment (LCA). In general, LCI analysis is classified into process, I-O and hybrid analysis based on different theory and system scope.

Process analysis is known as a bottom-up methodology. It is used to identify and trace the physical flow of goods and services for the particular processes, products or manufacturing chains [9]. In general, process analysis is defined as the EE analysis using high-resolution detailed site-specific data [10]. It most applicable and accurate for assessing products or processes that are highly energy intensive, and require large amounts of direct energy components [11]. However, process data is not always available to take advantage of detailed process analysis, due to lack of quantity and quality of process data locally and internationally [12]. According to Pomponi and Lenzen [13], process model induces systematic truncation errors due to the incomplete definition of system boundary.

To overcome the incompleteness of system boundary, I-O analysis is widely used. It is a top-down approach, which converts monetary values into physical values. According to Lenzen, et al. [14], this model uses a top-down linear macroeconomic approach to explain the complex inter-industry relationship in terms of monetary transactions in industrial structures. The use of I-O data increases the reliability of LCA by improving completeness and reliability of life cycle inventory analysis compared to the traditional method. Beside its advantages, I-O analysis has several limitations. Thus, using this method to replace process analysis does not always ensure the accuracy of the analysis being studied.

Hybrid analysis was developed to combine the advantages of the more accurate process analysis and the extended system boundaries of the I-O analysis [10]. This analysis comprises two form of distinct method, namely process-based hybrid analysis and I-O based hybrid analysis. Process-based hybrid analysis incorporates process data into I-O analysis during calculation of hybrid EE intensity of material and product (or building). This analysis is suited to analyse large typical product such as entire building [15]. Meanwhile, The I-O based hybrid analysis improves the completeness of the EE

IOP Conf. Series: Materials Science and Engineering **513** (2019) 012041 doi:10.1088/1757-899X/513/1/012041

intensities. Crawford [16] pointed out that the results obtained from the process-based hybrid analysis can be combined with further I-O values to extend the system boundary.

3. Model development

3.1. Hybrid based life cycle inventory model

Hybrid analysis integrates more reliable process data into the most comprehensive I-O data during calculation of the energy and carbon intensity. Crawford [17] suggested that in quantifying EE and EC intensity of a product, the first step is to calculate the EE and EC intensity of the basic materials. The hybrid EE and EC intensity of the basic material was determined by adding EE and EC intensity from detailed process analysis with indirect EE and EC intensity from comprehensive I-O analysis. The indirect EE and EC intensity of the material is calculated by subtracting direct EE and EC intensity from the total EE and EC intensity of material sector n from I-O data. This difference represents the indirect EE and EC intensity of the material unaccounted for by the process analysis due to truncation error or setting up the system boundary of material production. The Hybrid EE and EC intensities of basic material are then multiplied by the material's price as formulated by Equations 1 and 2 below.

$$\mathrm{HEI_m = EI_D + [(TEI_n - DEI_n) \times C_m]} \tag{1}$$

$$\mathrm{HECO_{2-e}I_m = ECO_{2-e}I_D + [(TCO_{2-e}I_n - DCO_{2-e}I_n) \times C_m]} \tag{2}$$

Where, HEIm is the EE intensity of the basic material using hybrid analysis; EID is the direct EE intensity of the basic material from process analysis; TEIn is the total EE intensity of the basic material sector using I-O analysis; DEIn is the direct EE intensity of the basic material sector using I-O analysis; HECO2-eIm is the EC intensity of the basic material using hybrid analysis; and Cm is the material price.

Total EE and EC intensity of a product was quantified using hybrid analysis. If the process data was unavailable, direct EE and EC intensity of the product were obtained from I-O analysis and was then multiplied by the product's price. The total EE and EC of the product were quantified by adding all EE and EC intensity of the basic materials with the direct EE and EC intensity of the product as formulated by Equations 3 and 4 below.

$$\mathrm{HEE_{pdt}} = \sum_{\mathrm{m=1}}^{\mathrm{M}} \mathrm{(HEI_m \times W_m \times Q_m) + DEI_{n,pdt} \times C_{pdt}} \tag{3}$$

$$\mathrm{HECO_{2-e,pdt}} = \sum_{\mathrm{m=1}}^{\mathrm{M}} \mathrm{(HECO_{2-e}I_m \times W_m \times Q_m) + DCO_{2-e}I_{n,pdt} \times C_{pdt}} \tag{4}$$

Where, HEEpdt is the total EE intensity of the product using hybrid analysis; Wm is the material wastage factor; Qm is the material quantity to produce the product; DEIn,pdt is the direct EE intensity of the product sector n from I-O analysis; Cpdt is the total price of the product; HECO2-e,pdt is the total EC intensity of the product using hybrid analysis; and DCO2-eIn,pdt is the direct EC intensity of the product sector n using I-O analysis.

3.2. Evaluation of I-O and Hybrid LCI model

The gap analysis was used to quantify differences between I-O and hybrid analysis by focusing on the final LCI results. It is used to assess the differences between models, as an evaluation of the completeness of each model. The gaps analysis as proposed by Crawford [18] was modified to calculate the gaps between I-O and hybrid analysis as given in Equation 5 below:

IOP Conf. Series: Materials Science and Engineering **513** (2019) 012041 doi:10.1088/1757-899X/513/1/012041

$$\text{GAP} = \frac{\text{HLCI} - \text{I} - \text{O LCI}}{\text{HLCI}} \times 100 \qquad (5)$$

Where, GAP is the percentage completeness of the I-O analysis when compared with hybrid analysis; HLCI is the EE and EC intensity of basic material and product (or material and product input) using hybrid analysis; and I-O LCI is the EE and EC intensity of basic material and product from I-O analysis.

The comparative analysis was previously conducted to evaluate the applicability of I-O data to the hybrid analysis [19]. Similarly, this analysis was also used in this research to evaluate the applicability of I-O and hybrid LCI model to quantify EE and EC intensities of individual materials and products, as well as material and product inputs to the whole building.

4. Result and Discussion

4.1. Evaluation of EE and EC intensities using gap analysis

Tables 1 presents the detailed proportion of EE and EC intensities of building materials and products based on process, I-O, and hybrid analysis. The gap analysis clearly pointed out that the EE and EC intensities of materials and products using I-O analysis can be higher or lower than hybrid analysis. For instance, EE and EC intensity of bituminous waterproofing coating using hybrid analysis were lower than I-O analysis, accounting for 34.29% and 104.63% respectively. The higher EE and EC intensity values using I-O analysis were due to variations in material and product prices and the small proportion of process data available and other variation associated with I-O analysis. The largest gaps in EE intensities in concrete and steel were found to be different with the gaps found in their EC intensities. These differences were due to the effect of carbon emission factors used to convert energy requirements to carbon emissions in EC analysis. While concrete and steel were the common materials used in building construction, using the I-O analysis to quantify both EE and EC intensities led to overestimation or underestimation of the total embodied impact of the building.

The results revealed that the largest gaps exist in EE intensities between I-O and hybrid analysis accounted for up to 65.90% for aluminium virgin, 55.06% for steel virgin, 50.77% for clay brick, 47.26% for structural steel, 45.83% for 40 MPa concrete, and 44.92% for Ordinary Portland Cement (OPC). Meanwhile, the largest gaps of EC intensities were also found to be 61.40% for 40 MPa concrete, 60.63% for OPC, 51.24% for aluminium virgin, 47.51% for clay brick, 47.27% for steel virgin, and 42.28% for medium hardwood. It is clearly shown that the top largest gaps in EE and EC intensities were the most commonly used materials and products in building construction. Asif, et al. [20] identified key materials in building construction include concrete, timber, glass, ceramic tiles, and aluminium. Lenzen and Treloar [21] revealed large differences in EE intensities of materials and products (e.g. softwoods, mineral wool insulations, plasterboards, and plastic products) when comparing between wood-framed and concrete-framed building. These gaps stemmed from major discrepancies of these materials due to the influence of a large proportion of indirect energy in the upstream boundary of material production (e.g. mineral wool insulation, plastic, and plasterboard); and different of production layer which are influence from economic structure (i.e. energy consumed by manufacturers of materials used in the buildings; and by the suppliers of these manufacturers).

Table 1. Detailed proportion of EE and EC intensities of building materials and products in Malaysia.

Material and product sector	Basic material and product	EE intensities (MJ/kg)			Gap[a] (%)	EC intensities (kg $CO_{2\text{-}e}$/kg)			Gap[a] (%)
		Process LCI	I-O LCI	Hybrid LCI		Process LCI	I-O LCI	Hybrid LCI	
Sand, aggregate, and stone	Granite aggregates	0.083	0.236	0.286	17.48	0.0052	0.0185	0.0206	9.82
	Normal river sand	0.008	0.201	0.181	-11.08	0.0051	0.0158	0.0182	13.15
	Polished granite slab	11.000	47.797	52.102	8.26	0.7000	3.7568	3.8125	1.46
Carpet	Nylon carpet tile	60.350	386.927	396.270	2.36	13.7000	38.3041	43.9257	12.80
Sawmill products	Heavy hardwood	10.400	15.072	23.586	36.10	0.8700	1.2260	1.8304	33.02
	Medium hardwood	10.400	11.270	20.259	44.37	0.8700	0.9167	1.5881	42.28
Wood products	HDF	16.000	34.150	43.775	21.99	1.0900	3.1199	3.2059	2.68
	MDF	11.000	17.776	25.457	30.17	0.7400	1.6240	1.8414	11.81
	Plywood	15.000	22.201	33.056	32.84	1.1000	2.0282	2.4755	18.07
Bitumen products	Bituminous waterproofing	51.000	146.425	109.034	-34.29	0.4900	9.3523	4.5704	-104.63
Paints	Oil-based paint	97.000	194.113	274.949	29.40	3.7600	15.8527	17.6166	10.01
	Water-based paint	59.000	182.869	226.642	19.31	2.5400	14.9344	15.5940	4.23
Plastic products	ABS pipe	95.300	156.191	232.099	32.70	3.7600	15.0199	15.2641	1.60
	HDPE pipe	84.400	114.490	184.675	38.00	2.5200	11.0097	10.9526	-0.52
	UPVC pipes	67.500	78.179	135.972	42.50	3.2300	7.5179	8.9882	16.36
	PE sheet membrane	83.100	111.963	181.162	38.20	2.5400	10.7668	10.7865	0.18
Sheet glass and glass products	Clear float glass	15.000	38.632	46.051	16.11	0.9100	3.4993	3.3567	-4.25
	Obscured glass	15.000	30.977	39.898	22.36	0.9100	2.8059	2.8719	2.30
	Tinted float glass	15.000	46.986	52.765	10.95	0.9100	4.2560	3.8858	-9.53
Clay and ceramic products	Ceramic tiles	12.000	37.029	41.049	9.79	0.7800	3.0204	2.8410	-6.31
	Clay bricks - common	3.000	2.406	4.888	50.77	0.2400	0.1963	0.3739	47.51
	Vitrified clay pipe	7.000	21.655	23.989	9.73	0.5000	1.7664	1.7053	-3.58
Cement	OPC	5.200	5.238	9.509	44.92	0.7400	0.4120	1.0466	60.63
Concrete and non-metallic mineral products	30 MPa concrete	1.098	1.057	1.944	45.62	0.1368	0.0770	0.1978	61.06
	35 MPa concrete	1.153	1.104	2.033	45.70	0.1438	0.0804	0.2072	61.21
	40 MPa concrete	1.220	1.161	2.143	45.83	0.1525	0.0845	0.2190	61.40
	50 MPa concrete	1.260	1.197	2.194	45.41	0.1565	0.0872	0.2237	61.00
	Fibre glass insulation	28.000	254.407	248.012	-2.58	1.3500	21.4743	19.1096	-12.37
	Rockwool insulation	17.568	138.833	137.632	-0.87	1.2240	11.7188	10.9157	-7.36
	Cellulose fibre ceiling	10.400	19.633	27.379	28.29	1.0900	1.6572	2.4606	32.65
	Plain gypsum board	6.750	18.268	22.548	18.98	0.3900	1.5420	1.6653	7.40
Iron and steel products	Galvanised iron pipe	40.000	52.232	83.629	37.54	3.0100	5.1267	6.8065	24.68
	Mild steel pipe	34.700	55.683	81.213	31.44	2.8700	5.4655	6.9174	20.99
	Stainless steel pipe	56.700	223.278	243.206	8.19	6.1500	21.9153	22.3791	2.07
	Steel virgin	35.400	25.473	56.677	55.06	2.8900	2.5002	4.7415	47.27
	Steel sheet decking	40.000	139.923	156.879	10.81	3.0100	13.7338	13.1804	-4.20
Non-ferrous metals	Aluminium virgin	218.000	107.002	313.756	65.90	12.7900	10.4162	21.3642	51.24
	Reflective foil insulation	217.000	172.263	371.158	53.59	12.8000	16.7690	26.6036	36.97
Structural metal products	Structural steel	38.000	38.373	72.753	47.26	3.0300	3.7575	6.2050	39.44

Note:
(a) Gap between I-O and hybrid analysis is calculated by: (HLCI – I-O LCI)/HLCI.

5. Conclusion

This research assessed the completeness and applicability of I-O and hybrid analysis as applied to the selected Malaysian building construction. The process, I-O, and hybrid analysis were evaluated at the basic materials and products level to identify the significant gaps or differences between analysis methods and further incorporated into the material and product inputs of a selected building. The gap analysis found that the largest differences in EE and EC intensities of the materials and products between I-O and hybrid analysis accounted for up to 65.90% and 62.08%, respectively. These largest differences were found in non-ferrous metals (aluminium), iron and steel products, and concrete which were considered as the key materials used in the building construction. Using I-O data as substitute for process data is considered better than not using any data at all. However, the use of I-O analysis is only limited to fill the gaps in upstream boundary of materials and products manufacturing. Therefore, this research initially quantified EE and EC intensities of the materials and products, before calculating the building's total embodied impact using actual quantities of materials, equipment, plants, etc. to overcome inherits problems such as underestimation of material, product and building's prices as used in the I-O analysis method.

References

[1] United Nations Environmental Programme 2005 *Building and Construction: Climate Neutral Network* United Nations Environmental Programme (UNEP).

[2] Mohd Safaai N S, Noor Z Z, Hashim H, Ujang Z, and Talib J 2011 Projection of CO2 emissions in Malaysia *Environmental Progress & Sustainable Energy*, **30** (4) 658-665.

[3] The Economic Planning Unit 2006 *Ninth Malaysia Plan 2006-2010* The Economic Planning Unit, Prime Minister's Department of Malaysia, Putrajaya, Malaysia.

[4] Treloar G J 1997 Extracting embodied energy paths from input-output tables: towards an input-output-based hybrid energy analysis method *Economic Systems Research* **9** (4) 375-375.

[5] The Institution of Structural Engineers 2011 *A Short Guide to Embodied Carbon in Building Structures* The Institution of Structural Engineers (IStructE), London, UK.

[6] Acquaye A A, Duffy A P, and Basu B 2011 Stochastic hybrid embodied CO2-eq analysis: An application to the Irish apartment building sector *Energy and Buildings* **43** (6) 1295-1303.

[7] Treloar G J 1998 *A Comprehensive Embodied Energy Analysis Framework*, PhD Thesis Faculty of Science and Technology, Deakin University.

[8] ISO 14040 2006 *ISO 14040: Environmental management - Life cycle assessment - Principles and framework* International Organization for Standardization (ISO), Switzerland.

[9] Crawford R H, Bontinck P A, Stephan A, Wiedmann T, and Yu M 2018 Hybrid life cycle inventory methods – A review *Journal of Cleaner Production* **172** 1273-1288.

[10] Lee C H, and Ma H W 2013 Improving the integrated hybrid LCA in the upstream scope 3 emissions inventory analysis *The International Journal of Life Cycle Assessment* **18** (1) 17-23.

[11] Wan Omar W M S, Doh J H, and Panuwatwanich K 2014 Variations in embodied energy and carbon emission intensities of construction materials *Environmental Impact Assessment Review* **49** 31-48.

[12] Wan Omar W M S, Doh J H, and Panuwatwanich K, and Miller D 2014 Assessment of the embodied carbon in precast concrete wall panels using a hybrid life cycle assessment approach in Malaysia *Sustainable Cities and Society* **10** (0) 101-111.

[13] Pomponi F, and Lenzen M 2018 Hybrid life cycle assessment (LCA) will likely yield more accurate results than process-based LCA *Journal of Cleaner Production* **176** 210-215.

[14] Lenzen M, Murray S A, Korte B, and Dey C J, Environmental impact assessment including indirect effects - a case study using input-output analysis *Environmental Impact Assessment Review* **23** (3) 263-282.

[15] Goggins J, Keane T, Kelly A, The assessment of embodied energy in typical reinforced concrete building structures in Ireland, Energy and Buildings, 42 (5) (2010) 735-744.
[16] Crawford R H, Validation of the use of input-output data for embodied energy analysis of the Australian construction industry, Journal of Construction Research, 6 (1) (2005) 71-90.
[17] Crawford R H 2011 *Life Cycle Assessment in the Built Environment* Spon Press London and New York.
[18] Crawford R H 2008 Validation of a hybrid life-cycle inventory analysis method *Journal of Environmental Management* **88** (3) 496-506.
[19] Crawford R H 2004 *Using Input-Output Data in Life Cycle Inventory Analysis*, PhD Thesis, Deakin University.
[20] Asif M, Muneer T, and Kelley R 2007 Life cycle assessment: A case study of a dwelling home in Scotland *Building and Environment* **42** (3) 1391-1394.
[21] Lenzen M, and Treloar G 2002 Embodied energy in buildings: wood versus concrete—reply to Börjesson and Gustavsson *Energy Policy* **30** (3) 249-255.

IOP Conf. Series: Materials Science and Engineering **513** (2019) 012042 doi:10.1088/1757-899X/513/1/012042

Double vertical carbon fiber reinforced polymer plates strengthened to reinforced concrete beams for six months saltwater exposure

Amiruddin Mishad[1*], Mohd Hisbany Mohd Hashim[2], Azmi Ibrahim[1] and Aidan Newman[1]

[1]Faculty of Civil Engineering, Universiti Teknologi MARA, Shah Alam, Selangor, Malaysia
[2]Faculty of Civil Engineering, Universiti Teknologi MARA, Pulau Pinang, Malaysia

[*]Corresponding author: amiruddinmishad@salam.uitm.edu.my

Abstract. Reinforced concrete structures might require rehabilitation after years exposed to extreme environment conditions. The use of Fiber Reinforced Polymer (FRP) material in construction industry has become popular due to its properties such as high tensile strength, durable and lightweight which can be used to improve mechanical behavior of structures. This research emphasis on the use of Carbon Fiber Reinforced Polymer (CFRP) to improve the flexural performance of reinforced concrete beam which is exposed to saltwater for 6 months. The purposes of this research are to determine the maximum flexural strength of normal reinforced concrete beam and reinforced beam strengthened with CFRP plate and to determine the effects of salt water on the flexural strength of concrete beam reinforced with CFRP plate. Two beam samples with dimension of 125mm x 300mm x 1800mm were prepared labelled as control beam and specimen beam respectively. The control beam was cure in fresh water while the specimen beam was cured in salt water for 28 days. All of the beams were tested using four point bending test. The control beam was tested until failure whereas the specimen was subjected to 30% of maximum load from control beam as pre-cracking load. The specimen beam was repaired by installing CFRP plate using near surface mounted (NSM) and is immerse in saltwater for another 6 months. After 6 months, the specimen beam was tested until failure. As a result, the flexural strength for normal beam and specimen beam is slightly the same. The specimen beam has form modes of failure known as concrete cover separation. It is also found that saltwater exposure did affect the performance of beam that has been repaired.

1. Introduction

The main significance of this research is to promote the use of Carbon Fiber-Reinforced Polymer (CFRP) in the construction industry. CFRP is one of the most suitable alternatives to replace the use of steel reinforcement bar in strengthening and repairing of new and old structure. CFRP not only can enhance the performance of the structure, but it can also extend the life of the structure providing long lasting age.

Research show that carbon fiber reinforced polymer (CFRP) has been used to strengthen reinforced concrete beam and enhance its flexural performance. Flexural failure occurs due to excessive load apply on the beam. It also happens due to the condition of the steel reinforced bar inside the beam. When the steel reinforcement bar is corroded causing the strength to decrease, it will result in the inability for the beam to sustain load. Alternative to that problem, CFRP plate is used so that it can improve the strength and load carrying capacity of the concrete beam. Other than that, CFRP is more durable compare to steel reinforcement bar since it can stand the adverse effects of environment. It has high corrosion resistance against salt water.

Meanwhile, this research can also help the construction industry to reduce the cost in strengthening, rehabilitation and maintaining of new and old structures. Since the cost of repairing and rehabilitation mainly depends on the degree of damage occurs, the introduction to CFRP can reduce the cost in terms of material since CFRP is cheaper than steel and maintenance since it requires less maintenance due to its durability. CFRP also have other good properties such as high strength and long lasting compare to steel.

Fiber reinforced polymer (FRP) or fiber reinforced plastic, is a non-metal reinforcement which is made of glass, carbon or aramid. FRP is a suitable materials used in construction, especially for structure exposed to extreme environment condition such as saltwater from the sea due to its durability and high resistance to corrosion. Recently, Fiber Reinforced Polymer (FRP) composites have developed as a potential answer for issues regarding infrastructure [1]. Plus, FRP can be used to repair and strengthen beams, slabs and columns even after the structure has been damaged due to excessive loading. Civil engineers have been attracted to FRP materials due to its electrochemical corrosion resistance, high weight-to-strength ratio and fabrication versatility [1].

FRP has also offer benefits to construction industry due to its ability to reduce crack and enhance the structures performance. Various researches have been conduct on the use of FRP materials in strengthening of structures. Therefore, FRP has turned out to be progressively noticeable in strengthening and rehabilitation of reinforced concrete structure [2].

Carbon Fiber Reinforced Polymer (CFRP) is the latest high performance composite material which serves numerous benefits such as extended breaking length, corrosion less and superb fatigue resistance [3].Besides that, the ultimate load carrying capacity of beam can also be increased by up to 53% [1].Furthermore, CFRP is a wide known material due to its superior properties such as durability, corrosion resistance, flexible application, easy handling and installation and high strength to weight ratio [4]. According to[2], by using CFRP in strengthening structure can be improved the performance of RC beams by increasing the load capacity as nearly as double of unstrengthen beams. Therefore, CFRP is the potential solution to overcome this problem. In addition, based on [4], both CFRP bars and strips are effective methods to improve the shear capacity of RC beams Reinforced concrete structures commonly lack in shear strength because of insufficient shear reinforcement of steel area due to excessive loads, corrosion and construction defects.

Near surface mounted (NSM) is one of the strengthening techniques that has been commonly used in strengthen reinforced concrete structures [5]. This technique requires the CFRP plate to be inserted into grooves on the beam surface and then bonded using epoxy adhesive. Application of near surface mounted CFRP strips is practical and effective for strengthening and repair damage concrete structures [1]. Shear failure in concrete structure normally occurs due to corrosion, overloading and lack of reinforcement in concrete structure itself. However, this issue can be overcome by using NSM techniques. NSM CFRP method gives more advantages compared to EBR method, which includes reduced of debonding risk, efficient FRP material utilization and a better external damage protection [6]. Bond behavior is the most essential issues in NSM techniques that require further studies. Bond behavior is important during the loading process of reinforced concrete element since it allows the transfer of stress between the concrete and the reinforcement [7].The bond behavior can affect the load carrying capacity and serviceability of the concrete structure.

2. Flexural strength test

Generally, the objective of using bending test machine is to determine the maximum flexural strength and the deflection of RC beam when subjected to loading. Rectangular RC beam (control beam) and RC beam strengthened with carbon fiber reinforced polymer (CFRP) plate are the specimens used in

IOP Conf. Series: Materials Science and Engineering **513** (2019) 012042 doi:10.1088/1757-899X/513/1/012042

this test. The design of reinforcement concrete beam is refer to [8]. The setup of this test is shown In Figure 1.

Figure 1. Setting up of four point bending test.

In this study, two RC beam specimens were prepared. Both of the RC beam specimens were in the same dimension, which is 150mm x 300mm and 1800mm length Figure 2.The beam specimens are also prepared using concrete grade of 30N/mm^2 and reinforced with steel bar of 12mm diameter and shear link of 6mm diameter.

One of the RC beam specimens is used as control specimen while the other is strengthened with two carbon fiber reinforced polymer (CFRP) plate in vertical position after pre-cracking. The (CFRP) plate used in this study has a dimension of 1.2mm thickness, 50mm width and 1800mm length. This CFRP plate is used to strengthen the RC beam using near surface mounted (NSM) method. The dimension and position of CFRP plate is shown in Figure 3.

The control beam in this study was undergone bending test until failure to determine the maximum applied load. The other beam specimen is exposed in saltwater with ratio of 1 liter of water equal to 40 gram of salt for 28 days. After 28 days, the beam will be subjected with 30 percent of maximum applied load of control beam specimen load until pre-cracked. As soon as the beam cracks, it will be repair and strengthen with carbon fiber reinforced polymer (CFRP) plate attached smooth groove line by using near surface mounted (NSM) method. The purpose of using salt water solution is to represent the condition of beam exposed to sea water. By doing this, the ability of CFRP plate and effectiveness of NSM method can be determined and further studied for future improvement. The four point bending test set up is shown in Figure 4.

In this research, the control specimen was installed with 3 concrete strain gauges (labelled strain gauge 6, 7 and 8) and 2 steel reinforcement strain gauges (labelled strain gauge 4 and 5). Whereas, specimen beam was installed with 3 concrete strain gauges and 2 steel strain gauges which was the same labelling as control beam, but with additional 3 CFRP strain gauges (labelled strain gauge 1, 2 and 3). Other than that, the beams were also installed with Linear Variable Displacement Transformer (LVDT) during the testing to measure the value of deflection. The location of the strain gauges and LVDT are shown in Figure 5.

IOP Conf. Series: Materials Science and Engineering **513** (2019) 012042 doi:10.1088/1757-899X/513/1/012042

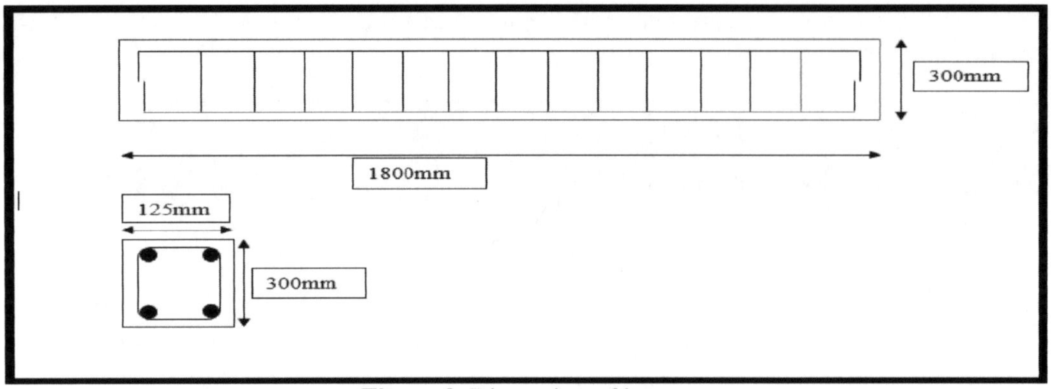

Figure 2. Dimension of beam.

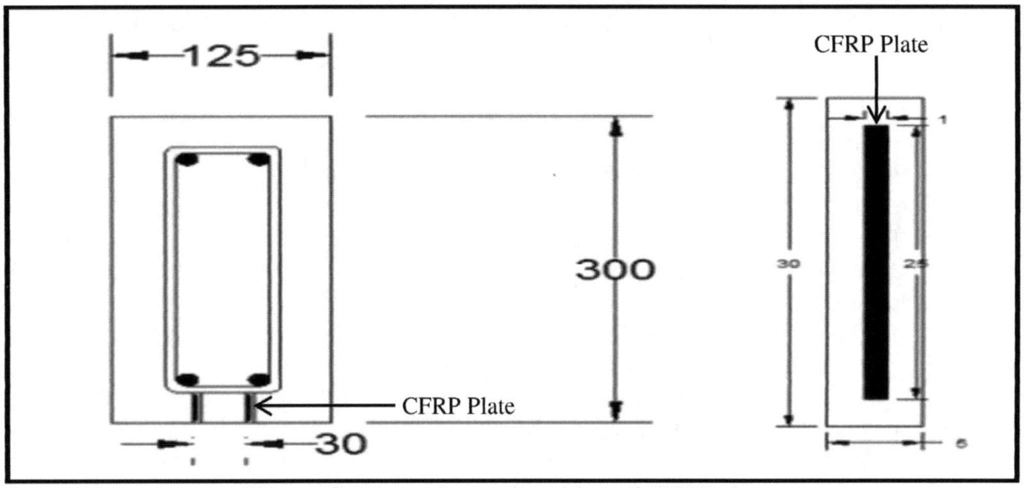

Figure 3. Dimension and position of CFRP plate.

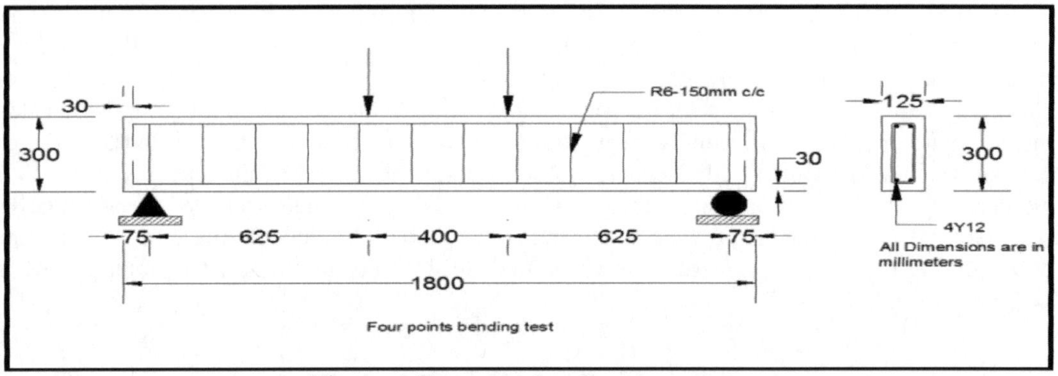

Figure 4. Four point bending test set up.

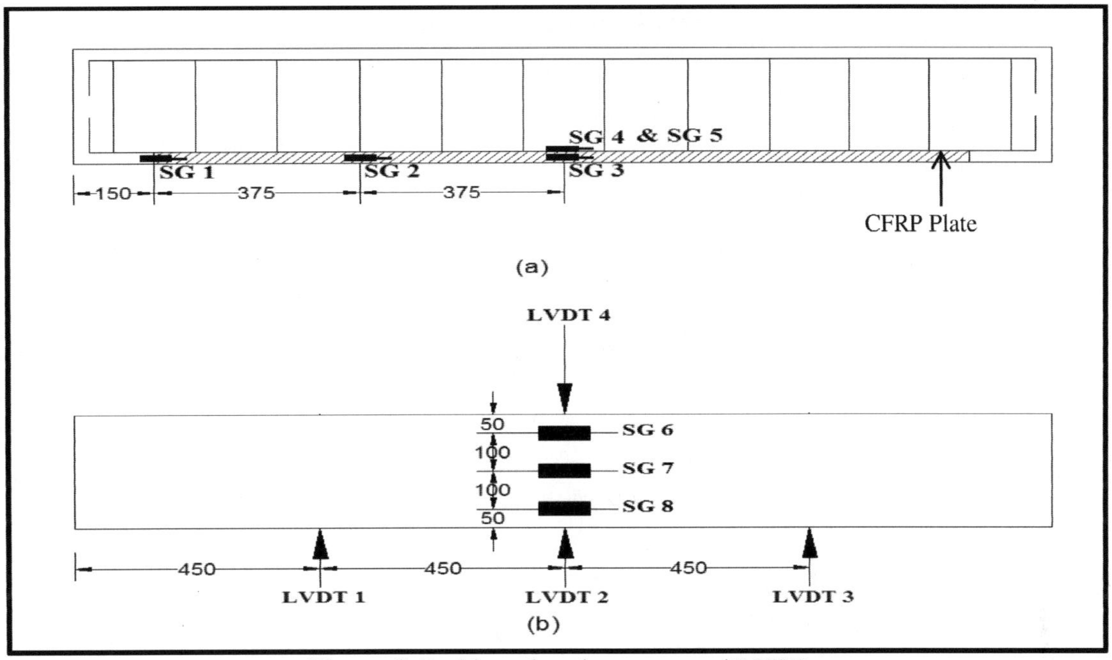

Figure 5. Position of strain gauges and LVDT.

3. Result and discussion

3.1. Load versus displacement

The result of bending test for control beam and beam strengthened with double vertical CFRP plate is shown in Table 1 and Figure 6. The flexural strength of reinforced concrete beam depends solely on the steel reinforcement performance to cater tensile stress produce during bending test. However, the performance can be affected due to the presence of saltwater. So, in order to improve the flexural strength of reinforced concrete beam that was exposed to saltwater, CFRP plate is used.

Table 1 shows that control beam has higher displacement value compare to the one strengthened with two vertical CFRP plate installed using NSM method. Figure 6 shows the load vs deflection graph of both control beam and the one strengthen with CFRP. The beam strengthened with CFRP tends to cater load which is more or less the same as control beam. This might be due to the fact that the beam has been exposed to saltwater for 6 months. It also gave a slightly lower deflection value at the ultimate load compare to control beam. Based on these observations, the strengthened beam has similar mechanical properties with the control beam, where the deflection value of both beam is almost the same. The ultimate load where both beam starts to fail is also slightly different. This might be due to the exposure of the strengthened beam to salt water for 6 months. This exposure might have cause the concrete to lose its compressive strength or even deterioration of the steel reinforcement inside the concrete. As a result, both beams have more or less the similar mechanical properties.

However, it is verified that CFRP can be used as a strengthening materials to improve the flexural performance of reinforced concrete even after it has been repaired due to pre-cracked loading and expose to salt water. It is proven that CFRP is able to resist the tensile force in the beam even though in this case, it does not increase the flexural strength of the beam itself.

IOP Conf. Series: Materials Science and Engineering **513** (2019) 012042 doi:10.1088/1757-899X/513/1/012042

Table 1. Load and deflection.

Type of Beam	Ultimate Load(KN)	Deflection (mm)
Control Beam	113.14	11.67
Double Vertical Plate	113.02	11.16

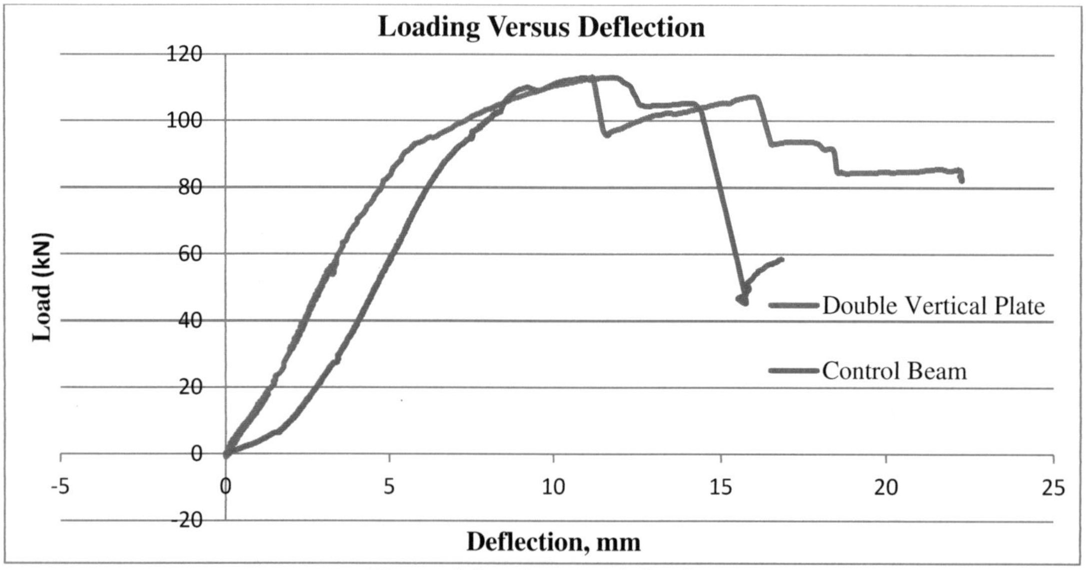

Figure 6. Load versus deflection.

3.2. Load versus strain

In this research, there are 3 concrete strain gauges used labelled strain gauge 6, 7 and 8 for the control beam. Based on Figure 7, strain gauge 6 give a strain reading of 0.001409 at ultimate load of 113.14kN where it starts to break. On the other hand, strain gauge 7 shows positive strain reading. However, most of the reading is observed to be between strain gauge 6 and strain gauge 8 reading. Strain gauge 7 give a strain reading of 0.002779 which breaks at ultimate load of 113.03kN. Whereas, strain gauge 8 give a strain reading of 0.004158 which breaks at ultimate load of 113.09kN. For steel reinforcement of control beam, two strain gauge are used which are strain gauge 4 and 5. It is observed that strain gauge 4 has a higher value compare to strain gauge 5, which is 0.001971 whereas strain gauge 5 has a reading of 0.001923. The ultimate load for strain gauge 4 and 5 is 112.98kN and 99.29kN respectively.

Whereas for the specimen beam, two steel strain gauge labelled strain gauge 4 and 5 were installed. In Figure 8, strain gauge 5 has a strain value of 0.00306 with an ultimate load of 85.96kN. Whereas, strain gauge 4 has a slightly lower strain reading compare to strain gauge 5. The strain value is 0.00305 which breaks at ultimate load of 86.07kN. On the other hand, 3 concrete strain gauges are labelled with strain gauge 6, 7 and 8. It is observed that the reading from strain gauge 6 has the highest strain value of 0.000549 compared to strain gauge 7 and 8. The ultimate load where the strain gauge breaks is 85.38kN. Strain gauge 7 gives a strain reading of 0.000085 with an ultimate load of 60.45kN. Whereas strain gauge 8 gives a reading of 0.000066 which breaks at ultimate load of 46.02kN.

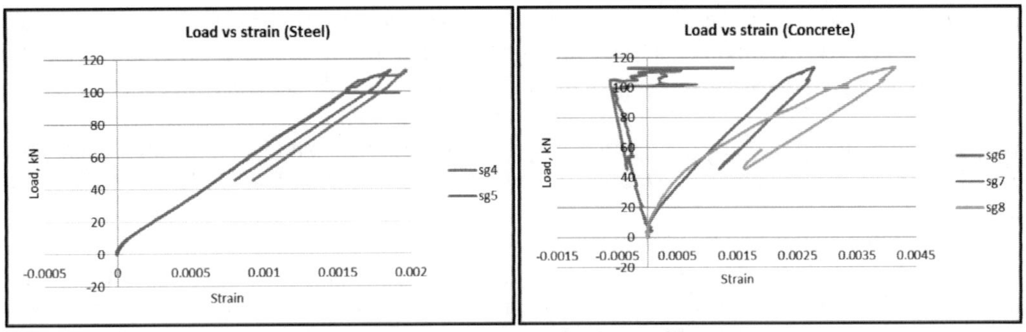

Figure 7. Load vs stress for control beam.

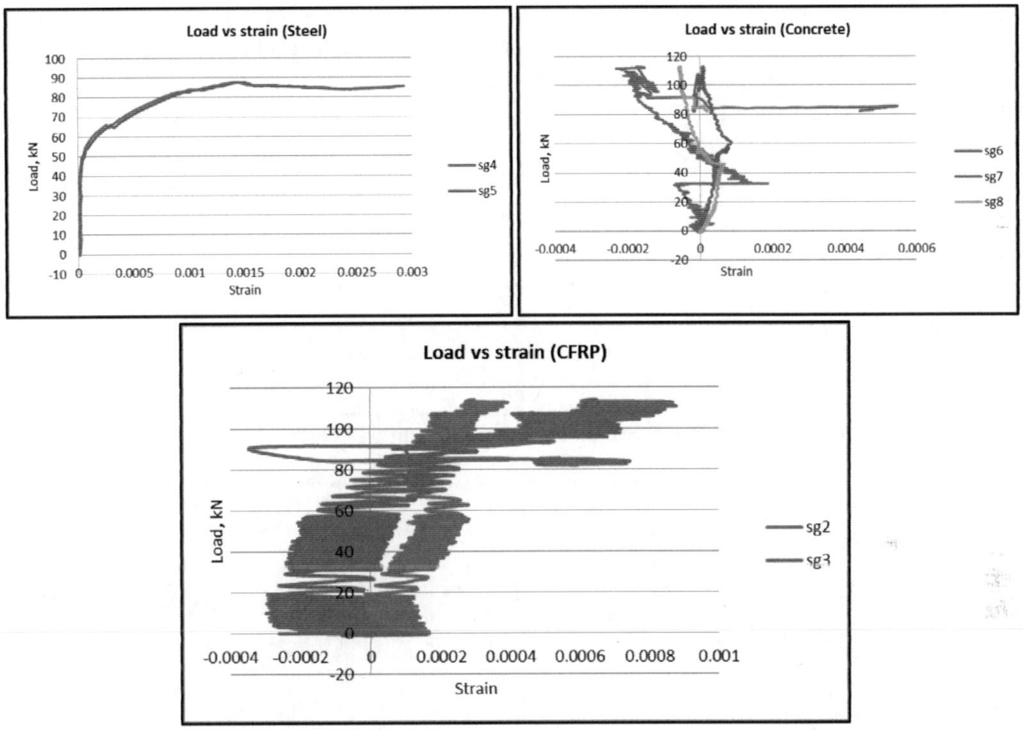

Figure 8. Load vs stress for beam with double vertical plate.

3.3. Stress versus strain

The result for control beam is shown in Figure 9. It is observed that the reading for strain gauge installed on steel reinforcement bar has a value of 0.001971 and 0.001923 for strain gauge 4 and 5 respectively. Strain gauge 4 breaks at ultimate stress of 394.2MPa whereas strain gauge 5 breaks at ultimate of 384.6MPa. It is observed that the maximum stress value is not over the maximum tensile strength of steel which is 460MPa. Strain reading for concrete shows that strain gauge 6 breaks at ultimate stress of 43.67MPa with a strain value of 0.001409. Strain gauge 7 breaks at ultimate stress of 86.15MPa with a slightly higher strain value of 0.002779. Whereas, strain gauge 8 breaks at ultimate load of 128.90MPa with the highest strain value among the three strain gauges, which is 0.004158. The maximum strain value of concrete between the three readings does exceed the maximum compressive strength of concrete that is 30MPa.

Figure 10 shows the result for specimen beam. For concrete strain gauge, strain gauge 6 has the highest strain value which is 0.000549 among the three. The ultimate stress when the strain gauge breaks is 17.02MPa. Whereas, strain gauge 8 has the smallest strain value which 0.000066 that breaks at ultimate stress of 2.06MPa. For strain gauge 7, it gives a strain reading of 0.00008508 with an ultimate stress of 2.63MPa. The maximum strain value of concrete between the three readings also exceeds the maximum compressive strength of concrete that is 30MPa. For CFRP strain gauges, researchers can observe that the strain gauge 3 breaks at ultimate stress of 144.67MPa with a strain value of 0.0008767. Compare to strain gauge 2, strain gauge 3 has a higher strain value. Whereas strain gauge 2, it gives a reading of 0.000393 of strain when it started to break at ultimate stress of 64.78MPa. Same as the control beam, the maximum stress value is not over the maximum tensile strength for the CFRP which is 3000N/mm^2. For steel reinforcement strain gauges, both strain gauges gave quite same value of the strain and stress. At ultimate stress of 611MPa, the steel bar starts to change from elastic state to plastic state with value of 0.00305 and 0.00306 for strain gauge 4 and 5. It is observed that the strain gauge 5 has slightly higher strain value compare to strain gauge 5.

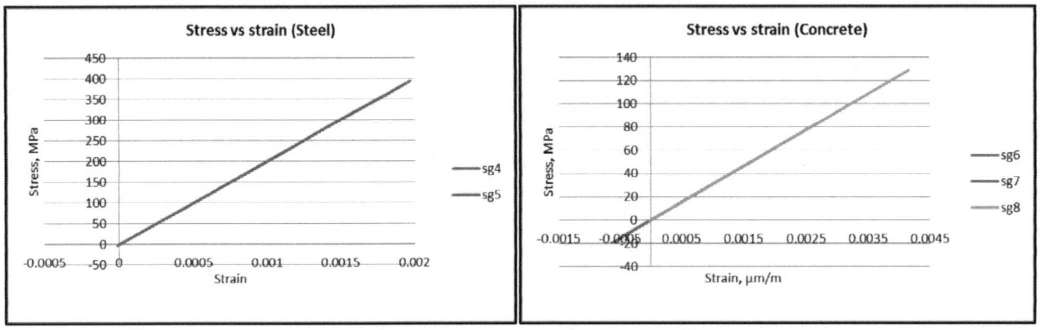

Figure 9. Stress vs strain for control beam.

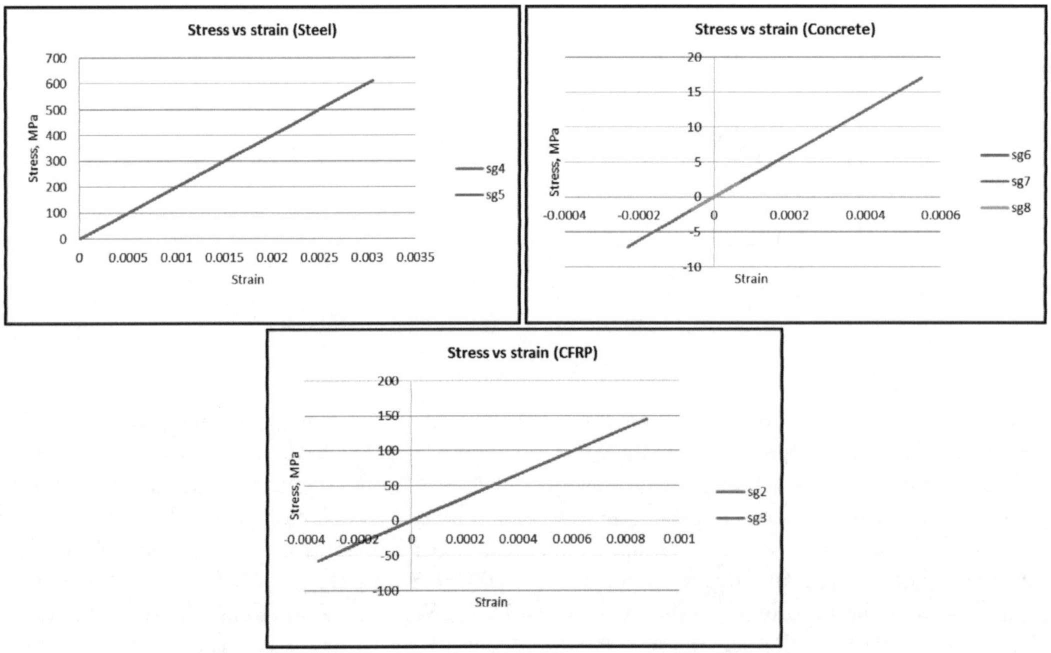

Figure 10. Stress vs strain for beam with double vertical plate.

IOP Conf. Series: Materials Science and Engineering **513** (2019) 012042 doi:10.1088/1757-899X/513/1/012042

4. Mode of failure

Modes of failure can be analyzed by observing the cracks on the beam. Crack started to occur when the loading on the beam increases. Since tensile stress is higher on the bottom of the beam, the cracks formed started from the bottom and continue to the top part of the beam which has less tensile stress.

As seen in Figure 12, the strengthened beam tends to have more cracks compare to the control beam seen in Figure 11. The exposure to saltwater may have reduced the concrete compressive strength which has caused the strengthened beam to have more cracks. Other than that, another reason the strengthened beam has more cracks compare to control beam is because it have been applied to pre-crack loading before.

As researchers can observe from both control beam and strengthened beam, researchers can identify that they have different modes of failure. Modes of failure for control beam are known as shear failure. This is because a large crack is formed on one side of the beam which is near to the support. Whereas for strengthened beam, it is found that it have undergone modes of failure known as concrete cover separation. It is observed that the CFRP plate is still intact and does not bend. However, it had caused the beam to crack as its concrete cover on the bottom. As conclusion, the CFRP has enhanced the beam but due to concrete ability to sustain tensile stress is low; it had caused these types of failure.

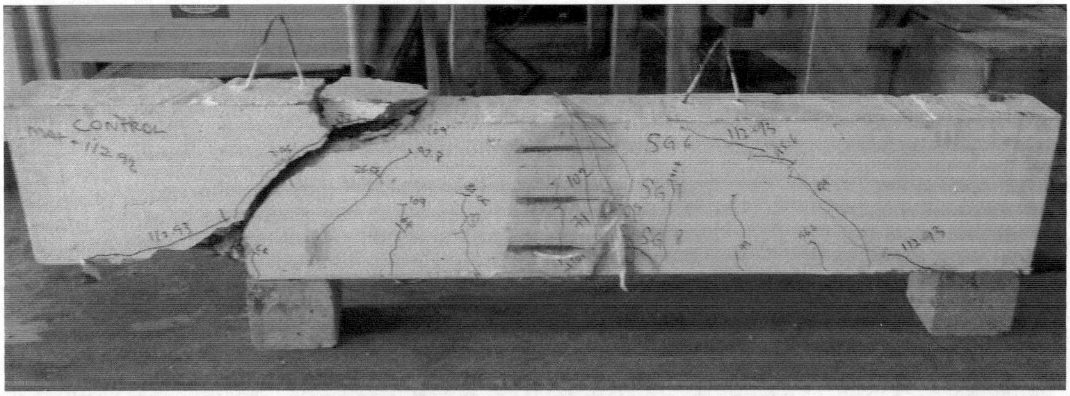

Figure 11. Failure mode of the control beam.

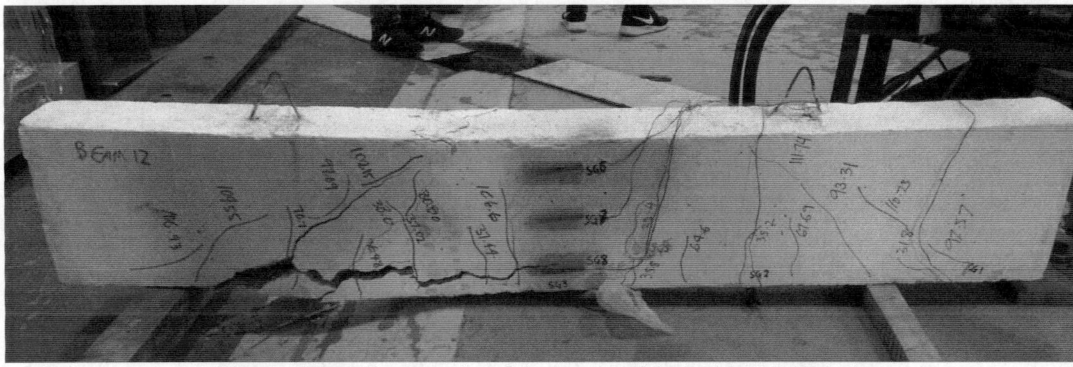

Figure 12. Failure mode of the beam with double vertical plate.

IOP Conf. Series: Materials Science and Engineering **513** (2019) 012042 doi:10.1088/1757-899X/513/1/012042

5. Conclusions

Flexural strength test method is used to measure the behavior of concrete beam subjected to bending force. In this research, the test is used to analyze the state of failure of concrete and beam to find out whether they undergone elastic deformation, plastic deformation, yield state or even fracture. Based on this research, findings are shown below:

a. At the same ultimate load, it is observed the deflection of control beam and beam strengthened with CFRP plate is slightly the same.
b. The strain produced by concrete for control beam is higher compare to the strengthened beam. Whereas the strain produced by steel reinforcement for control beam has almost the same value as strengthened beam.
c. The CFRP plate does not bend after testing, which result in the modes of failure known as concrete cover separating. This means that it can cater and improve the mechanical properties of the beam but is influence by the concrete compressive strength.

As conclusion, the strengthened beam could have higher flexural strength. However, due to pre-cracking load, the beam has slightly the same performance as control beam. Therefore, the use of CFRP can still produce high performance beam since it can still cater load even after subjected to pre cracking and repaired.

Acknowledgements

This work was supported in part by LESTARI under Grant Nos. 600-RMI/DANA 5/3/LESTARI (2/2016). The authors wish to express their appreciation to Universiti Teknologi MARA, providing the facilities assistance for successfully accomplishment of this research study.

References

[1] Hassan, T., & Rizkalla, S. (2003). Investigation of Bond in Concrete Structures Strengthened with Near Surface Mounted Carbon Fiber Reinforced Polymer Strips. Journal of Composites For Construction, 248.

[2] Soudki, K., El-Salakawy, E., & Craig, B. (2007). Behavior of CFRP Strengthened Reinforced Concrete Beams in Corrosive Environment. Journal of Composites For Construction, 291.

[3] Liu, Y., Zwingmann, B., & Schlaich, M. (2015). Carbon Fiber Reinforced Polymer for Cable. Polymers, 2097.

[4] Tanarslan, H. (2016). Effect Of CFRP Laminate And/Or CFRP Bars To The Shear.

[5] Mohd.Hashim, M. H., Mohd.Sam, A. R., Hussin, M. W., & Arshad, M. F. (2013). Structural Performance and Ductility of Fiber Reinforced Polymer Concrete Bonding System Under Tropical Climates. Jurnal Teknologi (Sciences & Engineering), 22.

[6] Ciobanu, P., Ţaranu, N., Popoaei, S., & Banu, D. (2012). Structural Responce OF Reinforced Strengthened in Flexure with Near Surface Mounted Fibre Experimental Setup. Faculty of Civil Engineering and Building Services, 14.

[7] Sena Cruz, J., Barros, J. A., Gettu, R., & Azevedo, A. F. (2006). Bond behavoir of near surface mounted CFRP laminate under monotonic and cyclic loading. Journal of Composite for Construction, 295-303.

[8] EN 1992-1-1 (2004) (English): Eurocode 2: Design of concrete structures - Part 1-1: General rules and rules for buildings [Authority: The European Union Per Regulation 305/2011, Directive 98/34/EC, Directive 2004/18/EC]

IOP Conf. Series: Materials Science and Engineering **513** (2019) 012043 doi:10.1088/1757-899X/513/1/012043

Developing a behavioural green index (BGI): introduce a framework and changing behaviour

Ram Ariff[1,3*], Z Sriyolja[1], A Wibawa[1], H Nofita[1], K Yahya[1] and S Sharif[2]

[1]School of Civil Engineering, Faculty of Engineering, Universiti Teknologi Malaysia, 81310, Johor
[2]School of Mechanical Engineering, Faculty of Engineering, Universiti Teknologi Malaysia, 81310, Johor
[3] School of Industrial Engineering, Faculty of Engineering, Universitas Pamulang, 15417, Indonesia

[*]Corresponding author: ramariff2017@gmail.com

Abstract. Some humans are the cause of environmental damage. Employee Green Behaviours must be able to maintain environmental quality because the construction industry is a contributor to high environmental damage in the world. Integrated control is needed in developing green behavioural awareness especially for human resources involved in building and housing projects. The purpose of this study is to develop a conceptual framework towards Behavioural Green Index (BGI) in construction. The methodology developed using frameworks studies and factors analysis. Factor analysis is carried out on factors that have been found to have influence on green behaviour such as external factors and internal factors. The results of this study is conceptual framework as inputs for developing a BGI that needs to be followed up by increasing green behaviour, reducing behavioural barriers, and supporting habitability.

1. Introduction

Kollmuss and Agyeman [37] concludes that many theoretical frameworks have been developed to explain the gap between environmental knowledge, environmental awareness and green behaviour.

IOP Conf. Series: Materials Science and Engineering **513** (2019) 012043 doi:10.1088/1757-899X/513/1/012043

Table 1. Research gap for green behaviour frameworks.

Frameworks	Factors	Research Gap
Early Models of pro-environment behaviour [11]	Environmental knowledge, environmental attitude and pro-environmental behaviour	There are commonalities, contradictions and omissions in the form of a framework. Some models have the same factors. Some models have contradictions in terms of perspective. Some models can be seen as complementary to one another. This study proposes a green behavior conceptual framework with the principle: a. the framework is structured based on a combination of existing frameworks b. the framework uses factors that have been found and adds to other literature for factors that do not yet exist in the framework c. the framework is integration from previous studies frameworks d. the framework is intended to obtain measurement and assessment of a construction project based on human capital
Theory of reasoned action [2]	Evaluative beliefs, normative beliefs, motivation to comply, attitude toward behaviour, relative importance of attitudinal and normative considerations, subjective norm, behaviour intention, behavior	
Models of predictors of environmental behaviour [31]	Personality factors (Attitudes, locus of control and personal responsibility), knowledge of issues, knowledge of action strategies and action skills (intention to act), situational factors, pro-environmental behavior	
Model of ecological behaviour [24]	Possibilities to act pro-environmentally, incentives for pro-environmental behaviour, perceived concequences of behaviour, environmental knowledge, environmental attitudes and values, pro-environmental behavior	
Barriers between environmental concern and action [6]	Type of barrier (individual barrier, individual in social context, social/ institutional barriers), environmental concern, individuality, responsibility, practicality, pro-environmental behavior	
Model of pro-environmental behaviour [37]	Internal factors (knowledge, feeling, values attitudes), external factors (infrastructure, political, social and cultural factors economic situation etc, pro-environmental behavior	

2. Theoretical study

Kollmuss and Agyeman [37] suggested that there are three factors important in green behaviour such as demographics, external factors and internal factors.

Table 2. The three factors important in green behaviour.

	Theory	Research Study
Green Behaviour (green behaviour **performance**) [45,46,47]	planned behaviour [2,57], risk, construction management [55], fundamental causes, construction economic, microeconomics, decision making process, psychology	How to make **green behaviour** on rational preferences, maximise outcomes and act independently to measure behaviour risk and root cases on construction project that can influence ecology based [15,16,37,60,61]
External Factors (green behaviour **policy**)	Sociological [27], infrastructure, economic [1], social culture [7], institutional	How to measure external factors that impact to behavioural change [33,63]. The **green habitability** is a enabler [23]
Internal Factors (green behaviour **competency**)	Motivation [40] [53], environmental knowledge [18,26,30,34,35], values [27,36,13,14], attitudes [44] [18,36], environmental awareness [53], perception of control [3,9,44], emotional involvement [13,14,30,36]	How to make green competency and to measure **barrier behaviour**, mental shortcut, emotional [25,28,29,51] [41,57]. Demographics is part of Internal Factors [64]

3. Methodology

Research methods used in this study using literature studies and factors analysis in the field of green behaviour, as illustrated in the **Figure 1** below.

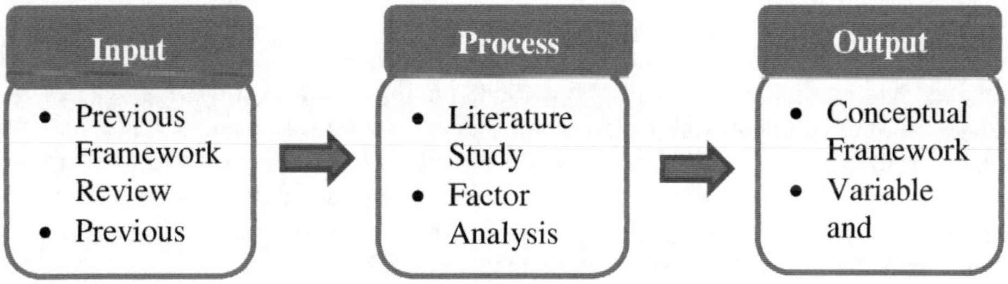

Figure 1. Research methodology.

4. Analysis

4.1. External factors or green habitability (Green behaviour policy)

Environmental policy is the commitment of an organization or government to the laws, regulations, and other policy mechanisms concerning environmental issue [19,20,21,65]. Lauritzen [39] state, "To be ecologically sustainable, we need to promote, influence, and change employee behaviors such that they are congruent with environmental sustainability goals of organizations". Employee Green Behaviours (EGBs) is scalable actions to contribute in environmental sustainability [40]. Policies relating to environmental impacts must touch the practical level in terms of the EGBs. Construction companies need to provide rules regarding green behavior on the project site. [42] suggested that company must propose a new company policy with addressing the environmental impact of the policy such as buy company supplies with thought for environmental impact, prioritize actions that would benefit the environment, reduce water consumption by turning off faucets when not in use, use inefficient work processes that waste natural resources and monitor the environmental impact of workplace processes.

4.2. Green behaviour performance (Green performance capability)

Lauritzen [39] suggested that behaviors can be counterproductive in that they actually detract from the organization's environmental performance. Green Performance factor are the working sustainably, avoiding harm, influencing others, conserving and taking initiative factors [57]. Meaningful outcomes could include performance of EGBs [12]. These behaviors can be performed as part of an employee's job duties and outside of an employee's job duties as organizational citizenship behaviors. Opposes the use of unethical practices to increase performance. Is fair and objective when evaluating EGBs performance and providing rewards. As a result all levels of employees must demonstrate the green behavior in their organizations and thrive for success. According to [8] initiatives made by employees have a significant impact on greening the organizations apart from the initiatives taken by the formal management systems, procedures or technologies. As a result of intensifying concern about green activities; a growing body of literature has been concerned with the motivation for adopting environmentally related standards and their impact on firms' performances, role of green psychological climate, organizational sustainability and employee green behavior [38,45,47]. EGBs can be part of any dimension of job performance and can be either required or discretionary, depending on the nature of the job [12].

According to [45] required EGB is delineated as green behavior performed within the context of employees' required job duties. Required EGB is also known as task related EGB which includes obliging to organizational policies, changing methods of work and creating sustainable products and processes [45]. The concept of required EGB is similar to task performance [45] which is defined as the activities formally identified as a part of the job of a person that contributes to the technical core of an organization [10]. Voluntary EGB has been defined as "green behavior involving personal initiative that exceeds organizational expectations" [45].

4.3. Internal factors or green barrier behaviour (Green competency)

[37] concluded that internal factors are motivation, environmental knowledge, values, attitudes, environmental awareness, emotional involvement, locus of control, responsibility and priorities. Recently human resources management functions such as recruitment, selection, training and performance evaluation are expected in considering environmental management issues [64]. Environmental protective acts with adequate ecological knowledge [62] and socio-economic behavior and skills are referred to in this paper as green competencies [32,42].

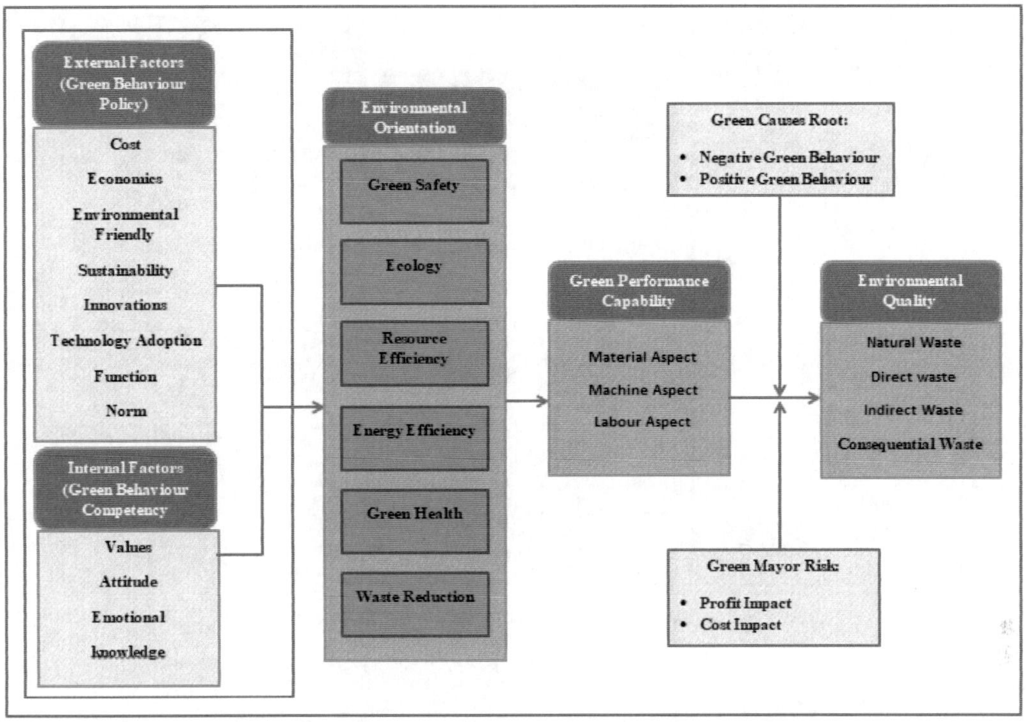

Figure 2. Behavioural green model of environmental quality.

5. Results

The conceptual model in this study builds on theories of green behaviour based on relevant previous research literature that has been described in the literature study. In this study will develop the theory of green behaviour which says that the green behaviour there are 3 dimensions in shaping the character that is 1) external factors (Green Habitability), 2) internal factors (Green Barrier) and 3) green performance (Green Behaviour). The theory is developed with green variables and indicators as a shaper of green behaviour in construction. This is because previous research on the development of the pro-environmental behaviour model does not explain that performance variables as something that should be integrated as micro implications or impacts that must be improved if they want to develop a green behaviour in the construction [58] [59]. However, various literatures prove that habitability and barrier behaviour cannot separately dimension from green performance. Next research will test in Indonesia construction [4] [5].

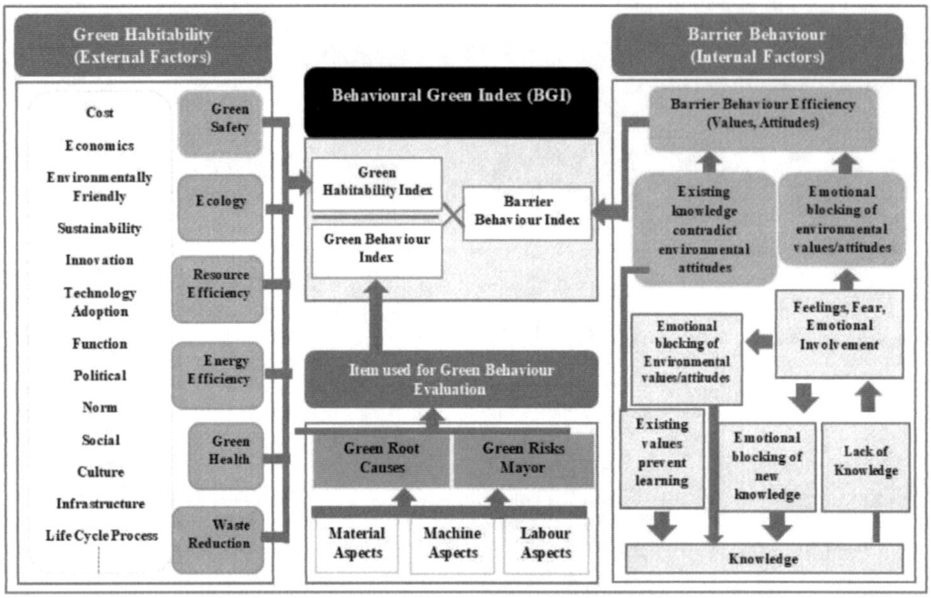

Figure 3. Conceptual framework of the Behavioural green index.

6. Discussion

Green behaviour are interconnected with the policy on macro level, performance on mezzo level and awareness on micro level.

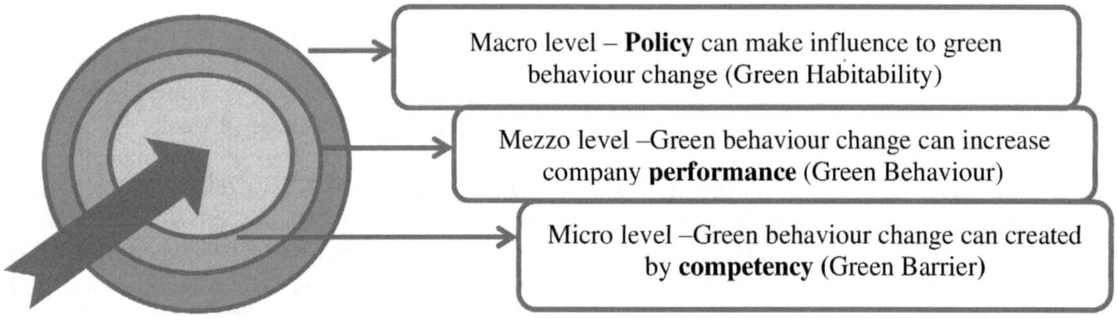

Figure 4. Green behaviour change model.

7. Conclusions

The results of this study indicate that in building a conceptual framework towards BGI, it is necessary policy, competency and performance as input at BGI [51]. In the policy there are four latent variables of environmental regulation policy, green behaviour or project policy, reward policy and punishment policy. For awareness there are five latent variables of barrier behaviour, mental shortcut, emotional, knowledge and values attitudes. Latent variables in performance include rational preferences, maximise outcomes, act independently to measure behaviour risk and root cases [39,52]. This conceptual framework will be developed for further research to find out the relationship between the integrated variables and know the method of development of Behavioural Green Index in the construction industry [43,46,48,49].

Acknowledgments
The authors wish to thank Universiti Teknologi Malaysia and Universitas Pamulang for the cooperation and assistance throughout conducting this research.

References
[1] Ackermann F 1997 *Why Do We Recyle? Markets, Values, and Public Policy* (Washington, DC,Island Press)
[2] Azjen I and Fishbein M 1980 *Understanding Attitudes and Predicting Social Behavior* (Englewood Cliffs, NJ, Prentice Hall)
[3] Allen J B and Ferrand J 1999 Environmental locus of control, sympathy, and proenvironmental behavior: a test of Geller's actively caring hypothesis, *Environment and Behavior.* **31**(3) 338–353
[4] Alwi 2014 *Waste in the Indonesian construction projects. In Proceedings The 1st International Conference of CIB W107 - Creating a sustainable Construction Industry in Developing Countries* 305-315, South Africa.
[5] Bambang and Elkhobar 2004 *Construction Waste Management in Construction Projects: Case Studies on Building Construction Projects in Jabotabek Area.*
[6] Blake J 1999 Overcoming the 'value–action gap' in environmental policy: tensions between national policy and local experience Local Environment **4**(3) 257–278.
[7] Boehmer-Christinsen S and Skea J 1991 *Acid Politics: environmental and energy policies in Britain and Germany* (New York, Belhaven Press)
[8] Boiral O and Paillé P 2012 Organizational Citizenship Behaviour for the Environment: Measurement and Validation. *Journal of Business Ethics.* **109**(4) 431-445
[9] Borden D and Francis J L 1978 Who cares about ecology? Personality and sex difference in environmental concern. *Journal of Personality.* **46** 190–203
[10] Borman W C and Motowidlo S J 1997 Task performance and contextual performance: The meaning for personnel selection research. *Human Performance* **10** 99-109
[11] Burgess J, Harrison C and Filius P 1998 Environmental communication and the cultural politics of environmental citizenship. *Environment and Planning A.* **30** 1445–1460
[12] Campbell J P and Wiernik B M 2015 The modeling and assessment of work performance. *Annu Rev Organ Psychol Organ Behav.* **2** 47–74
[13] Chawla L 1998 Significant life experiences revisited: a review of research on sources of pro-environmental sensitivity. *The Journal of Environmental Education.* **29**(3) 11–21
[14] Chawla L 1999 Life paths into effective environmental action. *The Journal of Environmental Education* **31**(1) 15–26
[15] Daily B F, Bishop J W and Govindarajulu N 2009 A conceptual model for organizational citizenship behavior directed toward the environment. *Business & Society* **48**(2) 243–256
[16] Defra (2008) *A framework for pro-environmental behaviours Department for Environmental and Rural Affairs London*
[17] Department for Environment Food and Rural Affairs (Defra) 2008 *A framework for pro-environmental behaviours.*
[18] Diekmann A and Franzen A 1996 *Einsicht in o ¨kologische Zusammenha ¨nge und Umweltverhalten in: R. Kaufmann-Hayoz & A. Di Giulio (Eds) Umweltproblem Mensch: Humanwissenschaftliche Zusammenha ¨nge zu umweltverantwortlichem Handeln* (Bern, Verlag Paul Haupt)
[19] Dolan at al. 2010 *Mindspace: Influencing behaviour through public policy* London: UK Cabinet Office Institute for Government
[20] Eisenberg N & Miller P 1987 The relation of empathy to prosocial and related behaviors. *Psychological Bulletin.* **101** 91–119
[21] European Commission 2012 *Science for Environment Policy, Future Brief: Green Behaviour,* Issue 4 October

IOP Conf. Series: Materials Science and Engineering **513** (2019) 012043 doi:10.1088/1757-899X/513/1/012043

[22] Festinger L 1957 *Theory of Cognitive Dissonance* (Stanford, CA, Stanford University Press)
[23] F Ferry, O Fadil, Y Khairulzan, and H Zaiton 2016 The Green Construction Site Index (GCSI): A Quantitative Tool used to Assess an Ongoing Project to Meet the Green Construction Concept. *International Journal Technology* **7**(4).
[24] Fietkau H J and Kessel H 1981 *Umweltlernen: Veraenderungsmoeglichkeiten des Umweltbewusstseins Modell-Erfahrungen* (Koenigstein, Hain)
[25] Fishbein M and Ajzen I 1975 *Belief, Attitude, Intention, and Behavior: an introduction to theory and research* (Reading, MA, Addison-Wesley)
[26] Fliegenschnee M & Schelakovsky M 1998 *Umweltpsychologie und Umweltbildung: eine Einfu ̈hrung aus humano ̈kologischer Sicht* (Wien, Facultas Universita ̈ts Verlag)
[27] Fuhrer U, Kaiser F G, Seiler J and Maggi M 1995 *From social representations to environmental concern: the in• uence of face to face versus mediated communication in: U. FUHRER (Ed.) Oekologisches Handeln als sozialer Prozess* (Basel, Birkhaeuser)
[28] Gigliotti L M 1992 Environmental attitudes: 20 years of change?. *The Journal of Environmental Education*. **24**(1) 15–26
[29] Gigliotti L M 1994 Environmental issues: Cornell students' willingness to take action. *The Journal of Environmental Education*. **25**(1) pp 34–42
[30] Grob A 1991 *Meinung Verhalten Umwelt* (Bern, Peter Lang Verlag)
[31] Hines J M, Hungerford H R and Tomera A N 1986–87 Analysis and synthesis of research on responsible pro-environmental behavior: a meta-analysis. *The Journal of Environmental Education*. **18**(2) 1–8
[32] Hungerford H R and Volk T L (1990) Changing learner behavior through environmental education. *The Journal of Environmental Education* **21**(3) pp 8–21
[33] Jackson 2005 *Morivating Sustainable Consumption: a review of evidence on customer behaviour and behavioural change* Sustainable Development Research Network DEFRA
[34] Kaiser F G, Woelfing S and Fuhrer U 1999 Environmental attitude and ecological behaviour. *Journal of Environmental Psychology*. **19** 1–19
[35] Kempton W Boster J S and Hartley J A 1995 *Environmental Values in American Culture* (Cambridge, MA, MIT Press)
[36] Lehmann J 1999 *Befunde empirischer Forschung zu Umweltbildung und Umweltbewusstsein* (Opladen, Leske und Budrich)
[37] Kollmuss and Agyeman 2002 Mind the Gap: why do people act environmentally and what are the barriers to pro-environmental behaviour*? Environmental Education Research* **8**(3)
[38] Lanfranchi J and Pekovic S 2014 How green is my firm? Workers attitudes and behaviors towards job in environmentally-related firms. *Ecological Economics*. **100** 16- 29
[39] Lauritzen E K 1994 *Economic and environmental benefits of recycling waste from the construction and demolition of buildings In: Waste Recycling* UNEP Industry and Environment 26–3
[40] Moisander J 1998 *Motivation for Ecologically Oriented Consumer Behavior*, Workshop Proceedings March The European Science Foundation (ESF) TERM (Tackling Environmental Resource Management Phase II 1998–2000) http://www.lancs.ac.uk/users/scistud/esf/ lind2.htm
[41] Mudgal, S, Lyons L, Bain J 2011 *Plastic Waste in the Environment – Revised Final Report for European Commission DG Environment BioIntelligence Service*
[42] Nachiappan et al. 2016 Green competence framework: evidence from China. *The International Journal of Human Resource Management*. **27** 151-172
[43] Nazech at al. (2008) *Issues on Construction Waste: The Need for Sustainable Waste Management*
[44] Newhouse N (1991) Implications of attitude and behavior research for environmental conservation. *The Journal of Environmental Education*. **22**(1) 26–32

IOP Conf. Series: Materials Science and Engineering **513** (2019) 012043 doi:10.1088/1757-899X/513/1/012043

[45] Norton T A, Parker S L, Zacher H, and Ashkanasy N M 2015 Employee green behavior: a theoretical framework, multilevel review, and future research agenda. *Organ Environ.* **28** 103–125
[46] Norton T A, Zacher H, Parker S L and Ashkanasy N M 2017 Bridging the gap between green behavioral intentions and employee green behavior: The role of green psychological climate. *Journal of Organizational Behavior.* **38**(7)
[47] Norton T Zacher H and Ashkanasy N 2014 Organisational sustainability policies and employee green behaviour: The mediating role of work climate perceptions. *Journal of Environmental Psychology.* **38** pp 49-54
[48] Ones D S and Dilchert S 2012a *Employee green behaviors* In S Jackson D Ones & S Dilchert (Eds.) *Managing human resources for environmental sustainability* 85-116
[49] Ones D S and Dilchert S 2012b Environmental sustainability at work: A call to action. *Industrial and Organizational Psychology* **5** 444–466
[50] Osbaldiston R and Schott J P 2012 *Environmental sustainability and behavioural science: Meta-analysis of pro-environmental behaviour experiments Environment and Behaviour* **44**(2) 257-299
[51] Oullier O and Saunero S 2011 *Policy Brief 216 : Green new incentives for ecological behaviour. Paris: Centre for Strategic Analysis (Centre d'analyse strategique)*
[52] Peter and Damar 2015 *Penyimpangan Perilaku Para Pelaku Jasa Konstruksi*
[53] Preuss S 1991 *Umweltkatastrophe Mensch. Ueber unsere Grenzen und Moeglichkeiten, oekologisch bewusst zu handeln* (Heidelberg, Roland Asanger Verlag)
[54] M. Redclift and T. Benton (Eds) Social Theory and the Global Environment (London, Routledge)
[55] Roseline 2016 Factors Influencing Waste Generation in the Construction Industry in Malaysia Procedia - *Social and Behavioral Sciences* **234** 11 – 18
[56] Ruben M and Carol Y Y 2015 Cultural antecedents of green behavioural intent: An environmental theory of planned behaviour. *Journal of Environmental Psychology.*
[57] Savage et al. 2011 *Social stratification, social capital and cultural practice in the UK*
[58] Sia A P, Hungerford H R and Tomera A N 1985–86 Selected predictors of responsible environmental behavior: an analysis .*The Journal of Environmental Education* **17**(2) 31–40
[59] Schwartz S H 1977 *Normative influences on altruism in: L Berkowitz (Ed.) Advances in Experimental Social Psychology* **10** (New York, Academic Press)
[60] Schwartz 2004 *The Paradox of Choice: Why More is Less* HarperCollins e-books
[61] Simon 1995 A Behavioral Model of Rational Choice. *The Quarterly Journal of Economics.* **69**(1) 99-118
[62] Skoyles E R and Skoyles J R 1987 *Waste prevention on site* London: Mitchell
[63] Steg and Vlek 2009 Encouraging pro-environmental behaviour. *Journal of Environmental Phychology* **29** 309-317
[64] Stern P S, Dietz T and Karlof L 1993 Values orientation, gender, and environmental concern. *Environment and Behavior.* **25**(3) 322–348
[65] Wolfgang 1983 The habitability framework: a conceptual approach towards linking human behaviour and physical environment. *Design Studies.* **4**(2) 84-91.